James G. Speight
Gas Engineering

Also of Interest

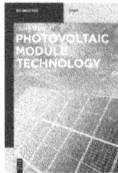

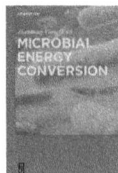

James G. Speight

Gas Engineering

Vol. 1: Origin and Reservoir Engineering

DE GRUYTER

Author
Dr. James G. Speight
2476 Overland Road,
Laramie, WY 82070-4808, USA
Tel: (307) 745-6069 Cell: (307) 760-7673
E-mail: JamesSp8@aol.com
Web page: https://www.drjamesspeight.com

ISBN 978-3-11-069089-7
e-ISBN (PDF) 978-3-11-069102-3
e-ISBN (EPUB) 978-3-11-069108-5

Library of Congress Control Number: 2021939882

Bibliographic information published by the Deutsche Nationalbibliothek
The Deutsche Nationalbibliothek lists this publication in the Deutsche Nationalbibliografie;
detailed bibliographic data are available on the Internet at http://dnb.dnb.de.

© 2021 Walter de Gruyter GmbH, Berlin/Boston
Cover image: nielubieklonu/iStock/Getty Images Plus
Typesetting: Integra Software Services Pvt. Ltd.
Printing and binding: CPI books GmbH, Leck

www.degruyter.com

Preface

The final three decades of the twentieth century saw not only perturbations of energy supply systems but also changes in attitudes of governments and voters alike toward environmental issues. Thus, environmental issues will be with us as long as there is manufacturing of consumer goods, and the use of fossil fuels for energy production. And it is this latter issue that is the subject of this text. The continued use of natural gas as a combustible fuel is a reality, and gas processing, although generally understandable using chemical and/or physical principles, still requires an attempt to alleviate some of the confusion that arises from uncertainties in the terminology.

This three-volume collection of books presents to the reader an understanding of the origin of gases, the properties of gases, and the uses of gases.

The primary aim of this first volume is to introduce the reader to the origins of natural gas and also contains chapters dealing with recovery, properties, and composition, including gas production from hydrocarbon-rich deep shale formations, known as shale gas, which is one of the most quickly expanding trends in onshore domestic gas exploration and presents the development of deep shale formations, typically located many thousands of feet below the surface of the Earth in tight, low permeability formations. The basic technology of reservoir engineering is presented using the simplest and most straightforward mathematical techniques. The book focuses on processes and, wherever possible, the advantages, limitations, and ranges of applicability of the processes are discussed so that the selection and integration into the overall gas plant can be fully understood. It is only through having a complete understanding of the technology that the engineer can hope to appreciate and solve complex reservoir engineering problems in a practical manner.

Volumes 2 and 3 are more specialized and describe the theory and practice of well testing and pressure analysis techniques, which are probably the most important subjects in the whole of reservoir engineering.

Volume 2 deals with the constituents of gas streams and the properties of the individual constituents. This volume also presents the chemistry and engineering aspects of the methods and principles by which the gas streams might be cleaned from their noxious constituents. The concept of gas condensate is also introduced and discussed, as well as the methods which can be applied to the analysis of gas streams and gas condensate.

Volume 3 presents a review of the uses of gas streams and their effects on the environment. This volume also introduces the concept of liquefied natural gas and the concept gas-to-liquids. Also the properties of gas streams as they are related to corrosion effects are also presented. The relationship of the properties of gas streams as they affect corrosion such as carburization and metal dusting as well as corrosion in steel and other materials used in refinery technology is also presented, and the book summarizes key findings into corrosion processes in gas-processing equipment as well as corrosion in offshore structures.

https://doi.org/10.1515/9783110691023-202

Each book contains copious references at the end of chapter, which include information from the open literature and meeting proceedings to give a picture of where the gas-processing technology stands as well as indicate some relatively new technologies that could become important in the future. Each book also contains a comprehensive glossary.

The books are written in an easy-to-read style and offer a ready-at-hand (one-stop shopping) guide to the many issues that are related to the engineering aspects of the properties and processing of natural gas and the effects of natural gas on various ecosystems as well as to pollutant mitigation and clean-up. The books present an overview with a considerable degree of detail on the various aspects of natural gas technology. Any chemistry presented in the books is used as a means of explanation of a particular point but is maintained at an elementary level.

Dr. James G. Speight,
Laramie, Wyoming, USA
May 2021

Contents

Chapter 1
History and background

1.1 Introduction

Natural gas (also called marsh gas and swamp gas in older texts) is a gaseous fossil fuel that is found in gas-bearing formations and in oil-bearing formations – coalbed methane is often referred to (incorrectly) as natural gas or as coal gas due to lack of standardization of the terminology (Levine, 1993; Speight, 2013, 2014). For clarification, natural gas is not the same as town gas (which is produced from coal), although the history of natural gas cleaning prior to sales has its beginnings in town gas cleaning (Speight, 2013, 2019a).

The term "natural gas" also includes gases and low-boiling liquids (often – in the current context – referred to as gas condensate or condensate) liquids from the recently developed shale formations (Kundert and Mullen, 2009; Aguilera and Radetzki, 2014; Khosrokhavar et al., 2014; Speight, 2017b) – biogas and landfill gas produced from various biological and non-biological sources are not included in the term "natural gas" (John and Singh, 2011; Ramroop Singh, 2011; Singh and Sastry, 2011). Thus, for the purposes of this book, any petroliferous natural gas is placed under the category of conventional gas while the non-petroliferous gases (such as biogas and landfill gas) are placed under the term "unconventional gas" (sometimes called non-conventional gas; Appendix A) (Speight, 2019b).

Thus, the terminology and definitions applied to natural gas (and, for that matter, to other gaseous products and gaseous fuels) are extremely important and have a profound influence on the manner by which the technical community and the public perceive a gaseous fuel. For the purposes of this book, natural gas and those products that are isolated from natural gas during recovery (such as natural gas liquids, gas condensate, and natural gasoline) are a necessary part of this text. In addition, gas in geopressurized zones, gas in tight formations (low-to-no permeability formations), and gas hydrates (which some observers prefer to categorize under the heading unconventional gas) are also (for the purposes of this book) included in the conventional gas group. Thus, the categorization of gaseous fuels is as follows listed alphabetically that than by occurrent or use:
Conventional natural gas
  Associated gas
  Non-associated gas
  Gas condensate (including natural gasoline and low-boiling naphtha)
  Gas from geopressurized zones
  Gas from tight formations (including shale gas)
  Gas hydrates

https://doi.org/10.1515/9783110691023-001

Unconventional gas
   Biogas
   Coalbed methane
   Coal gas
   Flue gas
   Landfill gas
   Manufactured gas
   Refinery gas
   Synthesis gas (syngas)

An alternate categorization of these gases is based on (i) the source of the gas and (ii) the method of production of the gas which also has some relationship to the composition of the gas. Thus:

Conventional natural gas
   Associated gas
   Non-associated gas
   Gas condensate
   Gas hydrates
   Gas in geopressurized zones
   Gas in tight formations (including shale gas)

Unconventional gas
   Coalbed methane
   Landfill gas

Manufactured gas
   Biogas
   Coal gas
   Flue gas
   Refinery gas
   Synthesis gas (syngas)

Landfill gas can be included in the unconventional gas category because, other than the burial of waste in the landfill, there are no external (manufacturing) forces that are used to produce the gas. However, to add confusion to this categorization, landfill gas is considered (by some observers) to fall into the biogas category. To mitigate any such confusion, it is recommended that a gas by categorized according to the source as well as method of production of the gas (as indicated above).

Nevertheless, whatever, the source or origin, natural gas and other fuel gases are vital components of the energy supply of the world and form a necessary supply chain for energy production:

reservoir gas → wellhead gas
wellhead gas → pre-treated gas
pre-treated gas → pipelined (transported) to gas processing plant → cleaned gas
(methane)
cleaned gas (methane) → transported/stored gas/sales gas

From a chemical standpoint, natural gas is a mixture of hydrocarbon compounds
and non-hydrocarbon compounds with minor-to-substantial amounts of carbon dioxide (CO_2), hydrogen sulfide (H_2S), mercaptan derivative (RSH, also called thiol derivatives), and (on occasion) helium (He) (Speight, 2012, 2014, 2019a). More generally,
"natural gas" is the generic term that is applied to the mixture of and gaseous hydrocarbon derivatives and low-boiling liquid hydrocarbon derivatives [typically up to
and including hydrocarbon derivatives such as n-octane, $CH_3(CH_2)_6CH_3$, boiling point
125.1–126.1 °C (257.1–258.9 °F) that are classified according to structure (Table 1.1;
Speight, 2014, 2019a).

Table 1.1: General classification system for hydrocarbon derivatives.

Hydrocarbons			
	Non-cyclic (open-chain) compounds	Saturated	Paraffins/alkanes
		Unsaturated	Alkene >C=C<
		Unsaturated	Alkyne -C≡C
	Cyclic (closed-chain or ring) compounds	Homocyclic	Alicyclic (non-aromatic)
			Cycloalkene
			Cycloalkyne
			Aromatic – benzenoid
			Aromatic – non-benzenoid
		Heterocyclic	

Typically, in field operations (Chapters 4 and 5), the composition of natural the gas
(which affects the specific gravity) can vary significantly because the product flowing out of the well can change with any variability in the production conditions as
well as the change of pressure as the gas is removed from the reservoir (Burruss and
Ryder, 2003; Mokhatab et al., 2006; Burruss and Ryder, 2014). When gas is removed
from the reservoir, the constituents of the gas that were in the liquid phase under
reservoir pressure reservoir can (will) revert to the gas phase as the reservoir pressure is reduced.

Thus, there is no guarantee that the natural gas from different sources – even though the elemental analysis may be similar – that the chemical composition of the gas will be the same. The composition varies depending on whether the gas is associated or non-associated with crude oil, or has or has not been processed in industrial plants. The basic composition includes methane, ethane, propane, and hydrocarbons with a heavier molecular weight (in smaller proportions). Normally it has low contaminant content, such as nitrogen, carbon dioxide, water, and sulfide derivatives. In fact, since the different gas supplies enter the natural gas system at different locations, the exact composition at any site will vary among the different regions. Furthermore, the system average heating value will depend on the mix of gas supplies and therefore can vary from the so-called typical values listed below. Hence the need for specifications for the gas that passes from the seller to the customer (Appendix C).

As a result, it should not be a surprise that at each stage of natural gas production (such as wellhead treating, transportation, and processing), analysis of the gas to determine the composition and properties by standard test methods is an essential part of the chemical and engineering aspects of natural gas including specification that must pass between the seller and the user (Appendix C). Use of analytical methods offers (Speight, 2018, 2019a) vital information related to the behavior of natural gas at each stage of the production (Table 1.2). The data produced from the test methods are the criteria by means of which the suitability of the gas for use and the potential for interference with the environment.

Table 1.2: General aspects of gas processing.

Feedstock	Intermediate product*	Final product	Use
Wellhead gas	Carbon dioxide removed Hydrogen sulfide removed Mercaptans removed		
Pre-treated gas		CH_4, methane C_2H_6, ethane C_3H_8, propane C_4H_{10}, butane C_5H_{12}+, pentane plus	Natural gas to consumer Petrochemical feedstock Liquefied petroleum gas (LPG) Liquefied petroleum gas (LPG) Natural gas condensate

*Preliminary treatment at the wellhead.
**Final treatment at a gas cleaning plant.

1.2 Reservoirs

Crude oil in the reservoir oil may be saturated with gas, the degree of saturation being a function, among other effects, of reservoir pressure and reservoir temperature. If the crude oil has dissolved in it all the gas that the crude oil is capable of holding under the reservoir conditions, the crude oil is referred to as saturated oil – any excess gas (undissolved gas) is present in the form of a gas cap. If there is less gas present in the reservoir than the amount that may be dissolved in oil under conditions of reservoir pressure and temperature, the crude oil is then termed undersaturated. The pressure at which the gas begins to come out of solution is the saturation pressure (more commonly referred to as the bubble-point pressure). In the case of saturated oil, the saturation pressure equals the reservoir pressure and the gas begins coming out of solution as soon as the reservoir pressure begins to decrease. In the case of undersaturated oil, the gas does not start coming out of solution until the reservoir pressure drops to the level of saturation pressure.

The evolution and the character of a specific reservoir is due to (i) the age of the reservoir, (ii) the depositional history of the reservoir, (iii) the temperature history of the reservoir, which increases with the depth of the reservoir, as well as (iv) the pressure history, of the reservoir, which also increases with the depth of the reservoir, are critical to the properties of the gas in a reservoir. Because of these effects, the organic debris and any evolved hydrocarbon derivatives underwent millions of years of natural transformation.

Generally, gas can be in the form of a gas cap on top of the oil zone or the gas can be dissolved in crude oil. As the depth of the reservoir increases, the amount of gas in the reservoir increases. Some of the most productive reservoirs occur at depths on the order of 10,000–12,000 feet in which crude oil coexists with substantial quantities of gas. At greater depths (e.g., on the order of and in excess of 17,000 feet), many reservoirs contain almost exclusively natural gas.

1.3 Formation

It is commonly accepted that natural gas, like crude oil, has been generated from organic debris that has been deposited in geologic time and have been embedded along with inorganic matter at a considerable depth below the surface of the Earth (Speight, 2014, 2017, 2018, 2019a, 2020a).

Natural gas (like crude oil) is a product of decomposed organic matter (often referred to as organic debris or detritus). The organic matter is the remains of ancient flora and fauna that was deposited over the past 550 million years. This organic debris is mixed with mud, silt, and sand on the sea floor, gradually becoming buried over time and is sealed off in an oxygen-free (anaerobic) environment. Exposure of the organic matter to pressure and an unknown temperature, the organic matter undergoes a decomposition process in which the products of the process (hydrocarbon derivatives and non-hydrocarbon derivatives) are formed.

Because natural gas and crude oil are often found with water in the reservoir, and because they are less dense than water, the gas and oil would rise vertically and escape to the atmosphere. However, if a horizontal barrier is encountered (cap rock), the migration of the gas and oil (or, at least, the product that are part way to gas and oil, sometimes referred to as proto-gas and protopetroleum) ceases and the gas and oil is confined (referred to as gas-in-place). Therefore, for natural gas to accumulate, there are two effects that are required to be present: (i) the source rock, which contains the compacted organic debris from which the natural gas is produced and (ii) a porous formation which becomes the reservoir in which the gas is stored. To become a reservoir, the formation must have an impermeable cap rock – the ceiling rock – and an impermeable basement rock which trap the gas in the porous formation.

It has been speculated, but not universally accepted (Speight, 2014, 2019a), that the deeper and hotter the source rock, the more likelihood of gas being produced. However, there is considerable discussion related to the heat to which the organic precursors have been subjected. Cracking temperatures (≥300 °C, ≥572 °F) are not by any means certain as having played a role in natural gas formation. Maturation of the organic debris through temperature effects occurred over geological time (millennia) and shortening the time to laboratory time and increasing the temperature to above and beyond the cracking temperature (at which the chemistry changes) does not offer conclusive proof of natural gas (and crude oil) formation involving high temperatures (Speight, 2014). Once the natural gas has formed, its fate depends on two critical characteristics of the surrounding rock: (i) porosity and (ii) permeability.

In order to gas and oil to accumulate, the reservoir formation must be of a suitable porosity – which refers to the amount of empty space contained within the grains of a rock. Highly porous rocks, such as a sandstone formation, typically has a porosity on the order of 5–25% v/v (percent volume of the rock), which gives the formation a substantial amount of space for the storage of fluids – the term "reservoir fluids" includes natural gas, crude oil, and water. On the other hand, the term "permeability" is a measure of the degree to which the pore spaces in a rock are interconnected and, therefore amenable to fluid flow. A highly permeable rock will permit gas and liquids to flow easily through the rock, while a low-permeability rock will not allow fluids to pass through. This latter property is characteristic of shale formations and tight formations (Chapter 2; Speight, 2017b).

After natural gas and crude oil form (or, at least the precursors to natural gas and crude oil form), they will tend to rise toward the surface through pore spaces in the rock because of its low density compared to the surrounding rock. Thus, at some point during or after the maturation process, the gas and crude oil migrated from the source rock either upward or sideways or in both directions (subject to the structure of the accompanying and overlying geological formations) through the underground sediments through fissures and faults until the gas enters a geological formation (*reservoir*) that retains or *traps* the gas through the presence of impermeable basement rock and cap rock. This has not occurred, there is the distinct likelihood that most of the natural gas would percolate through the surface formations and escape into the atmosphere.

As an additional note on the formation of natural gas and crude oil, it is rare that the source rock and the reservoir were one and the same and the reservoir may be many miles from the source rock. Thus, a natural gas field may have a series of layers of crude oil/gas and gas reservoirs in the subsurface. In some instances, the natural gas and crude oil parted company leading to the occurrence of reservoirs containing only gas (non-associated gas).

1.4 Conventional gas

Natural gas resources, like crude oil resources, are typically divided into two categories: (i) conventional gas and (ii) unconventional gas (Mokhatab et al., 2006; Islam, 2014; Speight, 2014; AAPG, 2015). For the purposes of this text, conventional gas includes natural gas from a variety of sources such as (alphabetically) associated gas, non-associated gas, gas in geopressurized zones, gas from tight formations (including shale gas), gas condensate, gas hydrates. On the other hand, the term "unconventional gas resources" includes (alphabetically) biogas, coalbed methane, coal gas, flue gas, landfill gas, manufactured gas, refinery gas, and synthesis gas (mixtures of carbon monoxide and hydrogen; Appendix A).

Conventional gas is typically found in reservoirs with a permeability greater than 1 milliDarcy (>1 mD) and can be extracted by means of traditional recovery methods. In contrast, unconventional gas is found in reservoirs with relatively low permeability (<1 mD) and hence cannot be extracted by conventional methods (Speight, 2016a, 2016b). Thus, conventional gas is produced geological formations and tend to be recovered using standard recovery methods that can be used to economically remove (but not always when low-permeability reservoirs are considered) the gas from the reservoir. However, some of these resources are trapped in reservoirs with poor-to-low permeability or low-to-no permeability and porosity and, as a result, it is difficult (if not impossible) for the natural gas to flow through the pores and into a production well. To be able to produce from these difficult reservoirs, specialized techniques and tools are used.

For example, the extraction of gas from geopressurized zones and from tight formations (including shale gas) often requires additional steps to produce the gas. For example, the recovery of gas from tight (low-to-no permeability) formations must include a hydraulic fracturing step in which fractures are created are created so that the gas will flow to the production well (Chapter 2, Chapter 6).

1.4.1 Associated gas

Associated natural gas (often referred to as dissolved natural gas) occurs either as free gas (the gas cap) in a crude oil reservoir or as gas in solution (dissolved gas) in the crude oil (Table 1.3). The gas that exists dissolved in the crude oil has a higher content of higher molecular weight hydrocarbon derivatives and the composition is variable (Table 1.4).

Table 1.3: Illustration of associated gas and non-associated gas.

Reservoir stratification	Gas type
Cap rock	
Natural gas	Associated gas
Crude oil	
Water	
Basement rock	
Cap rock	
Natural gas	
Crude oil (dissolved gas)	Associated gas
Water	
Basement rock	
Cap rock	
Natural gas	Non-associated gas
Water	
Basement rock	

Almost all crude oil reservoirs will produce some natural gas at the surface. However, the crude oil will not be shipped in a commercial pipeline or a tanker with gas still in the solution. The term "stock tank oil," which is used both as a measure for oil well performance and in commercial pricing of oil, means that all associated gas has been stripped from the liquid at one atmosphere pressure. The gas thus liberated is known as *associated gas*. Thus, associated or dissolved natural gas occurs either as free gas in a crude oil reservoir or as gas in solution in the crude oil. Gas that occurs as a solution in the crude oil is dissolved gas whereas the gas that exists in contact with the crude oil (gas cap) is associated gas.

Table 1.4: Composition of associated natural gas from a crude oil well.

Category	Component	Amount (%)
Paraffinic	Methane (CH_4)	70–98
	Ethane (C_2H_6)	1–10
	Propane (C_3H_8)	Trace–5
	Butane (C_4H_{10})	Trace–2
	Pentane (C_5H_{12})	Trace–1
	Hexane (C_6H_{14})	Trace–0.5
	Heptane and higher (C_7^+)	None–trace
Cyclic	Cyclopropane (C_3H_6)	Traces
	Cyclohexane (C_6H_{12})	Traces
Aromatic	Benzene (C_6H_6), others	Traces
Non-hydrocarbon	Nitrogen (N_2)	Traces–15
	Carbon dioxide (CO_2)	Trace–1
	Hydrogen sulfide (H_2S)	Trace occasionally
	Helium (He)	Trace–5
	Other sulfur and nitrogen	Trace occasionally
	Water (H_2O)	Trace–5

Crude oil cannot be produced without producing some associated gas, which consists of low-boiling hydrocarbon constituents that are emitted from solution in the crude oil as the pressure is reduced on the way to, and on, the surface. Designs for well completion and reservoir management protocols are used to minimize the production of associated gas to retain the maximum energy in the reservoir and thus increase ultimate recovery of the crude oil (Speight, 2014, Hsu and Robinson, 2017).

Crude oil in the reservoir with minimal or no dissolved associated gas (dead crude oil or dead oil) is rare and is often difficult to produce as there is little reservoir energy to drive the oil into the production well and to the surface. Thus, associated or dissolved natural gas occurs either as free gas or as gas in solution in the crude oil. Gas that occurs as a solution in the crude oil is dissolved gas whereas the gas that exists in contact with the crude oil is associated gas – the gas cap is an example of associated gas (Mokhatab et al., 2006 Hsu and Robinson, 2017; Speight, 2014, 2017a, 2019a).

After the production fluids are brought to the surface, the gas is treated to separate out the higher molecular weight natural gas liquids (NGLs) which are treated in a liquid petroleum gas (LPG) processing plant (refining plant) to provide propane and butane, either separately or as a mixture of the two. By definition, natural gas liquids include ethane, propane, butanes, and pentanes and higher molecular weight hydrocarbon derivatives (C_{6+}). While natural gas liquids are gaseous at underground pressure, these constituents condense at atmospheric pressure and turn into liquids. The composition of natural gas can vary by geographic region, the geological age of the deposit, the depth of the gas, and many other factors. Natural gas that contains a significant amount of natural gas liquids and condensates is referred to as wet gas, while natural gas that is primarily methane with little to no liquids in the gas is referred to as dry gas.

The liquids in a natural gas stream are hydrocarbon derivatives that are removed (condensed) as a liquid gas stream. The condensed liquids are maintained in the liquid state for storage, shipping, and consumption. There has been a movement to classify natural gas liquids on the basis of vapor pressure, for example: (i) low vapor pressure – condensate, (ii) high vapor pressure – liquefied natural gas and/or liquefied petroleum gas. However, the boundaries drawn in this manner are arbitrary and caution is advised when using such a classification scheme.

Mixtures of these higher molecular weight hydrocarbon derivatives are often referred to as gas condensate or natural gasoline and the mixture has the characteristics of low-boiling naphtha produced in a refinery by distillation and cracking processes (Parkash, 2003; Gary et al., 2007; Hsu and Robinson, 2017; Speight, 2014, 2017). The liquid petroleum gas stored ready for transport and the non-volatile residue (i.e., non-volatile under the conditions of the separation process), after the propane and butane are removed, is gas condensate (or, simply, condensate), which is mixed with the crude oil or exported as a separate product (low-boiling naphtha) (Mokhatab et al., 2006; Speight, 2014, 2019a).

Thus, in the case of associated gas, crude oil may be assisted up the wellbore by gas lift (Mokhatab et al., 2006; Speight, 2014, 2019a) in which the gas is compressed into the annulus of the well and then injected by means of a gas lift valve near the bottom of the well into the crude oil column in the tubing. At the top of the well the crude oil and gas mixture passes into a separation plant (consisting of high-pressure and low-pressure separators) in which the gas pressure is reduced considerably in two stages. The crude oil and water exits the bottom of the lower-pressure separator, from where it is pumped to tanks for separation of the crude oil and water. The gas produced in the separators is recompressed and the gas that comes out of solution with the produced crude oil (surplus gas) is then treated to separate out the NGLs that are treated in a gas plant to provide propane and butane or a mixture of the two (liquefied petroleum gas). The higher boiling residue, after the propane and butane are removed, is condensate, which is mixed with the crude oil or exported as a separate product. At each stage of this process (often referred to under the collective term wellhead processing), the composition of the gaseous and liquid products should be monitored to determine separator efficiency as well as for safety reasons (Colborn et al., 2011).

The gas itself is then dry and, after compression, is suitable to be injected into the natural gas system where it substitutes for natural gas from the non-associated gas reservoir. Pre-treated associated gas from other fields can also enter the system at this stage. Another use for the gas is as fuel for the gas turbines on site. This gas is treated in a fuel gas plant to ensure it is clean and at the correct pressure. The startup fuel gas supply will be from the main gas system, but facilities exist to collect and treat low-pressure gas from the various other plants as a more economical fuel source.

Other components such as carbon dioxide (CO_2), hydrogen sulfide (H_2S), mercaptans (thiols; RSH), as well as trace amounts of other constituents (such as helium) may also be present while methane (CH_4) and ethane (C_2H_6) constitute the bulk of the combustible components; However, there is no single composition of components which might be termed typical natural gas because of the variation in composition of the gas from different reservoirs, even from different wells from the same reservoir.

Throughout this text, the term "associated gas" is applied to natural gas that exists in a reservoir along with the water base and crude oil whereas non-associated gas is the gas that occurs in a (gas-and-water-only) reservoir where crude oil is not present. Thus:

Reservoir contents	Description	Type of gas
Gas	Gas above the oil (gas cap)	Associated gas
Crude oil Water	Gas dissolved in oil	Associated gas
Gas Water		Non-associated gas

Some observers have chosen to apply the term "non-associated gas" to the gas cap in a water-crude oil-gas reservoir. However, the definition does not allow for any changes that occur to the gas due to the back-and-forward dissolution and release of gaseous constituents from the crude oil to the gas cap and this form of the definition tends to assume that the gas and crude oil reach (and forever remain at) the equilibrium concentration of the gas in the oil.

1.4.2 Non-associated gas

In addition to the natural gas fund in crude oil reservoirs, there are also those reservoirs in which natural gas is the sole occupant and is referred to as non-associated gas. As with associated gas, the principal constituent of non-associated gas is methane – higher molecular weight hydrocarbon derivatives may also be present but in lower quantities than occur in associated gas. Carbon dioxide is also a common constituent of non-associated natural gas and trace amounts of rare gases, such as helium, may also occur, and certain natural gas reservoirs are a source of these rare gases.

Thus, non-associated gas (sometimes called well gas) is produced from geological formations that typically do not contain much, if any, crude oil, or higher boiling hydrocarbon derivatives (gas liquids) than methane. Non-associated gas occurs in reservoirs that contain low amounts (if any) of crude oil in the reservoir. Typically, these reservoirs are found at greater depth. If the gas brought to the surface through a well still remains gas, it is referred to as dry gas but if the surface pressures cause some liquid hydrocarbon derivatives to separate from the gas stream, the gas is referred to as wet gas. It must be emphasized that while non-associated gas reservoirs are likely to be found at greater depths, upward migration from the source rock, in geologic time, can result in shallow gas reservoirs.

The non-associated gas recovery system is somewhat simpler than the associated gas recovery system. Typically, non-associated gas flows up the production well under the reservoir energy and then through the wellhead control valves and along the flow line to the wellhead processing plant. At this stage, the first processing option is to reduce the temperature of the gas to a point dependent upon the pressure in the pipeline so that the higher molecular weight constituents which would exist as liquids at the temperature and pressure of the pipeline condense to a liquid phase and are removed in a separator. The temperature is reduced by expanding the gas through a Joule-Thomson valve, although other methods of removal do also exist (Mokhatab et al., 2006; Speight, 2014, 2019a). Briefly, the Joule-Thomson effect (also known as the Joule–Kelvin effect, the Kelvin–Joule effect, or the Joule-Thomson expansion) relates to the temperature change of a gas or liquid when it is forced through a valve while kept insulated so that no heat is exchanged with the environment.

Water in the gas stream must also be removed to mitigate the potential for the formation of gas hydrates (Gornitz and Fung, 1994; Collett, 2002; Buffett and Archer, 2004; Collett et al., 2009; Demirbaş, 2010a, 2010b, 2010c: Boswell and Collett, 2011; Chong et al., 2016) which would block flow lines and have the potential for explosive dissociation. One method for water removal from the gas stream, involves the injection of ethylene glycol (HOCH$_2$CH$_2$OH, also referred to as glycol) which combines with the water and is later recovered in a glycol plant (Mokhatab et al., 2006; Speight, 2014, 2019a). The treated gas passes from the top of the treatment vessel and into the pipeline. The water is treated in a glycol plant to recover the glycol and the fraction of the natural gas stream that has been isolated as natural gas liquids is sent, as additional feedstock, to the plant where liquefied petroleum gas is produced. Alternatively, the lower-boiling constituents of the natural gas liquids may be used as feedstock for the production of petrochemicals (Hsu and Robinson, 2017; Speight, 2014, 2019c).

Finally, one other aspect of gas processing that requires attention and is worthy of mention here is the removal of sulfur-containing derivatives from natural gas. The potential of sulfur-containing constituents, such as hydrogen sulfide (H_2S) and mercaptans (RSH), to corrode shipping equipment (such as pipelines) is high – especially in the presence of water. Once the hydrogen sulfide has been removed by a suitable wellhead treatment process – it is environmentally undesirable to flare the hydrogen sulfide, so where there are significant quantities in the gas stream, it is converted into elemental sulfur and used for the manufacturer of sulfuric acid and other products. The sulfur can be transported over long distances by being pumped as a liquid at a temperature on the order of 120 °C (248 °F) through an insulated pipeline, which is maintained at this temperature by a counter flow of hot pressurized water.

1.4.3 Gas in geopressurized zones

Geopressure (sometimes referred to as overpressure) occurs when the pore fluid pressure significantly exceeds the pressure predicted from the normal compaction of sediments with depth. Thus, geopressured gas reservoirs are abnormally pressured reservoirs. The initial pressure gradients in normally pressured reservoirs range between 0.43 psi/ft and 0.5 psi/ft. In geopressured reservoirs, initial pressure gradients are between 0.6 psi/ft and 1.0 psi/ft.

Thus, in the current context the term "geopressure" refers to the pressure of a reservoir fluid (including gas) that significantly exceeds hydrostatic pressure (which is on the order of 0.4–0.5 psi per foot of depth) and may even approach overburden pressure (on the order of 1.0 psi per foot of depth). Thus, geopressurized zones are natural underground formations that are under unusually high pressure for their depth. Typically, the hydrostatic pressure (which is on the order of 0.4–0.5 psi per foot of depth) may even approach overburden pressure (that is on the order of 1.0 psi per foot of depth) and, thus, the geopressurized zones are natural underground formations that are under unusually high pressure for their depth.

In normally pressured gas reservoirs, gas compressibility is dominant in comparison to water and formation (rock) compressibility as the source of energy for gas production. For this reason, water and formation compressibility are considered negligible in the development of the material balance equation for volumetric, normally pressured gas reservoirs. In geopressured reservoirs, water and formation compressibility can be almost as high as gas compressibility at initial pressures. Consequently, it is necessary to include water and formation compressibility in developing material balance equations for geopressured reservoirs.

These zones are formed by layers of clay that are deposited and compacted very quickly on top of more porous, absorbent material such as sand or silt. Water and natural gas that are present in this clay are squeezed out by the rapid compression of the clay and enter the more porous sand or silt deposits. Geopressured reservoirs frequently are associated with substantial faulting as well as complex stratigraphic structures and this form of geologic complexity can contribute to the uncertainty of in the estimates of reserves that are based on pressure/production performance. In addition, geopressurized zones are typically located at great depths, typically on the order of 10,000–25,000 feet below the surface of the Earth. The combination of all these factors complicates the extraction of natural gas from geopressurized zones. On the upside, since geopressured resources occur when there is an impermeable layer of sedimentary cap rock that traps a geothermal reservoir, the geopressured resources typically range from 90 to > 200 °C (195 to >390 °F), and the increase in pressure reduces the energy required to pump the resource making geopressured resources desirable.

The amount of natural gas in these geopressurized zones remains subject to considerable uncertainty although unproven estimates indicate that 5,000–49,000 trillion cubic feet (5,000–49,000 × 10^{12} ft^3) of natural gas may exist in these areas. Like gas hydrates, the gas in the geopressurized zones offers an opportunity for future supplies of natural gas.

1.4.4 Gas in tight formations

The term "tight formation" typically refers to a non-shale, sedimentary formation which has a relatively low (or low-to-no) permeability that contains natural gas and/or crude oil (Speight, 2017b). A tight reservoir (tight sands) is a low-permeability sandstone reservoir that produces primarily dry natural gas but (because of the low-to-no permeability) from which gas cannot be produced at economic flow rates unless the reservoir is stimulated by application of hydraulic fracturing and/or by the use of horizontal wellbores. This definition also applies low-to-no permeability carbonate reservoirs – shale gas reservoirs are also included in the tight formation term by some observers.

Typically, tight formations which formed under marine conditions contain lower amounts of clay minerals and are more brittle that non-tight formations and, thus, are more suitable for hydraulic fracturing than formations formed in fresh water which may contain higher amounts of clay minerals. The formations become more brittle with an increase in quartz (SiO_2) content and carbonate content (such as calcium carbonate, $CaCO_3$, or dolomite, $CaCO_3.MgCO_3$).

By way of explanation and comparison, in a conventional sandstone reservoir the pores are interconnected so that natural gas and crude oil can flow easily through the reservoir and to the production well. Conventional gas typically is found in reservoirs with permeability greater than 1 milliDarcy (mD) and can be extracted by use of traditional recovery techniques. In contrast, unconventional gas is found in reservoirs with relatively low permeability (less than 1 mD) and hence cannot be extracted via conventional methods. However, in tight sandstone formations, the pores are smaller and are poorly connected (if at all) by very narrow capillaries which result in low permeability and, hence, immobility of the natural gas. The tight gas is contained in lenticular or blanket reservoirs that are relatively impermeable, which occur downdip from water-saturated rocks and cut across lithologic boundaries. The reservoirs may contain a large amount of in-place gas but exhibit low recovery rates. Gas can be recovered from tight formations by creating downhole fractures with explosives or hydraulic pumping. The nearly vertical fractures provide a pressure sink and channel for the gas, creating a larger collecting area so that the gas recovery is at a faster rate.

In terms of chemical makeup, shale gas is typically a dry gas composed primarily of methane, but some formations do produce wet gas while crude oil from tight formations is typically more volatile than many crude oils from conventional reservoirs. The challenge in treating such gases is the low (or differing) hydrogen sulfide/carbon dioxide ratio and the need to meet pipeline specifications.

By way of recall, a specification is the collected data that give adequate control of natural gas (or condensate) behavior in a gas processing plant or refinery or for sales of the gas. More accurately, the specifications are derived from the set of tests and data limits applicable to the natural gas or to a finished product in order to ensure that every batch is of satisfactory and consistent quality at release for sales. The specifications should include all critical parameters in which variations would be likely to affect the safety and in-service use of the product.

1.4.5 Gas hydrates

The concept of natural gas from methane hydrates (also called gas hydrates, methane clathrates, natural gas hydrates, hydromethane, methane ice, fire ice) offers additional reserves of methane (Wang and Economides, 2009; Makogon, 2010). In terms of gas availability from this resource, one liter ($0.035 ft^3$) of solid methane hydrate can contains up to 168 liters ($0.59 ft^3$) of methane gas (Table 1.5).

Table 1.5: Density and molar volume of selected gas (hydrocarbon) hydrates.

Gas	Formula	Molecular weight	Density (g/cm³)	Mole volume (cm³/mol)
CH_4	$CH_4 \cdot 6H_2O$	124	0.910	136.264
C_2H_6	$C_2H_6 \cdot 7H_2O$	308	0.959	162.669
C_3H_8	$C_3H_8 \cdot 17H_2O$	350	0.866	404.157
iC_4H_{10}	$iC_4H_{10} \cdot 17H_2O$	354	0.91	403.996

The methane that is sequestered within the hydrate structure is generated by bacterial degradation of organic matter in an low oxygen (anaerobic) environment. Organic matter in the uppermost part of a sediment is first attacked by aerobic bacteria which generates carbon dioxide that escapes from the sediments into the water column wherein sulfate ($-SO_4$) derivatives are reduced to sulfide ($-S-$) derivatives. If the sedimentation rate is low (<1 cm per 1,000 years), the organic carbon content is low (<1%), and oxygen is abundant which allows the aerobic bacteria use all of the organic matter in the sediment. However, when the sedimentation rate is high and the organic carbon content of the sediment is high, the pore water in a sediment is typically anoxic at depths of less than one foot and methane is produced by anaerobic bacteria.

The three major conditions that promote hydrate formation are thus: (i) high gas pressure, (ii) low gas temperature, and (iii) the gas is at or below the water dew point with free water present (Sloan, 1998b; Collett et al., 2009). The hydrate derivatives form by migration of gas from deeper areas through geological faults, followed by precipitation (or crystallization) when there is contact of the rising gas stream with cold sea water. At high pressure, methane hydrates remain stable at temperatures up to 18 °C (64 °F) and the typical methane hydrate contains one molecule of methane for every six molecules of water that forms the ice cage. However, the methane (hydrocarbon)-water ratio is dependent on the number of methane molecules that fit into the cage structure of the water lattice.

Hydrate formation is favored by low temperature and high pressure – typically, gas hydrates form at temperatures on the order of 0 °C (32 °F) and elevated pressures (Sloan, 1998a; Lorenson and Collett, 2000). Chemically, gas hydrates are nonstoichiometric compounds with specific properties that differentiate the hydrates from ice and no actual chemical bond exists between guest and host molecules – the methane's relationship is that of a physical (cage) effect. Most methane hydrate deposits also contain lesser amounts of other hydrocarbon hydrates – these include ethane hydrate and propane hydrate. However, other nonpolar components such as argon (Ar) and ethyl cyclohexane ($C_6H_{11}C_2H_5$) can also form hydrates.

In the hydrate structure, methane is trapped within a cage-like crystal structure composed of water molecules in a structure that resembles ice (Buffett and Archer, 2004). Under the appropriate pressure, gas hydrates can exist at temperatures significantly above the freezing point of water, but the stability of the hydrate derivatives depends on pressure and gas composition and is also sensitive to temperature changes. For example, methane plus water at 600 psi forms hydrate at 5 °C (41 °F), while at the same pressure, methane with 1% v/v propane forms a gas hydrate at 9.4 °C (49 °F). Hydrate stability can also be influence by other factors, such as salinity.

Caution is advised when drawing generalities related to the formation and the stability of gas hydrates. Methane hydrates are also formed during natural gas production operations, when liquid water is condensed in the presence of methane at high pressure. Higher molecular weight hydrocarbon derivatives do not fit readily into the water cage structure and tend to destabilize the formation of hydrates. However, for this text, the emphasis is on methane hydrates (Collett, 2002; Collett et al., 2009; Demirbaş, 2010a, 2010b, 2010c; Chong et al., 2016).

As might be anticipated, the gas that is evolved from gas hydrates varies over a considerable range and is variously composed of the following hydrocarbon gases: methane (CH_4), ethane (C_2H_6), and the sum of the C_{3+} hydrocarbon derivatives composed of propane (C_3H_8), n-butane (n-C_4H_{10}), and isobutane (i-C_4H_{10}) (Lorenson and Collet, 2000). Other hydrocarbon derivatives such as the low boiling liquids are present in some sediments and include n-pentane (C_5H_{12}), iso-pentane (C_5H_{12}), neo-pentane (neo-C_5H_{12}), cyclopentane (cyclo-C_5H_{12}), n-hexane (n-C_6H_{14}), iso-hexane (i-C_6H_{14}), neo-hexane (neo-C_6H_{14}), n-heptane (n-C_7H_{16}), iso-heptane (i-C_7H_{16}), and methylcyclohexane (cyclo-$C_6H_{11}CH_3$) (Lorenson and Collet, 2000). Hydrogen sulfide (H_2S) has been detected in gas hydrates but because hydrogen sulfide is water-soluble, the gas hydrate dissociation measurements have the potential to be contaminated with some sediment and pore water, it is also possible that in some cases the hydrogen sulfide may not have existed within the has hydrate structure.

Gas production of from gas hydrates can be achieved by methods such as: (i) thermal stimulation, (ii) depressurization, or (iii) chemical inhibition. The gas content of gas hydrates typically involves the controlled decomposition of the hydrate in a dissociation chamber and allowed to decompose at room temperature. The gas pressure of the chamber will increase and reach a stable value over a period of time after which the hydrate gas is ready for analysis. The volume of water that remains in the dissociation chamber is then measured and stored in a glass ampoule for determination of the chemical composition and to ensure that no gases remain dissolved. Gas analysis will show that the gas released from the hydrate is predominantly methane with trace-to small-(typically < 5% v/v) of ethane, propane, isobutane, normal butane, nitrogen, carbon dioxide, and hydrogen sulfide (Gabitto and Tsouris, 2010).

To this point, the heat of dissociation (ΔH_d) is the enthalpy change to dissociate the hydrate phase to a vapor and aqueous liquid, with values given at temperatures just above the ice point. Also, the heat of dissociation is a function of the number of crystal hydrogen bonds (generally assumed to be the same as the hydration number). However, the value of the heat of dissociation is relatively constant for molecules which occupy the same cavity, within a wide range of component sizes. The enthalpy of dissociation may be determined via the univariant slopes of phase equilibrium lines (ln P versus 1/T) using the Clausius–Clapeyron relationship (Sloan, 2006):

$$(\Delta H\,d) = -zR_d(\ln P)/d(1/T)$$

Gas hydrates may formally be referred to as chemical compounds because they have a fixed composition at a certain pressure and temperature and form because of the Van der Waals attraction forces between the molecules. Covalent bonding is absent in the gas hydrates because during their formation there is no pairing of valence electrons and no spatial redistribution of electron cloud density.

1.4.6 Other definitions

In addition to the definitions presented above, there are several other definitions that have been applied to natural gas from conventional formations that can also be applied to gas from any source.

The terms "rich gas" and "lean gas," as used in the gas processing industry, are not precise indicators of gas quality but only indicate the relative amount of natural gas liquids in the gas stream. Thus, lean gas is gas in which methane is the predominant major constituent with other hydrocarbon constituents in the low minority while wet gas contains considerable amounts of the higher molecular weight hydrocarbon derivatives than dry gas (Table 1.6).

Table 1.6: Composition of dry gas, wet gas, and gas condensate.

Component	Dry gas	Wet gas	Condensate
Carbon dioxide (CO_2)	0.10	1.41	2.37
Nitrogen (N_2)	2.07	0.25	0.31
Methane (CH_4)	86.12	92.46	73.19
Ethane (C_2H_6)	5.91	3.18	7.80
Propane (C_3H_8)	3.58	1.01	3.55
n-Butane ($n\text{-}C_4H_{10}$)		0.24	1.45
iso-Butane ($i\text{-}C_4H_{10}$)	1.72	0.28	0.71
n-Pentane ($n\text{-}C_5H_{12}$)	–	0.08	0.68
iso-Pentane ($i\text{-}C_5H_{12}$)	0.50	0.13	0.64
Hexane isomers (C_6H_{14})	–	0.14	1.09
Heptane isomers-plus* ($\geq C_7H_{16}$)	–		

*Indicates higher molecular weight hydrocarbon derivatives.

Sour gas contains hydrogen sulfide and the equally odorous mercaptans whereas *sweet* gas contains little, if any, hydrogen sulfide or mercaptan. Residue gas is natural gas from which the higher molecular weight hydrocarbon derivatives have been extracted and casinghead gas is derived from crude oil but is separated at the separation facility at the wellhead. However, the term "residue" (as in *residue gas*) is used in relation to gas as a direct opposite as it applied to crude oil in a refinery (Parkash, 2003; Speight, 2014; Hsu and Robinson, 2017; Speight, 2017a, 2019a). In the refinery, the residue is the distillation residue of crude oil from which the lower molecular weight constituents have been removed. In natural gas technology, residue gas is natural gas from which the higher molecular weight constituents have been removed during gas processing operations (Appendix D) to leave methane (the lower-boiling constituent) as residue gas.

After production, the gas is stored in the reservoir formation in three different ways: (i) by adsorption, which refers to adsorbed gas that is physically attached (adsorption) or chemically attached (chemisorption) to organic matter or to clay minerals, (ii) non-adsorbed gas, which refers to free gas (also referred to as *non-associated gas*) that occurs within the pore spaces in the reservoir rock or in spaces created by the rock cracking (fractures or microfractures), and (iii) by solution, also referred to as associated gas which is gas that that exists in solution in liquids such as crude oil, heavy crude oil, and (in the current context) in the gas condensate that occurs in some tight reservoirs with the gas (Speight, 2014, 2019a).

The amount of adsorbed gas component (which is, typically, methane) usually increases with an increase in organic matter or surface area of organic matter and/or clay. On the beneficial side, a higher free-gas (non-associated) content in unconventional tight reservoirs generally results in higher initial rates of production because the free gas resides in fractures and pores and, when production is commenced, moves easier through the fractures (induced channels) relative to any adsorbed gas. However, the high, initial flow rate will decline rapidly to a low, steady rate as the non-associated gas is produced leaving the adsorbed gas to move to the well as it is slowly released from the shale.

1.5 Gas condensate

Natural gas condensate (sometimes referred to as condensate) is a mixture of low-boiling water-white or low-boiling straw-colored hydrocarbon liquids that is obtained when the vapors of these hydrocarbon constituents condenses either in the well or as the gas stream emits from the well – an older name for gas condensate is gas distillate (Mokhatab et al., 2006; Speight, 2014, 2019a).

Typically, the condensate is predominately pentane (C_5H_{12}) with varying amounts of higher-boiling hydrocarbon derivatives (up to C_8H_{18}) but relatively little methane or ethane; propane (C_3H_8) butane (C_4H_{10}) may be present in condensate by dissolution in the liquids. Other low-boiling hydrocarbon derivatives such as benzene (C_6H_6), toluene ($C_6H_5CH_3$), xylene isomers ($CH_3C_6H_4CH_3$), and ethyl benzene ($C_6H_5C_2H_5$) may also be present in the condensate (Mokhatab et al., 2006; Speight, 2014).

The condensate must be stabilized by removing the highly volatile constituents and, at the same time, increasing the amounts of the intermediate constituents (C_3 to C_5 hydrocarbon derivatives) and higher molecular weight constituents (C_6 and C_{6+} hydrocarbon derivatives) in the condensate. The stabilization process reduces the vapor pressure of the condensate by separation of any hydrocarbon gases (such as methane and ethane) from the higher-boiling (C_3+) hydrocarbon components. In this way, a vapor phase is not produced upon flashing the liquid to atmospheric storage tanks.

After degassing and dewatering in the production separation process, the pressurized liquid condensate enters the condensate stabilizer and flows through an exchanger in which hot, stabilized condensate is used to preheat the unstabilized condensate. In the next step, the unstable condensate flows to a line heater, where it is heated to stabilization temperature on the order of 95–120 °C (205–250 °F). The heated, unstabilized condensate is flash-distilled in the condensate separator at approximately 35–45 psi to remove low-density hydrocarbon vapors and any remaining water. The stabilized condensate flows through the plate exchanger for cooling and then to atmospheric storage.

One way to prevent the formation of condensate is to maintain the flowing well bottomhole pressure above the dew point pressure. This is often unsatisfactory because the drawdown (reservoir pressure minus flowing bottomhole pressure) may not be sufficient enough for the economic production rate. An alternative technique is to allow the formation of condensate, but occasionally to inject methane into the production well. The methane sweeps the liquid condensate into the reservoir after which the well is then put back in production. This approach is repeated several times in the life of the well (gas cycling). Another method is to inject both nitrogen and methane, which develops a miscible displacement process and results in high condensate recovery (Sanger and Hagoort, 1998).

1.5.1 Lease condensate

Lease condensate (having wide range of API gravity on the order of 45–75°) is produced at the wellhead as an unprocessed (unrefined) liquid except for stabilization at or near the wellhead. Lease condensate can be produced along with significant volumes of natural gas and is typically recovered at atmospheric temperatures and pressures from the wellhead gas production. These raw condensates come out of the ground as mixtures of various hydrocarbon compounds including NGLs, pentane derivatives (C5+), hexane derivatives (C6+), and depending on the condensate, varying amounts of higher molecular weight hydrocarbon derivatives in the C7, C8 range and even up to C10.

A lease condensate with a high API is typically clear or translucent in color and may contain substantial amounts of natural gas liquids (including ethane, propane, and butane) but not many of the higher molecular weight hydrocarbon derivatives.

The lower-API gravity condensates (API gravity on the order of 45°) level look more like crude oil and have much higher concentrations of the higher molecular weight hydrocarbon derivatives (C7, C8+).

The higher API gravity condensate can be difficult to handle because of the high vapor pressure and is usually stabilized at the wellhead. In this wellhead process (as compared to the more detailed processes), the condensate is passed through a stabilizer which may be nothing more than a large tank or a series of tanks that allows the high vapor pressure components (the NGLs) to vaporize and to be collected for processing. This process leaves a stabilized condensate that has a lower vapor pressure which is easier to handle, particularly when it must be shipped by truck or by rail.

1.5.2 Plant condensate

Plant condensate is a product of the processing of natural gas liquids and is similar to the product known as natural gasoline which is composed of hydrocarbon components that are similar to the hydrocarbon constituents in lease condensate, that is, pentane derivatives (C_{5s}), some hexane derivatives (C_{6s}), and small quantities of higher molecular weight hydrocarbon derivatives. Since plant condensate originates in a processing plant, the condensate is considered to be processed (manufactured) product rather than a naturally occurring product.

Furthermore, the quality of the natural gasoline (defined by the specifications) is in closer ranges than lease condensates. However, both products can be used interchangeably in some markets, such as a blend stock for crude oil and/or as a diluent for tar sand bitumen (Speight, 2014, 2017a).

1.5.3 Natural gasoline

Natural gasoline is a mixture of hydrocarbon derivatives (extracted from natural gas) that consists mostly of pentane (C5) derivatives and higher-boiling (C6+) hydrocarbon derivatives. In the beginning of the natural gasoline industry, the only use for natural gasoline was as motor fuel or as a blending agent in the production of motor fuel. This was followed by separation of the individual components of natural gasoline – namely isobutane, n-butane, and pentane derivatives as base stocks for reforming processes, alkylation processes, synthetic rubber products, and other petrochemicals (Speight, 2014, 2017a, 2019c).

On the other hand, liquefied refinery gases are mixtures of hydrocarbon derivatives – principally of propane, butane derivatives – which are recovered during the refining of crude oil. These materials are composed and are more likely to contain olefin derives such as propylene (C_3H_6, $CH_3CH=CH_2$) and butylene derivatives

(C_4H_{10}, $CH_3CH_2CH=CH_2$, $CH_3CH=CHCH_3$) which can be stored and handled at ambient temperatures and at moderate pressures.

1.5.4 Low-boiling naphtha

Low-boiling naphtha is a hydrocarbon product that is similar in properties to lease condensate and natural gasoline but it (low-boiling naphtha) is considered to be a refined product since it is produced by distillation. In fact, most naphtha is produced from crude oil in the first step of the refining process – distillation (Speight, 2014, 2017a). Low-boiling naphtha is composed predominantly of pentane (C5) and hexane (C6) derivatives as well as smaller amounts of higher molecular weight (C7+) hydrocarbon derivatives.

   Another source of naphtha is condensate which can be processed through a stand-alone, small crude oil distillation tower (sometime referred to as a splitter) which is effective for the separation of the naphtha constituents from the more volatile constituents of the condensate and from the higher-boiling hydrocarbon derivatives. The naphtha product can be used as a feedstock for refinery upgrading processes and also for petrochemical production (Speight, 2014; Speight, 2019c).

1.6 Uses

Natural gas itself is a versatile, clean-burning, and efficient fuel that is used in a wide variety of applications as well as for the production of a variety of chemicals, especially when the natural gas is used as the starting point for the production of synthesis gas (a mixture of hydrogen and carbon monoxide) and conversion of the synthesis gas to a variety of chemicals (Speight, 2017a, 2019a; Tables 1.7–1.9).

Table 1.7: Products from natural gas.

Natural gas processing	Non-hydrocarbon products
	Carbon dioxide
	Hydrogen
	Hydrogen sulfide
	Nitrogen
	Water
	Hydrocarbon products
	Ethane
	Butane
	Natural gasoline (condensate)

Table 1.8: Examples of routes to chemicals from natural gas via synthesis gas.

Starting material	Intermediate	Product
Natural gas (methane)	Synthesis gas	Oxo synthesis
		Alcohols
		Aldehydes
		Fischer–Tropsch synthesis
		Naphtha
		Diesel
		Kerosene
		Lubricants
		Waxes
		Carbonylation
		Formic acid
		Methanol
		Acetic acid
		Dimethyl ether
		Formaldehyde

Table 1.9: Natural gas liquids, uses, products, and consumers.

NGL	Chemical formula	Uses	Other uses
Ethane	C_2H_6	Ethylene production Power generation	Plastics Anti-freeze Detergents
Propane	C_3H_8	Heating fuel Transportation Petrochemical feedstock	Plastics
Butanes: n-butane and iso-butane	C_4H_{10}	Petrochemical feedstock Refinery feedstock Blend stock for gasoline	Plastics Synthetic rubber
Condensate	C_5H_{12} Higher-boiling hydrocarbons	Petrochemical feedstock Additive to gasoline Diluent for heavy crude oil	Solvents

More than 2,000 years ago, the Chinese discovered that the energy in natural gas could be harnessed and used as a heat source and from that time the use of natural gas has increased (Mokhatab et al., 2006; Speight, 2014). However, in the late nineteenth and in the early twentieth centuries, natural gas played a subsidiary role to coal gas insofar as coal gas was the fuel of choice for street lighting as well as for lighting in buildings (Mokhatab et al., 2006; Speight, 2013). However, as the twentieth century progressed and moved into the twenty-first century, the discovery of large reserves of natural gas in various countries (including gas in shale formations) as well as improved distribution of natural gas expanded the use of the gas into a wide variety of uses in homes, businesses, factories, and power plants (Nersesian, 2010; Hafner and Tagliapietra, 2013). Once the transportation of natural gas was possible over considerable distances, the increased use of natural gas led to new uses for natural gas. In fact, the fastest growing use of natural gas is for the generation of electric power and, to a large extent, has been a replacement fuel for many formerly coal-fired power plants and crude oil-fired power plants (Speight, 2021).

As the use of natural gas has increased and diversified, the need for an understanding of the composition and properties (and behavior) of the gas has also increased (Mokhatab et al., 2006; Speight, 2018). Natural gas has many applications: for domestic use, for industrial use, and for transportation. In addition, natural gas is also a raw material for many common products such as paints, fertilizer, plastics, antifreeze, dyes, photographic film, medicines, and explosives. Along with these newer uses, there has been an increased need not only for the compositional analysis of natural gas but also for analytical data that provide other information related to the behavior of natural gas.

Natural gas–fired power plants are currently among the cheapest power plants to construct which is a reversal of previous trends where operating costs were generally higher than those of coal-fired power plants because of the relatively high cost of natural gas. In addition, natural gas–fired plants have greater operational flexibility than coal-fired power plants because they can be fired up and turned down rapidly. Because of this, many natural gas plants in the United States were originally used to provide additional capacity (peak capacity, i.e., availability of the gas during peak-use periods) at times when electricity demand was especially high, such as the summer months when air-conditioning is widely used. During much of the year, these natural gas peak plants were idle, while coal-fired power plants typically provided base-load power. However, and natural gas is now increasingly used as for base-load power as well as for intermediate-load power source in many cities. Natural gas can also be used to produce both heat and electricity simultaneously (cogeneration or combined heat and power, CHP). Cogeneration systems are highly efficient, able to put 75–80% of the energy in gas to use. Trigeneration systems, which provide electricity, heating, and cooling, can reach even higher efficiencies using natural gas.

Natural gas also has a broad range of other uses in industry, not only as a source of both heat and power and as source of valuable hydrogen that is necessary for crude oil refining as well as for producing plastics and chemicals. Most hydrogen gas (H_2) production, for example, comes from reacting high temperature water vapor (steam) with methane – steam-methane reforming reaction followed by the water gas shift reaction:

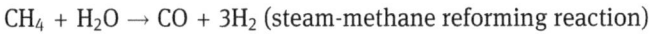

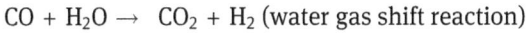

$CH_4 + H_2O \rightarrow CO + 3H_2$ (steam-methane reforming reaction)
$CO + H_2O \rightarrow CO_2 + H_2$ (water gas shift reaction)

Natural gas–fired plants have greater operational flexibility than coal plants because they can be fired up and turned down rapidly. Because of this, many natural gas plants were originally used to provide peaking capacity at times when electricity demand was especially high, such as the summer months when air-conditioning is widely used. During much of the year, the natural gas peak-plants were in low use or idle, while coal-fired power plants typically provided base load power. However, (i) with the current (and projected prolonged) plentiful supplies of natural gas, (ii) lower natural gas prices, and (iii) the projected environmental benefits of natural gas use vis-à-vis coal, natural gas is now increasingly used as a base and intermediate load power source in many places.

The integrated gasification combined cycle (IGCC) plant can be used as an example of the benefits of gas-fired power generation. The natural gas is combusted in a gas turbine unit that is connected to a generator after which the hot exhaust gases are then passed through a heat exchanger to generate steam for a steam turbine. By using this approach, a natural gas combined cycle (NGCC) power plant can reach efficiencies at least on the order of 50%, compared to a lower efficiency (30–35%) for a similar megawatt-size coal-fired power plant.

Furthermore, the hydrogen produced from natural gas can itself be used as a fuel. The most efficient way to convert hydrogen into electricity is by using a fuel cell, which combines hydrogen with oxygen to produce electricity, water, and heat. Although the process of reforming natural gas to produce hydrogen still has associated carbon dioxide emissions, the amount released for each unit of electricity generated is much lower than for a combustion turbine.

As part of the industrial use of natural gas, there is the need for analysis before the products (in this context, the gaseous products) are used by industrial and domestic consumers. Detection of even the slightest amounts of impurities can be an indication of process inefficiency and whether or not the gas is suitable for the designated use. In fact, one of the most important tasks in gas technology, especially in the context of crude oil–related natural gas, is the need for reliable values of the volumetric and thermodynamic properties for pure low-boiling hydrocarbon derivatives and their mixtures. These properties are important in the design and operation of much of the processing equipment (Poling et al., 2001).

For example, reservoir engineers and process engineers use pressure-volume-temperature (PVT) relationships and phase behavior of reservoir fluids (i) to estimate the amount of oil or gas in a reservoir, (ii) to develop a recovery process for a crude oil or gas field, (iii) to determine an optimum operating condition in a gas-liquid separator unit, (iv) to determine the need for a wellhead processing system to protect a pipeline from corrosion, and (v) to design suitable gas processing options. However, the most advanced design approaches or the most sophisticated simulation experiments cannot guarantee the optimum design or operation of a unit (or protection of a pipeline) if the physical properties are not known. For these reasons accurate knowledge of the properties of the gas is an extremely increasingly important aspect of gas technology.

Natural gas sees a broad range of other industrial uses, as a source of both heat and power and as an input for producing plastics and chemicals. For example, most of the hydrogen gas production comes from reacting high temperature water vapor (steam) with methane. The resulting hydrogen has wide use in crude oil refineries in order to produce marketable products from heavy crude oil, extra heavy cried oil and tar sand bitumen (Speight, 2014, 2017a) as well as to produce ammonia for fertilizer. Although the process of reforming natural gas to hydrogen still has associated carbon dioxide emissions, the amount released for each unit of electricity generated is much lower than for a combustion turbine.

Hydrogen produced from natural gas can itself be used as a fuel – the most efficient way to convert hydrogen into electricity is by using a fuel cell, which combines hydrogen with oxygen to produce electricity, water, and heat:

$$2H_2 + O_2 \rightarrow 2H_2O + \text{heat}$$

Compressed natural gas (CNG) has been used as a transportation fuel, mostly in public transit. The natural gas, which is compressed at over 3,000 psi to 1% of the volume that the gas would occupy at normal atmospheric pressure, can be burned in an internal combustion engine that has been appropriately modified. Approximately 0.1% v/v of the natural gas consumed in the United States has been used to power vehicles, representing the energy content of more than 5 million barrels of oil (US EIA, 2012).

Compared to gasoline, vehicles powered by CNG emit less carbon monoxide, nitrogen oxides (NO_x), and particulate matter. However, a disadvantage of CNG is the low energy density compared with the higher energy density of liquid fuels. A gallon of CNG has approximately one quarter of the energy in a gallon of gasoline and, therefore, vehicles powered by CNG require larger fuel tanks (compared to vehicles powered by liquid fuels).

Thus, a more suitable use for natural gas in the transportation sector may be as a resource to generate electricity for plug-in vehicles or hydrogen for fuel cell vehicles, which can reduce emissions savings on the order of 40% (or more).

NGLs, which are, by definition, hydrocarbon derivatives, also have use other than fuel components (Tables 1.7 and 1.8) because of the hydrocarbon constituents. Thus, there are many uses for natural gas liquids that span almost all sections of the industrial chemicals economy. Natural gas liquids are used feedstocks for petrochemical plants, burned for space heat and cooking, and blended into vehicle fuel.

The chemical composition of natural gas liquids from different sources is similar, yet their applications vary widely. Ethane occupies the largest share of the filed production of natural gas liquids and is used almost exclusively to produce ethylene, which is then converted into plastic products. By contrast, the majority of the propane, by contrast, is burned for heating, although a substantial amount is used as petrochemical feedstock. A blend of propane and butane, sometimes referred to as autogas (autogas), is a popular fuel in some parts of Europe, Turkey, and Australia. Natural gasoline (pentanes plus) can be blended into various kinds of fuel for combustion engines and is useful in energy recovery from wells and tar sand (oil sand) deposits.

A challenge with the use of natural gas liquids is that they are (i) expensive to handle, (ii) store, and (iii) transport compared to refined products. Natural gas liquids are highly flammable and require high pressure and/or low temperature to be maintained in the liquid state for shipping and handling. The flammability of these liquids necessitates the use of special trucks, ships, and storage tanks.

The last four decades of the twentieth century have seen not only perturbations of energy supply systems but also changes in attitudes of governments and voters alike toward environmental issues. Thus, environmental issues will be with us as long as there is manufacturing of e consumer goods, and the use of fossil fuels for energy production. It is this latter issue that is the subject of this text. The continued use of natural gas as combustible fuel is a reality and gas processing, although generally understandable using chemical and/or physical principles, still requires an attempt to alleviate some of the confusion that arises from uncertainties in the terminology.

Natural gas is a gaseous-phase fossil fuel which is colorless, odorless, shapeless, and lighter than air (Speight, 2018). When burned, natural gas produces approximately 1,000 Btu per scf (British thermal units per standard cubic foot, Btu/ft^3) and is used for domestic applications such as space heating, cooking, and, increasingly, to generate electricity. When compared with coal and crude oil, natural gas is a cleaner burning fuel and burns cleaner, more efficiently and with lower levels of potentially harmful by-products are released into the atmosphere. More important, there are large deposits of natural gas in the world (including the so-called tight gas – gas from low permeability formations).

Natural gas is the cleanest and most hydrogen rich of all hydrocarbon energy sources, and it has high energy conversion efficiencies for power generation. Of more significance is that gas resources discovered but as yet untapped remain plentiful. The sector is poised for considerable growth over the next two decades, and some believe that it may even overtake crude as the prime fuel within the next two to five decades. In fact, the trend toward natural gas becoming the premium fuel of the world economy is not now easily reversible. The key and the challenge for the energy industry is how the transition is to be managed. To develop a natural gas field, one of the first important steps is to understand the fundamentals of natural gas.

1.7 Energy security

Energy security is the uninterrupted availability of energy sources at an affordable price (IEA, 2018) or, put another way, a particular aspect of energy security is assuring access to a ready supply of energy (US DOE, 2017). Thus, energy security has many dimensions: long-term energy security mainly deals with timely investments to supply energy in line with economic developments and sustainable environmental needs. Short-term energy security focuses on the ability of the energy system to react promptly to sudden changes within the supply–demand balance. Lack of energy security is thus linked to the negative economic and social impacts of either (i) the unavailability of an energy source or (ii) the prices are overly are subject to the volatility of the prices of the energy source, or (iii) the prices overly volatile process of the energy source (Speight, 2011a; Speight and Islam, 2016; US DOE, 2017).

Historically, energy security for many countries was primarily associated with the supply of crude oil. Thus, using the international oil market as the example, prices are allowed to adjust in response to changes in supply and demand, and the risk of physical unavailability is limited to extreme events. However, in many instances, petro-politics (sometime called geo-politics) does play a role, thereby jeopardizing security of crude oil imports (Speight, 2011c).

In the United States and in many other energy-consuming countries, energy security has been an on-again-off-again political issue since the first Arab oil embargo in 1973. Since that time, the speeches of various governmental agencies have continued to call for an end to the dependence on energy sources but the governmental rhetoric of energy security and energy independence continues and meaningful suggestions of how to address this issue remain few and far between.

The past decade has yielded substantial change in the natural gas industry. Specifically, there has been rapid development of technology allowing the recovery of natural gas from tight formations such as shale formations and carbonate formations. In fact, in the United States, rapid growth in the production of natural gas from tight formations in North America has dramatically altered the global natural gas market landscape. Indeed, the emergence of shale gas (shale natural gas) is perhaps the most intriguing development in global energy markets.

As an example, beginning with the Barnett shale in northeast Texas, the application of innovative new techniques involving the use of horizontal drilling with hydraulic fracturing has resulted in the rapid growth in production of natural gas from shale (Speight, 2016a). In fact, the production of natural gas from shale formations in the United States has increased from virtually nothing in 2000 to (currently) more than 10 billion cubic feet per day (bcfd, 1×10^9 ft^3 per day) in 2010, and it is expected to more than quadruple by 2040, reaching 50% or more of total US natural gas production by the decade starting in 2030. Thus, natural gas stands to play an important role in the US energy mix for decades to come. Rising shale gas production has already delivered large beneficial impacts to the United States and since the US shale gas resources are generally located in close proximity to end-use markets where natural gas is utilized to fuel industry, generate electricity, and heat homes, shale gas offers both security of supply and economic benefits.

References

AAPG. 2015. Unconventional Energy Resources: 2015 Review. Natural Resources Research, 24(4): 443–508. American Association of Petroleum Geologists, Tulsa Oklahoma.

Aguilera, R.F., and Radetzki, M. 2014. The Shale Revolution: Global Gas and Oil Markets Under Transformation. Mineral Economics, 26(3): 75–84.

Boswell, R., and Collett, T.S. 2011. Current Perspectives on Gas Hydrate Resources. Energy Environ Sci. 4:1206–1215.

Buffett, B., and Archer, D. 2004. Global Inventory of Methane Clathrate: Sensitivity to Changes in the Deep Ocean. Earth and Planetary Science Letters, 227(3–4): 185.

Burruss, R.C., and Ryder, R.T. 2003. Composition of Crude Oil and Natural Gas Produced from 14 Wells in the Lower Silurian "Clinton" Sandstone and Medina Group, Northeastern Ohio and Northwestern Pennsylvania. Open-File Report 03–409, United States Geological Survey, Reston, Virginia.

Burruss, R.C., and Ryder, R.T. 2014. Composition of Natural Gas and Crude Oil Produced From 10 Wells in the Lower Silurian "Clinton" Sandstone, Trumbull County, Ohio. In: Coal and Petroleum Resources in The Appalachian Basin; Distribution, Geologic Framework, And Geochemical Character. L.F. Ruppert and R.T. Ryder (Editors). Professional Paper 1708, United States Geological Survey, Reston, Virginia.

Chong, Z.R., Yang, S.H.B., Babu, P., Linga, P., and Li, X.S. 2016. Review of Natural Gas Hydrates as an Energy Resource: Prospects and Challenges. Appl. Energy 162: 1633–1652.

Colborn, T., Kwiatkowski, C., Schultz, K., and Bachran, M. 2011. Natural Gas Operations from a Public Health Perspective. Hum. Ecol. Risk Assess., 17: 1039–1056.

Collett, T.S. 2002. Energy Resource Potential of Natural Gas Hydrates. AAPG Bulletin 86: 1971–1992.

Collett, T.S., Johnson, A.H., Knapp, C.C., and Boswell, R. 2009, Natural Gas Hydrates: A Review. In: Natural Gas Hydrates – Energy Resource Potential and Associated Geologic Hazards. T.S. Collett, A.H. Johnson, C.C. Knapp, and R. Boswell (Editors). AAPG Memoir No. 89, p. 146–219. American Association of Petroleum Geologists, Tulsa, Oklahoma.

Demirbaş A. 2010a. Methane from Gas Hydrates in the Black Sea. Energy Sources Part A, 32: 165–171.

Demirbaş A. 2010b. Methane Hydrates as Potential Energy Resource: Part 1-Importance, Resource and Recovery Facilities. Energy Convers. Manage., 51: 1547–1561.

Demirbaş A. 2010c. Methane Hydrates as Potential Energy Resource: Part 2 – Methane Production Processes from Gas Hydrates. Energy Convers. Manage. 51:1562–1571.

Gabitto, J., and Tsouris, C. 2010. Physical Properties of Gas Hydrates: A Review. Journal of
    Thermodynamics, Volume 2010, Article ID 271291. https://www.hindawi.com/journals/jther/
    2010/271291/citations/
Gary, J.G., Handwerk, G.E., and Kaiser, M.J. 2007. Petroleum Refining: Technology and Economics,
    5th Edition. CRC Press, Taylor & Francis Group, Boca Raton, Florida.
Gornitz, V., and Fung, I. 1994. Potential Distribution of Methane Hydrate in the World's Oceans.
    Glob. Biogeochem. Cycles, 8: 335–347.
Hafner, M., and Tagliapietra, S. 2013. The Globalization of Natural Gas Markets: New Challenges
    and Opportunities for Europe. Claeys & Casteels Law Publishers, Deventer, Netherlands.
    International Specialized Book Services, Portland Oregon.
Hsu, C.S., and Robinson, P.R. (Editors) 2017. Handbook of Petroleum Technology. Springer
    International Publishing AG, Cham, Switzerland.
IEA. 2018. https://www.iea.org/topics/energysecurity/; accessed May 21, 2108.
Islam, M.R. 2014. Unconventional Gas Reservoirs. Elsevier, Amsterdam, Netherlands.
John, E., and Singh, K. 2011. Production and Properties of Fuels from Domestic and Industrial
    Waste. In: The Biofuels Handbook. J.G. Speight (Editor). Royal Society of Chemistry, London,
    United Kingdom. Page 333–376.
Khosrokhavar, R., Griffiths, S., and Wolf, K-H. 2014. Shale Gas Formations and Their Potential for
    Carbon Storage: Opportunities and Outlook. Environmental Processes, 1(4): 595–611.
Kundert, D., and Mullen, M. 2009. Proper Evaluation of Shale Gas Reservoirs Leads to a More
    Effective Hydraulic-Fracture Stimulation. Paper No. SPE 123586. Proceedings. SPE Rocky
    Mountain Petroleum Technology Conference, Denver, Colorado. April 14–16.
Levine, J.R., 1993. Coalification: The Evolution of Coal as A Source Rock and Reservoir Rock for Oil
    and Gas. American Association of Petroleum Geologists, Studies in Geology, 38: 39–77.
Mokhatab, S., Poe, W.A., and Speight, J.G. 2006. Handbook of Natural Gas Transmission and
    Processing. Elsevier, Amsterdam, Netherlands.
Nersesian, R.L. 2010. Energy for the 21st Century 2nd Edition. M.E. Sharpe, Armonk, New York.
Parkash, S. 2003. Refining Processes Handbook. Gulf Professional Publishing, Elsevier,
    Amsterdam, Netherlands.
Poling, B.E., Prausnitz, J.M., and O'Connell, J.P. 2001. The Properties of Gases and Liquids
    5th Edition. McGraw-Hill New York.
Ramroop Singh, N. 2011. Biofuels. In: The Biofuels Handbook. J.G. Speight (Editor). Royal Society of
    Chemistry, London, United Kingdom. Page 160–198.
Sanger, P.J. and J. Hagoort. 1998. Recovery of Gas Condensate by Nitrogen Injection Compared With
    Methane Injection. SPE Journal 3(1): 26.
Singh, K., and Sastry, M.K.S. 2011. Production of Fuels from Landfills. In: The Biofuels Handbook.
    J.G. Speight (Editor). Royal Society of Chemistry, London, United Kingdom. Page 408–453.
Sloan, E.D. Jr. 1998a. Gas Hydrates: Review of Physical/Chemical Properties. *Energy & Fuels*, 12(2):
    191–196.
Sloan, E.D., Jr. 1998b. Clathrate Hydrates of Natural Gases 2nd Edition. Marcel Dekker Inc.,
    New York.
Sloan, E.D. Jr. 2006. Clathrate Hydrates of Natural Gases 3rd Edition. Marcel Dekker Inc., New York.
Speight, J.G. 2011b. An Introduction to Petroleum Technology, Economics, and Politics. Scrivener
    Publishing, Salem, Massachusetts.
Speight, J.G. 2011b. The Refinery of the Future. Gulf Professional Publishing, Elsevier, Oxford,
    United Kingdom.
Speight, J.G. 2011c. An Introduction to Petroleum Technology, Economics, and Politics. Scrivener
    Publishing, Salem, Massachusetts.

Speight, J.G. 2012. Crude Oil Assay Database. Knovel, New York. Online version available at: http://www.knovel.com/web/portal/browse/display?_EXT_KNOVEL_DISPLAY_bookid=5485&VerticalID=0

Speight, J.G. 2013. The Chemistry and Technology of Coal 3rd Edition. CRC Press, Taylor and Francis Group, Boca Raton, Florida.

Speight, J.G. 2014. The Chemistry and Technology of Petroleum 5th Edition. CRC Press, Taylor & Francis Group, Boca Raton, Florida.

Speight, J.G. 2015. Handbook of Petroleum Product Analysis 2nd Edition. John Wiley & Sons Inc., New York.

Speight, J.G., and Islam, M.R. 2016. Peak Energy – Myth or Reality. Scrivener Publishing, Beverly, Massachusetts.

Speight, J.G. 2016a. Handbook of Hydraulic Fracturing. John Wiley & Sons Inc., Hoboken, New Jersey.

Speight, J.G. 2016b. Introduction to Enhanced Recovery Methods for Heavy Oil and Tar Sands 2nd Edition. Gulf Publishing Company, Taylor & Francis Group, Waltham Massachusetts.

Speight, J.G. 2017. Handbook of Petroleum Refining. CRC Press, Taylor & Francis Group, Boca Raton, Florida.

Speight, J.G. 2017a. Handbook of Petroleum Refining. CRC Press, Taylor & Francis Group, Boca Raton, Florida.

Speight, J.G. 2017b. Deep Shale Oil and Gas. Gulf Professional Publishing, Elsevier, Oxford, United Kingdom.

Speight, J.G. 2018. Handbook of Natural Gas Analysis. John Wiley & Sons Inc., Hoboken, New Jersey.

Speight. J.G. 2019a. Natural Gas: A Basic Handbook 2nd Edition. Gulf Publishing Company, Elsevier, Cambridge, Massachusetts.

Speight, J.G. 2019b. Biogas: Production and Properties. Nova Science Publishers, New York.

Speight, J.G. 2019c. Handbook of Petrochemical Processes. CRC Press, Taylor & Francis Group, Boca Raton, Florida.

Speight, J.G. 2020a. Shale Oil and Gas Production Processes. Gulf Publishing Company, Elsevier, Cambridge, Massachusetts.

Speight, J.G. 2021. Coal-Fired Power Generation Handbook Second Edition. Scrivener Publishing, Beverly, Massachusetts.

US DOE. 2017. Report to Congress. Valuation of Energy Security for the United States. United States Department of Energy, Washington, DC. January. https://www.energy.gov/sites/prod/files/2017/01/f34/Valuation%20of%20Energy%20Security%20for%20the%20United%20States%20%28Full%20Report%29_1.pdf

US EIA. 2012. Natural Gas Consumption by End Use. United States Energy Information Administration. Washington, DC. https://www.eia.gov/dnav/ng/ng_cons_sum_dcu_nus_a.htm

Wang, X., and Economides, M. 2009. Advanced Natural Gas Engineering. Gulf Publishing Company, Elsevier BV, Amsterdam, Netherlands.

Chapter 2
Reservoir fluids and reservoirs

2.1 Introduction

Natural gas, along with crude oil, is the product of decomposed organic matter, typically from ancient marine microorganisms, that were deposited (along with mud, silt, and sand on the sea floor) over the past 550 million years (Table 2.1) and gradually becoming buried over time. As the organic matter decomposed the end result was the formation of natural gas and/or crude oil which migrated from the original source beds to more porous and permeable rocks, such as sandstone and siltstone, where the natural gas and the crude oil finally became trapped in the rock formation that was to become the reservoir.

By way of clarification, sandstone is a sedimentary rock composed of sand-size grains of mineral, rock, or organic material which also contains a cementing material that binds the sand grains together and may contain a matrix of silt-size or clay-size particles that are of a size that they can occupy the spaces between the sand grains (Table 2.2). Sandstone is composed primarily mostly of quartz (SiO_2) sand, but it can also contain varying amounts of feldspar, and sometimes silt and clay minerals. Sandstone that contains more than 90% quartz is called quartzose sandstone and the color of sandstone varies, depending on the composition. Chemically, the composition of sandstone is variable over fairly narrow limits, namely, silica (SiO_2, 93–94% w/w), aluminum oxide (Al_2O_3, approximately 1.5% w/w), iron oxide (Fe_2O_3, approximately 1.5% w/w), lime (CaO, up to 1% w/w), and mixed sodium and potassium oxides (Na_2O/K_2O, approximately 1% w/w).

On the other hand, siltstone, also known as aleurolite, is a clastic sedimentary rock that is composed mostly of silt (Marshak, 2012). It is a form of mudrock with a low content of clay minerals which can be distinguished from shale by its lack of fissility. The varying types of mudrocks include siltstone, claystone, mudstone, slate, and shale.

Clastic rocks are composed of fragments, or clasts, of pre-existing minerals and rock. A clast is a fragment of geological detritus – chunks and smaller grains of rock broken off other rocks by physical weathering.

The word silt does not refer to a specific substance but is used to describe loose granular particles in a specific size range (0.00015 and 0.0025 in, 0.0039 and 0.063 mm) in diameter. The particles are intermediate in size between coarse clay on the small side and fine sand on the large side. Silt does not have a definite composition and is usually a mixture of clay minerals, feldspars, and quartz. The small-size fraction of silt is mostly clay, while the coarse-size fraction is mostly grains of feldspar and quartz.

https://doi.org/10.1515/9783110691023-002

Table 2.1: The geologic timescale.*

Era	Period	Epoch	Duration (10^6 years)	Years ago (10^6 years)
Cenozoic	Quaternary	Holocene	10,000 to present	
		Pleistocene	2	0.01
	Tertiary	Pliocene	11	2
		Miocene	12	13
		Oligocene	11	25
		Eocene	22	36
		Paleocene	71	58
Mesozoic	Cretaceous		71	65
	Jurassic		54	136
	Triassic		35	190
Paleozoic	Permian		55	225
	Carboniferous		65	280
	Devonian		60	345
	Silurian		20	405
	Ordovician		75	425
	Cambrian		100	500
Precambrian			3,380	600

*The numbers are approximate (±5%) due to variability of the data in literature sources; nevertheless, the numbers do give an indication of the extent of geologic time.

Table 2.2: Grain sizes of the various sediments.

Sediment type	Size range
Sand	
Very coarse	1–2 mm
Coarse	0.5–1 mm
Medium	0.25-0-0.5 mm
Fine	0.125–0.25 mm
Very fine	0.05–0.125 mm
Silt	
Coarse	0.01–0.05 mm
Fine	0.005–0.01 mm
Clay	<0.005 mm

Feldspar is the name given to a group of minerals distinguished by the presence of alumina and silica (SiO_2) in their chemistry. This group includes aluminum silicates of soda, potassium, or lime. It is the single most abundant mineral group on Earth.

A series of reservoirs within a common rock structure or a series of reservoirs in separate but neighboring formations is commonly referred to as a field – more specifically gas field or an oil field. These fields are often found in a single geologic environment known as a sedimentary basin or province. When a reservoir contains natural gas or crude oil (commonly referred to as a hydrocarbon reservoir, although non-hydrocarbon constituents are typically present in the reservoir) is identified, it is also important to identify the types of fluids that are present, along with their main physicochemical characteristics. Generally, that information is obtained by performing a pressure–volume–temperature analysis on a fluid sample taken from the reservoir.

The term "reservoir fluid" is used to describe the contents of a reservoir. Typically, there are four main types of reservoir fluids: (i) crude oil, (ii) condensate, which is a retrograde gas, and (iii) gas, which is often subdivided into wet gas, and dry gas. The term "black oil" has also been used to describe a reservoir fluid but the term is of little value in describing the properties of the oil since color is not a good descriptor of any particular property. Each of the fluid types enumerated above requires different approaches when analyzing the reservoir, so it is important to identify the correct fluid type early on in the reservoir's life.

Geologic formations that contain natural gas and crude oil include clastic rocks (also called detrital rocks which are rocks composed of fragments of older rocks or minerals), chemical rocks (formed by chemical precipitation of minerals), and organic rocks (formed by biological debris from shells, plant material, and skeletons). The three most common sedimentary rock types encountered in natural gas fields and in crude oil fields are (i) shale rocks, (ii) sandstone rocks, and (iii) carbonate rocks. The classification of these types of rock primarily depends on characteristics such as grain size and composition, porosity (pore space within and between grains), and cementitious character (the manner in which the rock grains are held together), each of which can influence natural gas production and crude oil production (Bustin et al., 2008).

Historically, the majority of the natural gas and crude oil produced (as an example) in the United States have been withdrawn from sandstone reservoirs and from carbonate reservoirs.

Carbonate reservoirs are characterized by extremely heterogeneous porosity and permeability. These heterogeneities are caused by the wide spectrum of environments in which the carbonate minerals are deposited and by subsequent diagenetic alteration of the original rock fabric. Both reservoir pressure and temperature affect the type, magnitude and severity of diagenesis. Moreover, increasing temperatures increases the solubility of minerals and causes the pore waters to become saturated, thereby increasing precipitation and formation of cementitious materials.

However, over the past three decades, the production of natural gas and crude oil from shale formations and other tight rock formations has increased dramatically. Shale is a fine-grained, clastic sedimentary rock, composed of mud that is a mix of flakes of clay minerals and small fragments of other minerals, especially quartz (SiO_2) and calcite (a rock-forming mineral with a chemical formula of $CaCO_3$). Shale is characterized by breaks along thin laminae or parallel layering or bedding less than 1 cm in thickness, called fissility.

The natural gas produced in the source rock migrates (in the partially mature or mature from) to the reservoir rock thus offering differentiation between the original state and the final maturation state of the natural gas and crude oil (Speight, 2014, 2019). Typically, source rocks are sedimentary rocks in which natural gas and crude oil commence formation from organic debris (Chapter 1). After forming in the source rock, gas and the protopetroleum (the collection of precursors to crude oil) as well as any already-formed hydrocarbon derivatives – which can vary from simple structures (such as methane, CH_4) to more complex organic hydrocarbon derivatives (Table 2.3) – will migrate into the reservoir rock (Speight, 2014, 2019).

Table 2.3: Constituents of natural gas.

Name	Formula	Vol.%
Methane	CH_4	85+
Ethane	C_2H_6	4
Propane	C_3H_8	1–5
Butane	C_4H_{10}	1–2
Pentane[+]	C_5H_{12}	1–5
Carbon dioxide	CO_2	1–2
Hydrogen sulfide	H_2S	1–2
Nitrogen	N_2	1–5
Helium	He	<0.5

Pentane[+]: pentane and higher molecular weight hydrocarbons up to approximately C_{10}, including benzene and toluene.

2.2 Reservoir fluids

The term "reservoir fluid" is commonly used in a collective manner to include any organic material fluid that exists in a reservoir, which includes gases, liquids, and solids – water may also be included in this terminology and is an important aspect of the reservoir fluid category. Nevertheless, water notwithstanding, the specific types of fluids of interest are (i) natural gas, (ii) crude oil, which includes gas condensate and paraffin wax, (iii) heavy oil, (iv) extra heavy oil, and (v) tar sand bitumen. However, the focus of this text is predominantly on the first three categories of fluids.

The term "fluid" as used by reservoir engineers can mean the contents of the reservoir that are in the gaseous state or in the liquid state and the fluids exist as: (i) a solid phase, of which wax is an example; (ii) a liquid phase, of which crude oil is the example; and (iii) a gaseous phase, of which natural gas is the example. Gas and liquids coexist in a given reservoir. More specifically in the current context, a reservoir typically contains three main fluids: (i) natural gas, (ii) crude oil, and (iii) water with minor constituents being acid gases (carbon dioxide and hydrogen sulfide). These components will vary greatly in combination and proportion within each reservoir and, in the case of heavy crude oil, the amount of gas will be substantially less than would be found in a conventional oil reservoir. Reservoir fluids vary greatly in composition and chemical properties because of several factors (Table 2.4).

At this point another term – irreducible water saturation– should also be acknowledged. Irreducible water saturation (sometimes called critical water saturation) is the maximum water saturation that a formation with a given permeability and porosity can retain without producing water – this water is held in place by capillary forces and will not flow.

Table 2.4: Examples of factors that affect fluids distribution in the reservoir.

Factor	Comment
Depth	The difference in the density of the fluids results in their separation over time due to gravity (i.e., differential buoyancy).
Fluid composition	Important control on its pressure–volume–temperature properties, which define the relative volumes of each fluid in a reservoir. It also affects distribution through the wettability of the reservoir rocks.
Reservoir temperature	Exerts a major control on the relative volumes of each fluid in a reservoir.
Fluid pressure	Exerts a major control on the relative volumes of each fluid in a reservoir.
Fluid migration	Different fluids migrate in different ways depending on their density, viscosity, and the wettability of the rock and the mode of migration helps to define the distribution of the fluids in the reservoir.
Trap type	The effectiveness of the hydrocarbon trap also has a control on fluid distribution (e.g., cap rocks may be permeable to gas but not to oil).
Rock structure	The microstructure of the rock can preferentially accept some fluids and not others through the operation of wettability contrasts and capillary pressure. In addition, the common heterogeneity of rock properties results in preferential fluid distributions throughout the reservoir in all three spatial dimensions.

Thus, using the term "organic" as the keyword, typically, there are four main types of reservoir fluids: (i) crude oil, (ii) condensate, sometimes referred to as retrograde gas, (iii) wet gas, and (iv) dry gas. Heavy crude oil, extra heavy crude oil, and tar sand bitumen are often excluded in this list but acknowledgment of the present of these materials in some reservoirs or deposits is necessary for completion.

On occasion the terms "black oil" and "volatile oil" may be added to this list, but these terms only serve to added confusion to the list. "Black oil" is a somewhat ill-defined (even meaningless) term that offers little more description than the color of the oil and the terms "heavy crude oil" and "extra heavy crude oil" are more pertinent descriptors, while the term "volatile oil" offer little more in terms of the description of the oil (Speight, 2014, 2017). Each of the fluid types listed above (i.e., crude oil, condensate, sometimes referred to as retrograde gas, wet gas, and dry gas) requires different approaches when analyzing a reservoir, so it is important to identify the correct relevant fluid type in the early stages of reservoir delineation.

To produce a wellhead product from a reservoir there has to be flow of the fluids to the wellbores through the heterogeneous porous media of the reservoir. Fluid movements within the reservoir are governed by the local fluid potential gradients (Whitson and Belery, 1994) and reservoir effective permeability, the injection and production points, and the fluid viscosities. However, fluid flow is also governed by the character of the fluids in the reservoir.

In reservoirs that contain natural gas and crude oil, it is often assumed (with reasonable justification) that natural gas, being the least dense, is found at the top of the reservoir, with the crude oil beneath the natural gas, typically followed by water at the base of the reservoir. Thus, water is also a reservoir fluid but more specifically, the water that occurs in natural gas and crude oil reservoirs is usually a *brine* which consists of dissolved sodium chloride (NaCl) as well as salts (minerals) which include calcium (Ca), magnesium (Mg), sulfate (SO_4), bicarbonate (HCO_3), iodide (I), and bromide (Br). Under reservoir conditions, the brine that is sharing pore space with hydrocarbons always contains a limited amount of solution gas (predominantly methane) but increasing salinity decreases gas in solution. Reservoir brines exhibit only slight shrinkage (<5%) when produced to the surface.

The water found in the oil and gas zones (interstitial water, the compositional analysis of which can often provide information related to the mineralogy of the reservoir from the presence of water-soluble minerals) and usually occurs as: (i) collars around grain contacts, (ii) a filling of pores with unusually small throats connecting with adjacent pores, or, to a much smaller extent, (iii) as wetting films on the surface of the mineral grains when the rock is preferentially wet by water.

The water (brine) produced from the reservoir along with natural gas and crude oil is brought to the surface with the crude oil. Because the water has been in contact with the oil, it contains some of the chemical characteristics of the formation and the oil itself. Oil and gas wells produce more water than oil (7 bbl/bbl-of-crude oil in some fields). The composition (salt content) of co-produced water determines the need for anti-scaling additives. There are strict regulations to limit disposal and beneficial use options as well as environmental impacts that pertain to oil field waters.

However, rather than discrete zones of natural gas, crude oil, and water, it is more likely than a reservoir will consist of several boundary zones, such as (i) an oil-in-water zone, (ii) a water in oil zone, (iii) a gas in oil zone, (iv) a gas in water zone, as well as various zones of mixed composition. It is safe to assume that in the reservoir, zones are not always ordered according to density. Thus, the distribution of the fluids in a reservoir rock is not always dependent on the density of the reservoir fluids as well as on the properties of the rock.

If the pores of the reservoir rock are of uniform size and evenly distributed, there is: (i) an upper zone where the pores are filled mainly by gas (the gas cap), (ii) a middle zone in which the pores are occupied principally by oil with gas in solution, and (iii) a lower zone with its pores filled by water. A certain amount of water (approximately 10 to 30%) occurs along with the oil in the middle zone. There is a transition zone from the pores occupied entirely by water to pores occupied mainly by oil in the reservoir rock, and the thickness of this zone depends on the densities and interfacial tension of the oil and water as well as on the sizes of the pores. Similarly, there is some water in the pores in the upper gas zone that has at its base a transition zone from pores occupied largely by gas to pores filled mainly by oil.

Typically, fluids from a crude oil reservoir are brought to the surface as a mixture of natural gas, crude oil, and water, which is then sent to a surface production facility before they can be disposed or sold to an industrial consumer (e.g., a refinery). A surface production facility is the system in charge of the separation of the well stream fluids into its three single phase components – oil, gas and water – and of their transport and processing into marketable products and/or their disposal in an environmentally acceptable manner. Once separated, the oil, natural gas, and water follow different paths. Water is typically re-injected for reservoir pressure maintenance operations. The oil usually goes through a process of dehydration, which removes basic sediments, and hydrocarbon fluids are assumed to comprise two components – stock tank oil and surface gas.

The fluids, particularly natural gas and crude oil, which exist in a reservoir (and which vary widely in properties), must be determined early after the discovery of the reservoir. Fluid type is a critical consideration in the decisions that must be made related to producing the fluids. Furthermore, fluid properties play a key role in the design and optimization of injection/production strategies and surface facilities for efficient reservoir management and longevity. Inaccurate fluid characterization will lead to uncertainty in the amount of the resource that is in-place as well as predictions of recovery efficiency. Prior to production, determination of the fluid properties may only represent laboratory (hence, ex-situ) properties but once production commences, variations in fluid composition because of pressure changes and flow throughout the reservoir will become apparent from which the in-place properties can be assessed accurately as well as a measure of reservoir longevity can be assessed.

Moreover, reservoir fluids vary greatly in composition – in some fields, the fluid is in the gaseous state and in others it is in the liquid state but generally gases and liquids frequently coexist in a reservoir – in some reservoirs (or deposits) solids may exist as a wax or as a tar mat (Wilhelms and Larter, 1994a, 1994b; Zhang and Zhang, 1999; Speight, 2014). The rocks which contain these reservoir fluids also vary considerably in composition and can influence the physical properties and the flow properties. Other factors, such as producing area, height of the fluid column, natural fracturing, or faulting, and water production, also serve to distinguish one reservoir from another and which also affect the choice of the production method.

In fact, in keeping with understanding the nature of the reservoir fluids, understanding the elastic properties of reservoir rocks is crucial for exploration and successful production of natural gas and crude oil from tight shale and tight formation reservoirs. In the case of the static and dynamic elastic properties of shale from Barnett, Haynesville, Eagle Ford, and Fort St. John shale formations, the matter is not so straightforward since the elastic properties of these rocks vary significantly between reservoirs (and even within a reservoir) due to the wide variety of minerals that form the reservoir rock as well as the microstructures exhibited by these shale reservoirs and tight reservoirs. For example, the static (Young's modulus) and dynamic (P-wave and S-wave moduli) elastic parameters generally decrease monotonically with the content of the clay minerals plus any kerogen. However, the elastic properties of the shale formations are strongly anisotropic (the properties are not identical in all directions) since the degree of anisotropy correlates with the organic content of the shale as well as the minerals that constitute the amount and type of clay minerals that constitute the shale (Tables 2.5 and 2.6) (Hillier, 2003; Bergaya et al., 2011). This is not (and should not be) uprising considering the complex and varying composition of clay minerals (Sone and Zoback, 2013a, 2013b).

2.3 Sediments

A sediment is the matter that settles to the bottom of a liquid and, thus, sediments (sedimentary rocks, sedimentary strata) are types of rock that are formed by the deposition of material within bodies of water through the process of sedimentation – the process by which mineral matter and/or organic detritus (i.e., organic particles) settle and accumulate in layers (strata).

The geologic age of any sediment is an important determinant of the potential of the sediment to natural gas and/or crude oil. While many rocks of different ages produce natural gas (and crude oil), the areas of prolific production include formations that are from several different geologic periods (Table 2.1): (i) the Devonian period, approximately 405–345 million years ago, (ii) the Carboniferous period, approximately

Table 2.5: General types of clay minerals.

Group	Minerals in group
Kaolin[1]	Kaolinite
	Dickite
	Halloysite
	Nacrite (polymorphs of $Al_2Si_2O_5(OH)_4$)
Smectite[2]	Montmorillite
	Nontronite
	Beidellite
	Saponite
Illite[3]	Illite
	Clay-micas
Chlorite[4]	Considerable chemical variation throughout this group

[1] Also called china clay, soft white clay that is an essential ingredient in the manufacture of china and porcelain and is widely used in the making of paper, rubber, paint, and many other products.
[2] The name used for a group of phyllosilicate mineral species, the most important of which includes montmorillonite.
[3] A group of closely related non-expanding clay minerals. Illite is a secondary mineral precipitate, and an example of a phyllosilicate, or layered alumino-silicate.
[4] The name of a group of common sheet silicate minerals that form during the early stages of metamorphism.

Table 2.6: Chemical formulas of clay minerals.

Group	Layer type	Layer charge	Chemical formula
Kaolinite	1:1	<0.01	$[Si_4]Al_4O_{10}(OH)_8 \cdot nH_2O$ (n = 0 or 4)
Illite	2:1	1.4–2.0	$M_x[Si_{6.8}Al_{1.2}]Al_3Fe \cdot 025Mg_{0.75}5O20(OH)_4$
Vermiculite	2:1	1.2–1.8	$M_x[Si_7Al]AlFe \cdot 05Mg0.5O_20(OH)_4$
Smectite	2:1	0.5–1.2	$M_x[Si_8]Al_{3.2}Fe_{0.2}Mg_{0.6}O_20(OH)_4$
Chlorite	2:1:1	Variable	$(Al(OH)_{2.55})4[Si_{6.8}AlO_{1.2}]Al_{3.4}Mg_{0.6})20(OH)_4$

345–280 million years ago, (iii) the Permian period, approximately 280–225 million years ago, and (iv) the Cretaceous period, approximately 136–71 million years ago. During these periods, organic-rich materials accumulated with the sedimentary rocks and, over geologic time (millions of years), chemical changes (induced by pressure from the overlying sediments and any resulting heat from increasing pressure) transformed the original organic detritus into what eventually became natural gas and crude oil.

2.3.1 Rock types

There are three kinds of rocks (i) igneous rocks, (ii) metamorphic rocks, and most important in the current context (iii) sedimentary rocks. Igneous rocks form when molten rock (magma or lava) cools and solidifies. Metamorphic rocks are formed when existing rocks are changed by heat, pressure, or reactive fluids, such as hot, mineral-laden water. Sedimentary rocks originate when particles settle out of water or air, or by precipitation of minerals from water and the result is a stratified rock formation.

In the current context, the four examples of sedimentary rocks are: (i) breccia, which consists of angular fragments cemented together in which the angular grains on the order of 2–64 mm in size; (ii) conglomerate, which is a clastic sedimentary rock made up of rounded clasts that are greater than 2 mm in diameter and the grains are predominately rounded and on the order of 64 mm to >256 mm in size; (iii) sandstone, in which the grains range from 2 mm to 1/16 mm (cf., shale formations are composed of particles less than 1/16 mm in size, (iv) siltstone, and (v) shale. In the current context, sandstone and shale are the most important sedimentary rocks. Sandstone is the most commonly encountered reservoir rock for natural gas, and crude oil and sandstone formations are created by larger sediment particles, and are typically deposited in river channels, deltas, and shallow sea environments.

Conglomerate formations are typically consolidated gravel deposits with variable amounts of sand and mud between the particles (sometimes referred to as pebbles). Conglomerate formations accumulate in stream channels, along the margins of mountain ranges, and on beaches and are composed largely of angular pebbles (breccias) and some (tillites) are formed in glacial deposits. On the other hand, sandstone formations are composed essentially of cemented sand and comprise approximately one-third of all sedimentary rocks. The most abundant mineral in sandstone is quartz (SiO_2), along with lesser amounts of calcite ($CaCO_3$), gypsum ($CaSO_4 \cdot 2H_2O$), and various iron compounds, such as the iron oxides (FeO, Fe_2O_3). These formations tend to be more porous than shale formations, and consequently make excellent reservoir rocks – as long as impermeable basement rocks and cap rocks (such as shale formation) are present. The third most abundant formations, carbonate formations, are created by the accumulation of shells and skeletal remains of water-dwelling organisms in marine environments.

Chemical and organic sedimentary rocks are the other main group of sediments and are formed by weathered material in solution precipitating from water or as biochemical rocks made of dead marine organisms. Chemical sediments are deposited from the water in which the material is dissolved – examples include limestone ($CaCO_3$), dolomite ($CaCO_3.MgCO_3$), and rock salt ($NaCl$) – and these types of deposits are referred to as inorganic chemical sediments. Chemical sediments that have been deposited by or with the assistance of floral or faunal materials are classed as organic sediments or biochemical sediments.

Organic-rich sedimentary rocks are a specific type of sedimentary rock that contains significant amounts (>3% w/w) of organic carbon. The most common types include coal, lignite, or oil shale – the latter contains an organic compound frequently referred to as kerogen.

2.3.2 Characteristics

Sedimentary rocks possess (and display) definitive physical characteristics which include (i) stratification, (ii) cross-bedding, (iii) graded bedding, (iv) texture, (v) ripple marks, (vi) mud cracks, (vii) concretions, (viii) fossils, and (ix) color.

Stratification, probably the most characteristic feature of sedimentary rocks, is their tendency of the rocks to occur in beds (strata), which are formed when geological agents such as wind, water, or ice gradually deposit sediment. Cross-bedding (cross-stratification) occurs to sets of beds that are inclined relative to one another – the beds are inclined in the direction that the wind or water was moving at the time of deposition and are common in beach deposits, sand dunes, and river-deposited sediment and enable determinations to be made related to the origin and formation of ancient sediments. Graded bedding occurs as a result of a reduction in velocity (typically in a stream bed) and (i) larger or denser particles are deposited followed by (ii) deposition of smaller particles. This results in the bedding showing a decrease in grain size from the bottom of the bed to the top of the bed (fine sediment particles at the top of the bed and coarse sediment particles at the bottom of the bed). Fossils are the remains of once-living organisms that have been preserved in the crust of the Earth and give clues to the relative age of the sediment and can be important indicators of past climates.

Finally, the minerals in a sediment can impart color to the sediment. With the exception of gray and black, which mostly result from partially decayed organic matter, most rock colors are the result of the presence of iron-containing minerals. Ferric iron (Fe^{3+}) produces red, purple, and yellow colors – minerals such as hematite (iron oxide, Fe_2O_3, also spelled hematite) and limonite, which is an iron ore consisting of a mixture of hydrated ferric oxide-hydroxides in varying composition [frequently written as $FeO(OH){\cdot}nH_2O$], produce a pink or red color, while the ferrous iron (Fe^{2+}) produces green colors in sediments. Dark gray to black colors mean anoxic conditions (depleted of oxygen) which are common in muddy ocean bottoms where there are both high amounts of organic matter and low levels of inflow of oxygenated water through the sediment and the interstitial water (water between the sediment grains) is oxygen free several inches below the surface. These conditions could occur in a mean deep water environment or in a swamp environment.

2.3.3 Composition

A sediment is composed of three basic components (i) grains, (ii) matrix, and (iii) cementitious materials. Grains (sometimes referred to as framework grains) refer to the larger, solid components in the sediment which form the basic small-scale units of a sandstone reservoir. The original grain composition is controlled by the composition of the sediment source (i.e., the origin and history of the sediment) as well as the physical and chemical processes which are prevalent when the sediment is created and transported to the geologic basin. Often referred to as detrital grains, the grain composition of most sandstone reservoirs consists primarily of quartz, feldspars – a group of minerals distinguished by the presence of silica (SiO_2) and alumina (Al_2O_3) – and rock fragments (Berg, 1986). Following deposition and burial, the framework grains (typically referred to as authigenic grains) – minerals and other materials formed in their present position – are often altered by the physical effects of compaction as well as various chemical processes (diagenesis). The original grain composition governs the type and severity of diagenesis (Rushing et al., 2008).

Clay minerals vary widely in the structure or morphology of both the individual and aggregate particles, and the presence of these minerals in the pore of sediments can significantly reduce both permeability and primary porosity (Neasham, 1977a, 1977b; Wilson and Pittman, 1977). This effect of clay minerals reinforces the importance of a comprehensive program to identify clay type, clay origin, and the factors controlling the occurrence of the clay minerals.

In terms of tight sand formations, a major component common for many tight gas sands is the grain cement (cementitious material) which typically refers to any mineral that forms during diagenesis and is precipitated after deposition of both grains and matrix components (Berg, 1986).

2.3.4 Texture

Sediment texture refers to the size, shape, and arrangement of materials that is derived from processes of weathering, transportation, deposition, and diagenesis. Sediment texture is an important aspect of the character of a sandstone reservoir and includes grain size, sorting, packing, shape, and grain orientation. The texture of the sediment not only affects properties of the sediment at the time of deposition but also can impact the rate, magnitude, and the extent of mineral diagenesis.

The texture in sediment and sedimentary rocks is dependent on the processes that occur during each stage of formation which also includes: (i) the nature of the source materials, (ii) the nature of wind and water currents present, (iii) the distance that materials were transported as well as the timer spent in the transportation process, (iv) any biological activity, and (v) exposure to various chemical environments.

Each of these aspects can affect the ability of the sediment to hold (store) natural gas (and crude oil).

Depending on the sediment type, the presence of smaller matrix materials (such as clay minerals and shale minerals) in clean coarse-grained sands will tend to reduce both permeability and primary porosity. Other textural traits include grain shape and orientation. Grain shape is usually expressed as sphericity (a measure of the deviation of a grain from a perfect sphere) and roundness (a measure of the roundness of the grain edges) (Berg, 1986). On the other hand, grain orientation refers to the preferred direction of the long axes of the grain.

2.3.5 Structure

Sediment structure (including identification of bed geometry, bedding planes, contacts between beds, and bedding plane orientation) is an important element of the depositional process since the type of structure may help in identifying the original depositional environment. In fact, an understanding of sediment structure is an important component in optimizing field development activities since bed geometry and dimensions may impact both the vertical and lateral continuity of the sediment (i.e., the reservoir) which would, in turn, dictate well spacing and the type of wellbore architecture which is important in terms of predicting the long-term production behavior of a reservoir. For example, significant vertical heterogeneity may determine the effectiveness of horizontal wellbores for the recovery the natural gas (or crude oil).

Other aspects of sediment structure are (i) ripple marks, which provide information related to the conditions when the sediment was originally deposited, (ii) mud cracks, which also provide other signs that are a guide to the environment in which the sediment was formed, and (iii) concretions, which are an indication that the softer environmentally prone rock was eroded away leaving the harder concretions intact.

Ripple marks in a sediment are characteristic of deposition of the inorganic materials in shallow water and are caused by forces such as wave-related forces or wind-related forces that leave ripples of sand as typified by the marks seen on beach sand or on the bottom of a shallow stream. Mud cracks which result from the drying out of wet sediment on the bottom of dried-up lakes, ponds, or stream beds may be many-sided (polygonal) shapes that present a honeycomb-type appearance on the surface of the beach sand or the stream sand. Concretions are spherical or flattened masses of rock enclosed in some (but not all) shale formations or in some limestone formations and which are generally harder than the rock enclosing them.

2.4 Reservoir evaluation

A reservoir is a subsurface rock structure, with sufficient size and closure that contains a three-dimensional network of interconnected void (pore) space and is underlain (the basement rock) by a non-porous (non-permeable) rock formation and overlain (the cap rock) by a non-porous (non-permeable) rock formation. Some reservoirs occur at depths up to two miles under the surface while others may be as much as five miles under the surface of the Earth. In addition, there are also sub-sea reservoirs, that is, reservoirs that are located under the floor of the ocean (Speight, 2015a).

Natural gas and crude oil reservoirs are typically classified as (i) conventional reservoirs and (ii) unconventional reservoirs. In the former type of reservoir – the conventional reservoir – the natural gas and crude oil are trapped by an underlying rock formation (the basement rock) and by an overlying rock formation (the cap rock) with lower permeability than the reservoir rock. In the second type of reservoir – the unconventional reservoir – the reservoir rock typically has high porosity and low-to-no permeability in which the natural gas and crude oil are trapped in place without the need for a cap rock or basement rock (GAO, 2012; Speight, 2014, 2019).

Whatever the location of the reservoir, the natural gas must be extracted by the application of methods that assure maximum recovery of the gas. More recently, substantial reserves of natural gas have been discovered in tight formations, such as shale formation where the permeability is low-to-zero. Furthermore, the methods selected to recover the natural gas will vary with the reservoir and even from wells drilled across a gas reservoir due to differences such as (i) reservoir thickness, (ii) reservoir extent, (iii) reservoir pressure, (iv) reservoir depth, (v) mineralogy of the reservoir, (vi) gas composition, and (vii) water content, all of which are required in order for the project to proceed with the potential for success (Speight, 2014, 2019).

The geologic age (Table 2.1) of any sediment is an important determinant of the potential of the sediment to contain crude oil and natural gas. While many rocks of different ages produce oil and natural gas, the areas of prolific production include formations that are from several specific geological periods which are (i) the Devonian period, approximately 405–345 million years ago, (ii) the Carboniferous period, approximately 345–280 million years ago, (iii) the Permian, approximately 280–225 million years ago, and (iv) the Cretaceous period, approximately 136–71 million years ago (Table 2.1). During these periods, organic-rich materials accumulated with the sediments and, over geologic time (millions of years), chemical changes (induced by pressure from the overlying sediments and any resulting heat from increasing pressure) changed the original organic detritus, thereby (eventually) producing natural gas and crude oil.

Moreover, each reservoir will exhibit individual properties that are specific to that reservoir (site-specific properties). Indeed, within a reservoir, these properties may even change with longitudinal extent and with vertical height of the reservoir

(Hunt, 1996; Dandekar, 2013; Speight, 2014). Thus, reservoir evaluation is an important aspect of oil and gas production.

2.4.1 Reservoir geology

An understanding of the geology of a reservoir is essential to reservoir development, oil and gas production, and management, including reservoir longevity and environmental management. Furthermore, reservoir evaluation includes both the external geology of the reservoir (the forces responsible for the formation of the reservoir) and the internal geology of the reservoir (the nature of the rocks that constitute the reservoir). These aspects are even more important when hydraulic fracturing methodology is to be applied to the reservoir (Speight, 2014, 2016a). In addition, the efficient extraction of natural gas (and crude oil) requires that the reservoir be visualized in three dimensions (Solano et al., 2013).

An important geologic aspect of the reservoir is the external geometry of the reservoir, defined by seals that inhibit the further migration of the natural gas and crude oil. Migration will cease, and a hydrocarbon reservoir will form, only where hydrocarbons encounter a trap, which are composed of a suitable gas-holding or oil-holding rock with the following types of seals: (i) top, (ii) lateral, and (iii) bottom seals. In addition, the geometry of traps can be: (i) structural, (ii) sedimentary, and (iii) diagenetic (Hunt, 1996; Dandekar, 2013; Speight, 2014, 2019).

Another important geologic aspect of the reservoir is the internal architecture which involves the lateral distribution of depositional textures is related to depositional environments, and the vertical stacking of textures is described by stratigraphy, which is the geological study of the following aspects of rock strata: (i) form, (ii) arrangement, (iii) geographic distribution, and (iv) chronologic succession. Diagenesis, which refers to changes that happen to the sediment after deposition, can also control the lateral continuity and vertical stacking of reservoir rock types. This phenomenon is an important aspect of carbonate reservoirs, in which the conversion of limestone ($CaCO_3$) to dolostone ($CaCO_3 \cdot MgCO_3$) and the dissolution of carbonate have a large effect on internal reservoir architecture (Tucker and Wright, 1990; Blatt and Tracy, 1996).

In general, reservoir rocks tend to show greater variations in permeability than in porosity and in addition, these two properties, as measured on core samples from reservoir rocks, are not always identical with the values indicated for the bulk rock in the underground formation because of the non-representative nature of many core sample. Porosity is typically on the order of 5–30%; while permeability is typically on the order of 0.005 Darcy (5 milliDarcys) and several Darcys (several thousand milliDarcys) (Kovscek, 2002).

2.4.2 Rock types

The identification and comparison of different rock types, such as (i) depositional rock types, (ii) petrographic rock types, and (iii) hydraulic rock types, is a fundamental aspect of reservoir evaluation.

Depositional rock types are collections of rocks that are grouped according to similarities in composition, texture, sedimentary structure, and stratigraphic sequence as influenced by the environment at the time of deposition. Petrographic rock types are grouped according to pore-scale, microscopic imaging of the current pore structure, as well as the rock texture and composition, clay mineralogy, and diagenesis. Hydraulic rock types are rock types that are also defined on the basis of pore scale and quantify the physical flow and storage properties of the rock relative to the native fluid(s) as controlled by the dimensions, geometry, and distribution of the current pore and pore throat structure (Rushing et al., 2008).

Each rock type represents different physical and chemical processes affecting the properties of the rock during the depositional and paragenetic cycles which represent an equilibrium sequence of mineral phases. It is used in studies of igneous and metamorphic rock genesis and importantly in studies of the hydrothermal deposition of minerals and rock alteration associated with ore mineral deposits.

Since most tight gas sands have been subjected to post-depositional diagenesis, a comparison of all three rock types will allow an assessment of the impact of diagenesis on rock properties. If diagenesis is minimal, the depositional environment (and depositional rock types) as well as the expected rock properties derived from those depositional conditions will allow prediction of the quality and properties of the reservoir rock properties.

One of the most difficult parameters to evaluate in tight gas reservoirs is the drainage area size and shape of a typical well. Knowledge of the depositional system and the effects of diagenesis on the rock is needed to estimate the drainage area size and shape for a specific well. In some tight gas reservoirs, the average drainage area of a well largely depends on the number of wells drilled, the size of the fracture treatments pumped on the wells, and the time frame being considered. In lenticular or compartmentalized tight gas reservoirs, the average drainage area is a function of the average sand-lens size or compartment size, and may not be a strong function of the size of the fracture treatment.

A main factor controlling the continuity of the reservoir is the depositional system. Generally, reservoir drainage per well is small in continental deposits and larger in marine deposits. Fluvial systems tend to be more lenticular whereas barrier-strand-plain systems tend to be more blanket and continuous. To date, most of the more successful tight gas plays are those in which the formation is a thick, continuous, marine deposit.

2.4.3 Temperature

Reservoir temperatures may vary up to 90 °C (194 °F) or even higher, while surface conditions are on the order of 20–40 °C (68–104 °F). Pressure can vary from its atmospheric value (14.7 psi, or lower in the case of vacuum distillation) to a number in the thousands of psi. Within such an ample range of conditions, hydrocarbon derivatives in the reservoir fluids undergo transformations and exist as a single phase (gas, liquid, or solid) or co-exist in several forms (liquid-plus-gas, solid-plus-liquid, vapor-plus-solid, or even in liquid-plus-liquid combinations). Understanding the methods by which the hydrocarbon fluids interact with and react to their thermodynamic surroundings is essential to understanding the pressure–temperature diagram or the pressure–temperature envelope. Each envelope represents a thermodynamic boundary separating the two-phase conditions (inside the envelope) from the single-phase region (outside). The correct identification of the type of hydrocarbon fluid is critical to the design and development of the production strategy for the field under consideration.

2.4.4 Heterogeneity

In addition to understanding the petrophysics of the reservoir, gas recovery requires an understanding of displacement and flow through porous media, which can be complicated (Grattoni and Dawe, 2003; Dawe, 2004; Maxwell and Norton, 2012). Reservoir heterogeneity, which is typically due to lateral and vertical variability in permeability and porosity within the same sand body, is mostly caused by post-depositional diagenesis. Most diagenetic processes do not cause completely isolated reservoir compartments – but such processes may yield complex and/or poor quality flow paths, which may result in low productivity for a given reservoir system.

Physically, natural gas (and crude oil) reservoirs are always the homogeneous porous media that is often envisaged on paper and using data from laboratory simulations. Heterogeneity means that a specific property of interest to reservoir development will vary along vertical and longitudinal axes within the reservoir (Grattoni and Dawe, 2003; Dawe, 2003, 2004). For example, well log and core analysis reports indicate that reservoirs are typically heterogeneous and the rock properties (such porosity and pore saturation) vary within the reservoir. In addition, permeability heterogeneity (compared to a homogeneous system) causes variations in the fluid movements (Anderson, 1986; Caruana and Dawe, 1996a, 1996b; Grattoni and Dawe, 2003; Dawe, 2003, 2004).

In summary, the heterogeneity of a reservoir, which is typically manifested by lateral and vertical variability in permeability and porosity within the same sand body, is predominantly caused by post-depositional diagenesis. Most diagenetic processes do not cause completely isolated reservoir compartments, but such processes may yield complex and/or poor quality flow paths, which may result in low productivity for a given reservoir system.

2.4.5 Porosity and permeability

Porosity and permeability – which are the result of both depositional and diagenetic factors – are related properties of any rock or loose sediment and both properties are necessary to make a productive natural gas well and (Alreshedan and Kantzas, 2015). Although a relationship between porosity and permeability is often difficult to interpret (Speight, 2014, 2019) there are wide variations in the permeability of the reservoir rock (Table 2.7). Nevertheless, there are trends that show a very general relationship between the porosity and the permeability (Figure 2.1).

Table 2.7: Examples of permeability limits for various reservoirs.

Permeability limits
1 nanodarcy   Shale reservoir
1 microdarcy   Tight reservoir
1 millidarcy   Conventional reservoir
1 darcy

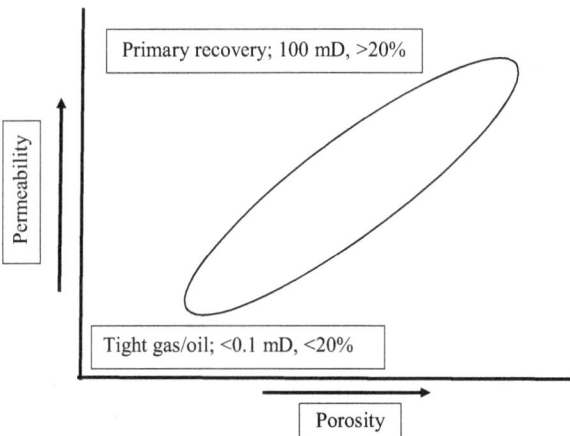

Figure 2.1: General trends in the relationship between porosity and permeability.

The characterization of porosity and permeability is of fundamental importance for the proper evaluation of a reservoir. At the microscopic scale, porosity and permeability are highly dependent on the geometry of the pores and pore throats within volumetrically finite homogeneous systems. These microscopic, locally homogeneous domains are usually found as layered sediments and/or clusters which confer different degrees of heterogeneity to the reservoir. Thus, porosity is the proportion of void space to the total volume of rock and is a measure of the ability of the rock to hold natural gas (and other fluids). Mathematically, porosity (ø) is the open space in a rock divided by the total rock volume and is normally expressed as a percent of the total rock which is taken up by pore space. Thus:

$$ø = V_V/V_T$$

In this equation, V_V is the void volume and V_T is the total volume.

As an example, sandstone may have 8% porosity – the other 92% is solid space-filling rock. In newly deposited sand formations and poorly consolidated sandstone formations, grain size correlates well with pore size and is a primary factor in controlling the permeability. Also, the throat size of the pores controls the permeability of the rock, thereby affecting the gas flow rate and porosity controls the distribution of the natural gas in the formation.

Porosity and permeability generally decrease with increasing depth (thermal exposure and effective pressure); however, a significant number of deep (approximately 13,000 feet) sandstone reservoirs worldwide are characterized by anomalously high porosity and permeability (Bloch et al., 2002). Anomalous porosity and permeability can be defined as being statistically higher than the porosity and permeability values occurring in typical sandstone reservoirs of a given lithology (composition and texture), age, and burial/temperature history.

In conventional reservoirs, pore space (pore volume, porosity) can vary from fairly large, visible openings to microscopic pores, and generally comprises less than 30% v/v of the reservoir rock volume. Also, in conventional reservoirs, natural gas and crude oil tend to move relatively easily through the formation until they are trapped against a rock formation that prevents further flow downward (impermeable basement rock) or upward (impermeable cap rock). In reservoirs that consist of tight formations, the natural gas (and crude oil) is often found within the same rocks where the gas and oil were generated (the source rock) trapped in the pore spaces, natural fractures, and within the organic matter itself.

On the other hand, in tight reservoirs, porosity is commonly less than 10% v/v of the reservoir rock. However, regardless of the total porosity volume, if these pores are not efficiently connected one to the other (to give permeability), natural gas and crude oil cannot migrate. Thus, the higher the permeability, the greater the amount of fluid that can flow through the rock. Conventional reservoirs may have a permeability in the range of tens to hundreds of milliDarcys. Tight reservoirs usually have

permeability from 0.1 to 0.001 milliDarcys, and shale reservoirs are even less permeable – in the 0.001–0.0001 millidarcy range.

In pore network modeling, the complex pore structure in a rock is represented
by a network of pore bodies (void spaces) and pore–throats (narrow paths that connect pore bodies) with simplified geometries. When this type of model is successfully initiated, single and multiphase flow calculations can be performed. At the
early stages of network modeling, capillary pressure and relative permeability curves
of drainage using two-dimensional (2D) regular lattice networks where the radii can
be assigned after which 3D pore network models can be produced that realistically
represent the porous media.

Permeability k is defined by the following equation:

$$k = q/A = q/\mu(-dp/dL)$$

In this equation, q is the volume flow rate (cm³/s), A is the cross-sectional area
(cm²), μ, is the fluid viscosity (centipoises, cp), k is the permeability (darcy), and –
(dP/dL) is the pressure drop per unit length (atm/cm).

Permeability is measured by passing a fluid of known viscosity through a core
plug of measured dimensions and then measuring flow rate and pressure drop. The
pressure drop is measured by a manometer, and the flow rate either by a calibrated
orifice or, with very slow rates, by the rate of movement of a soap bubble in a calibrated glass tube.

However, as a word of caution, reservoirs rocks have connate water in them,
and the clay minerals and other minerals are adjusted to the presence of water or
brine. Cleaning and drying the core prior to porosity and permeability measurements can alter the permeability results when air is used for measurement. The use
of water foreign to the native water can alter the results, fresh water usually giving
the lowest permeability. The ionic content and pH of the water are believed to be
prime variables.

The choice of drilling fluids and the use of air drilling or cable tool completions
are influenced by the necessity of keeping non-connate water (connate water is the
natural water retained in a reservoir after gas or oil have entered the closure by
water displacement) away from some reservoir rocks. However, it may be ascer-
tained that the chemistry of a given reservoir system permits contact with aqueous
solutions without reduction of the permeability.

In complex sandstones, it is recommended to create first a 3D image-based representation of the pore space that should capture the statistics of the real rock.
This can be generated using a direct imaging technique such as micro-CT scanning
or by various process/object-based modeling approaches. The CT scan (a computed
tomography scan, formerly known as a computed axial tomography, or CAT scan) is
an imaging technique that uses computer-processed combinations of multiple X-ray
measurements taken from different angles to produce cross-sectional images (virtual images) of a reservoir thereby allowing the engineer to obtain an "inside" view

of the reservoir. Subsequently, using various image-based network extraction techniques, an equivalent pore network is then extracted from the 3D image to estimate the single and multiphase fluid flow properties.

In summary, the success or failure of a recovery process (especially a hydraulic fracture treatment) will depend on the quality of the candidate well selected for the treatment (Speight, 2016a). Evaluation and selection of a suitable candidate reservoir for stimulation is a move in the depiction of success, while choosing a poor candidate normally results in failure. To select the best candidate for stimulation, the design engineer must consider many variables, of which the most critical parameters for hydraulic fracturing are: (i) formation permeability, (ii) the in situ stress distribution, (iii) viscosity of the reservoir fluid, (iv) reservoir pressure, (v) reservoir depth, (vi) the condition of the wellbore, and (vii) prior stimulation of, or damage to, the reservoir.

2.4.6 Reservoir morphology

The term "reservoir morphology" is used to define (i) the dimensions, (ii) the geometry, orientation, (iii) the heterogeneity, and (iv) the continuity of the reservoir as developed by depositional and post-depositional processes. Both the quality and quantity of the reservoir rock are controlled by primary and secondary depositional environments and processes. Also, understanding reservoir morphology will affect the optimum well spacing applicable to successful field development. Moreover, quantification of the morphology of the reservoir will assist in the definition of the reservoir architecture and compartments and ultimately to determine the original reservoir volume. For example, the gas-in-place volumes and producing characteristics for a blanket sand reservoir will be much different than for a lenticular sand reservoir.

Depositional environment and post-depositional diagenesis both have a significant bearing on morphology, including reservoir compartmentalization and heterogeneity. Reservoir compartments refer to intervals or sections of the sand deposits that are mostly or completely isolated (i.e., not in pressure communication) from other parts of the reservoir. Compartments may be created by significant changes in the depositional environment or by post-depositional processes (such as diagenesis and/or tectonic activity creating sand pinch-outs, and no-flow barriers).

In a reservoir, the quality and quantity of the sand are controlled by primary and secondary depositional environments and processes. Quantification of the reservoir morphology is an aid to defining the reservoir architecture and compartments, and, ultimately, to determine the original reservoir volume. For example, the gas-in-place volumes and producing characteristics for a blanket sand reservoir will be much different than for a reservoir characterized as a lenticular sand reservoir.

Reservoir morphology also affects the optimum well spacing to for field development. Depositional environment and post-depositional diagenesis both have a significant bearing on morphology, including reservoir compartmentalization and heterogeneity. Reservoir compartments refer to intervals or sections of the sand deposits that are mostly or completely isolated (i.e., not in pressure communication) from other parts of the reservoir. Compartments may be created by significant changes in the depositional environment or by post-depositional processes (such as diagenesis and/or tectonic activity creating sand pinch-outs, and no-flow barriers).

The final analysis includes an economic evaluation of whether to complete an oil or gas well and once completed, an ongoing analysis of how to produce the well most effectively. These interpretations and analyses are affected by geological complexity of the reservoir, rock quality, reservoir heterogeneity, and, from a logistical standpoint, the areal extent and location of the project of interest. In the early stages of development, the purpose of formation evaluation is to define reservoir thickness and areal extent, reservoir quality, reservoir fluid properties, and ranges of rock properties. The key rock properties are porosity, permeability, oil, gas, and water saturations.

2.5 Tight formations

The term "tight formation" refers to a formation consisting of low-to-no permeability rock such as shale formations and carbonate formations that (in the present context) contain natural gas (and crude oil). Natural gas resources that occur in shale formations and in other tight formations (e.g., carbonate formations) represent a significant portion of the North American current natural gas (and crude oil) resource base and these systems represent an important source for future reserve growth and production (Moore et al., 2016).

Reservoirs with an estimated in situ permeability of 0.1 milliDarcy (Moslow, 1993). Thus, for all practical purposes, a tight reservoir is generally recognized as any low-to-no permeability formation in which special well completion techniques (such as horizontal drilling and hydraulic fracturing) are required to stimulate production. These types of reservoirs are characterized by complex geological and petrophysical systems as well as reservoir heterogeneity and, in fact, low-permeability oil reservoirs typically exhibit a high degree of heterogeneity. Site-specific variations of porosity, permeability, and pore geometry are variably affected by the compositional nature of the sediments and the depositional environment in which they formed, as well as the evolving diagenetic and tectonic history of the reservoir rocks (Solano et al., 2013). In addition, and unlike conventional reservoirs, tight reservoirs typically exhibit storage and flow characteristics which are uniquely tied to depositional and diagenetic processes.

Generally, a tight (low-to-no permeability) reservoir is a layered system and, in a clastic depositional system, the layers are composed of: (i) shale, (ii) mudstone, (iii) and siltstone while in a carbonate system the layers are composed predominantly of: (i) limestone, $CaCO_3$, (ii) dolomite, $CaCO_3.MgCO_3$, (iii) possibly halite, NaCl, or anhydrite, $CaSO_4$, and some shale. In order to optimize the development of a tight reservoir, a multi-disciplinary approach must fully characterize all the layers of rock above, within, and below the pay zones in the reservoir.

2.5.1 Shale formations

Shale formations (also referred to as shale plays) are a worldwide occurrence (Ma et al., 2016; Moore et al., 2016), and shale formations can serve as pressure barriers in basins, as top seals, and as reservoirs in shale gas plays. More technically, shale is a fissile (capable of splitting into thin sheets along bedding), terrigenous (of a marine deposit consisting of material eroded from the land) sedimentary rock in which the particles are mostly of the size found in silt and clay size. In many basins, the fluid pressure of the aqueous system becomes significantly elevated, leading to the formation of a hydro-fracture, and fluid bleed-off. However, the occurrence of a natural hydro-fracture is an unlikely process in the circumstances that exist in most basins.

A significant factor associated with tight gas reservoirs is the low productivity, which is emphasized in the case of gas condensate fluids that exhibit complex phase behavior and complex flow behavior due to the appearance of condensate banking in the near-well region. Thus, it is necessary to have a thorough understanding of the manner by which the condensate accumulation influences the productivity.

More specific to this text, four general types of shale formation have been defined as being predominant in shale formations and have been given the simple designations: Type 1, Type 2, Type 3, and Type 4, without any order of preference but more on the basis of composition and behavior. Type 1 shale is a fractured organic mudstone with high carbonate content – an example is the Barnett Shale – and primary typically involves a mix of gas released from fractures and micropores through gas desorption from the organic material and clay minerals. On the other hand, Type 2 shale has laminated sands embedded in organic-rich shale – an example is the Bakken formation, which is primarily is an oil resource play. Type 3 shale is an organic-rich black shale, such as the Marcellus Shale, and production occurs through gas desorption – the gas has been observed to carry with it gas condensate. Finally, Type 4 shale (an example is the Niobrara Shale formation that is situated in northeastern Colorado, parts of adjacent Wyoming, Nebraska, and Kansas) is a combination of the other three type of shale formations and production is through desorption, matrix structures, and fractures.

A key factor that controls the deliverability of a gas condensate is the relative permeability, which is influenced directly by the condensate accumulation that not only reduces both the gas and liquid relative permeability but also changes the phase composition of the reservoir fluid, hence changes the phase diagram of reservoir fluid and varies the fluid properties (Wheaton and Zhang, 2000. Pedersen and Christensen, 2006). In addition, changing the manner in which the well is brought into flowing condition can affect the liquid dropout composition and can therefore change the degree of productivity loss.

The Barnett Shale (located in the Bend Arch-Fort Worth Basin, Texas, which consists of sedimentary rocks dating from the Mississippian period) was the first major natural gas field developed in a shale reservoir rock. Producing gas from the Barnett Shale was a challenge because the pore spaces in shale are small and the gas has difficulty moving through the shale and into the well.

Although the interstitial spaces in a shale formation are small, they can take up a significant volume of the rock. This allows the shale to hold significant amounts of water, natural gas, or crude oil but not be able to effectively transmit them because of the low-to-no permeability. The limitations can be overcome by using horizontal drilling and hydraulic fracturing which creates artificial porosity and permeability within the rock. Some of the clay minerals that occur in shale have the ability to absorb or adsorb large amounts of water, natural gas, ions or other substances which can enable the shale to selectively (even tenaciously) to hold or release fluids.

2.5.2 Sandstone and carbonate formations

The low-to-no permeability associated with tight formations is attributed to a wide distribution of small pores as well as a complex system of pore throats connecting those pores. Furthermore, both small pore systems and pore throat systems can result from several processes which are (i) the deposition of fine to fine grained sediments, (ii) the presence of various types of dispersed shale minerals and clay minerals in the pores, and/or (iii) the post-depositional diagenesis that alter the original pore structure (Rushing et al., 2008). Therefore, successful exploitation of a tight gas reservoir requires an understanding the rock pore structure and the properties as well as the various processes that affect the properties.

The development of tight reservoirs depends upon several factors (i) the provenance, (ii) the mineralogy, (iii) the grain size, (iv) the grain sorting, (v) the flow regime, (vi) the sedimentary depositional environment, (vii) the lithification, and (viii) the diagenesis which involves compaction, cementation, and dissolution followed by tectonics and development of fractures. In addition, the regional and local tectonic play an important role in the evaluation of the tight sand reservoirs. For example, the pressure and the thermal gradient are affected by the tectonics and are also an important aspect of the evolution of these types of reservoirs.

In tight reservoirs, the typical drainage area of a well largely depends on (i) the number of wells drilled, (ii) the size of the fracture treatments, and (iii) the time frame being considered. In lenticular or compartmentalized tight reservoirs, the drainage area is usually a function of the sand lens size or compartment size, and may not be a strong function of the size of the fracture treatment. A main factor controlling the continuity of the reservoir is the depositional system since reservoir drainage per well is generally small in continental deposits and larger in marine deposits. Fluvial systems tend to be more lenticular whereas barrier-strand plain systems tend to be more continuous.

2.5.3 Development and production

Whether the reservoir is a tight shale reservoir or carbonate reservoir or a sandstone reservoir, the development of the reservoir requires a vertical well drilled and completed and which must be stimulated to produce natural gas at commercial gas flow rates and volumes. Typically, hydraulic fracturing is required to produce natural gas economically from a tight reservoir (Speight, 2016a). Horizontal wells and/or multilateral wells can be used to provide the stimulation required for commercial production of natural gas from some naturally fractured tight reservoirs. Moreover, to optimize the development of a tight reservoir, the number of wells drilled must be optimized along with the necessary drilling and completion procedures for each well.

Because of this, effective reservoir exploitation requires a comprehensive description and characterization program to quantify gas-in-place and to identify those reservoir properties which can affect production. In terms of reservoir evaluation, reservoirs are generally evaluated on the basis of (i) rock types, (ii) structural types, (iii) heterogeneity, and (iv) porosity and permeability, which are obtained through core analysis.

2.6 Reservoir types

The reservoirs containing natural gas are a geologically varied group and involve geological feature such as – alphabetically – (i) an anticline (ii) a fault, (iii) a pinchout, (iv) a salt dome, and (v) an unconformity (Speight, 2014, 2019). In each case, this list of reservoir types contains a geological feature that prevents the gas from escaping from the formation. However, this list does not include reservoirs that are classed as tight reservoirs in which the permeability of the reservoir formation is low-to-none.

An anticline is a structural trap formed by the folding of rock strata into an arch-like shape. The rock layers in an anticlinal trap were originally laid down horizontally and then earth movement caused it to fold into an arch-like shape (the anticline).

In the case of the fault reservoir, a fault is a break or planar surface in brittle rock across which there is observable displacement. Depending on the relative direction of displacement between the rocks, or fault blocks, on either side of the fault, its movement is described as normal, reverse, or strike-slip. In a normal fault, the hanging wall moves down relative to the footwall along the dip of the fault surface, which is steep, from 45° to 90°. A reverse fault forms when the hanging wall moves up relative to the footwall parallel to the dip of the fault surface. A strike-slip fault in which the block across the fault moves to the right is described as a dextral strike-slip fault. Some fault surfaces contain relatively coarse rubble that can act as a conduit for migrating oil or gas, whereas the surfaces of other faults are smeared with impermeable clay minerals or broken grains that can act as a fault seal.

A pinch-out reservoir is a type of stratigraphic trap the termination of which occurs by thinning or tapering out of the reservoir formation against a nonporous sealing rock. This creates a favorable geometry to trap hydrocarbons, particularly if the adjacent sealing rock is a source rock such as a shale formation. The geology of the reservoir involved a reduction in bed thickness that results from the onlapping stratigraphic sequences.

A salt dome reservoir is a type of structural dome that is formed when a thick bed of evaporite minerals (mainly salt, NaCl, or halite – the mineral name for salt and also known as rock salt). It is found at depth intrudes vertically into surrounding rock strata, forming a diapir, which is a domed rock formation in which a core of rock has moved upward to pierce the overlying strata.

Unconformity reservoirs are those types of traps that resulted from the truncation of reservoir rocks and the subsequent sealing of the subcrop by an unconformable, relatively impermeable, fine-grained, rock unit. The source rocks may be within the pre-unconformity sequence, or in the immediate post-unconformity cap rocks. The timing of secondary migration is not, earlier than the time of sealing of the subcrop.

In keeping with understanding the nature of the reservoir fluids, understanding the variable properties of reservoir rocks is crucial for exploration and successful production of natural gas from tight shale and tight formation reservoirs. However, in the case of the static and dynamic elastic properties of shale from Barnett shale formation (located in the Bend Arch-Fort Worth Basin, Texas), Haynesville shale formation (that underlies large parts of southwestern Arkansas, northwest Louisiana, and East Texas.), Eagle Ford shale formation (located in South Texas), and Fort St. John shale formation (located in the Western Canada Sedimentary Basin), the matter is not so straightforward since the elastic properties of these rocks vary significantly not only between reservoirs but also within a reservoir due to the wide variety of minerals that form the reservoir rock as well as the microstructures exhibited by these reservoirs.

For example, the static (Young's modulus) and dynamic (P-wave and S-wave moduli) elastic parameters generally decrease monotonically with the content of the clay minerals plus any kerogen. The Young's modulus equation is:

$$E = \sigma/\varepsilon$$

E is Young's modulus, σ is the force per unit area, and ε is the axial strain (or proportional deformation).

However, the elastic properties of the shale formations are strongly anisotropic (the properties are not identical in all directions) since the degree of anisotropy correlates with the organic content of the shale as well as the minerals that constitute the amount and type of clay minerals that constitute the shale (Hillier, 2003; Bergaya et al., 2011). This is not (and should not be) surprising when the complex and varying composition of clay minerals are taken into consideration (Sone and Zoback, 2013a, 2013b).

More generally, the production of natural gas (and crude oil) occurs from two classes of rock which are (i) source rock and (ii) reservoir rock, although it is generally believed that at some time during the history of the formation of the natural gas (and crude oil) there has been migration of the gas (and oil) or a precursor to gas (and oil) from the source rock to the reservoir rock (Speight, 2014, 2019), thus differentiating between the original and final maturation state of the natural gas (and crude oil).

In addition, if the reservoir temperature is above the critical temperature (T_c) the reservoir is classified a natural gas reservoir and there are various types of gas reservoirs (i) the dry gas reservoir in which the gases remain in the vapor state, (ii) the wet gas reservoir in which the gases remain in the vapor state with a small amount of liquid, and (iii) the gas condensate reservoir which, on occasion, is referred to as a natural gas liquids reservoir in which it is a low-density mixture of hydrocarbon liquids that are present as gaseous components in the raw natural gas, (iv) the associated gas reservoir in which the gases are produced as a by-product of the production of crude oil, and (v) the non-associated gas reservoir in which the gas is not associated with crude oil and is produced as a primary product of the reservoir (natural gas condensate or natural gas liquids notwithstanding).

The reservoirs of interest in the present context are three basic types: (i) natural gas reservoirs, which do not contain crude oil; (ii) gas condensate reservoirs, which contain natural gas and natural gas liquids – also often referred to as natural gasoline; and (iii) crude oil reservoirs, which also contain natural gas (i.e., associated natural gas).

2.6.1 Natural gas reservoirs

In general, if the reservoir temperature is above the critical temperature (T_c), the reservoir is classified a natural gas reservoir and there are three types of gas reservoirs: (i) the dry gas reservoir in which the gas mixture remains in the vapor state, (ii) the wet gas reservoir in which the gas mixture remain in the vapor state but there is a small amount of liquid, and (iii) the gas condensate reservoir in which there is a low-density mixture of hydrocarbon liquids that are present as gaseous components in the raw natural gas. Also there are the reservoirs with associated gas and non-associated gas – the associated gas is produced as a by-product of the production of crude oil and the non-associated gas that is not associated with crude oil and the gas is produced as a primary product of the reservoir.

The principal constituent of natural gas is methane, but other hydrocarbon derivatives (especially when the gas occurs in a crude oil reservoir), such as ethane, propane, butane, and even higher molecular weight hydrocarbons up to and including octane (C_8H_{18}), may also be present in varying amounts. However, the higher molecular weight constituents are typically in lower proportions in the gas from natural gas reservoirs than found in natural gas from crude oil reservoirs. Carbon dioxide (CO_2) and hydrogen sulfide (H_2S) are also common constituents of natural gas as well as trace amounts of rare gases, such as helium, may also occur – in fact, certain natural gas reservoirs are a source of this rare gas.

Since natural gas has a low density, once formed it will rise toward the surface of the earth through loose, shale-type rock and other material. Most of this methane will simply rise to the surface and dissipate into the air. However, a great deal of this methane will rise up into geological formations that trap the gas under the ground. These formations are made up of layers of porous, sedimentary rock with a denser, impermeable layer of rock on top. This impermeable rock traps the natural gas under the ground. If these formations are large enough, they can trap a great deal of natural gas underground, in what becomes the reservoir.

There are a number of different types of these formations, but the most common is created when the impermeable sedimentary rock appears as dome-shaped formation that traps all of the natural gas (Speight, 2014, 2019). There are a number of ways that this sort of dome may be formed. For instance, faults are a common location for natural gas and crude oil to exist. A fault occurs when the normal sedimentary layers sort of split vertically, so that impermeable rock shifts down to trap natural gas in the more permeable limestone or sandstone layers (Speight, 2014, 2019).

Once a potential natural gas deposit has been located by a team of exploration geologists and geophysicists, it is up to a team of drilling experts to actually dig down to where the natural gas is thought to exist. Thus, natural gas that is trapped under in a sub-surface formation can be recovered by drilling into and through the impermeable rock. Gas in these reservoirs are typically under pressure, allowing the gas to escape from the reservoir on its own. The decision of whether or not to drill a well into a reservoir depends on several factors, not the least of which are the economic characteristics of the potential natural gas reservoir.

If a new well, once drilled, does in fact come in contact with a natural gas reservoir, it is developed to allow for the extraction of this natural gas, and is termed a development well or productive well. At this point, with the well drilled and hydrocarbon derivatives present, the well may be completed to facilitate its production of natural gas. However, if the exploration team was incorrect in the estimation of the existence of marketable quantity of natural gas at a well site, the well is termed a dry well, and production does not proceed. Then, in order to successfully bring natural gas to the surface, a hole must be drilled through the impermeable rock to release the natural gas that is under pressure.

The exact placement of the drill site depends on a variety of factors, including (i) the nature of the potential formation to be drilled, (ii) the characteristics of the subsurface geology, and (iii) the depth and size of the target formation. After the geophysical team identifies the optimal location for a well, it is necessary for the drilling company to ensure that they complete all the necessary steps to ensure that they can legally drill in that area. This usually involves securing permits for the drilling operations, establishment of a legal arrangement to allow the natural gas company to extract and sell the resources under a given area of land, and a design for gathering lines that will connect the well to the pipeline. There are a variety of potential owners of the land and mineral rights of a given area.

In order to combat the presence of sulfur compounds in natural gas (many pipeline owners specify maximum sulfur content), scrubbers are installed, usually at or near the wellhead. The scrubbers serve primarily to remove sand and other large-particle impurities such as hydrogen sulfide. Heaters also ensure that the temperature of the gas does not drop too low. With natural gas that contains even low quantities of water, natural gas hydrates tend to form when the temperature decreases. These hydrates are solid or semi-solid compounds, resembling ice-like crystals (Berecz and Balla-Achs, 1983; Gudmundsson et al., 1998; Sloan, 1997, 2000). When gas hydrates accumulate, the passage of natural gas through valves and gathering systems can be seriously impeded. To reduce the occurrence of hydrates, small natural gas–fired heating units are typically installed along the gathering pipe wherever it is likely that hydrates may form.

A reservoir containing wet gas (Table 2.8) with a large amount of valuable natural gas liquids (any hydrocarbon derivatives other than methane such as ethane, propane, and butane) and even light crude oil (>30° API) and condensate has to be treated carefully. When the reservoir pressure drops below the critical point for the mixture, the liquids may condense into a liquid phase out and remain in the reservoir. Thus, it is necessary to implement a cycling process in which the wet gas is produced to the surface and the natural gas liquids are condensed as a separate stream and the gas is compressed and injected back into the reservoir to maintain the pressure.

Table 2.8: Composition of dry gas, wet gas, and condensate.

Component	Dry gas	Wet gas	Condensate
CO_2	0.10	1.41	2.37
N_2	2.07	0.25	0.31
C_1	86.12	92.46	73.19
C_2	5.91	3.18	7.80
C_3	3.58	1.01	3.55
i-C_4	1.72	0.28	0.71
n-C_4	–	0.24	1.45
i-C_5	0.50	0.13	0.64
n-C_5	–	0.08	0.68
C_6	–	0.14	1.09
C_7+	–	0.82	8.21

The most common reservoir rocks are sandstone (SiO_2), limestone ($CaCO_3$), and dolomite (a mixed calcium-magnesium carbonate mineral) ($CaCO_3.MgCO_3$). The four basic elements of a reservoir system include: (i) the source rock, which is the rock containing the organic material that converted into natural gas and/or crude oil; (ii) a migratory pathway, which is a route for the partially formed or completely formed natural gas and crude oil to migrate from the source rock to the reservoir; (iii) the reservoir rock, which is rock of suitable porosity that can store and suitable permeability that can allow the fluid to move through the reservoir to a production well; and (iv) a seal, which is an impermeable cap rock that prevents the upward escape of the natural gas and crude oil to the surface of the Earth.

Since natural gas has a low density, once formed it will rise toward the surface of the earth through loose, shale-type rock and other material. Most of this methane will simply rise to the surface and dissipate into the air. However, a great deal of this methane will rise up into geological formations that trap the gas under the ground. These formations are made up of layers of porous, sedimentary rock with a denser, impermeable layer of rock on top. This impermeable rock traps the natural gas under the ground. If these formations are large enough, they can trap a great deal of natural gas underground, in what is known as a reservoir.

There are a number of different types of these formations, but the most common is created when the impermeable sedimentary rock forms a *dome* shape, like an umbrella that catches all of the natural gas that is floating to the surface. There are a

62 —— Chapter 2 Reservoir fluids and reservoirs

number of ways that this sort of dome may be formed. For instance, faults are a common location for oil and natural gas deposits to exist. A fault occurs when the normal sedimentary layers sort of split vertically, so that impermeable rock shifts down to trap natural gas in the more permeable limestone or sandstone layers. Essentially, the geological formation which layers impermeable rock over more porous, oil- and gas-rich sediment has the potential to form a reservoir.

To successfully bring natural gas to the surface, a hole must be drilled through the impermeable rock to release the natural gas that is under pressure.

With natural gas trapped under the earth in this fashion, it can be recovered by drilling a hole through the impermeable rock. Gas in these reservoirs is typically under pressure, allowing it to escape from the reservoir on its own.

Once a potential natural gas deposit has been located by a team of exploration geologists and geophysicists, it is up to a team of drilling experts to actually dig down to where the natural gas is thought to exist. Although the process of digging deep into the Earth's crust to find deposits of natural gas that may or may not actually exist seems daunting, a number of innovations and techniques have been developed that increase the efficiency of drilling for natural gas, with a concurrent decrease in cost. The advance of technology has also contributed greatly to the increased efficiency and success rate for drilling natural gas wells.

The exact placement of the drill site depends on a variety of factors, including the nature of the potential formation to be drilled, the characteristics of the subsurface geology, and the depth and size of the target deposit. After the geophysical team identifies the optimal location for a well, it is necessary for the drilling company to ensure that they complete all the necessary steps to ensure that they can legally drill in that area. This usually involves securing permits for the drilling operations, establishment of a legal arrangement to allow the natural gas company to extract and sell the resources under a given area of land, and a design for gathering lines that will connect the well to the pipeline. There are a variety of potential owners of the land and mineral rights of a given area.

If the new well, once drilled, does in fact come in contact with a natural gas reservoir, it is developed to allow for the extraction of this natural gas, and is termed a *development well* or *productive well*. At this point, with the well drilled and hydrocarbons present, the well may be completed to facilitate its production of natural gas. However, if the exploration team was incorrect in its estimation of the existence of marketable quantity of natural gas at a well site, the well is termed a *dry well*, and production does not proceed.

In order to combat the presence of sulfur compounds in natural gas (many pipeline owners specify maximum sulfur content), scrubbers are installed, usually at or near the wellhead. The scrubbers serve primarily to remove sand and other large-particle impurities such as hydrogen sulfide. Heaters also ensure that the temperature of the gas does not drop too low.

With natural gas that contains even low quantities of water, natural gas hydrates have a tendency to form when temperatures drop. These hydrates are solid or semi-solid compounds, resembling ice like crystals. If the hydrates accumulate, they can impede the passage of natural gas through valves and gathering systems. To reduce the occurrence of hydrates, small natural gas–fired heating units are typically installed along the gathering pipe wherever it is likely that hydrates may form.

2.6.2 Gas condensate reservoirs

Natural gas condensate is a low-density mixture of hydrocarbon liquids that are present as gaseous components in the raw natural gas produced from many natural gas fields (Table 2.9). Some gas species within the raw natural gas will condense to a liquid state if the temperature is reduced to below the hydrocarbon dew point temperature at a definitive pressure.

Table 2.9: Density of various hydrocarbon liquids*.

Hydrocarbon (phase)	Formula	Molecular weight	Density
*n-Alkanes**			
Pentane	C_5H_{12}	72	0.626
Hexane	C_6H_{14}	86	0.661
Heptane	C_7H_{16}	100	0.684
Octane	C_8H_{18}	114	0.703
Nonane	C_9H_{20}	128	0.718
Decane	$C_{10}H_{18}$	142	0.730
*Aromatics**			
Benzene	C_6H_6	78	0.877
Toluene	C_7H_8	92	0.867

*Possible constituents of gas condensate and natural gasoline listed in order of molecular weight.

There are many reservoirs that produce gas condensate, and each reservoir will produce gas condensate with its own unique composition. However, in general, gas condensate has a specific gravity on the order of ranging from 0.5 to 0.8 and is composed of higher molecular weight hydrocarbon derivatives – typically up to C8 hydrocarbons but, in some cases, up to the C_{12} hydrocarbons. Propane (C_3H_8) and butane (C_4H_{10}), normally gases at standard temperature and pressure may also occur in gas condensate as gas gases soluble in the liquid hydrocarbon derivatives.

The C8 to C_{12} hydrocarbon derivatives include the higher-boiling hydrocarbon compounds such as cyclohexane derivatives and aromatic derivative such as benzene, toluene ($C_6H_5CH_3$), xylene isomers (ortho-, meta-, and para-$CH_3C_6H_4CH_3$), and ethyl benzene ($C_6H_5C_2H_5$). In addition, the gas condensate may contain additional impurities such as hydrogen sulfide, thiol derivatives (also called mercaptans, RSH), carbon dioxide, cyclohexane (C_6H_{12}), and low molecular weight aromatics such as benzene (C_6H_6), toluene ($C_6H_5CH_3$), ethylbenzene ($C_6H_5CH_2CH_3$), and xylene derivatives ($H_3CC_6H_4CH_3$) (Mokhatab et al., 2006; Speight, 2014, 2019).

The composition of the gas condensate can have a major influence on the production of natural gas and condensate at the wellhead. For example, when condensation occurs in the reservoir, the phenomenon known as condensate blockage can halt flow of the liquids to the wellbore. Hydraulic fracturing is the most common mitigating technology in siliciclastic reservoirs (reservoirs composed of clastic rocks), and acidizing is used in reservoirs composed of carbonate mineral rocks (generally referred to as carbonate reservoirs) (Speight, 2016a).

Briefly, clastic rocks are composed of fragments, or clasts, of pre-existing minerals and rock. A clast is a fragment of geological detritus, chunks and smaller grains of rock broken off other rocks by physical weathering. The geological term "clastic" is used with reference to sedimentary rocks as well as to particles in sediment transport whether in suspension or as bed load, and in sedimentary deposits.

Reservoirs with the relative amount of at least 5–10 g of condensate per cubic meter are usually referred to as gas condensate reservoirs. Gas condensate reservoirs can be confined in any formations that from suitable traps and arbitrarily defined in two ways: (i) primary gas condensate reservoirs formed at the depths in excess of 10,000 feet below the surface that are separate from crude oil accumulations, and (ii) secondary gas condensate reservoirs formed through partial vaporization of the constituents of the crude oil. There are saturated gas condensate fields (the formation pressure is equal to the pressure of initial condensation) and non-saturated gas condensate fields (the initial condensation pressure is lower than the formation pressure) according to the thermo-baric conditions.

A gas-condensate reservoir (also called a dew-point reservoir) is a reservoir in which condensation causes a liquid to leave the gas phase. The condensed liquid remains immobile at low concentrations. Thus, the gas produced at the surface will have a lower liquid content, and the producing gas-oil ratio therefore rises. This process of *retrograde* condensation continues until a point of maximum liquid volume is reached. The term *retrograde* is used because generally vaporization, rather than condensation, occurs during isothermal expansion. After the dew point is reached, because the composition of the produced fluid changes, the composition of the remaining reservoir fluid also changes.

A retrograde-condensate gas reservoir initially contains a single-phase fluid, which changes to two phases (condensate and gas) in the reservoir when the reservoir pressure decreases. Additional condensate forms with changes in pressure and temperature in the tubing and during lease separation. From a reservoir standpoint, dry and wet gas can be treated similarly in terms of producing characteristics, pressure behavior, and recovery potential. Studies of retrograde-condensate gas reservoirs must consider changes in condensate yield as reservoir pressure declines, the potential for decreased well deliverability as liquid saturations increase near the wellbore, and the effects of two-phase flow on wellbore hydraulics.

Typically, a gas condensate reservoir will have a reservoir temperature located between the critical point and the cricondentherm on the reservoir fluid pressure–temperature diagram. This is one way of identifying a gas condensate reservoir – any other definition (such as condensate–gas ratio or molecular weight of the C^{7+} fraction or the API gravity of the C_{7+} fraction), may leave gaps in knowledge of the behavior of the reservoir and the condensate (Thomas et al., 2009).

Drip gas, so named because it can be drawn off the bottom of small chambers (called drip chambers or drips) sometimes installed in pipelines from gas wells, is another name for natural gas condensate, a naturally occurring form of gasoline obtained as a by-product of natural gas extraction. Drip gas is defined in the United States Code of Federal Regulations as consisting of butane, pentane, and hexane derivatives. Within set ranges of distillation, drip gas may be extracted and used to as a cleaner and solvent as well as a lantern and stove fuel. Accordingly, each type of condensate (including drip gas, natural gasoline, and casing-head gas) requires a careful compositional analysis for an estimation of the potential methods of preliminary purification at the wellhead facilities prior to transportation through a pipeline to a gas processing plant or a refinery (Speight, 2014, 2017, 2019).

Because gas condensate is typically liquid in ambient conditions and also has a low viscosity, it is often used as a diluent for highly viscous heavy crude oil that cannot otherwise be efficiently transported by means of a pipeline. In particular, condensate (or low-boiling naphtha from a refinery) is frequently mixed with bitumen from tar sand (called oil sand in Canada) to create the blend known as *Dilbit*. However, caution is required when condensate having an unidentified composition is blended with heavy oil, extra heavy oil, and/or tar sand bitumen since the potential for incompatibility of the blended material may become a reality.

This is especially true if the condensate is composed predominantly of n-alkane hydrocarbon derivatives of the type (pentane, C_5H_{12}, and heptane, C_7H_{16}, as well as other low-boiling liquid alkane derivatives). These hydrocarbon derivatives are routinely used in laboratory deasphalting and in commercial deasphalting units in which the asphaltene fraction is produced as a solid insoluble as a solid product from the heavy oil or bitumen feedstock (Speight, 2014, 2015b).

Condensate differs substantially from conventional crude oil insofar as (i) the color of crude oil typically varies from dark green to black – typically, condensate is almost colorless, (ii) crude oil typically contains some naphtha which is often incorrectly referred to as gasoline – condensate is almost equivalent in boiling range to the low-boiling naphtha fraction of crude oil, (iii) crude oil usually contains dark-colored, high-molecular-weight non-volatile components – condensate does not contain any dark-colored, high-molecular-weight non-volatile components, and (iv) the

API gravity of crude oil, which is measure of the weight per unit volume or density, is commonly less than 45° – condensate typically has an API gravity on the order of 50° and higher (Speight, 2014, 2015b, 2019).

Furthermore, although the difference between retrograde-condensate and wet gas is notable, there is much less distinction between wet gas and dry gas. For both wet gas and dry gas, reservoir engineering calculations are based on a single-phase reservoir gas. The only issue is whether there is a sufficient volume of produced liquid to be considered in such calculations as material balance or wellbore hydraulics. Retrograde systems require more-complex calculations using equation of state that depend upon data produced in an analytical or research laboratory.

2.6.3 Crude oil reservoirs

In reservoirs that contain natural gas and crude oil, the natural gas (being the least dense) is found closest to the surface, with the oil beneath it, typically followed by a certain amount of water. In order to process and transport associated dissolved natural gas, it must be separated from the oil in which it is dissolved. This separation of natural gas from petroleum is most often done using equipment installed at or near the wellhead.

The actual process (and equipment) that is used to separate natural gas from crude oil can vary widely. Although dry pipeline quality natural gas is virtually identical across different geographic areas, raw natural gas from different regions will vary in composition (Table 2.3) and therefore separation requirements may emphasize or de-emphasize the optional separation processes. In many instances, natural gas is dissolved in crude oil in the reservoir primarily due to the formation pressure.

When this natural gas and crude oil is produced, it is possible that it will separate without application of a separation process. In these cases, separation of natural gas and crude oil is relatively easy, and the two hydrocarbon sources are sent separate ways for further processing. When mechanical separation is required, the most basic type of separator is known as a conventional separator which consists of a simple closed tank, where the force of gravity serves to separate the heavier liquids (the crude oil) or the higher molecular weight constituents of the gas stream. In some cases, specialized equipment may be necessary to separate the natural gas and the crude oil. An example of this type of equipment is the low-temperature separator which is most often used for wells producing high-pressure gas along with condensate and/or light crude oil. These separators use pressure differentials to cool the wet natural gas and to separate the condensate form the crude oil.

In the process, wet gas enters the separator, after being cooled by passage through a heat exchanger system. The gas stream then travels through a high-pressure liquid knockout pot, which serves to remove any liquids into a low-temperature separator. The gas stream is then sent into the low-temperature separator through a choke mechanism, which expands the gas as it enters the separator. This rapid expansion of the gas allows for the lowering of the temperature in the separator. After liquid removal, the dry gas then travels back through the heat exchanger and is warmed by the incoming wet gas. By varying the pressure of the gas in various sections of the separator, it is possible to vary the temperature, which causes the crude oil and some water to be condensed out of the wet gas stream. This basic pressure–temperature relationship can work in reverse as well, to extract gas from a liquid oil stream.

In addition, the presence of impurities in gas streams – coupled with shipping container regulations – may require some pre-transportation processing at the wellhead. Typical pre-transportation processing would include the removal of acid gas (hydrogen sulfide, H_2S, and carbon dioxide, CO_2).

References

Alreshedan, F., and Kantzas, A. 2015. Investigation of Permeability, Formation Factor, and Porosity Relationships for Mesaverde Tight Gas Sandstones Using Random Network Models. Journal of Petroleum Exploration and Production. Technology. DOI 10.1007/s13202-015-0202-x; published on line September 22, 2015. http://download.springer.com/static/pdf/29/art%253A10.1007%252Fs13202-015-0202-x.pdf?originUrl=http%3A%2F%2Flink.springer.com%2Farticle%2F10.1007%2Fs13202-015-0202-x&token2=exp=1460646961~acl=%2Fstatic%2Fpdf%2F29%2Fart%25253A10.1007%25252Fs13202-015-0202-x.pdf%3ForiginUrl%3Dhttp%253A%252F%252Flink.springer.com%252Farticle%252F10.1007%252Fs13202-015-0202-x*~hmac=342abf10be838dd3f869a372bd753a11debc3698f1c525ec0a689af6b4ff2677; accessed March 10, 2016.

Anderson, W.G., 1986. Wettability literature survey: Part 1. Rock-Oil-Brine Interactions and the Effects of Core Handling on Wettability. Journal of Petroleum Technology, October, pages 1125–1144

Berecz, E. and Balla-Achs, M. 1983. Gas Hydrates. Elsevier BV, Amsterdam, Netherlands.

Berg, R.R. 1970. Method for Determining Permeability from Reservoir Rock Properties. Transactions of the GCAGS 20: 303. Gulf Coast Association of Geological Societies, Houston, Texas.

Berg, R.R. 1986. Sandstone Reservoirs. Prentice-Hall, Pearson Education Group, Upper Saddle River, New Jersey.

Bergaya, F., Theng, B.K.G., and Lagaly, G. 2011. Handbook of Clay Science. Elsevier, Amsterdam, Netherlands.

Blatt, H., and Tracy, R.J. 1996. Petrology: Igneous, Sedimentary, and Metamorphic 2nd Edition. W.H. Freeman and Company, Macmillan Publishers, New York.

Bloch, S., Lander, R.H., and Bonnell, L. 2002. Anomalously High Porosity and Permeability in Deeply Buried Sandstone Reservoirs: Origin and Predictability. AAPG Bulletin, 86(2): 301–328. American Association of Petroleum Geologists, Tulsa, Oklahoma.

Bustin, R.M., Bustin, A.M.M., Cui, X., Ross, D.J.K., and Murthy Pathi, V.S. 2008. Impact of Shale Properties on Pore Structure and Storage Characteristics. Paper No. SPE 119892. Proceedings.

SPE Conference on Shale Gas Production, Fort Worth, Texas, November 16–18. Society of Petroleum Engineers, Richardson, Texas.

Caruana A and Dawe R A, 1996a. Effect of Heterogeneities on Miscible and Immiscible Flow Processes in Porous Media, Trends in Chemical Engineering. **3**: 185–203.

Caruana A and Dawe R A, 1996b, Flow Behavior in the presence of Wettability Heterogeneities, Transport in Porous Media, **25**, 217–233.

Dandekar, A.Y. 2013. Petroleum Reservoir Rock and Fluid Properties 2nd Edition. CRC Press, Taylor & Francis Group, Boca Raton, Florida.

Dawe R.A. 2003. Reservoir Engineering, 2000. In: Dawe RA (Editor). Modern Petroleum Technology, Upstream Volume. Institute of Petroleum, John Wiley and Sons, Chichester, England. Chapter 7. Page 207–282.

Dawe, R.A. 2004. Miscible Displacement in Heterogeneous Porous Media. Proceedings. Sixth Caribbean Congress of Fluid Dynamics, University of the West Indies, Augustine, Trinidad. January 22–23.

GAO. 2012. Information on Shale Resources, Development, and Environmental and Public Health Risks. Report No. GAO-12-732. Report to Congressional Requesters. United States Government Accountability Office, Washington, DC. September.

Grattoni C A and Dawe R A, 2003. Consideration of Wetting and Spreading in Three-Phase Flow in Porous Media. In: Progress in Mining and Oilfield Chemistry, Volume 5. Recent Advances in Enhanced Oil and Gas Recovery. I Lakatos (Editor). Akad. Kiado, Budapest, Hungary.

Gudmundsson, J.S., Andersson, V., Levik, O.I., and Parlaktuna, M. 1998. Hydrate Concept for Capturing Associated Gas. Proceedings. SPE European Crude oil Conference, The Hague, Netherlands, 20-22 October 20–22.

Hillier, S. 2003. Clay Mineralogy. In: Encyclopedia of Sediments and Sedimentary Rocks. G.V. Middleton, M.J. Church, M. Coniglio, L.A. Hardie, and F.J. Longstaffe (Editors). Kluwer Academic Publishers, Dordrecht, Netherlands. Page 139–142.

Hunt, J.M. 1996. Petroleum Geochemistry and Geology 2nd Edition. W.H. Freeman and Co., New York.

Kovscek, A.R. 2002. Heavy and Thermal Oil Recovery Production Mechanisms. Quarterly Technical Progress Report. Reporting Period: April 1 through June 30, 2002. DOE Contract Number: DE-FC26-00BC15311. July.

Ma, Y.Z., Moore, W.R., Gomez, E., Clark, W.J. and Zhang, Y. 216. Tight Gas Sandstone Reservoirs, Part 1: Overview and Lithofacies. In: Unconventional Oil and Gas Resources Handbook. *Y.Z. Ma, S. Holditch, and J.J. Royer* (Editors). Elsevier, Amsterdam, Netherlands. Chapter 14.

Marshak, S. 2012. Essentials of Geology 4th Edition. W.W. Norton & Company, New York.

Maxwell, S., and Norton, M. 2012. The Impact of Reservoir Heterogeneity on Hydraulic Fracture Geometry: Integration of Microseismic and Seismic Reservoir Characterization. Proceedings. AAPG Annual Convention and Exhibition, Long Beach, California. April 22–25. http://www.searchanddiscovery.com/documents/2012/40993maxwell/ndx_maxwell.pdf; accessed April 15, 2015.

Mokhatab, S., Poe, W.A., and Speight, J.G. 2006. Handbook of Natural Gas Transmission and Processing Elsevier, Amsterdam, Netherlands.

Moore, W.R., Ma, Y.Z., Pirie, I., and Zhang, Y. 2016. Tight Gas Sandstone Reservoirs, Part 2: Petrophysical Analysis and Reservoir Modeling. In: Unconventional Oil and Gas Resources Handbook. *Y.Z. Ma, S. Holditch, and J.J. Royer* (Editors). Elsevier, Amsterdam, Netherlands. Chapter 15.

Moslow, T.F. 1993. Evaluating Tight Gas Reservoirs. In: Development Geology Reference Manual. American Association of Petroleum Geologists, Tulsa, Oklahoma.

Neasham, J.W. 1977a. Applications of Scanning Electron Microscopy to the Characterization of Hydrocarbon-Bearing Rocks. Scanning Electron Microscopy, 7: 101–108.

Neasham, J.W. 1977b. The Morphology of Dispersed Clay in Sandstone Reservoirs and Its Effect on Sandstone Shaliness, Pore Space and Fluid Flow Properties. Paper No. SPE 6858. Proceedings. 52nd Annual Fall Technical Conference and Exhibition of the Society of Petroleum Engineers. October 9–12. Society of Petroleum Engineers, Richardson, Texas.

Pedersen, K.S., ands Christiansen, P.L. 2006. Phase Behavior of Petroleum Reservoir Fluids. CRC Press, Taylor & Francis Group, Boca Raton, Florida.

Radlinski, A.P., Ioannidis, M.A., Hinde, A.L., Hainbuchner, M., Baron, M., Rauch, H., and Kline, S.R. Angstrom-to-millimeter Characterization of Sedimentary Rock Microstructure. Journal of Colloidal and Interface Science, 274: 607–612.

Rushing, J.A., Newsham, K.E., and Blasingame, T.A. 2008. Rock Typing – Keys to Understanding Productivity in Tight Gas Sands. Paper No. SPE 114164. Proceedings. 2008 SPE Unconventional Reservoirs Conference, Keystone, Colorado. February 10–12. Society of Petroleum Engineers, Richardson, Texas.

Sloan, E.D. 1997. Clathrates of Hydrates of Natural Gas. Marcel Dekker Inc., New York.

Sloan, E.D. 2000. Clathrates Hydrates: The Other Common Water Phase. Ind. Eng. Chem. Res. 39: 31123–3129.

Solano, N.A., Clarkson, C.R., Krause, F.F., Aquino, S.D. and Wiseman, A. 2013. On the Characterization of Unconventional Oil Reservoirs. CSEG Recorder, 38(4): 42–47. http://csegre corder.com/articles/view/on-the-characterization-of-unconventional-oil-reservoirs; accessed July 20, 2015.

Sone, H., and Zoback, M.D. 2013a. Mechanical Properties of Shale-Gas Reservoir Rocks – Part 1: Static and Dynamic Elastic Properties and Anisotropy. Geophysics, 78(5): D381-D392.

Sone, H., and Zoback, M.D. 2013b. Mechanical Properties of Shale-Gas Reservoir Rocks – Part 2: Ductile Creep, Brittle Strength, and Their Relation to The Elastic Modulus. Geophysics, 78(5): D393-D402.

Speight, J.G. 2014. The Chemistry and Technology of Petroleum 5th Edition. CRC Press, Taylor & Francis Group, Boca Raton, Florida.

Speight, J.G. 2015a. Handbook of Offshore Oil and Gas Operations. Gulf Professional Publishing Company, Elsevier, Oxford, United Kingdom.

Speight, J.G. 2015b. Handbook of Petroleum Product Analysis 2nd Edition. John Wiley & Sons Inc., Hoboken, New Jersey.

Speight, J.G. 2016a. Handbook of Hydraulic Fracturing. John Wiley & Sons Inc., Hoboken, New Jersey.

Speight, J.G. 2016b. Enhanced Recovery Methods for Heavy Oil and Tar Sands 2nd Edition, Gulf Publishing Company, Elsevier, Oxford, United Kingdom.

Speight, J.G. 2017. Handbook of Crude oil Refining. CRC Press, Taylor & Francis Group, Boca Raton, Florida.

Speight, J.G. 2019. Natural Gas: A Basic Handbook 2nd Edition. Gulf Publishing Company, Elsevier, Cambridge, Massachusetts.

Thomas, F.B., Bennion, D.B., and Andersen, G. 2009. Gas Condensate Reservoir Performance. Journal of Canadian Crude oil Technology, 48(7): 18–24.

Tucker, M.E., and Wright, V.P. 1990. Carbonate Sedimentology. Blackwell Scientific Publications, Wiley-Blackwell, John Wiley & Sons Inc., Hoboken, New Jersey.

Wheaton, R., and Zhang, H. 2000. Condensate Banking Dynamics in Gas Condensate Fields: Compositional Changes and Condensate Accumulation Around Production Wells. Paper No. 62930. Proceedings. SPE Annual Technical Conference and Exhibition, Dallas, Texas. October 1–4. Society of Petroleum Engineers, Richardson, Texas.

Whitson, C.H., and Belery, P. 1994. Compositional Gradients in Crude oil Reservoirs. Paper No. SPE28000. Proceedings. Centennial Crude oil Engineering Symposium. University of Tulsa, Tulsa, Oklahoma. August 29–31. Society of Crude oil Engineers, Richardson, Texas.

Wilhelms, A., and Larter, S.R. 1994a. Origin of Tar Mats in Petroleum Reservoirs. Part I: Introduction and Case Studies. Marine and Petroleum Geology, 11(4): 418–441.

Wilhelms, A., and Larter, S.R. 1994b. Origin of Tar Mats in Petroleum Reservoirs. Part II: Formation Mechanisms for Tar Mats. Marine and Petroleum Geology, 11(4): 442–456.

Wilson, M. D., and E. D. Pittman, 1977, Authigenic Clays in Sandstones: Recognition and Influence on Reservoir Properties and Paleoenvironmental Analysis. Journal of Sedimentary Petrology, 47: 3–31.

Zhang, M., and Zhang, J. 1999. Geochemical Characteristics and Origin of Tar Mats from the Yaha Field in the Tarim Basin, China. Chinese Journal of Geochemistry, 18(3): 250–257.

Chapter 3
Reservoir engineering

3.1 Introduction

Reservoir engineering is a branch of petroleum engineering that applies scientific principles to the fluid flow through porous medium during the development and production of natural gas and crude oil from reservoirs in order to maximize resource recovery (Baker et al., 2015; Terry and Rogers, 2015; Satter and Iqbal, 2016; Ahmed, 2019). More specifically, reservoir engineering is the application of engineering principles for evaluating and managing reservoirs. This discipline is devoted to evaluating field performance and through reservoir modeling studies, reservoir engineers endeavor to increase hydrocarbon production and maximize exploration and production assets. Because of this, reservoir engineering covers a broad range of subjects including the occurrence of fluids in a gas or oil-bearing reservoir, movement of the reservoir fluids or injected fluids, and evaluation of the factors governing the recovery of oil or gas. The objectives of a reservoir engineer are to maximize producing rates and to ultimately recover oil and gas from reservoirs in the most efficient manner possible.

The initial amount of fluid in a reservoir is an important number. Typically, the symbol N (from the word *naphtha*) represents the initial volume of oil in the reservoir expressed as a standard surface volume, such as the stock-tank barrel (STB). G and W are initial reservoir gas and water, respectively. As these fluids are produced, the subscript p is added to indicate the cumulative oil (Np), gas (Gp), or water (Wp) produced.

Reservoir engineering involves more than the application of applied reservoir mechanics and put simply, the objective of reservoir engineering is optimization of recovery of the resource. To obtain optimum recovery of the resource – in this case natural gas and/or crude oil – the engineer or the engineering team must (i) identify and define all individual properties of the reservoir and the reservoir fluids, (ii) deduce the performance of the reservoir, (iii) prevent drilling of unnecessary wells, (iv) initiate operating controls at the proper time, and last but not by any means the least (v) consider all important economic factors, including any applied taxes from local, state, or federal authorities.

Thus, early and accurate identification and definition of the reservoir system (including the reservoir fluids) is essential to any effective reservoir engineering program (Chapters 1 and 2). Conventional geologic techniques seldom provide sufficient data to identify and define each individual reservoir and the reservoir engineer must supplement the geologic study with engineering data and tests to provide the necessary information. The most successful efforts are those that apply extensive efforts to understand the reservoir and acquire as many facts related to the reservoir

https://doi.org/10.1515/9783110691023-003

system as possible thereby mitigating the number of assumptions related to the reservoir system.

In the context of this book, a reservoir can contain natural gas alone or natural gas and crude oil. In the latter case, the crude oil may be saturated with gas – the degree of saturation is (among other parameters) a function of reservoir pressure and reservoir temperature. If all of the natural gas capable of dissolving in the crude oil has dissolved under the reservoir conditions, the crude oil is referred to as saturated (crude) oil. The excess gas is present in the reservoir in the form of a free gas cap. If there is less gas present in the reservoir than the amount that may be dissolved in oil under conditions of reservoir pressure and temperature, the oil is then termed undersaturated. The pressure at which the gas begins to come out of solution is called the saturation pressure or the bubble-point pressure (Appendix C). In the case of saturated crude oil, the saturation pressure is equal to the reservoir pressure and the gas begins to emerge from solution as soon as the reservoir pressure begins to decrease. In the case of undersaturated crude oil, the gas does emerge from the solution until the reservoir pressure drops to the level of saturation pressure.

As the gas emerges for the solution, the viscosity of the crude oil increases and the density (specific gravity of the crude oil) decreases – to allay any confusion, the API gravity of the crude oil increases as the density of the crude oil decreases.

The predictability of such phenomena can be determined by the consideration of the physical properties of the reservoir fluids in the laboratory, either from bottom-hole samples or from recombined surface separator samples. Some of the correlations on which the predictability is based involve determination of (i) the bubble-point pressure, (ii) the formation volume factors of bubble-point liquids, (iii) the formation volume factors of gas plus liquid phases, and (iv) the density of a bubble-point liquid as empirical functions of gas-oil ratio, gas gravity, oil gravity, pressure, and temperature.

Thus, the purpose of reservoir engineering is to provide the facts, information, and knowledge necessary to control operations to obtain the maximum possible recovery from a reservoir at the least possible cost. Above all, this must be a group effort in which members of the necessary disciplines and sub-disciplines must work together.

This chapter presents the basic fundamentals useful to reservoir engineering. Moreover, the topics are introduced at a level that can be understood by engineers and geologists who may not be expert in this field. But as a start, the necessary definitions are presented in the next section.

3.2 Definitions

A reservoir is a porous and permeable underground formation containing an individual and separate natural accumulation of producible natural gas and/or crude oil (Chapter 2). The natural gas and the crude oil are confined within the reservoir by impermeable rock and/or water barriers and the reservoir is characterized by a single natural pressure system. In most situations, reservoirs are classified by a local (or national) regulatory agency as (i) natural gas reservoirs, (ii) gas condensate reservoirs, and (iii) crude oil reservoirs (Chapter 2). In the absence of a regulatory authority, the classification is based on the occurrence of the contents of the reservoir as determined by the operator or recovery personnel.

Volumetric gas reservoirs are considered as totally isolated, closed systems with approximately constant hydrocarbon pore volume. For volumetric gas reservoirs, it is assumed that the reservoir does not receive significant pressure support or fluid from outside sources, such as water influx from aquifers or neighboring shale (non-reservoir) layers. In practice, there are relatively few gas reservoirs that are truly volumetric as defined. The classification of a gas reservoir as volumetric allows the application of simplifying assumptions for the assessment of its in-place hydrocarbon volumes and prediction of reservoir performance. In this section, volumetric calculations for dry gas, wet gas, and retrograde gas condensate reservoirs are presented. This is then followed by material balance on volumetric gas reservoirs.

Natural gas (by definition and by properties) is a mixture of hydrocarbon derivatives and varying quantities of nonhydrocarbon derivatives that exist either in the gaseous phase or in solution in crude oil in a natural underground reservoir. Furthermore, natural gas may be sub-classified into two predominant types as follows (i) associated gas, commonly known as gas-cap gas, that overlies the crude oil thereby being in contact with the crude oil in the reservoir – dissolved gas is natural gas that is in solution by dissolution in the crude oil in the reservoir, and (ii) nonassociated natural gas, which is gas that is in a reservoir that does not contain significant quantities of crude oil (Chapter 1). Dissolved gas and associated gas may be produced concurrently from the same wellbore. In such situations, it is not feasible to measure the production of dissolved gas and associated gas separately; therefore, production is reported under the heading of associated-dissolved or casinghead gas. Reserves and productive capacity estimates for associated and dissolved gas also are reported as totals for associated gas and dissolved gas combined.

Natural gas condensate (also referred to as gas condensate, or condensate) is a mixture of low-density-low-boiling hydrocarbon liquids that are present as gaseous components in the raw natural gas produced from many natural gas fields (Chapters 1 and 2). Some gas species within the raw natural gas will condense to a liquid state if the temperature is reduced to below the hydrocarbon dew point temperature at a definitive pressure.

Reservoirs with the relative amount of at least 5–10 g of condensate per cubic meter (5–10 g/m^3) are usually referred to as gas condensate reservoirs. These types of reservoir can be located in any suitable formation that form a suitable trap and arbitrarily defined in two ways: (i) a primary gas condensate reservoir that is formed at the depths in excess of 10,000 feet below the surface and are separate from crude oil accumulations, and (ii) a secondary gas condensate reservoir formed through partial vaporization of the constituents of the crude oil. There are saturated gas condensate reservoirs in which the formation pressure is equal to the pressure of initial condensation and non-saturated gas condensate reservoirs in which the initial condensation pressure is lower than the formation pressure.

Similarly, the term "natural gas liquids (NGLs)" is applied to a hydrocarbon liquid stream that consists of higher molecular weight hydrocarbon derivatives that exist in the reservoir as constituents of natural gas, but which are recovered as liquids in field (well-head) separators or gas-processing plants (Table 3.1). Also, the term "natural gasoline" is used in the gas processing industry to refer to a mixture of liquid hydrocarbon derivatives (predominantly pentane derivatives) (C_5H_{12}), including n-pentane ($CH_3CH_2CH_2CH_2CH_3$), iso-pentane [$(CH_3)_2CHCH_2CH_3$], and higher molecular weight hydrocarbon derivatives. In terms of physical properties (and, quite often, constituents) there is little difference between natural gas condensate, NGLs, and natural gasoline.

Table 3.1: Constituents of natural gas.

Constituent	Formula	% v/v
Methane	CH_4	>85
Ethane	C_2H_6	3–8
Propane	C_3H_8	1–5
n-Butane	C_4H_{10}	1–2
iso-Butane	$C4H_{10}$	<0.3
n-Pentane	C_5H_{12}	1–5
iso-Pentane	C_5H_{12}	<0.4
Hexane, heptane, octane*	C_nH_{2n+2}	<2
Carbon dioxide	CO_2	1–2
Hydrogen sulfide	H_2S	1–2
Oxygen	O_2	<0.1
Nitrogen	N_2	1–5
Helium	He	<0.5

*Hexane (C_6H_{14}) and higher molecular weight hydrocarbon derivatives up to octane as well as benzene (C_6H_6) and toluene ($C_6H_5CH_3$).

On the other hand, crude oil is a mixture of higher-boiling (than condensate) hydrocarbon derivatives that exist in the liquid phase in a natural underground reservoirs and which remains liquid at atmospheric pressure after passing through surface facilities. From a chemical standpoint crude oil is an extremely complex mixture of hydrocarbon compounds, usually with minor amounts of nitrogen-, oxygen-, and sulfur-containing compounds as well as trace amounts of metal-containing compounds.

For statistical purposes, volumes reported as crude oil reserves include (i) liquids technically defined as crude oil; (ii) small amounts of hydrocarbon derivatives that existed in the gaseous phase in a natural underground reservoir but which are liquid at atmospheric pressure after being recovered from a well (casinghead) gas in lease separators; and (iii) small amounts of nonhydrocarbon derivatives produced with the oil.

3.3 Reservoir evaluation

Reservoir evaluation involves a careful assessment of (i) the structure of the reservoir, such as the types of rock(s) from which the reservoir is composed and (ii) the temperature of the reservoir. However, these two parameters are not the only parameters used to evaluate a reservoir and there are other parameters that are (at least) equally important (Chapter 2). However, each rock type represents different physical and chemical processes affecting rock properties during the depositional and paragenetic cycles.

A natural gas reservoir and/or a crude oil reservoir is a subsurface collection (sometime referred to as a pool, which leaves open the possibility of misinterpretation) of hydrocarbons and hydrocarbonaceous derivatives in porous or fractured rock formations. However, natural gas and crude oil reservoirs are typically classified as (i) conventional reservoirs and (ii) unconventional reservoirs. In the former type of reservoir – the conventional reservoir – the natural gas and crude oil are trapped by an underlying rock formation (the basement rock) and by an overlying rock formation (the cap rock) with lower permeability than the reservoir rock. In the second type of reservoir – the unconventional reservoir – the reservoir rock typically has high porosity and low permeability in which the natural gas and crude oil are trapped in place without the need for a cap rock or basement rock.

In the case of tight formations (Chapter 2), most tight gas sands have been subjected to post-depositional diagenesis, a comparison of all the rock types will allow an assessment of the impact of diagenesis on rock properties. If diagenesis is minor, the depositional environment (and depositional rock types) as well as the expected properties of the rocks derived from those depositional conditions will be allowable predictors of rock quality. However, if the reservoir rock has been subjected to significant diagenesis, the original rock properties present at deposition must be anticipated to be different to the current properties. More specifically, use of the depositional environment and the associated rock types (in isolation) to guide field development activities should be used with caution and have the potential to result in ineffective exploitation of the reservoir.

In a conventional natural gas (and crude oil) reservoir, natural gas and crude oil are fairly mobile and easily move through the permeable formation because of buoyancy – the gas and crude oil are less dense than the water in the same formation and therefore rise until they are trapped against an impermeable rock (the cap rock seal) that prevents further movement. This leads to localized collections (often referred to as pools) of natural gas and crude oil while the rest of the formation is filled with water. However, both biogenic and thermogenic shale gas remain where it was first generated and can be found in three forms: (i) free gas in the pore spaces and fractures; (ii) adsorbed gas, where the gas is electrically stuck to the organic matter and clay; and (iii) a small amount of dissolved gas that is dissolved in the organic matter.

Typically, in the so-called conventional reservoir, an impermeable shale formation is either the basement rock or the cap rock of a sandstone formation thereby preventing any fluids (gases or liquids) within the sandstone from escaping or migrating to other formations. When a significant amount of organic matter has been deposited with the sediments, the shale rock can contain organic solid material (kerogen). The properties and composition of shale place it in the category of sedimentary rocks known as mudstones. Shale is distinguished from other mudstones because it is laminated and fissile – the shale is composed of many thin layers and readily splits into thin pieces along the laminations.

The evaluation of any reservoir, including reservoirs that have low-to-no permeability – such as formations and form tight formations – should always begin with a thorough understanding of the geologic characteristics of the formation. The important geologic parameters for a trend or basin are: (i) the structural and tectonic regime, (ii) the regional thermal gradients, and (iii) the regional pressure gradients, (iv) the depositional system, (v) the genetic facies, (vi) textural maturity, (vii) mineralogy, (viii) diagenetic processes, (ix) reservoir dimensions, and (x) the presence of natural fractures, all of which can affect drilling, evaluation, completion, and stimulation. Without understanding the above-listed factors can lead to guesswork in determining reservoir behavior, performance, and longevity.

An important geologic aspect of the reservoir is the external geometry of the reservoir, defined by seals that inhibit the further migration of the natural gas and crude oil. Migration will cease, and a hydrocarbon-containing reservoir will form, only where the hydrocarbon derivatives encounter a trap, which is composed of a suitable gas-holding or oil-holding rock with the following types of seals: (i) top, (ii) lateral, and (iii) bottom seals. In addition, the geometry of traps can be (i) the structural properties or configuration, (ii) the sedimentary properties or stratification, and (iii) the diagenetic properties.

Another important geologic aspect of the reservoir is the internal architecture, which involves the lateral distribution of depositional textures is related to depositional environments, and the vertical stacking of textures is described by stratigraphy, which is the geological study of the following aspects of rock strata such as

(i) form, (ii) arrangement, (iii) geographic distribution, and (iv) chronologic succession. Diagenesis, which refers to changes that happen to the sediment after deposition, can also control the lateral continuity and vertical stacking of reservoir rock types. This phenomenon is an important aspect of carbonate reservoirs, in which the conversion of limestone to dolostone and the dissolution of carbonate have a large effect on internal reservoir architecture. Other parameters include the amount (the reserves) of natural gas in the reservoir and the ability of the developer to recover these reserves.

Also, reservoir rocks tend to show greater variations in permeability than in porosity and in addition, these two properties, as measured on core samples from reservoir rocks, are not always identical with the values indicated for the bulk rock in the underground formation because of the non-representative nature of many core sample. Generally, porosity is on the order of 5–30%; while permeability falls between 0.005 darcy (5 millidarcy) and several darcies (several thousand millidarcies).

One of the most difficult parameters to evaluate in tight gas reservoirs is the drainage area size and shape of a typical well. In tight reservoirs, months or years of production are normally required before the pressure transients are affected by reservoir boundaries or well-to-well interference. As such, it may be necessary to estimate the drainage area size and shape for a typical well in order to estimate reserves. Knowledge of the depositional system and the effects of diagenesis on the rock are needed to estimate the drainage area size and shape for a specific well. In a blanket sand tight gas reservoir, the typical drainage area of a well largely depends on (i) the number of wells drilled, (ii) the size of the fracture treatments pumped on the wells, and (iii) the time frame being considered. In lenticular (or compartmentalized) tight gas reservoirs, the average drainage area is typically a function of the average sand-lens size or compartment size and may not be a reliable on the size of the hydraulic fractures. The process of hydraulic fracturing is a stimulation process that induces hydraulic fractures into a formation to provide flow paths of the natural gas to the wellbore (Chapter 6).

In the hydraulic fracturing process, a fracturing fluid is pumped (injected) into the formation at a calculated pressure and rate that has the tendency to create fracture in a low-to-no permeability reservoir. The process can enhance production of natural gas by elevation the production rate. The refracturing treatment should be applied on a regular basis while still preserving gas production.

Thus, an understanding of the geology of a reservoir is essential to reservoir development, oil and gas production, and management, including reservoir longevity and environmental management. Furthermore, reservoir evaluation includes both the external geology of the reservoir (the forces responsible for the formation of the reservoir) and the internal geology of the reservoir (the nature of the rocks that constitute the reservoir). These aspects are even more important when hydraulic fracturing methodology is to be applied to the reservoir. In addition, the efficient extraction of crude oil and natural gas requires that the reservoir be visualized in three dimensions, which can only be adequately provided through a variety of scientific and geological studies.

Thus, reservoir evaluation, as applied to natural gas reservoirs, consists of (i) the quantitative and qualitative interpretation of core sample, (ii) geophysical well logs, (iii) mud logs, (iv) flow tests, (v) pressure tests, and (vi) samples of reservoir fluids. The goal of the interpretation is to provide information concerning reservoir lithology, fluid content, storage capacity, and producibility of oil or gas reservoirs and only after this is known can estimates of the resource base be made.

3.3.1 Coring and core analysis

Coring is a drilling-technical operation, which allows integral sampling of the drilled rock – a so-called core. Based on technical equipment used for coring, and based on a chosen coring technology, a mechanically undamaged core (or in physically unaffected condition) can be recovered. Individual techniques and technologies allow core-sampling of various types of rocks. There is equipment that can be used for coring, however, basic coring methods are: (i) side coring and (ii) full face coring.

Side cores are taken by logging cable using explosives (logging inside the well). The process is carried out only rarely if additional information on the rock side wall (and its condition) is needed. Fool face coring is the most common coring method and the goal is to obtain a core and the coring method is chosen based on what information on deposit or rock is needed. In a wider context, the laboratory analyses of core samples produce information, which cannot be obtained from other sources (or it would not be efficient). The information obtained by coring is much more accurate than the data obtained from fragments or logging measurements. Thus, the objective of coring and core analysis is to reduce uncertainty in reservoir evaluation by providing data representative of the reservoir at in-situ conditions. The quality and reliability of core data have become more important with the ever-increasing need to optimize field development.

Routine or conventional core analyses refer to common procedures that provide information on porosity, permeability, resident fluids, lithology, and texture of crude oil reservoirs. There is a variety of the types of analyses that are obtained and the results of each analysis can be used as needed (Table 3.2). Specialized core analyses (Table 3.3) are also necessary and are important for specific applications. Routine core analyses can be performed on whole cores or on small plugs that are cut from a larger core. With the exception of petrographic analyses (thin sections, x-ray; scanning electron microscopy, etc.), special core analyses are normally performed one with core plugs. After a well is drilled and logs are available to identify zones of interest, small portions of the reservoir can be obtained with percussion sidewall **or** sidewall drilled cores. Sidewall cores **are** less expensive and are valuable for petrographic analyses, but are generally not suitable for special core analyses.

Table 3.2: Routine core analysis tests.*

Type of analysis	Use of data
Gamma-ray log	Core and log depth
Grain density	Log interpretation and lithology
Lithology	Rock type
	Fractures
	Laminations, shale content
	Log interpretation, recovery forecasts, and capacity estimates
	Shale content
	Vugs**
Porosity	Volume and storage determinations.
Permeability – horizontal and vertical	Defines flow capacity
	Crossflow
	Gas/water coning
	Relative profile capacity of different zones
Saturations	Completion interval
	Fluid contacts
	Presence of hydrocarbons
	Type of recovery

*Listed alphabetically and not in order of preferences.
**Small unfilled cavities in a rock formation.

Table 3.3: Special core analysis tests.*

Type of test	Use of data
Acoustic velocity	Log and seismic interpretation
Air, water, and liquid permeability	Completion
	Injection fluids
Capillary pressure	Defines irreducible fluid content, contacts
Electrical	Log interpretation
Rock compressibility	Volume change caused by pressure change
Permeability/porosity vs pressure	Corrects to reservoir conditions

Table 3.3 (continued)

Type of test	Use of data
Petrographic studies	
Diagenesis	Origin of oil and source bed studies
Mineral	Used in log interpretation
Sieve analysis	Selection of screens, sand grain size
Wettability	Used in capillary pressure interpretation
	Recovery analysis-relative permeability
Electrical	
Formation factor	Used in log interpretation
Resistivity index	Used in log interpretation
Relative permeability	Effective permeability to each fluid when two or more fluids flow simultaneously; relative permeability enables the calculation of recovery versus saturation and time while values from flood-pot test give only end-point results

*Listed alphabetically and not in order of preference.

3.3.1.1 Well coring

Well coring refers to the process of obtaining representative samples of the productive formation in order to conduct a variety of laboratory testing. Various techniques are used to obtain core samples: conventional diamond-bit coring, rubber-sleeve coring, pressure coring, sidewall coring, and recovery of cuttings generated from the drilling operation. Conventional coring is normally done in competent formations to obtain full-diameter cores. Rubber sleeve-coring improves core recovery in softer formations. Pressure coring, although relatively expensive, is used to obtain cores that have not lost any fluids during lifting of the core to the surface.

A common problem with all of these techniques is to decide when to core. In many instances, cores from the interval of interest are not obtained because of abrupt stratigraphic changes. A second problem is that, typically, nonproductive intervals of the desired strata are obtained. These intervals did not initially contain a significant amount of hydrocarbon.

The importance of not altering wettability with drilling mud filtrate has been discussed in this chapter in the section entitled "Wettability and Contact Angle." Preventing wettability changes in core material, after it has been recovered at the surface, can be equally important **so** that subsequent laboratory measurements are representative of formation conditions.

Cores obtained with drilling muds that minimize wettability alteration, and that are protected at the well-site to prevent evaporation **or** oxidation, are called preserved cores. They are also referred to as fresh cores or native-state cores.

Cores that are cleaned with solvents and resaturated with reservoir fluids are called restored-state cores or extracted cores. The restoring process is often performed on non-preserved or weathered cores, but the same technique could apply to cores that had been preserved.

Two methods of preserving conventional cores, immediately after they have been removed from the core barrel, will prevent changes in wettability for several months. One method consists of immersing the core in deoxygenated formation brine or suitable synthetic brine (i.e., drilling mud filtrate) and keeping the samples in suitable containers that can be sealed to prevent leakage and the entrance of oxygen. In the second method, the cores are wrapped in Saran or polyethylene film and aluminum foil and then coated with wax or strippable plastic. The second method is preferred for cores that will be used for laboratory determination of residual oil content, but the first method may be preferred for laboratory displacement tests. Plastic bags are often recommended for short-term (24 days) storage of core samples. However, this method will not ensure unaltered rock wettability. Air-tight metal cans are not recommended because of the possibility of rust formation and potential leakage.

Cores taken with a pressure core barrel are often frozen at the well-site for transportation to the laboratory. (Cores are left in the inner core barrel.) Normally, the inner barrel containing the cores is cut into lengths convenient for transport. Because of the complexity of the operation, the pressure core barrel is not used as extensively as the conventional core barrel. An alternate procedure involves bleeding off the pressure in the core and core barrel while the produced liquids are collected and measured.

Depending on the type of core testing to be done, core samples may be tested as received in the laboratory or they may be cleaned to remove resident fluids prior to analysis. The best practices for core handling start during retrieval while extracting from the barrel, followed by correct procedures for transportation and storage. Appropriate preservation measures should be adopted depending on the objectives of the scientific investigation and core coherency, with respect to consolidation and weathering. It is particularly desirable to maintain a constant temperature of 1–4 °C (33–39 °F) and a consistent relative humidity of >75% to minimize any micro fracturing and internal moisture movement in the core. While core re-sampling, it should be ensured that there is no further core compaction, especially while using a hand corer. Analysis of the depressurized core is done by conventional techniques. Fluids collected from the barrel during depressurization are proportionately added to the volumes of liquid determined from core analysis. In this manner a reconstructed reservoir core saturation is provided.

Drill core shale samples are critical for paleo-environmental studies and potential and the samples need to be preserved carefully to maximize the retention of reservoir parameters. However, the samples are susceptible to alteration due to cooling and depressurization during retrieval to the surface, resulting in volume expansion and formation of desiccation and microfractures. This leads to inconsistent measurements of different critical attributes, such as porosity and permeability (Basu et al., 2020).

3.3.1.2 Core analysis

The objective of coring and core analysis is to reduce uncertainty in reservoir evaluation by providing data representative of the reservoir at in-situ conditions. Core derived data are integrated with other field data to minimize reservoir uncertainties that cannot be addressed by other data sources such as well logging, well testing or seismic. The quality and reliability of core data have become more important with the ever-increasing pressure to optimize field development.

Obtaining and analyzing cores are crucial to the proper understanding of any complex reservoir system. To obtain the data needed to understand the fluid flow properties, the mechanical properties and the depositional environment of a specific reservoir require that cores be cut, handled correctly, and tested in the laboratory using modern and sophisticated laboratory methods. Of primary importance is measuring the rock properties under restored reservoir conditions. The effect of net overburden pressure (NOB) must be reproduced in the laboratory to obtain the most accurate quantitative information from the cores.

After acquiring the cores in the field, it is important to handle the core properly and suggestions are (i) the core should not be hammered out of the barrel but should be pumped out, (ii) once the core is laid out on the pipe racks, it should be wiped with rags to remove the mud without washing with water and then described as quickly as possible, (iii) the bedding features, natural fractures, and lithology should be described for each foot length of the core, (iv) permanent markers should be used to label the depth of the core and clearly mark the up direction on the core, (v) the core should be wrapped in heat shrinking plastic, then sealed in paraffin as quickly as possible in preparation for transportation to the core analysis laboratory, and (vi) precautions should be taken to minimize and induced changes to the core properties while retrieving and describing the core in the field.

Once in the laboratory, the core is unwrapped and slabbed, and plugs are cut for testing. Normally, a core plug should be cut every foot in the core, trying to properly sample all the rock – not just the cleaner pay zones. Routine core analyses can be run on these core plugs. Once the routine core analyses are completed, additional core plugs are cut for special core analyses. Sometimes samples of whole core are used for testing. Both the routine and the special core analyses are required to calibrate the open-hole logging data, and to prepare the data sets required to design the optimum completion. The core plugs must also be treated with care. For example, if a core plug from a shale-containing sand is placed in a standard oven, it is likely that the clay minerals in the pores will be altered as they dry out. A more accurate core analysis can be achieved if the core plugs are dried in a humidity-controlled oven in which the free water is evaporated, but the treatment should not be so severe as to affect the bound clay water.

Routine core analyses should be run on core plugs cut every foot along the core. Routine core analyses should consist of measurements of: (i) grain density, (ii) porosity and permeability to air, both unstressed and stressed, (iii) cation exchange capacity, and (iv) fluid saturations analysis. In addition, each core plug should be described in detail to understand the lithology and grain size and to note any natural fractures and other details that could be of importance to the geologist, petrophysicist, or engineer.

The porosity is used to determine values of gas in place and to develop correlations with permeability. The grain density should be used to determine how to correlate the density log values and to validate any calculation of lithology from log data. The cation exchange capacity can be used to determine how much electric current can be transmitted by the rock rather than the fluid in the pore space. The cation exchange capacity must be measured in the laboratory, using samples of rock, and is a function of the amount and type of clay in the rock. Saturation analysis measures the amount of water, oil, and gas in the core plugs in the laboratory. Saturation analysis can be misleading in rocks that are cored with water-based mud because of mud filtrate invasion during the coring process and problems that occur with core retrieval and handling prior to running the laboratory tests. However, the values of water saturation from the core analysis of cores cut with an oil-based mud can be used to calibrate the log data and to estimate values of gas in place in the reservoir.

The measurements of porosity and permeability are a function of the net stress applied to the rock when the measurements are taken. For low porosity rock, it is important to take measurements at different values of net stress to fully understand how the reservoir will behave as the gas is produced and the reservoir pressure declines. In fact, a challenge in the development of resources from low permeability formations is the ability to accurately evaluate rock properties (such as permeability, porosity, and capillary pressure). This information is necessary to quantify the resource potential of the formation and also to predict production behavior. However, due to the complex structure of the pore network, simple relationships relating permeability to porosity are not representative since the low permeability of tight formations renders the standard steady-state techniques, which are applicable in evaluating conventional natural gas and crude oil reservoirs, difficult to implement in terms of producing reliable and meaningful data.

Furthermore, the measurement of permeability and porosity of conventional, high permeability reservoirs are typically performed at low pressure in the laboratory and are not representative of in-situ conditions, which means that the effect of the overburden stress is largely ignored. As stress is increased, the high aspect ratio of the pore structures dominating conductive pathways are compressed and ultimately closed off, restricting fluid flow and increasing tortuosity of flow pathways. Thus, after the values of porosity and permeability are measured in the laboratory, the values should be correlated to the conditions in the tight formation (Thomas et al., 1972; Jones et al., 1980; Soeder and Randolph, 1987; Guha et al., 2013). In addition, however, it should be remembered that these estimates are from routine core analyses, which means the core has been tested dry with no water in the core. If similar measurements are made at connate water saturation, the permeability in the core is further reduced, maybe by a factor of 2 or even an order of magnitude in some cases. As such, in tight gas reservoirs, it is often found that in situ permeability to gas is 10–100 times lower than gas permeability measured at ambient conditions on dry core plugs cut from whole core. If cores come from a percussion sidewall device, the core plugs are typically altered, and the values of permeability under unstressed conditions can be even more optimistic.

Special core analysis testing is performed when specifically required. Visual inspection and some petrographic studies are frequently done. For sandstones and conglomerates, particle size is often obtained by disaggregating and sieving reservoir rock material. Fractions of the various sizes of grains are determined can be described according to a standardized nomenclature (Table 3.4). Larger grain size is associated with higher permeability, and small grain sizes include silt and clay fractions that are associated with lower permeabilities.

Table 3.4: Particle size definitions.

Material	Particle size, micron	US standard sieve mesh no.
Coarse sand	>500	<35
Medium sand	250–500	35–60
Fine sand	125–250	60–1 20
Very fine sand	62.5–125	120–230
Coarse silt	31–62.5	NA
Medium silt	15.6–31	NA
Fine silt	7.8–1 5.6	NA
Very fine silt	3.9–7.8	NA
Clay	<2	

3.3.1.3 Tight reservoirs

In order to fully understand the properties of tight gas formations, special core analyses must be performed on selected core plugs to measure values of gas permeability versus water saturation, resistivity index, formation factor, capillary pressure, acoustic velocity, and the rock mechanical properties (Guha et al., 2013). The values of resistivity index and formation factor are used to better analyze the porosity and resistivity logs. The acoustic velocity can be used to better estimate porosity and to determine how to estimate the mechanical properties of the rock from log data. The mechanical properties are measured and correlated to log measurements and lithology. The capillary pressure measurements and the gas permeability versus water saturation relative permeability measurements are required to properly simulate fluid flow in the reservoir and to design hydraulic fracture treatments.

It is important to choose the correct core samples for conducting the special core analyses. Special core analysis tests are expensive and require weeks or months of special laboratory measurements. As such, the core samples must be chosen with care to provide the optimum data for designing the well completion and the well stimulation treatment and forecasting future gas recovery. A good way to select the core samples for special core analysis testing is to: (i) form a team of geologists, engineers, and petrophysicists, (ii) lay out the core, (iii) have the routine core analysis and log analysis available, (iv) determine how many rock types or lithology types that are contained in the core are important to the completion and stimulation process, and (v) pick three to six locations for each rock type or lithology where core plugs are cut for testing.

In addition to the composition of the shale (i.e., shale type), the depth of the formation is a major parameter since depth also influences thermal maturity, bottomhole temperatures, formation pressure, and overall formation behavior. For example, the Haynesville shale (a Jurassic Period rock formation that underlies large parts of southwestern Arkansas, northwest Louisiana, and East Texas) which lies at depths of 10,500–13,000 feet below the surface of the land, which is reflected in high-pressure and high-temperature conditions whereas the Marcellus shale (which is buried thousands of feet beneath the surface of the Earth and stretches from northern New York state south through Pennsylvania to West Virginia and west to parts of Ohio) is relatively shallow and is only mildly over-pressured in the most productive areas. Pore pressure can also be used as an indicator of reservoir quality – higher pressures typically reflect a high degree of gas generation and storage. Furthermore, in order to better understand the effect of hydraulic stimulation, it is necessary to understand the stress state in conjunction with the elastic and strength properties of the rock mass. Such a study may commence with determination of the brittleness index (BI) of the rocks.

Specialized methods to characterize pore structure, storage capacity and flow characteristics of tight reservoirs need further development. The techniques currently applied to assess the properties of tight formations are methods that were originally developed for conventional reservoir rocks and coal seams using the assumption that the same transport and storage mechanisms occur in tight reservoirs containing natural gas and crude oil. However, tight formations have a rock composition and structure that is different to coal and conventional reservoirs and many of the techniques employed to characterize porosity are not always suitable for tight formations. In addition, most methods for characterization require pre-drying of the samples, which can cause the clay minerals to shrink and, hence, alter the rock structure. As a consequence, the sample and the properties determined therefrom may not to be representative of the in-situ formation. Whether or not the errors are small enough to be discounted, in reality small errors in porosity (which is related to gas storage capacity) can eventually lead to errors in the capacity of the reservoir to hold natural gas and crude oil because of the large areal extent of many tight reservoirs. Understanding the limitations of current characterization techniques and the adjustments that are required to calibrate these measurements to the realities of the tight formation important for accurate assessment of any natural gas or crude oil resource base.

Finally, a valuable property of shale formations is the BI, which is a measure of the ability of the rock to crack or fracture. A high BI is typically associated with high quartz content or high carbonate (dolomite) content, as illustrated by the Barnett shale play and the Woodford shale formation (located in the geologic basins of Oklahoma, Anadarko Shelf, Arbuckle Uplift, Ardmore Basin, Arkoma Basin, and the Ozark Uplift at depths from 7,500 to 8,500 feet, and is generally 50 to 300 feet thick). On the other hand, the BI decreases with increasing clay and organic matter, as is in the Marcellus shale formation. Thus, the BI serves as a guide for placement of perforations, isolation points, and fracture stages (Wylie, 2012).

However, the BI should be used according to the relevant definition (Herwanger et al., 2015) within which the brittleness of rocks is characterized from (i) the elastic properties, (ii) the petrophysical properties, and (iii) the strength properties. The definition that is predominant in the geophysical literature relates to a high brittleness for rocks that exhibit a high Young's modulus E and low Poisson ratio v (Rickman et al., 2008).

The Young's modulus or the modulus of elasticity in tension is a mechanical property that measures the tensile stiffness of a solid material and quantifies the relationship between tensile stress (force per unit area) and axial strain (proportional deformation) in the linear elastic region of a material.

Young's modulus:

$$E = \sigma/\varepsilon$$

In this equation, E is Young's modulus, σ is the force per unit area, and ε is the axial strain (or proportional deformation).

The Poisson ratio is a measure of the Poisson effect, which is the phenomenon in which a material tends to expand in directions perpendicular to the direction of compression. Conversely, if the material is stretched rather than compressed, it usually tends to contract in the directions transverse to the direction of stretching.

Poisson ratio:

$$\nu = d\varepsilon_{trans}/d\varepsilon_{axial}$$

In this equation, $d\varepsilon_{trans}$ is the transverse strain and $d\varepsilon_{axial}$ is the axial strain.

There is also a BI that is related to a specific combination of Lamé parameters λ and μ (Goodway et al., 2010). Another category of definitions of BI is based on mineral content of the rocks while definition of BI can also be based on strength parameters to derive a BI from a combination of uniaxial compressive strength σ_c and tensile strength σ_t. Whatever system is used, the BI is essentially a lithology (or mineralogy) indicator and the user should make sure to provide the method employed to determine the index thereby mitigating any possible confusion.

3.3.2 Pressure transient testing of wells

Production rates depend on the effectiveness of the well completion (skin effect), the reservoir permeability, the reservoir pressure, and the drainage area. By way of clarification, a transient event is a short-lived burst of energy in a system caused by a sudden change of state. The source of the transient energy may be an internal event or a nearby event. The energy then couples to other parts of the system, typically appearing as a short burst of oscillation. The transient analysis calculates the response over a period of time defined by the user and the accuracy of the analysis is dependent on the size of internal time steps, which together make up the complete simulation time. Thus, transient well tests measure changes in reservoir pressure associated with changes in well rates. The pressure transient testing (often referred to as PTT) method relies on measuring the changes in pressure at the wellbore as a function of time that accompany fluid flow rate changes in the wellbore.

The PPT of wells is a means of assessing reservoir performance by measuring flow rates and pressures under a range of flowing conditions and then applying the data to a mathematical model. In most well tests, a limited amount of fluid is allowed to flow from the formation being tested. The pressure-transient analysis (often referred to as PTA) is the analysis of the resulting pressure and flow data set, and involves matching a model (either analytical or numerical) of the well and reservoir to the data.

PTA is a valuable method for determining the reservoir characteristics required to forecast production rates. Transient pressure data are generated by changing the producing rate and observing the change in pressure with time. The transient period should not exceed 10% of the previous flow or shut-in period. There are a number of methods to generate the transient data available to the reservoir engineer.

Single-well tests such as buildup, falloff, drawdown, injection, and variable rate describe the isotropic reservoir adjacent to the test well while multiple well tests such as long-term interference or short-term pulse describe the characteristics between wells. Buildup and falloff tests are most popular because the zero flow rate is readily held constant. Drawdown and injection tests are run less frequently due to problems with maintaining a constant rate. Variable rate tests are useful when wellbore storage is a problem. Multi-well testing for characterizing anisotropic reservoirs has been popularized by the increased use of sophisticated simulation software.

3.3.3 Reserves

The term "reserves" refers to the estimated volumes (in the overall context) of natural gas, gas condensate, NGLs, crude oil, and any other associated substances anticipated to be commercially recoverable and marketable under existing economic conditions. Various terms are used to describe the reserves of natural gas but caution is advised as many of these terms may be a best estimate or even speculative (Appendix B) (Terry and Rogers, 2015). More specifically in the current context, the reserves of natural gas refer to the reserves of natural gas which, based on geological surveys and engineering calculations, are thought to exist to a high degree of certainty. Recently, the reserves of natural gas have been increased by inclusion of the gas that occurs in shale reservoirs and in other tight reservoirs.

In the current context, natural gas reserves are those volumes, which are expected to be produced and that may have been reduced by onsite usage, by removal of nonhydrocarbon constituents, condensate, or NGLs. The reserves may be attributed to either natural reservoir or improved recovery methods, which includes all of the methods for supplementing natural reservoir energy to increase ultimate recovery from a reservoir. Such methods include (i) pressure maintenance, (ii) cycling, (iii) waterflooding, (iv) thermal methods, (v) chemical flooding, and (vi) the use of miscible and immiscible displacement fluids. Thus, natural gas reserves may be attributed to either natural reservoir or improved recovery methods. However, it is preferable (to avoid any confusion) the reserves attributed to improved recovery methods are usually distinguished from those attributed to primary and secondary recovery methods.

The decision to drill a well is based upon the potential reserves that can be recovered which means that an engineer must be able to predict reserves before a well is drilled. Typically, the methods of estimating the reserves of natural gas (or crude oil) arise using the method of analogy, which involves information from known reserves in similar formations and locations. In fact, the analogy method is the only method, which can be used without specific well information such as porosity, reservoir thickness, and water saturation (Lyons and Plisga, 2005). Because analogy employs no specific information related to a well, it may be the least accurate method of determining the reserves of natural gas. The best analogy can be made by taking the median ultimate recovery of a number of wells that are closest and have the same formation and characteristics expected in the proposed well. When ultimate recovery data are not available, volumetric, decline curve, or other methods of estimating ultimate recovery may be used.

Thus, if a well is drilled after reserves are determined by analogy, factual information becomes available and reserves can then be determined volumetrically. From log analysis the porosity, water saturation, and productive formation thickness are estimated. A reasonable drainage area is assigned and total natural gas in place is then calculated. When enough wells have been drilled to delineate the field, a subsurface geological contour map showing the subsea sand top and bottom depth, oil-water contact, and gas-oil contact can be prepared. From this map the total areas in acre-feet of each contour are planimetered and graphed as the abscissa against the subsea depth as the ordinate which ultimately result in an estimation of the gross gas-bearing reservoir volume in acre-feet.

For more accuracy, the gross reservoir volume that is productive can be estimated from core data and/or electric logs. If there is no subsurface contour map available or if the reservoir is very heterogeneous, an isopach map should be constructed (by contouring net reservoir thickness) or an isovol map (which indicates resource thickness) is constructed by contouring the value of net pay height multiplied by porosity (and by one minus the water saturation) should be constructed. The isopach map works well when the reservoir is uniform and when porosity and water saturation are relatively constant but when the water saturation and porosity vary widely from well to well, the isovol map is more useful.

Generally, reserve estimates are based on interpretation of geologic and/or engineering data available at the time of the estimate. Future economic conditions may lead to different estimates of recoverable volumes, which are not considered reserves under existing economic conditions constraints but may be identified as future resources. The term marketable means that facilities to process and transport reserves to market are operational at the time of the estimate, or that there is a commitment to install such facilities in the near future. Reserve estimates should be continually revised (i) as reservoirs are in production, (ii) as additional geologic and/or engineering data become available, or (iii) as economic conditions change.

Natural gas is a resource and, in particular, a resource is the entire commodity that exists in the sediments and strata whereas the reserves represent that fraction of a commodity that can be recovered economically. However, the use of the term reserves as being descriptive of the resource is subject to much speculation. For example, reserves are classified as (i) proved, (ii) unproved, probable, possible, and undiscovered (Appendix B). The differences between the data obtained from the various estimates can be considerable, but it must be remembered that any data related to the reserves of natural gas will always be open to questions related to the degree of certainty. Thus, in reality, proven reserves may be a small part of the total hypothetical and/or speculative amounts of a natural gas resource.

At some time in the future, certain resources may become reserves because of improvements in recovery techniques, which may either make the resource accessible or to cause an increase in the efficiency of resource recovery. In addition, other uses may also be found for a commodity, and the increased demand may result in an increase in price. Alternatively, a large deposit may become exhausted and unable to produce any more of the resource thus forcing production to focus on a resource that is lower grade but has a higher recovery cost.

However, no matter how well the reserves estimates are calculated, all such reserve estimates involve some degree of uncertainty and depend predominantly on the amount and reliability of geologic and engineering data available at the time of the estimate and, especially, the interpretation of these data. The relative degree of uncertainty may be conveyed by placing reserves in one of two classifications, either proven (proved) or unproven (unproved). These latter types of reserves i.e., (unproven reserves) are less certain to be recovered than proven reserves and may be subclassified as probable reserves or possible reserves to denote progressively increasing uncertainty.

3.3.3.1 Reserve estimates and material balance

The first action in a field development program is to calculate reserves by use of a material balance equation, which is the simplest expression of the conservation of mass in a reservoir. A material balance, also called a mass balance, is an application of conservation of mass to the analysis of (in this case) a reservoir and the resource.

Material balance (and energy balance) is (are) based on a conservation law, which is stated generally in the form

$$\text{input} + \text{source} = \text{output} + \text{sink} + \text{accumulation}.$$

The individual terms can he plural and can be rates as well as absolute quantities.

By accounting for material leaving the reservoir system, mass flows can be identified, which might have been unknown, or difficult to measure without this technique. Thus, the equation mathematically defines the different producing mechanisms and effectively relates the reservoir fluid and rock expansion to the subsequent fluid withdrawal. The equation is derived on the assumption that the reservoir is a homogeneous vessel with uniform porosity, permeability, and fluid properties. The equation accounts for all quantities of materials that enter **or** leave the vessel. The simplest form of the equation is that initial volume is equal to the volume remaining plus the volume removed. As material is withdrawn from a constant-volume reservoir the pressure declines and remaining material expands to fill the reservoir.

Reservoirs that contain only unassociated natural gas contain a mixture of gaseous hydrocarbon derivatives, which may be dry, wet, or condensate gas (Chapter 1). Gas reservoirs may be volumetric with no water influx. Applying the mass balance principle to a closed reservoir with 100% gas, we may derive the general equation:

$$GB_{g1} = (G - G_p)B_{g2}$$

In this equation, G is gas initially in place, G_p is the cumulative gas production, and B_g is the formation-volume-factor for gas.

In the case of a gas reservoir with water encroachment and water production, the material balance equation is

$$GB_{g1} = (G - G_p)B_{g2} + (W_e - W_p)$$

In this equation, W_e is the cumulative aquifer influx, and W_p is the cumulative water produced.

In either equation, the gas formation volume factor can be obtained, as a function of pressure and temperature.

Reserve estimates are based on interpretation of geologic and/or engineering data available at the time of the estimate. Existing economic conditions are prices, costs, and markets prevailing at the time of the estimate. Other assumed future economic conditions may lead to different estimates of recoverable volumes; these volumes are not considered reserves under existing economic conditions constraints, but may be identified as resources.

The most widely used method of estimating reserves is the production rate decline-curve, which involves extrapolation of the trend in performance. Decline analysis is a reservoir engineering empirical technique that extrapolates trends in the production data from oil and gas wells. The purpose of a decline analysis is to generate a forecast of future production rates and to determine the expected ultimate recoverable (EUR) reserves.

If a continuously changing continuous function is plotted as the dependent variable against an independent variable, a mathematical or graphical trend can be established. Extrapolation of that trend can then permit a prediction of future performance. Although decline-curve analysis is empirical, if care is taken to ensure that production rates are not being affected by, for example, mechanical degradation of equipment or the plugging of the formation by fines or paraffin derivatives, the method is reasonably accurate.

The conventional analysis of production decline curves for natural gas production consists of plotting the log of flow rate versus time on semilog paper. In cases for a decline in rate of production, the data **are** extrapolated into the future to provide an estimate of expected production and reserves. The empirical relationships for the analysis of production decline curves in which a decline rate, a, was defined as the fractional change in the flow rate, q, with time, t:

$$a = (dq/dt)/q$$

If the time is days, the flow rate in this equation is expressed in terms of stock tank barrels per day in the case of oil and *scf* per day for gas. Other consistent units of flow rate and time can be used.

Three types of decline can occur: a constant percentage or exponential decline, a hyperbolic decline, and a harmonic decline. For the semilog plot, the exponential decline is a straight line whereas the slopes of the hyperbolic and harmonic decline curves decrease with time. For the exponential decline, the drop in production per unit time is a constant fraction of the production rate. For a hyperbolic decline, the decrease in production per unit time as a fraction of the production rate is proportional to a fractional power (between 0 and 1) of the production rate. For a harmonic decline, the decrease in production per unit time as a fraction of the production rate is directly proportional to the rate. An equation is presented for the hyperbolic decline that will reduce to the other types under certain circumstances (Slider, 256):

$$a/a_1 = q^n/q_1{}^n$$

In this equation, a is the decline rate when the production rate is q, and a_1 and q are the decline rate and production at an initial time. The exponent, n, is a number between, but not including zero and one for the hyperbolic decline. When n is zero, the decline rate is constant, which is the exponential decline. When n is one, the decline rate is proportional to the rate which is the harmonic decline.

3.3.3.2 Methods for estimating residual gas

The economics of primary and secondary recovery processes are usually sufficiently attractive to permit considerable error in the estimation of recoverable reserves of natural gas. However, for tertiary recovery the amount of natural gas remaining in a reservoir and its distribution must be known with reasonable confidence. This requires (i) reliable estimate of residual content is extremely important to technical evaluation of field tests, (ii) the high front-end costs of tertiary processes are such that overestimates of residual oil saturation could have disastrous economic consequences. Thus, a well-planned effort to measure residual natural gas is a necessity before any tertiary recovery application. Under favorable circumstances, accuracies of the reservoir pore space can be achieved.

There is no absolute measure of natural gas saturation for a reservoir. When evaluating a tertiary prospect, a combination of methods should be used, which provide information on both amount and distribution. Evaluation will normally begin with material balances using information that is already available.

Comparisons of material balance with other methods of determining residual natural gas may show unacceptable scatter, and on average, the material balance gives saturations, which are too high.

In addition to material balance, other estimates of residual natural gas from resistivity logs and laboratory waterflood tests may also be available. However, in general, none of the conventional methods of determining residual oil saturation-analysis of conventional cores, laboratory displacement tests, conventional logging, material balance are considered sufficiently reliable in and can provide useful guides as to whether a tertiary prospect should be investigated further.

Pressure coring and the sponge core technique provide information on the vertical distribution of residual oil and also have the advantage that the core analysis procedures directly demonstrate the presence of natural gas. Because of the possibility of flushing, results may tend to be conservatively low. For pressure coring, the time taken to obtain results tends to be longer than other methods because of transportation and the specialized core analysis work that is needed. After determining residual oil, the extracted cores are used to obtain needed information on reservoir properties, in particular vertical heterogeneity, and they also can be used in laboratory displacement tests. The use of the sponge coring technique has increased because of the lower coring costs, reduced analysis costs, and, since larger diameter cores are obtained, core plugs can be obtained for subsequent conventional or specialized core testing.

Results with the tracer test will be conservatively low because they are usually weighted toward the more permeable zones, where residual oil will tend to be lower. On balance this is probably an advantage because these zones will also tend to be swept more readily during a tertiary process. The tracer test *can* be used in old wells, but it is important that the well has not been fractured or stimulated severely.

Of the various log-inject-log procedures, the pulsed-neutron capture method is the most widely tested and has the advantage that it can be used in cased holes. Problems can arise with borehole rugosity effects (a measure of small-scale variations of amplitude in the height of a surface) and high values of residual oil if displacements are incomplete during the log-inject-log procedure. Results can be affected by flushing because of the small depth of investigation and the method can give accurate results under favorable circumstances. Stabilization of capture-cross-section values for the log-inject-log procedure can take unexpectedly long times and can present problems in interpretation. The nuclear magnetic log has been rated highly as to accuracy, but cannot be used in cased holes and it still has limited commercial availability. Log-inject-log resistivity measurements can give good results under favorable circumstances but are not applicable to cased holes.

3.4 Recovery

Natural gas exists in nature under pressure in rock reservoirs in the crust of the Earth, either in conjunction with and dissolved in crude oil (associated gas) or without crude oil (non-associated gas). Reservoirs containing natural gas vary in size can be anything from a few hundred yards to miles across and tens to hundreds of yards thick, with the gas trapped (similar to crude oil traps) against an impermeable lower barrier (which can be rock or water) and an impermeable upper layer (Speight, 2014, 2019).

Typically, natural gas is produced from the reservoir in a manner similar to crude oil (when it occurs in conjunction with crude oil). However, extracting natural gas from deposits deep underground is dependent upon several factors in the underground environment including the (i) pressure of the gas in the reservoir, (ii) the composition of the gas, (iii) the porosity of the reservoir rock, and (iv) the permeability of the reservoir rock. Primary recovery relies on underground pressure to drive the gas though the production well to the surface. When the pressure decreases due to production of the gas, the remaining gas is brought to the surface using artificial lift technologies, such as the horsehead pump at the surface or a downhole pump at the bottom of the production well.

However, gas fields are managed differently from oil fields whereas crude oil is relatively easy to transport in bulk volumes long distances, whereas this is difficult for gas, unless a very expensive liquid natural gas (LNG) plant is built or there is an extensive regional gas pipeline network. A gas field will typically be developed once a gas sales contract has been made to supply the gas to customers living close to the gas field. The contract will involve a commitment to supply a daily volume of gas over a certain period of time. Thus, there is a requirement to be reasonably sure of what a gas field will produce before it starts production. It will be necessary to test every well preproduction to get a good knowledge of well productivity and flow rates.

Recovery factors are higher for gas than they are for crude oil, commonly in the range of 50–80% v/v for natural gas. The recovery factor for gas fields is dependent on factors such as the *abandonment pressure*, the *initial pressure*, and the type of reservoir drive mechanism. Recovery can also be sensitive to the engineering of the surface plant. In big gas fields, the installation of compression equipment can lead to higher recoveries.

A key property of a gas is the compressibility (Appendix C) – gas compresses readily with increasing pressure and. Conversely, with decreasing pressure, the gas will expand. The measure of how much a gas will expand between the reservoir and surface conditions is the gas expansion factor. A typical value for this is 200. Expansion is the main mechanism by which gas is produced to the surface. Once the pressure drops to reduced levels, then surface flow rates may be too low to be profitable. This is the abandonment pressure, which effectively defines the economic limit of flow from a gas field.

One difference between gas and oil is that, as the pressure decreases, oil with its limited compressibility will stay trapped in a reservoir dead end, whereas gas may not remain trapped for too long. The gas will expand on decreasing pressure and a significant proportion of the trapped volume will eventually escape around the edges of the dead end. This is an important actor contributing to high recovery factors in gas fields.

Another factor is that gas has a lower viscosity than oil and will flow through low-permeability rocks that would not produce oil. Hence, gas can be produced economically from poorer quality reservoir rocks. A wider spectrum of rock types will produce gas by comparison to oil.

A strong water drive is unfavorable to gas recovery as water breakthrough to the production wells will make flow rates sluggish and uneconomic at higher pressures than with closed gas reservoirs. The intensity of the water drive can be a major factor behind the ultimate recovery of gas; a slower encroachment of water will result in higher recoveries. Permeability is also a critical factor in gas reservoirs. Higher permeability results in a high flow rate for a given pressure drop. Thus, the abandonment pressure can be lower for a high permeability gas reservoir.

Gas reservoirs without an aquifer have been called volumetric reservoirs (or depletion reservoirs). This is because the elementary physics of the gas laws allow the volume of gas in the reservoir to be calculated once a certain amount of gas has been produced. Recoveries are higher (on the order of 80–90%) as pressure depletion is much more efficient than a water drive regime, because the wells do not load up with water.

A material balance technique used by reservoir engineers to estimate gas volumes is the *P/Z plot*. This is based on the gas law relationship whereby if the volume of a gas is reduced within a closed system, then the pressure will drop in a predictable manner. Two parameters are cross-plotted on a graph: pressure, divided by the gas deviation (dimensionless compressibility) factor Z, used in the equation for the nonideal gas law; and cumulative gas production. In closed system reservoirs, the values will plot on a straight-line trend. The trend is extrapolated to the abandonment pressure to estimate the contactable volume of gas. For example, in the Novillero gas field in the Veracruz basin of Mexico, the P/Z plot extrapolates to the base line at a total gas volume of close to 1,400 MM m³ (49.4×10^9 ft³). If an aquifer is present, the system cannot be considered closed, and the P/Z plot will deviate from a straight-line trend.

Pressures can fall below the dew point in the reservoir, with liquid condensate dropping out near the well bore. This *condensate banking* can lead to a reduction in the gas relative permeability, and gas flow rates can drop off significantly as a result. Therefore, there is an incentive with a gas condensate reservoir to keep the pressure above the dew point so as to prevent this from happening. A typical strategy is to recycle gas back into the reservoir to maintain pressure.

Once a potential natural gas reservoir has been located, the decision of whether or not to drill a well depends on a variety of factors, not the least of which is the economic characteristics of the reservoir. After the decision to drill has been made, the exact placement of the drill site depends on a variety of factors, including the nature of the potential formation to be drilled, the characteristics of the subsurface geology, and the depth and size of the target deposit. During this time, it is also necessary for the drilling company to ensure that they complete all the necessary steps to ensure that they can legally drill in that area. This usually involves securing permits for the drilling operations, establishment of a legal arrangement to allow the natural gas company to extract and sell the resources under a given area of land, and a design for gathering lines that will connect the well to the pipeline.

If the new well does in fact come in contact with a natural gas reservoir, the well is developed to allow for the extraction of the natural gas and is termed a development well or productive well. At this point, with the well drilled and hydrocarbons present, the well may be completed to facilitate its production of natural gas. However, if the exploration team was incorrect in its estimation of the existence of marketable quantity of natural gas at a well site (and this does happen), the well is termed a dry well, and work on the well is terminated.

Gas wells are technically similar to crude oil wells with casing, tubing, and a wellhead and controls at the top (Rojey et al., 1997; Arnold and Stewart, 1999). Conventional wells use casings *telescoped* inside each other that is given pressure resistance by cement. However, with *expanded tubulars* each tubular casing is expanded against the previous one by pumping a tool called a *mandrel* down the casing. Thus, the well can either be thinner and cheaper for a given size of final tubing, or alternatively, using the conventional 20-inch diameter external casing, there is space to install wider and higher capacity tubing.

When gas is contained in a deep low-to-no permeability (tight) reservoir, it is often economic to drill large-diameter wells with horizontal sections through the reservoirs to collect more gas. In addition, gas production can be increased by fracturing the reservoir rock using over-pressure and by pumping in ceramic beads that maintaining the integrity of the fractures. It is also attractive to install smart wells that have measurement gauges permanently installed downhole. This eliminates the need for lowering gauges down the wells during production. With the high gas flow rates, sand can often be pulled into the well with the gas and therefore wearing parts are hardened against sand erosion.

If the natural gas is to be recovered from a crude oil reservoir (as is often the case), the production methods differ somewhat because of the presence of crude oil hydrocarbons in the natural gas. Thus, once a crude oil reservoir is discovered and assessed, production engineers begin the task of maximizing the amount of oil or gas that can ultimately be recovered from it. However, before a well can produce crude oil or gas the borehole must also be stabilized with casing that is cemented in place. The casing also serves to protect any fresh water intervals that the well passes through, so that oil cannot contaminate the water. A small-diameter tubing string is centered in the wellbore and held in place with packers. This tubing will carry the crude oil and natural from the reservoir to the surface.

Reservoirs are typically at elevated pressure because of underground forces. To equalize the pressure and avoid the wasteful *gushers* of the early 1900s (and the Hollywood movies), a series of valves and equipment is installed on top of the well. This wellhead (also called a *Christmas tree*) regulates the flow of hydrocarbons out of the well.

Early in its production life, the underground pressure (often referred to as the *reservoir energy*) will push the crude oil and natural gas up the wellbore to the surface and depending on reservoir conditions, this *natural flow* may continue for many years. When the pressure differential is insufficient for the crude oil and natural gas to flow naturally, mechanical pumps must be used to bring the products to the surface (*artificial lift*).

Most wells produce crude oil and natural gas in a predictable pattern (*decline curve*) in production increases for a short period, then peaks, and follow a long, slow decline. The shape of this decline curve, how high the production peaks, and the length of the decline are all driven by reservoir conditions. There are two steps that can be taken to influence the decline curve of a well: (i) perform a periodic *work over* which cleans out the wellbore to help the crude oil or natural gas move more easily to the surface, or (ii) fracture or treat the reservoir rock with acid around the bottom of the wellbore to create better pathways for the crude oil and natural gas to move through the subsurface to the producing well.

When natural gas, crude oil, and water are produced from a well, the mixture is separated at the surface. However, although natural gas wells usually do not produce crude oil, they do produce varying amounts of liquid hydrocarbon derivatives (NGLs) that are removed in the field or at a gas processing plant (which may remove other impurities as well). NGLs often have significant value as petrochemical feedstocks. Once water is produced, onshore natural gas is usually transported through a pipeline, although alternate methods may be more appropriate depending on the location of the well and the destination of the gas.

Just as many crude oil reservoirs require enhancement to continue oil production, gas reservoirs also require application of enhanced production methods to maximize the recovery of the gas but caution is advised (Table 3.5). As an example, the injection of carbon dioxide can be used to enhance the gas production from the existing reservoir. The principle of enhanced gas recovery (EGR) is that the injected fluid (in this example, carbon dioxide) forms a front that pushes the natural gas to the production wells. However, the injected carbon dioxide may follow the path of lowest resistance, namely the depleted sand layers of high permeability, instead of creating a wide front pushing forward the gas toward a producing well.

Thus, extracting natural gas from deposits deep underground is not merely a matter of drilling and completing a well. Any number of factors in the underground environment – such as the porosity of the reservoir rock – can impede the free flow of product into the well (Sim et al., 2008; Godec et al., 2014). In the past, it was common to recover as little a minority of the gas in the reservoir.

Natural gas reservoirs may be found beneath impermeable barriers of considerable thickness. Thus, the encountered pressure upon entering a gas reservoir may be quite large, a combination of both hydrostatic pressure and the weight of impermeable overburden. Anticipation of such large pressure is essential for both blowout prevention, and the eventuality of a gas kick, which is a sudden influx of

Table 3.5: Enhanced methods applied to recovery of natural gas.*

Alkaline or caustic flooding
Carbon dioxide flooding
Chemical flooding
Enriched gas drive
Flue gas injection
High-pressure gas drive
Hydrocarbon gas Injection
Inert gas (nitrogen) injection
Miscible solvent (LPG or propane) flooding
Polymer-augmented waterflooding
Steamflooding
Surfactant flooding
Low tension waterflooding
Micellar/polymer (microemulsion) flooding

*Listed alphabetically rather than the preference of efficiency of the method.

reservoir gas (or reservoir fluids) into the drilling fluid column. This often occurs in gas wells and results in an increase in the annular pressure compared with the shut-in drill pipe pressure. Typically, this would require weighing the drilling mud further in order to circulate the gas kick out and also to prevent further gas influx.

Predicting reservoir pressure ahead of entering a formation of interest is important. Assuming that a barrier is at a depth H_a and the depth below the barrier is H_b then the expected pressure upon entering the formation just below the barrier would be

$$p = 0.465H_a + 1.1(H_b - H_a)$$

In this equation, 0.465 psi/ft is the reservoir fluid gradient and 1.1 psi/ft is the lithostatic or overburden gradient.

A drill stem test (DST) is designed to determine the productivity and fluid properties prior to completion of the well and although a DST can be performed in uncased (open) hole, or in cased hole (perforation tests), the open hole test is more common. The tool assembly, which consists of a packer, a test valve, and an equalizing valve, is lowered on the drill pipe to a position opposite the formation to be tested. The packer expands against the hole to segregate the mud-filled annular section from the interval of interest, and the test valve allows formation fluids to enter the drill pipe during the test. The equalizing valve allows pressure equalization after the test so the packer can be retrieved. By closing the test valve, a buildup in pressure is obtained; by opening the test valve, a decline in pressure is obtained. During the DST, both pressures and flow rates are measured as a function of time.

In summary, a DST can provide a valuable indication of commercial productivity from a well, provided engineering judgment and experience are properly utilized.

Unconventional natural gas is the gas contained in geologic formations, which have not been exploited traditionally by the gas industry and include shale formation, tight gas sands, formations with active aquifers, and coal bed methane. A number of these geologic formations are known to contain sizable quantities of natural gas. However, EGR is a recent consideration because of the large storage capacity of depleted gas pools. Considering that any additional gas recovered as a result of carbon dioxide injection might offset the cost of the storage of carbon dioxide, a number of experimental and simulation studies have been undertaken to examine displacement efficiency of reservoir gas by gas injection (Mamora and Sea, 2002; Seo and Mamora, 2003; Pooladi-Darvish et al., 2008).

EGR is used to describe the recovery of unconventional, deep, or otherwise difficult-to-recover natural gas (Guo et al., 2014). The term refers to the technologies and processes that bare employed to recover natural gas from reservoirs that are difficult to access by conventional methods. As an example, in order to enhance the recovery of natural gas from a gas reservoir, nitrogen is injected into the reservoirs to artificially increase pressure, which results in more yield from the well (often referred to as EGR). This process can also be accomplished with other inert gases such as carbon dioxide (CO_2), which sourced from underground wells is an economical alternative.

However, nitrogen can be recovered virtually anywhere from atmospheric air – via an air separation plant. Physically, there are also differences between the two gases: (i) nitrogen requires less compression that carbon dioxide and (ii) greater amounts of carbon dioxide are needed to create high pressure in the reservoirs. Thus, the EGR process using compressed nitrogen is more energy-efficient than when using carbon dioxide. Another option to maintain the pressure in the reservoir involves injection of part of the recovered natural gas.

The principles of EGR can also be applied to the recovery of coalbed methane, which, in the context of this book, is an unconventional source of methane gas (Appendix A). Carbon dioxide has a greater adsorption affinity onto coal than methane and if carbon dioxide is pumped into a coal seam toward the end of a coalbed methane production project, it displaces any remaining methane at the adsorption sites (allowing methane recovery jointly with carbon dioxide storage). However, the low permeability of coal seams means that a large number of wells may be needed to inject sufficient amounts of carbon dioxide to make methane recovery economical. Moreover, the methane in coal represents only a small proportion of the energy value of the coal, and the remaining coal could not be mined or gasified underground without releasing the carbon dioxide to the atmosphere. Also, methane is a far more potent greenhouse gas than carbon dioxide and the necessary precautions would have to be in place to ensure no methane leakage to the atmosphere occurred.

3.5 Storage and storage facilities

As a brief introduction, natural gas that is scheduled to be transported and stored must meet specific quality measures so that the pipeline network (or grid) can provide uniform quality gas suitable for sales (Creti, 2009).

More specifically, natural gas storage serves two primary functions: to meet seasonal demands for gas (base-load storage); and to meet short-term peaks in demand (peaking storage). Peaks in natural gas demand can range from a few hours to a few days, typically during unusually cold winter weather. To ensure that adequate natural gas supplies are available to meet seasonal customer requirements, underground natural gas storage facilities are filled during low utilization periods in what is commonly called the injection season, typically in the spring and summer months of any year. Natural gas that is placed into storage is ultimately moved to markets to supplement domestic production and imports during what is referred to as the withdrawal season between the autumn-winter peak usage months.

Wellhead natural gas will contain other hydrocarbons, inert gases, and contaminants, which must be removed before the natural gas can be safely delivered to the high-pressure, long-distance pipelines that transport natural gas to consumers. Natural gas processing can be complex and usually involves several processes, or stages, to remove oil, water, hydrocarbon gas liquids, and other impurities such as sulfur, helium, nitrogen, hydrogen sulfide, and carbon dioxide (Appendix D). The composition of the wellhead natural gas determines the number of stages and the process required to produce pipeline-quality dry natural gas. These processes or stages may be integrated into one unit or operation, be performed in a different order or at alternative locations (lease/plant), or not be required at all.

The stages of natural gas processing/treatment are: (i) gas-oil-water separators, in which pressure relief causes a natural separation of the liquids from the gases in the natural gas; (ii) condensate separator, in which gas condensate and/or natural gasoline is removed from the natural gas stream at the wellhead with separators much like gas-oil-water separators, (iii) dehydrator, in which water is removed to reduce the potential for corrosion in the pipeline as well as the formation of undesirable hydrates and water condensation in pipelines; (iv) contaminant removal units, in which nonhydrocarbon gases such as hydrogen sulfide, carbon dioxide, water vapor, helium, nitrogen, and oxygen are removed from the natural gas stream; (v) nitrogen rejection unit, in which nitrogen is removed and the gas stream is further dehydrated using molecular sieve beds; (vi) methane separator, which methane is removed from the gas stream can occur as a separate operation by cryogenic processing and absorption methods; and (vii) fractionator, in which the gas stream is separated into component hydrocarbons (ethane, propane, butane) using the varying boiling points of the individual hydrocarbons. Each of these unit operations is described in more detail elsewhere in this text (Chapter 4). Also, there is the need to determine the purity of the gas streams as well as the efficiency of each process unit by means of gas stream analysis, usually by means of online or offline monitoring of the product streams by means of, for example, a technique such as gas chromatography (Speight, 2018).

As simple as it may seem to many observers, the storage of natural gas is not only a matter of putting the gas into a container without the potential for any adverse consequences. However, there are several analytical measures that must be taken to ensure that the gas is not only stored safely but also recoverable safely and in a usable form.

Traditionally, natural gas has been a seasonal fuel and demand is usually higher during the winter, partly because it is used for heat in residential and commercial settings. Therefore, the natural gas that reaches its destination is not always needed right away and, fortunately, natural gas can be stored for an indefinite period of time. Storage used to serve only as a buffer between transportation and distribution, to ensure adequate supplies of natural gas were in place for seasonal demand shifts, and unexpected demand surges. Now, in addition to serving those purposes, natural gas storage is also used by industry participants for commercial reasons; storing gas when prices are low, and withdrawing and selling it when prices are high, for instance. The purpose and use of storage has been closely linked to the regulatory environment of the time.

Typically, in the United States interstate pipeline companies, intrastate pipeline companies, independent storage providers, and local distribution companies are the primary owners and operators of underground natural gas storage facilities. Storage facility owners and/or operators may or may not own the stored natural gas, which can be held under lease with natural gas shipment companies, local distribution centers, or end-users that also own the gas. The entity that owns or operates an underground facility will generally determine how the facility's storage capacity will be used. The individual States primarily regulate natural gas storage facilities involved in intrastate and interstate commerce. Thus, there are relevant legal issues that must be addressed (Burt, 2016).

Furthermore, because of the geographic and geologic diversity in gas resources and storage operations in the United States, there is no single approach that can be taken without the application of analytical test methods and consideration of the data relating to (i) the potential storage facility, such as a depleted reservoir, (ii) the composition and properties of the gas to be stored, and (iii) whether the gas is adequate for each and every storage container; in this case, the word container includes natural container such as underground natural storage facilities as described below.

Storage for base load requirements (base load storage) is gas that is used to meet seasonal demand increases and the facilities are capable of holding enough natural gas to satisfy long-term seasonal demand requirements. Typically, the turnover rate for natural gas in these facilities is a year; natural gas is generally injected during the summer (non-heating season) and withdrawn during the winter (heating season). These reservoirs are larger, but their delivery rates are relatively low, meaning the natural gas that can be extracted each day is limited. Instead, these facilities provide a prolonged, steady supply of natural gas. Depleted gas reservoirs are the most common type of base load storage facility.

Stored natural gas plays a vital role in ensuring that any excess supply delivered during the summer months is available to meet the increased demand for gas during the winter months. In addition, the recent trend toward electricity generation using natural gas as the fuel had caused demand to increase during the summer months due to the need for electricity to power air conditioners and the like. Natural gas in storage also serves as insurance against any unforeseen accidents, natural disasters, or other occurrences that may affect the production or delivery of natural gas. In addition, natural gas storage plays a vital role in maintaining the reliability of supply needed to meet the demands of consumers. The increased use of natural gas and the need to provide a plentiful supply of this fuel are causing an expansion of the need for natural gas storage facilities so that this domestically produced gaseous fuel will be readily available through the utility infrastructure. Storage is the main link between the three segments of the natural gas industry that are involved in delivering natural gas from the wellhead to the consumer: (i) the production company explores, drills, and extracts the natural gas from the ground, (ii) the transportation company operates the pipelines that link the gas fields to major consuming areas, and (iii) the distribution company – in the form of local utility companies that deliver natural gas to the customer.

Any underground storage facility is reconditioned before injection, to create a sort of storage vessel underground. Natural gas is injected into the formation, building up pressure as more natural gas is added. In this sense, the underground formation becomes a sort of pressurized natural gas container. As with newly drilled wells, the higher the pressure in the storage facility, the more readily gas may be extracted. Once the pressure drops to below that of the pressure at the wellhead, there is no pressure differential left to push the natural gas out of the storage facility. This means that, in any underground storage facility, there is a certain amount of gas that may never be extracted. This is known as physically unrecoverable gas; it is permanently embedded in the formation.

In addition to this physically unrecoverable gas, underground storage facilities contain what is known as base gas or cushion gas. This is the volume of gas that must remain in the storage facility to provide the required pressurization to extract the remaining gas. In the normal operation of the storage facility, this cushion gas remains underground. However, a portion of it may be extracted using specialized compression equipment at the wellhead.

Commonly assessed risks or threats include, but are not limited to the following: (i) the physical properties of the casing, such as diameter, weight, and grade, (ii) presence of atmospheric or external corrosion at or near the surface, (iii) known metal loss indications from casing inspection surveys, (iv) the presence of annulus pressure or flow, (v) water production in the well, (vi) the presence of corrosive hydrogen sulfide and bacteria, (vii) or naturally corrosive zones, (viii) the flow potential of the well, and (ix) the work history of the well and its reliability in providing natural gas service. Each category requires the production of data by application of the relevant test methods. In addition, there are several volumetric measures that are used to quantify the fundamental characteristics of an underground storage facility and the gas contained within it. Indeed, it is important to distinguish between the characteristic of a facility, such as the *capacity* of the facility and the characteristic of the natural gas within the facility, such as the actual *inventory level*.

By way of definition, the injection capacity (or injection rate) is the complement of the deliverability or withdrawal rate and is the amount of natural gas that can be injected into a storage facility on a daily basis (or unit-time basis). As with deliverability, injection capacity is usually expressed in millions of standard cubic feet per day (MMScf/day), although dekatherms/day is also used. A further definition that is necessary at this point, a dekatherm (dth) is a unit of energy used primarily to measure natural gas and which is equal to 10 therms or 1,000,000 British thermal units (MMBtu) or, using the SI system, 1.055 gigajoules (GJ). The dekatherm is also approximately equal to one thousand cubic feet (Mcf, Mft^3) of natural gas or exactly one thousand cubic feet of natural gas with a heating value of 1,000 Btu/ft^3.

The total natural gas storage capacity of an underground storage facility is the maximum volume of natural gas that can be stored in the facility in accordance with the design of the facility, which comprises (i) the physical characteristics of the facility, such as if it is a depleted reservoir, (ii) the installed equipment, and (iii) the operating procedures that are particular to the site. Following from this, the total gas in storage is the volume of natural gas in the underground facility at a particular time. Thus, the injection capacity of a storage facility is also variable and is dependent on factors comparable to those that determine the deliverability of the natural gas. In contrast, or as can be anticipated, the injection rate varies inversely with the total amount of gas in storage: the injection rate is at its lowest when the facility is full, but the injection rate increases as gas is withdrawn.

The base gas (or cushion gas) is the volume of natural gas intended as permanent inventory in a storage facility to maintain adequate pressure and deliverability rates throughout the withdrawal season. On the other hand, the working gas is the volume of gas in the storage facility above the level of base gas and is, simply, the natural gas that is available for withdrawal and sales. In relation to the definition of working gas, the working gas capacity is the total gas storage capacity minus base gas. The deliverability of the gas from the storage facility (also referred to as the deliverability rate, withdrawal rate, or withdrawal capacity) is often expressed as a measure of the amount of gas that can be delivered (withdrawn) from a storage facility on a daily basis.

Gas deliverability is usually expressed in terms of million cubic feet per day (MMcf/day). Occasionally, deliverability is expressed in terms of equivalent heat content of the gas withdrawn from the facility, most often in dekatherms per day (one Therm is equal 100,000 Btu, which is roughly equivalent to 100 cubic feet of natural gas; a dekatherm is the equivalent of approximately one thousand cubic feet, Mcf). The deliverability of a given storage facility is variable, and it depends on factors such as, but not limited to: (i) the amount of natural gas in the storage facility, (ii) the pressure within the storage facility, (iii) the compression capability available to the storage facility, and (iv) the configuration and capabilities of surface facilities associated with the underground storage facility. In general, the deliverability rate of a storage facility will vary directly with the total amount of natural gas in the facility. Thus, gas deliverability of the gas is at its highest when the facility is full and declines as working gas is withdrawn.

In addition, the mineralogy of the underground storage facility is important since the present of certain minerals may enhance the adsorption (and non-recovery) of certain constituents of the gas, which can result in changes in the composition of the gas. Also, other mineralogical constituents of the storage facility may cause adverse chemical changes to the gas. Thus, in addition to use of geological, geophysical, and core analysis methods, presence of discontinuities in the caprock (e.g., because of presence of faults or fractures) is often tested using long water pump tests. In these tests, water is either withdrawn or injected into the target formation, and pressure (or water level) is monitored in the formation (aquifer) above the caprock.

Because of the potential variability in gas composition (Mokhatab et al., 2006; Speight 2017; Speight, 2014, 2019) and mineralogical characteristics of the storage facility, none of the aforementioned measures for any given storage facility are fixed or absolute. The injection rate and the withdrawal rate can (will) change as the level of natural gas varies within the facility. In practice, a storage facility may be able to exceed certificated total capacity in some circumstances by exceeding certain operational parameters. The total capacity of the storage facility can also vary, temporarily or permanently, as its defining parameters vary. Measures of base gas, working gas, and working gas capacity also can also change from time to time. These changes occur, for example, when a storage operator reclassifies one category of natural gas to the other, often as a result of new wells, equipment, or operating practices (such a change generally requires approval by the appropriate regulatory authority and is subject to data derived by chemical and physical analysis of the gas). Finally, storage facilities can withdraw base gas for supply to market during times of particularly heavy demand, although by definition, this gas may not be intended for that use.

Natural gas may be stored in several different ways. In the modern world, natural gas is most commonly held in inventory underground under pressure in three main types of facilities. These underground facilities are (i) depleted reservoirs in oil and/or natural gas fields, (ii) aquifers, and (iii) salt cavern formations. Other potential types of underground storage facilities such as an abandoned mine or a hard rock cavern are not included here being the less likely potentially usable storage facilities. Natural gas is also stored in liquid or gaseous form in above-ground storage tanks, which were the conventional method of storing coal gas in the early-to-mid-twentieth century (Speight, 2013), and such storage facilities may still be seen at some sites.

More pertinent to underground storage facilities, each type of facility has its own physical characteristics (porosity, permeability, retention capability) and economics (site preparation and maintenance costs, deliverability rates, and cycling capability), which govern its suitability for storage applications. In relation to the type of facility, two important characteristics of an underground storage facility are (i) the capacity to hold natural gas for future use and (ii) the rate at which gas inventory can be withdrawn (the deliverability rate).

It is important to recognize that no matter which type of underground storage method is used, the caverns need to be prepared to receive the natural gas and be suited for the task (Table 3.6). This often includes adding pipes and valves, as well as sealing any cracks that might have occurred with the original drilling. These costs, gas accessibility, ease of gas extraction, or proximity to a gas distribution center must be given due consideration in order to evaluate the ability of the reservoir or the cavern as a viable option for natural gas storage.

Table 3.6: Examples of general comments on underground storage facilities.

Facility	Comment
Depleted reservoirs	Most common; has slower injection rates and withdrawal rates. Conversion of a natural gas field from production to storage duty takes advantage of existing wells, gathering systems, and pipeline connections.
Aquifers	Usually used only in areas where there are no nearby depleted reservoirs. Single withdraw period (winter) is used to meet peak load requirements as well. Least desirable and most expensive type of natural gas storage facility.
Salt Caverns	Very high withdrawal and injection rates. More expensive, but faster and more flexible. The safest form of natural gas storage if originally designed for that purpose.

3.5.1 Depleted reservoirs

Most existing natural gas storage in the United States is in depleted natural gas reservoirs or depleted crude oil reservoirs that are close to consumption centers. A benefit of the use of such reservoirs is that conversion of a gas field or crude oil field from production to storage duty takes advantage of existing wells, gathering systems, and pipeline connections. Because of their wide availability, depleted crude oil reservoirs and depleted natural gas reservoirs are the most commonly used underground of all storage facilities.

By definition, depleted reservoirs are those formations that have already been tapped of all their recoverable natural gas and/or crude oil. This leaves an underground formation with porosity and permeability that is capable of holding natural gas. In addition, using an already developed reservoir for storage purposes allows the use of the extraction and distribution equipment left over from when the field was productive such as the existing wells, gathering systems, and pipeline connections as well as the proximity of consumption centers.

Having this extraction network in place reduces the cost of converting a depleted reservoir into a storage facility. The factors that determine whether or not a depleted reservoir will make a suitable storage facility are both geographic and geologic. Geographically, depleted reservoirs must be relatively close to consuming regions and they must also be close to transportation infrastructure, including pipelines, and distribution systems.

To be suitable for storage of natural gas, a depleted reservoir formation must have high porosity and high permeability. The porosity of the formation determines the amount of natural gas that it may hold, while the permeability of the formation determines the rate at which natural gas flows through the formation, which in turn determines the rate of injection and withdrawal of working gas. However, a depleted reservoir where withdrawal of the crude oil or natural gas has caused a decrease in the porosity and/or a decrease in the permeability may not be acceptable for natural gas storage. In certain instances, the formation may be stimulated to increase permeability (Speight, 2016).

In some cases, and in order to maintain pressure in depleted reservoirs, approximately 50% v/v of the natural gas in the formation must be kept as cushion gas. However, a depleted reservoir that has already been used as a source of filled with natural gas may not require the injection of what will become physically unrecoverable gas (cushion gas) since that gas already exists in the formation.

In regions without depleted reservoirs, like the upper Midwest of the United States, one of the other two storage options is required – either an aquifer or a salt cavern. Of the approximately 400 active underground storage facilities in the United States, the majority (ca. 79%) are depleted natural gas or depleted crude oil reservoirs. Conversion of a crude oil or a natural gas reservoir from production to storage takes advantage of existing infrastructure such as wells, gathering systems, and pipeline connections. Depleted crude oil and natural gas reservoirs are the most commonly used underground storage sites because of their relatively wide availability.

The injected gas volume in a depleted gas reservoir can be calculated by using the gas formation volume factor for the initial gas-in-place calculation of a producing field. If it is assumed that the reservoir pore volume is constant, the initial gas-in-place in the depleted gas reservoir in standard conditions is G_i and the total gas volume in storage facility is G, then the cumulative injected gas volume, G_s is

$$G_s = G - G_i$$

By employing the formation volume factors at initial and final conditions, the equation becomes

$$G = G_i(B_{gi}/B_g) - G_i = G_i(B_{gi}/B_g) - 1$$

The term G_i is the residual gas in a depleted gas reservoir that will be used for storage, or the initial gas in a storage field after the seasonal withdrawal and at the beginning of the resumption of injection. It can be calculated by using Eq. (1.13). Thus, assuming that the temperature is constant and substituting the already derived terms, it becomes

$$Gs = G_i[(p/Z/p_i/Z_i) - 1] = G_i(p_i/Z_i)(p/Z - p_i/Z_i)$$

In these equations, the subscript i stands for the initial conditions of the gas storage. The pressures are measured when the storage is at its maximum and minimum capacities. The pressures measured are then near the maximum and minimum pressures. The final equation (immediately above) is valid when there is no active water drive.

3.5.2 Aquifers

An aquifer is an underground highly porous and permeable rock formation that can act as natural water reservoirs. Natural aquifers may be suitable for gas storage if the water-bearing sedimentary rock formation is overlaid with an impermeable cap rock. Also, the aquifer should not be a part of drinking water system of aquifers and make up only approximately 10% of the natural gas storage facilities.

An aquifer is suitable for gas storage if the water bearing sedimentary rock formation is overlain with an impermeable cap rock. Storage is created by injecting gas and displacing the water. In some cases, an aquifer may be reconditioned and used as a facility for natural gas storage.

Typically, an aquifer is costlier to develop than a depleted reservoir and, consequent, these aquifer-derived storage facilities are used only in areas where there are no depleted reservoirs that are nearby and cheaper to develop. As might be anticipated since aquifers have not always been explored to the same extent as crude oil or natural gas reservoirs, seismic testing must be performed, much the same as is performed for the exploration of potential natural gas formations and application of text method to determine the mineralogy of the aquifer would also be a benefit, perhaps even a necessity. The area of the formation, the composition and porosity of the formation itself, and the existing formation pressure as well as the capacity for gas storage must all be determined prior to development of the aquifer for gas storage. Typically, an aquifer is suitable for gas storage if the water-bearing sedimentary rock formation is overlaid with an impermeable cap rock. Although the geology of aquifers is similar to depleted production formations, their use for natural gas storage usually requires more base gas (cushion gas) and allows less flexibility in has injection and gas withdrawal.

Furthermore, in order to develop a natural aquifer into an effective natural gas storage facility, all of the associated infrastructure must also be developed. This includes installation of wells, extraction equipment, pipelines, dehydration facilities, and possibly compression equipment. Since aquifers are naturally full of water, in some instances powerful injection equipment must be used, to allow sufficient injection pressure to push down the resident water and replace it with natural gas.

While natural gas being stored in aquifers has already undergone all of its processing, upon extraction from a water bearing aquifer formation the gas typically requires further dehydration prior to transportation, which would, more than likely, require the installation of specialized equipment at or near to near the wellhead. In addition, the aquifer may not have the same natural gas retention capabilities as a depleted crude oil reservoir or a natural gas reservoir and a part of the injected natural gas that may escape from the formation and must be gathered and extracted by *collector wells* that are specifically designed to collect any gas that has escaped from the primary aquifer formation.

In addition, an aquifer formation may typically require cushion gas than is required in a depleted crude oil reservoir or a depleted natural gas reservoir. Since there is no naturally occurring gas in the aquifer formation to begin with, a portion of the natural gas that is injected will ultimately prove to be physically unrecoverable. While it is possible to extract cushion gas from depleted reservoirs, extracting cushion gas from an aquifer formation could have negative effects, among which formation damage might be extensive. As such, most of the cushion gas that is injected into an aquifer formation may remain unrecoverable, even after the storage facility is shut down and abandoned.

Nevertheless, in spite of these issues, in some areas of the United States, most notably the Midwestern United States, natural aquifers have been converted to storage reservoirs for natural gas. Deliverability rates may be enhanced by the presence of an active water drive, which supports the reservoir pressure through the injection and production cycles.

3.5.3 Salt caverns

Essentially, salt caverns are formed out of existing salt deposits. These underground salt deposits may exist in two possible forms: salt domes and salt beds. Salt domes are thick formations created from natural salt deposits that, over time, leach up through overlying sedimentary layers to form large dome-type structures. They can be as large as a mile in diameter, and 30,000 feet in height. Typically, salt domes used for natural gas storage are between 6,000 and 1,500 feet beneath the surface, although in certain circumstances they can come much closer to the surface. Salt beds are shallower, thinner formations. These formations are usually no more than 1,000 feet in height. Because salt beds are wide, thin formations, once a salt cavern is introduced, they are more prone to deterioration, and may also be more expensive to develop than salt domes.

Once a suitable salt dome or salt bed deposit is discovered, and deemed suitable for natural gas storage, it is necessary to develop a "salt cavern" within the formation. Essentially, this consists of using water to dissolve and extract a certain amount of salt from the deposit, leaving a large empty space in the formation. This is done by drilling a well down into the formation and cycling large amounts of water through the completed well. This water will dissolve some of the salt in the deposit, and be cycled back up the well, leaving a large empty space that the salt used to occupy – this process is known as "salt cavern leaching." Some of the salt is dissolved leaving a void and the water, now saline, is pumped back to the surface. The process continues until the cavern is the desired size. Once created, a salt cavern offers an underground natural gas storage vessel with very high deliverability.

Salt cavern leaching is used to create caverns in both types of salt deposits and can be quite expensive. However, once created, a salt cavern offers an underground natural gas storage vessel with very high deliverability. In addition, cushion gas requirements are the lowest of all three storage types, with salt caverns only requiring approximately 33% of total gas capacity to be used as cushion gas.

These underground salt caverns offer another option for natural gas storage. These formations are well suited to natural gas storage in that salt caverns, once formed, allow little injected natural gas to escape from the formation unless specifically extracted. The walls of a salt cavern also have the structural strength of steel, which makes it very resilient against degradation over the life of the storage facility. Salt caverns allow very little of the injected natural gas to escape from storage unless specifically extracted. The walls of a salt cavern are strong and impervious to gas over the lifespan of the storage facility. Salt caverns provide high withdrawal rates and high injection rates relative to the working gas capacity. Base gas requirements are relatively low.

Salt cavern storage facilities are primarily located along the Gulf Coast, as well as in the northern states, and are best suited for peak load storage. Salt caverns are typically much smaller than depleted gas reservoirs and aquifers, in fact underground salt caverns usually take up only one one-hundredth of the acreage taken up by a depleted gas reservoir. As such, salt caverns cannot hold the volume of gas necessary to meet base load storage requirements. However, deliverability from salt caverns is typically much higher than for either aquifers or depleted reservoirs. Therefore, natural gas stored in a salt cavern may be more readily (and quickly) withdrawn, and caverns may be replenished with natural gas more quickly than in either of the other types of storage facilities. Moreover, salt caverns can readily begin flowing gas on as little as one hour's notice, which is useful in emergency situations or during unexpected short-term demand surges. Salt caverns may also be replenished more quickly than other types of underground storage facilities.

Typically, a salt cavern will provide a high withdrawal rate and a high injection rate relative to the working gas capacity of the cavern. Most salt cavern storage facilities have been developed in salt dome formations located in the Gulf Coast states of the United States. Salt caverns have also been formed (by a leaching process) in bedded salt formations in Northeastern, Midwestern, and Southwestern states.

Salt formation storage facilities make up approximately 10% of the natural gas storage facilities. These subsurface salt formations provide very high withdrawal and injection rates.

3.5.4 Other types of storage

3.5.4.1 Gas holders
In addition to underground storage of natural gas, there are also facilities for above-ground storage of the gas. This may be the situation if there are no regional underground storage facilities or if it is more convenient, because of the lower volume if gas to be stored, and the need for the gas is immediate. In the case of above-ground storage, the gas is stored in specially fabricated tanks above ground, which do allow for easy access to the gas and complete control of gas extraction from storage. However, while the costs for above-ground storage options are typically less than underground, tanks can store only a fraction of the natural gas that underground caverns can. Another style of above-ground storage is the transportable tank, which are typically used to maximize the amount of LNG being moved. Transportable tanks can be loaded onto train cars, used with eighteen-wheeler trucks for road transport, or moved onto barges for an overseas journey.

A gas holder (also called a gasometer) is a large container in which natural gas (or town gas) can be stored at or near atmospheric pressure and at ambient temperature. Gas holders were initially developed in the early part of the twentieth century for the storage of coal gas (Speight, 2013). The volume of a gas holder follows the quantity of stored gas, with pressure coming from the weight of a movable cap. Typical volumes for large gas holders are approximately 1,800,000 cubic feet, with 200 feet diameter structures. Gas holders tend to be used for balancing purposes (making sure gas pipes can be operated within a safe range of pressures) rather than for actually storing gas for later use.

A gas holder can provide on-site storage for purified gas and can act as a buffer by removing the need for continuous gas processing. The weight of the gas holder lift (cap) controlled the pressure of the gas in the mains and provided back pressure for the gas-making plant. A water-sealed gas holder consisted of two parts: a deep tank of water that was used to provide a seal, and a vessel that rose above the water as the gas volume increased. In the well-known telescoping type of gas holder, the

tank floats in a circular or annular water reservoir, held up by the roughly constant pressure of a varying volume of gas, the pressure determined by the weight
of the structure, and the water providing the seal for the gas within the moving
walls.

The rigid waterless gas holder is a design that neither expands nor contracts.
The modern version of the waterless gas holder is the dry-seal type (membrane
type) of gas holder that consists of a static cylindrical shell, within which a piston
rises and falls. As it moves, a grease seal, tar/oil seal or a sealing membrane which
is rolled out and in from the piston keeps the gas from escaping.

The benefit of using a gas holder (although possibly limited in storage capacity)
is that the gas holder can store gas at district pressure and can provide extra on-site
gas very quickly and at peak times. Furthermore, the gas holder is the only storage
method that can maintain the gas at the required pressure, which is the pressure
required in local gas lines, and thus the gas holder may hold a large advantage over
other methods of storage.

3.5.4.2 Cryogenic storage

Liquefied natural gas (LNG) is natural gas stored as a super-cooled (cryogenic)
liquid – the temperature required to condense natural gas depends on its precise
composition, but it is typically between –120 and –170 °C (–185 and –275 °F). Large
cryogenic tanks are needed to store the LNG which, typically, these may be 230 feet
in diameter, 145 feet high and hold over 26,400,000 gallons of LNG. At the consumer
end of the transportation process, an infrastructure for handling the reprocessing of
vast quantities of natural gas from LNG is required, which is also expensive and vulnerable to sabotage.

The advantage of LNG is that it offers an energy density comparable to petrol
and diesel fuels, extending range and reducing the frequency of refueling. The disadvantage, however, is the high cost of cryogenic storage on vehicles and the major
infrastructure requirement of dispensing station for the LNG, production, plants,
and transportation facilities. For transportation, the LNG is loaded onto double-
hulled ships, which are used for both safety and insulating purposes. Once the ship
arrives at the receiving port, the LNG is typically off-loaded into well-insulated storage tanks. Regasification is used to convert the LNG back into its gas form, which
enters the domestic pipeline distribution system and is ultimately delivered to the
end-user. The current largest specially built refrigerated tankers can carry 135,000 m^3
LNG, which is approximately 4,767,480 cubic feet of gas.

Within the United States and many other countries, LNG must meet heating
value specifications and the gas can contain only moderate quantities of NGLs. If
LNG is shipped with NGLs, the NGLs must be removed upon receipt or blended with
lean gas or nitrogen before the natural gas can enter the transportation system
(especially the pipeline system).

3.5.5 Storage measures

The total gas storage capacity is the maximum volume of gas that can be stored in an underground storage facility by design. It is determined by the physical characteristics of the reservoir and installed equipment. The total gas volume in storage is the volume of storage in the underground facility at a particular time.

The base gas (also known as the cushion gas) is the volume of gas intended as permanent inventory in a storage reservoir to maintain adequate pressure and deliverability rates throughout the withdrawal season. It contains two elements: (i) recoverable base gas – the portion of the gas that can be withdrawn with current technology, but it is left in the reservoir to maintain the pressure and (ii) non-recoverable base gas – the portion of the gas that cannot be withdrawn with the existing facilities both technically and economically. The working gas capacity is the total gas storage capacity minus base gas, i.e., the volume of gas in the reservoir above the level of base gas. So, for a given storage capacity, the higher the base gas is, the lower the working gas will be, the less efficient the storage will be.

The injection volume is the volume of gas injected into storage fields during a given period. The deliverability or the deliverability rate (also known as the withdrawal rate or the withdrawal capacity) is a measure of the amount of gas that can be delivered or withdrawn from a storage facility on a daily basis with the unit of MMscf/day, same as that for production rate. Occasionally, it is expressed in terms of equivalent heat content of the gas withdrawn from the facility such as dekatherms per day. A therm is roughly equivalent to 100 scf of natural gas; a dekatherm is approximately 1 Mscf. In general, the deliverability rate of a facility varies directly with the total amount of gas in the reservoir; it is at its highest when the reservoir is most full and declines as working gas is withdrawn.

The injection capacity or injection rate is the amount of gas that can be injected into a storage facility on a daily basis. As with deliverability, injection capacity is usually expressed in MMscf/day, although dekatherms per day is also used. By contrast, the injection rate varies inversely with the total amount of gas in storage; it is at its lowest when the reservoir is most full and increases as working gas is withdrawn.

These measures for any given storage facility are not necessarily absolute and are subject to change or interpretation. In the following sections, natural gas storage is viewed in terms of a depleting or increasing pressure in a closed reservoir without active water drive. If the reservoir pressure is supported by active water movement, equations have to be modified.

The initial gas-in-place in the depleted gas reservoir in standard conditions is G_i, and the total gas volume in storage facility is G, then the cumulative injected gas volume, Gs is (if is assumed that the reservoir pore volume is constant):

$$G_s = G - G_i$$

If the formation volume factors at initial and final conditions are employed, the equation is

$$G = G_i \left(B_{gi}/B_s \right) - G_i = G_i (Bgi/Bs - 1)$$

The G_i is the residual gas in a depleted gas reservoir that will be used for storage, or the initial gas in a storage field after the seasonal withdrawal and at the beginning of the resumption of injection. It can be calculated by using the following equation, assuming the temperature is constant:

$$G_s = G_i (p/Z/p_i/Z_i - 1) = G_i (p_i/Z_i)(p/Z - p_i/Z/)$$

In these equations, the subscript i stands for the initial conditions of the gas storage. The pressures are measured when the storage is at its maximum and minimum capacities. The pressures measured are then near the maximum and minimum pressures and the final equation (above) is valid when there is no active water drive.

3.5.6 Losses during gas storage

Gas loss in gas storage is a serious issue. It happens when the cap rock does not seal well, cement around the wellbore is flawed, or there is a communication between the storage and other reservoirs. Once gas loss is happening, the storage deliverability or withdrawal rate will decline from year to year, and the operator will have to bear with high cost or even the risk of not meeting the peak demand. The annual losses can be up to 0.5 Bcf (Wang and Economides, 2009). Therefore, gas storage must be monitored carefully in order to determine the magnitude of such loss, the root cause, and remedy it as soon as it is detected.

For gas storage that is converted from depleted gas reservoir with no water drive, the gas flows to the wells primarily by gas expansion and a procedure can be used to determine the gas loss. There are several ways to determine the reservoir pressure. One way is to conduct regular (e.g., semiannual) pressure build-up tests similar to pressure surveys done in gas production fields. Another way is to monitor the bottomhole pressure in observation wells. Ordinarily, these pressure surveys are conducted in the fall and spring when reservoir pressure is near maximum and minimum for the total gas volume calculation. During the withdrawal season, fluctuation can happen as the demands from pipeline systems can be different (Mayfield, 1981).

The total gas in storage or gas-in-place can be plotted along with the determined p/Z. If there is no gas loss, all data points should fall on the same line after repeated cycles of injection and withdrawal. If the slope of the line becomes smaller, this is likely to mean that the storage increases because of gas migration or leakage. When there is gas loss, parallel lines would appear from year to year and are shifted toward a larger gas volume at a given p/Z. The difference between these lines is gas loss.

3.6 Transportation

The efficient and effective movement of natural gas from the production site to the consumers requires an extensive and elaborate transportation system. In many instances, natural gas produced from at the wellhead will have to travel a considerable distance to reach its point of use. The typical transportation system for natural gas consists of a complex network of pipelines, designed to transport natural gas quickly and efficiently from its origin to areas of high natural gas demand. Transportation of natural gas is closely linked to storage insofar as if the natural gas being transported is not required when it reaches the end of the pipeline (or transportation system) it can be put into storage facilities to be released at a time when it is needed.

Natural gas, as it is used by consumers, is much different from the natural gas that is brought from underground up to the wellhead. In some cases, constituents of the gas may remain in the reservoir because of various property constraints or because of interactions with the reservoir rock. Furthermore, although the processing (treating, refining) of natural gas is, in many respects, less complicated than the processing and refining of crude oil, gas processing is equally as necessary before its use by end users (Parkash, 2003; Mokhatab et al., 2006; Gary et al., 2007; Speight, 2007; Riazi et al., 2013; Speight, 2014; Faramawy et al., 2016; Hsu and Robinson, 2017; Speight, 2017). The natural gas used by consumers is composed almost entirely of methane.

However, natural gas produced at the wellhead, although still composed primarily of methane, is by no means as pure and includes several varieties of impurities such as higher molecular weight hydrocarbons, carbon dioxide, and hydrogen sulfide (including mercaptans). Raw natural gas comes from three types of wells: oil wells, gas wells, and condensate wells. Natural gas that comes from oil wells is typically termed *associated gas,* which can exist separate from oil in the formation (free gas) that occur with crude oil in the reservoir or gas that dissolved in the crude oil (dissolved gas). Natural gas from gas and condensate wells, in which there is little or no crude oil, is termed *non-associated gas* (Chapter 1).

The natural gas used by consumers is composed almost entirely of methane but the natural gas that is produced at the wellhead, although still composed primarily of methane, is by no means as pure. Whatever the source of the natural gas (Chapter 1), it commonly exists in mixtures with other hydrocarbon derivatives – principally ethane, propane, butane, and pentanes. In addition, raw natural gas contains water vapor, hydrogen sulfide (H_2S), carbon dioxide, helium, nitrogen, and other compounds.

Natural gas processing (Chapter 4) consists of separating all of the various hydrocarbons and fluids from the pure natural gas, to produce pipeline-quality dry natural gas. Major transportation pipelines usually impose restrictions on the make-up of the natural gas that is allowed into the pipeline. That means that before the natural gas can be transported it must be purified. The associated hydrocarbons, which include ethane, propane, butane, iso-butane, and natural gasoline, are often referred to as NGLs can be very valuable by-products of natural gas processing. These NGLs are sold separately and have a variety of different uses, including enhancing oil recovery in oil wells, providing raw materials for oil refineries or petrochemical plants, and as sources of energy. If the natural gas has a significant helium content, the helium may be recovered by fractional distillation. Natural gas may contain as much as 7% helium and is the commercial source of this noble gas (Ward and Pierce, 1973).

While some of the needed processing can be accomplished at or near the wellhead (field processing), the complete processing of natural gas takes place at a processing plant (Mokhatab et al., 2006; Speight, 2014, 2019). The extracted natural gas is transported to these processing plants through a network of gathering pipelines, which are small-diameter, low-pressure pipes. In addition to processing performed at the wellhead and at centralized processing plants, some final processing is also sometimes accomplished at straddle extraction plants, which are located on major pipeline systems. Although the natural gas that arrives at these straddle extraction plants is already of pipeline quality, in certain instances there still exist small quantities of NGLs, which are extracted at the straddle plants.

Major transportation pipelines usually impose restrictions on the make-up of the natural gas that is allowed into the pipeline. Natural gas processing consists of separating all of the various hydrocarbons and fluids from the pure natural gas, to produce what is known as "pipeline quality" dry natural gas. Associated hydrocarbons (NGLs) can be valuable by-products of natural gas processing. These liquids include ethane, propane, butane, isobutane, and natural gasoline and are sold separately. NGLs have a variety of different uses; including enhancing oil recovery in oil wells and provide raw materials for oil refineries or petrochemical plants, and as sources of energy.

Thus, the raw natural gas must be purified to meet the quality standards specified by the major pipeline transmission and distribution companies. These quality standards vary from pipeline to pipeline and are usually a function of a design of the pipeline system and the markets that are served by the natural gas. In general, the standards specify that the natural gas should:

1. Be within a specific range of heating value (caloric value). For example, in the United States, the heating value should be on the order of 1,035 ±5% Btu per cubic foot of gas at 1 atmosphere and 15.6 °C (60 °F).
2. Be delivered at or above a specified hydrocarbon dew point temperature (below which some of the hydrocarbons in the gas might condense at pipeline pressure forming liquid slugs that could damage the pipeline). The dew-point adjustment serves the reduction of the concentration of water and heavy hydrocarbons in natural gas to such an extent that no condensation occurs during the ensuing transport in the pipelines.

3. Be free of particulate solids and liquid water to prevent erosion, corrosion or other damage to the pipeline.
4. Be dehydrated (water removed) sufficiently to prevent the formation of methane hydrates within the gas processing plant or subsequently within the sales gas transmission pipeline. A typical water content specification in the United States is that gas must contain no more than seven pounds of water per MMScf of gas (Table 3.7).
5. Contain no more than trace amounts of components such as hydrogen sulfide, carbon dioxide, mercaptans, and nitrogen. The most common specification for hydrogen sulfide content is 4 ppm of hydrogen sulfide per 100 cubic feet of gas. The specifications for the allowable amount of carbon dioxide typically limit the content to less than 3% by volume.
6. Maintain mercury at less than detectable limits (approximately 0.001 parts per billion, ppb, 0.001×10^{12}) by volume primarily to avoid damaging equipment in the gas processing plant or the pipeline transmission system from mercury amalgamation and embrittlement of aluminum and other metals.

Table 3.7: Examples of pipeline specifications for natural gas.*

Components (mol%)	Minimum	Maximum
Methane	75	10
Ethane		
Propane		5
Butanes		2
Pentanes plus		0.5
Nitrogen and other inerts		3–4
Carbon dioxide		3–4
Trace components		
Hydrogen sulfide		0.25–1.0 g/100 ft^3
Mercaptan sulfur		0.25–1.0 g/100 ft^3
Total sulfur		5–20 g/100 ft^3
Water vapor		7.0 lbs/mm ft^3
Oxygen		0.2–1.0 ppm v/v
Heating value	950	1,150 Btu/ ft^3

*May vary from state to state (within the United States and also from country to country).

Furthermore, there is need for caution when condensate mixtures are to be transported. The transport of gas-condensate mixtures of various compositions has been found to be accompanied by a slight increase in viscosity in the coldest period when ground temperatures at depth of a condensate pipeline reach −4 to 0 °C (25–32 °F). The decrease in the temperature of reservoir oil fluids under study to minus −30 to 10 °C (−22 to 50 °F) is accompanied by a sharp increase in all structural and rheological parameters of the mixture. As a result, cloud and pour point of a mixture falls, its amount decreases, the structure of paraffin deposits changes (Loskutova et al., 2014).

3.6.1 Pipelines

After extraction from the reservoir, natural gas it must be transported to different places to be processed, stored, and then finally delivered to the end consumer and this can occur by means of a pipeline or ship. Thus, once natural gas is extracted from onshore and offshore sites, it is transported to consumers, typically by way of pipelines. Before it reaches the pipelines, however, it needs to be purified into the state it will be in when it enters homes and businesses. This requires the separation of various hydrocarbons and fluids from the pure natural gas to produce "pipeline quality" dry gas. Restrictions are placed on the quality of natural gas that is allowed to enter pipelines. Natural gas can also be stored in large underground areas because demand is higher in different seasons of the year. From the large pipelines, the gas goes into smaller pipelines called mains, and then further into even smaller pipes called services that lead directly into homes and buildings to be heated. Natural gas can also be cooled to a very cold temperature and stored as a liquid. Changing the phase of the natural gas from gas to liquid allows for easier storage because it takes up less space. Then, when it needs to be distributed, it is returned to its original state and sent through pipelines.

There are essentially three major types of pipelines along the natural gas transportation route: (i) the gathering pipeline system, (ii) the transmission pipeline system, sometimes referred to as the interstate pipeline system, and (iii) the distribution system.

Gathering systems, primarily made up of small-diameter, low-pressure pipelines, move raw natural gas from the wellhead to a natural gas processing plant or to an interconnection with a larger mainline pipeline. Transmission pipelines are typically wide-diameter, high-pressure transmission pipelines that transport natural gas from the producing and processing areas to storage facilities and distribution centers. Compressor stations (or pumping stations) on the pipeline network keep the natural gas flowing forward through the pipeline system. Local distribution companies deliver natural gas to consumers through small-diameter, lower pressure service lines. Whatever the pipeline system, there is the need to determine if the natural gas received at the wellhead has a high sulfur content and a high carbon dioxide content (sour gas), a specialized sour gas gathering pipe must be installed. Sour gas is extremely corrosive and dangerous, thus its transportation from the wellhead to the sweetening plant must be done carefully (Speight, 2014b).

Thus, an issue that always arises at the wellhead when natural gas is transported by pipeline is the degree of processing at the wellhead to remove potential corrosive contaminants that would seriously affect the integrity (corrosivity) of the pipeline. While carbon dioxide and hydrogen sulfide are often considered to be non-corrosive in the dry state, the presence of water in the natural gas can render these two gases extremely corrosive (Speight, 2014b). This emphasizes the need for compositional analysis of the natural gas as it exits the production well. As a result of the compositional analysis, the pipeline operators can make the decision related to the extent of the wellhead treating such as separation of (i) hydrocarbon gas liquids, (ii) nonhydrocarbon gases, and (iii) water from the natural gas before the (treated) gas is delivered into a mainline transmission system.

Major transportation pipelines usually impose restrictions on the make-up of the natural gas that is allowed into the pipeline and natural gas must be processed to produce "pipeline quality" dry natural gas. Some field processing can be accomplished at or near the wellhead; however, the complete processing of natural gas takes place at a processing plant, usually located in a natural gas producing region. Thus, from the wellhead, natural gas is transported to processing plants through a network of small-diameter, low-pressure gathering pipelines, which may consist of a complex gathering system can consist of thousands of miles of pipes, interconnecting the processing plant to upward of 100 wells in the area.

Thus, it is essential that the composition of the gas be known so that natural-gas processing can begin at the wellhead. The composition of the raw natural gas extracted from producing wells depends on the type, depth, and location of the underground deposit and the geology of the area. A natural gas processing plant is a facility designed to clean raw natural gas by separating impurities and various non-methane hydrocarbons and fluids to produce what is known as *pipeline quality* dry natural gas. A gas processing plant is also used to recover NGLs (condensate, natural gasoline, and liquefied petroleum gas (LPG)) and sometimes other substances such as sulfur-containing constituents and should be (at least) checked for sulfur content and for residues to ensure that the LPG (Table 3.8) meets the specification (Speight, 2015, 2018).

Four basic types of LPGs are provided to cover the common use applications, particularly LPGs consisting of propane, propene (propylene), butane, and mixtures of these materials that are intended for use as domestic, commercial and industrial heating, and engine fuels. However, care must be taken in the sampling of the liquefied gases to ensure that the sample is representative otherwise the for the test results may not be significant (Speight, 2018). All four types of LPGs should conform to the specified requirements for vapor pressure, volatile residue, residue matter,

Table 3.8: Properties of liquefied petroleum gas.

Constituent	Propane	Butane
Formula	C_3H_8	C_4H_{10}
Boiling point, °F	−44°	32°
Specific gravity of the gas (air = 1.00)	1.53	2.00
Specific gravity of the liquid (water = 1.00)	0.51	0.58
Lbs./gallon: liquid @ 60 °F	4.24	4.81
BTU/gallon: gas @ 60°F	91,690	102,032
BTU/lb: gas	21,591	21,221
BTU/ft^3: gas @ 60 °F	2,516	3,280
Ft^3 of vapor @ 60 °F./gal. of liquid, 60 °F	36.39	31.26
Ft^3 of vapor @ 60 °F./lb. of liquid, 60 °F	8.547	6.506
Latent heat of vaporization @ boiling point BTU/gal	785.0	808.0
Combustion data		
Flash point, °F	−155	−76
Auto-ignition temperature, °F	878	761
Maximum flame temperature in air, °F	3,595	3,615
Flammability limits, % v/v of gas in air mixture		
Lower limit – %	2.4	1.9
Upper limit – %	9.6	8.6
Octane number (iso-octane = 100)	100+	92

relative density, and corrosion. There is also a further series of standard test methods that can be used to provide information related to the composition and properties of domestic and industrial fuel gases (Speight, 2018).

While some of the needed processing can be accomplished at or near the wellhead (field processing), the complete processing of natural gas takes place at a processing plant, usually located in a natural gas producing region. The extracted natural gas is transported to these processing plants through a network of gathering pipelines, which are small-diameter, low pressure pipes. A complex gathering system can consist of thousands of miles of pipes, interconnecting the processing plant to upward of 100 wells in the area. In addition to processing done at the wellhead and at centralized processing plants, some final processing is also sometimes accomplished at straddle extraction plants. These plants are located on major pipeline systems. Although the natural gas that arrives at these straddle extraction plants is already of pipeline quality, in certain instances there still exist small quantities of NGLs, which are extracted at the straddle plants.

The actual practice of processing natural gas to pipeline dry gas quality levels can be quite complex, but usually involves four main processes to remove the various impurities: (i) oil and condensate removal, (ii) water removal, (iii) separation of NGLs, (iv) hydrogen sulfide removal, and (v) carbon dioxide removal (Chapter 4) (Mokhatab et al., 2006; Speight, 2007, 2014). If mercury is present in the gas, typically as trace amounts, opinions differ whether or not the mercury should be removed at the wellhead. The presence of mercury can cause corrosion of aluminum heat exchangers as well as cause environmental pollution. If needed, there are two forms of removal processes: (i) regenerative processes and (ii) non-regenerative processes. The regenerative process uses sulfur-activated carbon or alumina, while non-regenerative processes use silver on a molecular sieve (Mokhatab et al., 2006; Speight, 2014a, 2019).

In addition, hydrogen sulfide and carbon dioxide can be removed by olamine scrubbing (Mokhatab et al., 2006; Kidnay et al., 2011; Speight, 2014, 2019) and heaters and scrubbers are installed, usually at or near the wellhead. The scrubbers serve primarily to remove sand and other large-particle impurities. The heaters ensure that the temperature of the gas does not drop too low. With natural gas that contains even low quantities of water, natural gas hydrates (NGHs) tend to form when temperatures drop. These hydrates are solid or semi-solid compounds, resembling ice like crystals and when gas hydrates accumulate, they can impede the passage of natural gas through valves and gathering systems. After processing, the pipeline quality natural gas is injected into gas transmission pipelines and transported to the end users. This often involves transportation of the gas over hundreds of miles, as the location of gas production is generally not the location where the gas is used.

Nevertheless, whatever of the extent of the gas processing operations (at the wellhead or in a processing facility) it is essential that the composition of the gas be known through analysis in order to ensure a (relatively) pure product for transportation and that the gas meets sales specification goals.

Pigging is the process of forcing a solid object (the pig) through a pipeline and the process involves inserting the pig, via a pig launcher, into the pipelines and removing it by use of a pig receiver. Pigging is used to perform any of the following functions, which are: (i) to provide a barrier between liquid products that use the same pipeline, (ii) to check wall thickness and find damaged sections of lines, (iii) remove debris such as dirt and wax from lines, (iv) to provide a known volume for calibrating flow meters, (v) to coat the inner pipe walls with inhibitors, and (vi) to remove condensed hydrocarbon liquids and water in two phase pipelines. For field operations, the last function is the most important.

Gathering systems typically are in the two phase stratified flow region, where the liquid flow rate is much slower than the gas flow rate. Thus, liquid accumulates in low spots in the line. Field operations must follow a rigorous pigging schedule to prevent the plant from being hit by large slugs of liquid that would flood inlet receiving and carry liquids into the gas-processing units. Fortunately, plant operators usually know when a "killer pig" is coming, and they draw down liquid levels in inlet receiving. To protect the plant from large liquid surges, operators respond by shutting in gas, which shuts down field compressors and upsets the plant with the potential for producing off-spec products.

3.6.2 Liquefied natural gas

Contrary to some loose and inaccurate definitions, LNG is not the same as LPG (often referred to as *propane*), which is determined by a standard test method (Speight, 2015, 2018; ASTM 2021). LNG is the liquid form of natural gas and is used principally for transporting natural gas to markets, where it is regasified and distributed as pipeline natural gas. The temperature required to condense natural gas depends on its precise composition, but it is typically between –120 and –170 °C (–184 and –274 °F). The advantage of LNG is that it offers an energy density comparable to petrol and diesel fuels, extending range and reducing refueling frequency. The disadvantage is the high cost of cryogenic storage on vehicles and the major infrastructure requirement of LNG dispensing stations, production plants and transportation facilities.

LNG is composed predominantly of methane since the liquefaction process requires the removal of the non-methane components like carbon dioxide, water, butane, pentane and heavier components from the produced natural gas. LNG is odorless, colorless, non-corrosive, and non-toxic. When vaporized it burns only in concentrations of 5–15% when mixed with air. In terms of composition (Table 3.1), LNG is predominantly methane and it is not surprising that the density of LNG is close to (but not exactly equal to) the density of methane.

If gas is produced at lower pressures than typical sales pipeline pressure (approximately 700–1,000 psi), it is compressed to sales gas pressure. Transport of sales gas is done at high pressure to reduce pipeline diameter. Pipelines may operate at very high pressures (>1,000 psig) to keep the gas in the dense phase thus preventing condensation and two-phase flow. Compression typically requires two to three stages to attain sales gas pressure. As stated previously, processing may be done after the first or second stage, prior to sales compression.

Compression is used in all aspects of the natural gas industry, including gas lift, reinjection of gas for pressure maintenance, gas gathering, gas processing operations (circulation of gas through the process or system), transmission and distribution systems, and reducing the gas volume for shipment by tankers or for storage. In recent years, there has been a trend toward increasing pipeline-operating pressures. The benefits of operating at higher pressures include the ability to transmit larger volumes of gas through a given size of pipeline, lower transmission losses due to friction, and the capability to transmit gas over long distances without additional boosting stations. In gas transmission, two basic types of compressors are used: reciprocating and centrifugal compressors. Reciprocating compressors are usually driven by either electric motors or gas engines, whereas centrifugal compressors use gas turbines or electric motors as drivers.

Thus, when natural gas is cooled to a temperature of approximately –160 °C (approximately –260 °F) at atmospheric pressure, it condenses to a liquid (LNG). One volume of this liquid takes up approximately 1/600th the volume of natural gas. LNG weighs less than one-half that of water, approximately 45% as much. Neither LNG, nor its vapor, can explode in an unconfined environment. Since LNG takes less volume and weight, is presents more convenient options for storage and transportation.

The task of gas compression is to bring gas from a certain suction pressure to a higher discharge pressure by means of mechanical work. The actual compression process is often compared to one of three ideal processes: (i) isothermal, (ii) isentropic, and (iii) polytropic compression.

Isothermal compression occurs when the temperature is kept constant during the compression process. It is not adiabatic because the heat generated in the compression process must be removed from the system. The compression process is isentropic or adiabatic reversible if no heat is added to or removed from the gas during compression and the process is frictionless. The polytropic compression process is, like the isentropic cycle, reversible but it is not adiabatic. It can be described as an infinite number of isentropic steps, each interrupted by isobaric heat transfer. This heat addition guarantees that the process will yield the same discharge temperature as the real process.

LNG can be used in natural gas vehicles, although it is more common to design vehicles to use compressed natural gas (CNG). The relatively high cost of production of LNG and the need to store the liquid in expensive cryogenic tanks have prevented its widespread use in commercial applications. Prior to and during transportation, CNG is stored on the vehicle in high-pressure tanks, usually on the order of 3,000–3,600 psi.

The production of LNG process involves removal of certain components, such as dust, acid gases (such as hydrogen sulfide and carbon dioxide), helium, water vapor, and higher molecular weight hydrocarbon derivatives, which could cause difficulty downstream of the wellhead. The natural gas is then condensed into a liquid at close to atmospheric pressure by cooling it to approximately –162 °C (–260 °F); maximum transport pressure is set at around 4 psi. One volume of this liquid takes up approximately 1/600th the volume of natural gas. LNG weighs less than one-half that of water and it is colorless, odorless, and non-corrosive. When vaporized, LNG burns only in concentrations of 5–15% v/v when mixed with air. Since LNG takes less volume and weight that natural gas, liquefaction enhances ease of storing and transporting. Hazards include flammability after vaporization into a gaseous state, freezing and asphyxia.

Facilities for liquefying natural gas require complex machinery with moving parts and special refrigerated ships for transporting the LNG to market. The costs of building LNG plant have lowered over the past 25 years because of greatly improved thermodynamic efficiencies so that LNG is becoming a major gas export method worldwide and many plants being extended, or new ones built in the world.

3.6.3 Liquefied petroleum gas

LPG is the term applied to certain specific hydrocarbons and their mixtures, which exist in the gaseous state under atmospheric ambient conditions but can be converted to the liquid state under conditions of moderate pressure at ambient temperature. Typically, fuel gas with four or less carbon atoms in the hydrogen-carbon combination have boiling points that are lower than room temperature and these products are gases at ambient temperature and pressure. LPG also called liquid petroleum gas (LP gas), also referred to as simply propane or butane, is flammable mixtures of hydrocarbon gases used as fuel in heating appliance and vehicles. Propylene and butylene derivatives as well as various other hydrocarbons are usually also present in small concentrations.

LPG is prepared by at a crude oil refinery or at a gas processing plant and is almost entirely derived from fossil fuel sources being (i) manufactured during the refining of crude oil or (ii) extracted from crude oil or natural gas streams as they emerge from the ground. As its boiling point is below room temperature, LPG will evaporate quickly at normal temperature and pressure (STP) and is usually supplied in pressurized steel vessels. The pressure at which LPG becomes liquid is the vapor pressure which varies depending on composition and temperature. LPG is heavier than air and, unlike methane, will flow along floors and tend to settle in low spots, such as basement or depressions in the earth. This can lead to explosion if the mixture of the LPG and air is within the explosive limits and there is an ignition source. Also, LPG can cause suffocation due to when it displaces air, causing a decrease in oxygen concentration.

LPG is a hydrocarbon mixture containing propane ($CH_3CH_2CH_3$) and butane ($CH_3CH_2CH_2CH_3$). To a lesser extent, *iso*-butane [$CH_3CH(CH_3)CH_3$] may also be present. The most common commercial products are propane, butane, or a specific of the two gases (Table 3.8) and are generally extracted from natural gas or crude oil. The propane and butane can be derived from natural gas or from refinery operations but, in this latter case, substantial proportions of the corresponding olefins will be present and need to be separated. The hydrocarbons are normally liquefied under pressure for transportation and storage.

Propylene and butylene isomers result from cracking other hydrocarbons in a crude oil refinery and are two important chemical feedstocks. The presence of propylene and butylenes in LPG used as fuel gas is not critical. The vapor pressures of these olefins are slightly higher than those of propane and butane and the flame speed is substantially higher, but this may be an advantage since the flame speeds of propane and butane are slow. However, one issue that often limits the amount of the olefins in LPG is the propensity of the olefins to form soot.

As already noted, the compositions of natural, manufactured, and mixed gases can vary so widely, no single set of specifications could cover all situations (Chapter 3). The requirements are usually based on performances in burners and equipment, on minimum heat content, and on maximum sulfur content. Gas utilities in most states come under the supervision of state commissions or regulatory bodies and the utilities must provide a gas that is acceptable to all types of consumers and that will give satisfactory performance in all kinds of consuming equipment.

The specific gravity of product gases, including LPG, may be determined conveniently by a number of methods and a variety of instruments (Speight, 2015, 2018; ASTM 2021). This test method covers the determination of relative density of gaseous fuels, including LPGs, in the gaseous state at normal temperatures and pressures. The methods specified as sub-sections of the test method are sufficiently varied in nature so that one or more may be used for laboratory, control, reference (quality control), gas measurement, or in fact, for any purpose in which it is desired to know the relative density of gas or gases as compared to the density of dry air at the same temperature and pressure.

The *heat value* of a fuel gas is generally determined at constant pressure in a flow calorimeter in which the heat released by the combustion of a definite quantity of gas is absorbed by a measured quantity of water or air.

The lower and upper limits of *flammability* indicate the percentage of combustible gas in air below which and above which flame will not propagate. When flame is initiated in mixtures having compositions within these limits, it will propagate and therefore the mixtures are flammable. A knowledge of flammable limits and their use in establishing safe practices in handling gaseous fuels are important, such as (i) when purging equipment is used in gas service, (ii) in controlling the atmospheric composition in a factory or in a mine, or (iii) in handling liquefied gases.

Many factors enter into the experimental determination of flammable limits of gas mixtures, including the diameter and length of the tube or vessel used for the test, the temperature and pressure of the gases, and the direction of flame propagation-upward or downward. For these and other reasons, great care must be used in the application of the data. In monitoring closed spaces where small amounts of gases enter the atmosphere, often the maximum concentration of the combustible gas is limited to one-fifth of the concentration of the gas at the lower limit of flammability of the gas-air mixture.

LPG can be transported in a number of ways, including by ship, rail, tanker trucks, intermodal tanks, cylinder trucks, pipelines, and local gas reticulation systems. However, a challenge with LPG is that it can vary widely in composition, leading to variable engine performance and cold starting performance. At normal temperatures and pressures, LPG will evaporate, and, because of this tendency, LPG is stored in pressurized steel bottles. Unlike natural gas, LPG is heavier than air, and thus will flow along floors and tend to settle in low spots, such as basements or, if outside, in depressions in the earth. Such accumulations can cause explosion hazards and are the reason that LPG fueled vehicles are prohibited from indoor parking area in some jurisdictions.

In addition, LPG is available in different grades (usually specified as: (i) commercial propane, (ii) commercial butane, (iii) commercial propane-butane (P-B) mixtures, and (iv) special duty propane). During the use of LPG, the gas must vaporize completely and burn satisfactorily in the appliance without causing any corrosion or producing any deposits in the system.

Commercial propane consists predominantly of propane and/or propylene while *commercial butane* is mainly composed of butanes and/or butylenes. Both must be free from harmful amounts of toxic constituents and free from mechanically entrained water (that may be further limited by specifications). *Commercial propane-butane* mixtures are produced to meet sales specifications such as volatility, vapor pressure, specific gravity, hydrocarbon composition, sulfur and its compounds, corrosion of copper, residues, and water content. These mixtures are used as fuels in areas and at times where low ambient temperatures are less frequently encountered. Analysis by gas chromatography is possible (Speight, 2015, 2018; ASTM 2021). *Special duty propane* is intended for use in spark-ignition engines and the specification includes a minimum *motor octane number* to ensure satisfactory antiknock performance. Propylene ($CH_3CH=CH_2$) has a significantly lower octane number than propane, so there is a limit to the amount of this component that can be tolerated in the mixture. Analysis by gas chromatography is possible (Speight, 2015, 2018; ASTM 2021).

LPG and LNG can share the facility of being stored and transported as a liquid and then vaporized and used as a gas. In order to achieve this, LPG must be maintained at a moderate pressure but at ambient temperature. The LNG can be at ambient pressure but must be maintained at a temperature of roughly −1 to 60 °C (30 to 140 °F). In fact, in some applications it is actually economical and convenient to use LPG in the liquid phase. In such cases, certain aspects of gas composition (or quality such as the ratio of propane to butane and the presence of traces of heavier hydrocarbons, water and other extraneous materials) may be of lesser importance compared to the use of the gas in the vapor phase.

For normal (gaseous) use, the contaminants of LPG are controlled at a level at which they do not corrode fittings and appliances or impede the flow of the gas. For example, hydrogen sulfide (H_2S) and carbonyl sulfide (COS) should be absent.

Organic sulfur to the level required for adequate odorization, or *stenching*, is a normal requirement in LPG, dimethyl sulfide (CH_3SCH_3) and ethyl mercaptan (C_2H_5SH) are commonly used at a concentration of up to 50 ppm. Natural gas is similarly treated possibly with a wider range of volatile sulfur compounds.

The presence of water in LPG (or in natural gas) is undesirable since it can produce hydrates that will cause, for example, line blockage due to the formation of hydrates under conditions where the water *dew point* is attained. If the amount of water is above acceptable levels, the addition of a small methanol will counteract any such effect.

In addition to other gases, LPG may also be contaminated by higher boiling constituents (residua) such as the constituents of middle distillates to lubricating oil. These contaminants become included in the gas during handling and must be prevented from reaching unacceptable levels. Olefins and especially diolefins are prone to polymerization and should be removed.

Control over the amount of residue in LPG is essential in end-use applications of LPG. In fact, an oily residue in LPG is a contamination that can lead to problems during production, transportation, storage, or when in use. For example, when the liquefied gas is used for automotive fuel application, a residue can lead to troublesome deposits that will accumulate and corrode or plug the LPG fuel filter, the low-pressure regulators, the fuel mixer, or the control solenoids. LPG can be contaminated with oily residues during its production or transport. Transport contamination can be a result of shared pipelines, valves, and trucks used for the distribution of other products. Production sources such as the desulfurization process may contribute sulfur absorbent oil to the LPG stream. Commercial LPG, especially for automotive applications, should comply with current fuel specifications.

Fuel specifications for oily residue in LPG use a method known as the oil stain method (Speight, 2015, 2018; ASTM 2021). In the method 100 milliliters of the LPG are evaporated and the remaining volume of residue is read from the glass evaporation tube. In addition, the residue is dissolved in a solvent and the resulting solution is slowly dripped on the adsorption paper. The size and persistence of the stain which remains on the paper after the solvent evaporates is the other, empirical, quantification of the oily residue in the LPG sample. The accuracy of both quantifications is subject to question.

On the other hand, the liquefied has injector method, in which a dedicated sampler (the liquefied gas injector) is used to inject the liquefied gas at room temperature directly on to the gas-chromatographic column. This test method is based on evaporation of large sample volumes followed by visual or gravimetric estimation of residue content. In addition, this method provides enhanced sensitivity in measurements of heavier (oily) residues, with a quantification limit of 10 mg/kg total residue. This test method gives both quantitative results and information related to contaminant composition such as boiling point range and fingerprint, which can be useful in tracing the source of a particular contaminant. The method covers the determination, by gas chromatography, of soluble hydrocarbon materials, sometimes called "oily residue," which can be present in LPG and which are substantially less volatile than the LPG itself. Also, the method offers quantitative data over the range of 10–600 mg/kg (ppm w/w), the residue with a boiling point between 174 °C (345 °F) and 522 °C (970 °F; decane to tetracontane, i.e., C_{10} to C_{40}) in LPG. Higher boiling materials, or materials that adhere permanently to the chromatographic column, will not flow though the column and, therefore, will not be detected.

Finally, most methods for the compositional analysis of LPG recommend the use of a liquid sampling valve for sample introduction into the split inlet of the gas chromatographic instrument.

3.6.4 Compressed natural gas

CNG (methane stored at high pressure) is a fuel, which can be used in place of gasoline, diesel fuel, and LPG. The composition and properties of compressed natural are dependent on the properties and composition of the original natural gas feedstock and test methods for application to the compressed natural should be designed based on the test methods applied to the gaseous (non-compressed) feedstock.

CNG is often confused with LNG – the main difference between the two (CNG and LNG) is that CNG is stored at ambient temperature and high pressure, while LNG is stored at low temperature and approximately ambient pressure. In their respective storage conditions, LNG is a liquid while CNG is a supercritical fluid (a fluid at a temperature and pressure above the critical point where distinct liquid and gas phases do not exist). Also, CNG does not require an expensive cooling process and cryogenic tanks but CNG does require a much larger volume to store the energy equivalent of gasoline and the use of very high pressures (3,000–4,000 psi). As a consequence of this, LNG is often used for transporting natural gas over large distances, in ships, trains, or pipelines, where the gas is converted into CNG before distribution to the consumers.

The combustion of CNG produces fewer undesirable gases than many other fuels. Also, CNG is safer than other fuels in the event of a spill because natural gas is lighter than air and disperses quickly when released. CNG may be found above oil deposits or may be collected from landfills or wastewater treatment plants where it is known as biogas. CNG is produced by compressing natural gas to less than 1% of the volume it occupies at standard atmospheric pressure. It is stored and distributed in hard containers at a pressure on the order of 2,900–3,600 psi, usually in cylindrical-shaped or spherical-shaped containers. Also, gas can be transported in containers at high pressures, typically 1,800 psig for a rich gas (significant amounts of ethane, propane, etc.) to roughly 3,600 psig for a lean gas (mainly methane). Gas at these pressures is termed *compressed natural gas*.

CNG has been used as a transportation fuel, mostly in public transit as an alternative to conventional fuels (gasoline or diesel). CNG, which is compressed at over 3,000 psi to one percent of the volume the gas would occupy at normal atmospheric pressure, can be burned in an internal combustion engine that has been appropriately modified. Compared to gasoline, CNG vehicles emit far less carbon monoxide, nitrogen oxides (NOx), and particulates. The main disadvantage of CNG is its low energy density compared with liquid fuels. A gallon of CNG has only a quarter of the energy in a gallon of gasoline. CNG vehicles therefore require big, bulky fuel tanks, making CNG practical mainly for large vehicles such as buses and trucks. The filling stations can be supplied by pipeline gas, but the compressors needed to get the gas to 3,000 psig can be expensive to purchase, maintain, and operate.

An alternative approach has dedicated transport ships carrying straight long, large diameter pipes in an insulated cold storage cargo package. The gas has to be dried, compressed, and chilled for storage onboard. By careful control of temperature, more gas should be transported in any ship of a given payload capacity, subject to volume limitation and amount and weight of material of the pipe (pressure and safety considerations). Suitable compressors and chillers are needed but would be much less expensive than a natural gas liquefier, and would be standard, so that costs could be further minimized. According to the proposers, the terminal facilities would also be simple and hence would be of low cost.

3.6.5 Gas-to-solid

Gas can be transported as a solid, with the solid being the gas hydrate (Børrehaug and Gudmundsson, 1996; Gudmundsson, 1996; Gudmundsson, and Børrehaug, 1996; Gudmundsson et al., 1997; Gudmundsson et al., 1998; Taylor et al., 2003). NGH is the product of mixing natural gas with liquid water to form a stable water crystalline ice-like substance. The transport of NGHs is (with due consideration of the safety issues) a viable alternative to LNG or pipelines for the transportation of natural gas from source to demand. Consideration of the various aspects leads to the conclusion that transporting natural gas as the solid gas hydrates can be done at a higher temperature and at a lower pressure than the transportation of LNG and the risk of ignition in transport is much lower remembering that gas hydrates may decompose explosively with simultaneous ignition.

The gas-to-solids process involves three stages: (i) production, (ii) transportation, and (iii) regasification. NGHs are created when certain small molecules, particularly methane, ethane, and propane, stabilize the hydrogen bonds within water to form a three-dimensional cage-like structure with the gas molecule trapped within the cages. A cage is made up of several water molecules held together by hydrogen bonds and the solid gas hydrate has a snow-like appearance. Hydrates are formed from natural gas in the presence of liquid water provided the pressure is above and the temperature is below the equilibrium line of the phase diagram of the gas and liquid water.

For the most part, in the crude oil industry and in the natural gas industry, gas hydrates are a pipeline nuisance and safety hazard, and require considerable care by the operators to ensure that they do not form as they can block pipelines if precautions, such as methanol injection, are not taken. On the other hand, vast quantities of gas hydrate have been found in permafrost and at the seabed in depths below 1,500 feet (500 meters), and if properly exploited could become the major energy source in the next 30 years.

For gas transport, NGHs can be deliberately formed by mixing natural gas and water at approximately 1,175–1,500 psi and 2–10 °C (35–50 °F). If the slurry is refrigerated to approximately –15 °C (5 °F), it decomposes very slowly at atmospheric pressure, so that the hydrate can be transported by ship to market in simple insulated under near adiabatic conditions – conditions that occur without transfer of heat or matter between a thermodynamic system and its surroundings.

At the market, the slurry is melted back to gas and water by controlled warming for use after appropriate drying in electricity power generation stations or other requirements. The hydrate mixture yields up to 5,600 ft^3 (approximately 160 m^3) of natural gas per ton of hydrate, depending on the manufacturing process. The manufacture of the hydrate could be carried out using mobile equipment for onshore and ship for offshore using a floating production, storage and offloading vessel with minimal gas processing (cleaning, etc.) prior to hydrate formation, which is attractive commercially.

The water can be used at the destination if there is water shortage, or returned as ballast to the hydrate generator and, since it is saturated with gas, will not take more gas into solution. Process operability of continuous production of hydrate in a large-scale reactor, long-term hydrate storage and controlled regeneration of gas from storage has all been demonstrated.

The hydrate mixture can be stored at normal temperatures (0 to –10 °C; 32–14 °F) and pressures (10–1 atmosphere) where 1 m^3 of hydrate should contain approximately 160 m^3 gas per m^3 of water. This concentration of gas is attractive as it is easier to produce, safer and cheaper to store compared to the 200 m^3 per 1 m^3 of compressed gas (high pressure ca. 3,000 psig) or the 637 m^3 gas per 1 m^3 of LNG (low temperatures of –162 °C, –260 °F).

Gas storage in hydrate form becomes especially efficient at relatively low pressures where substantially more gas per unit volume is contained in the hydrate than in the free state or in the compressed state when the pressure has dropped. When compared to the transportation of natural gas by pipeline or as LNG, the hydrate concept has lower capital and operating costs for the movement of quantities of natural gas over adverse conditions.

Thus, gas hydrate is an effective means of gas storage and gas transport as it eliminates low temperatures and the necessity of compressing the gas to high pressures.

To recall, dry hydrate pellets yield approximately 160 volumes of gas at standard conditions from 1 volume of hydrate compared to the approximately 637 volumes of gas per volume of LNG. This is a considerable volume penalty (and hence transport cost) if considered in isolation, with the cheaper ships for hydrate transport the process could be economic.

3.6.6 Gas-to-power

Electric power can be an intermediate product, such as in the case of mineral refining in which electricity is used to refine bauxite into aluminum; or it can be an end product that is distributed into a large utility power grid. Thus, the concept of gas-to-power (GTP, sometimes referred to as *gas-to-grid*, GTG) is not a new thought but is certainly worthy of consideration in this section.

Currently, much of the transported gas destination is fuel for electricity generation. Electricity generation at or near the storage facility and transportation by cable to the destination(s) (GTP) is possible. Thus, for instance offshore or isolated gas could be used to fuel an offshore power plant (may be sited in less hostile waters), which would generate electricity for sale onshore or to other offshore customers. Unfortunately, installing high-power lines to reach the shoreline appear to be almost as expensive as pipelines, so that gas-to-power could be viewed as defeating the purpose of an alternative cheaper solution for transporting gas. There is significant energy loss from the cables along the long-distance transmission lines, more so if the power is AC rather than DC; additionally, losses also occur when the power is converted to DC from AC and when it converted from the high voltages used in the transmission to the lower values needed by the consumers.

Some observers consider having the energy as gas at the consumers' end gives greater flexibility and better thermal efficiencies, because the waste heat can be used for local heating and desalination. This view is strengthened by the economics as power generation uses approximately 1 million scf/day of gas for every 10 MW of power generated, so that even large generation capacity would not consume much of the gas from larger fields, and thus not generate large revenues for the gas producers. Nevertheless, gas-to-power has been an option much considered in the United States for getting energy from the Alaskan gas and oil fields to the populated areas.

There are other practical considerations to note such as if the gas is associated gas, then if there is generator shutdown and no other gas outlet, the whole oil production facility might also have to be shut down, or the gas released to flare. Also, if there are operational problems within the generation plant the generators must be able to shut down quickly (on the order of sixty seconds, or less) to keep a small incident from escalating to a major incident. Additionally, the shutdown system itself must be safe so that any plant that has complicated processes that requires a purge cycle or a cool-down cycle before it can shut down is clearly unsuitable. Finally, if the plant cannot shut down easily and/or be able to start up again quickly (perhaps in an hour), operators will be hesitant to ever shut down the process, for fear of financial retribution from the power distributors.

3.6.7 Gas-to-liquids

In gas-to-liquids (GTL) transport, there are two basic technologies for the conversion of gas to liquids: (i) direct conversion of natural gas to liquid fuels and (ii) indirect conversion via synthesis gas (after referred to syngas because it is used to synthesize products without the need for additional reactants). The direct conversion avoids the production of synthesis gas and several direct conversion processes have been developed.

The direct conversion of methane to higher hydrocarbons may result from a number of reactions:

Dehydrogenation

$$2CH_4 \rightarrow C_2H_6 + H_2, \qquad\qquad \Delta G^{\circ}(500\,^{\circ}C) = +35.6\,kJ/mol \qquad (3.1)$$

Oxidative coupling

$$2CH_4 + O_2 \rightarrow C_2H_4 + 2H_2O, \qquad \Delta G(500\,^{\circ}C) = -374.2\,kJ/mol \qquad (3.2)$$

$$2CH_4 + 0.5O_2 \rightarrow C_2H_6 + H_2O, \qquad \Delta G(500\,^{\circ}C) = -169.3\,kJ/mol \qquad (3.3)$$

Partial oxidation

$$CH_4 + 0.5O_2 \rightarrow CH_3OH, \qquad \Delta G(500\,^{\circ}C) = -86.1\,kJ/mol \qquad (3.4)$$

$$CH_4 + 0.5O_2 \rightarrow CH_2O + H_2, \qquad \Delta G(500\,^{\circ}C) = -83.7\,kJ/mol \qquad (3.5)$$

Oxydehydrochlorination

$$CH_4 + 0.5O_2 + HCl \rightarrow CH_3Cl + H_2O, \quad \Delta G(500\,^{\circ}C) = -119.9\,kJ/mol \qquad (3.6)$$

Complete oxidation

$$CH_4 + 2O_2 \rightarrow CO_2 + 2H_2O, \qquad \Delta G(500\,^{\circ}C) = -792.9\,kJ/mol \qquad (3.7)$$

In the dehydrogenation reaction (3.1), hydrogen is removed from two molecules of methane that are assembled to produce ethane thermally. It is the most direct reaction, but unfortunately its free energy is so positive, that the reaction is not feasible at reasonable temperatures. The use of oxygen makes the removal of hydrogen from methane and successive coupling of C–C bonds easier, as in the oxidation reactions ((3.2)–(3.6)). However the compete oxidation reaction (3.7) inadvertently dominates the process when oxygen is used, producing undesirable products. Selective acceleration of the oxidation reactions ((3.2)–(3.6)) has been achieved using various catalysts and, as a result, selectivity (i.e., percentage of useful products in the product mix) up to 20% has been achieved at 40% conversion.

As shown above, partial oxidation (POx) is another proven method for production of synthesis gas. In the process, natural gas reacts with pure oxygen at a temperature above 1,000 °C (>1,830 °F). The overall reaction is noncatalytic and slightly exothermic:

$$CH_4 + 0.5O_2 \rightarrow CO + 2H_2, \quad \Delta H(1,000\,°C) = -35.67\,kJ/mol$$

As this reaction illustrates, the resulting ratio of hydrogen to carbon monoxide (H_2: CO) for the partial oxidation process is on the order of 2, which is in the middle of the desired range for the Fischer–Tropsch synthesis (1.8–2.3). However, this ratio is not always achieved because of the occurrence of reverse water gas shift reaction:

Water gas shift reaction

$$CO + H_2O \rightarrow CO_2 + H_2$$

Reverse water gas shift reaction

$$CO_2 + H_2 \rightarrow CO + H_2O$$

This reaction consumes hydrogen to form carbon monoxide thereby lowering the hydrogen-to-carbon monoxide ratio to less than 2. To avoid the consumption of hydrogen and the formation of carbon monoxide, steam is added to the feed, in a process known as autothermal reforming (Speight, 2014, 2017).

In the autothermal reforming process, a mixture of methane and oxygen (or air) enters the burner, where part of the methane is combusted:

$$CH_4 + 2O_2 \rightarrow CO_2 + 2H_2O$$

The hot mixture of combustion gases and unburned methane passes through a packed catalyst bed (usually nickel, Ni), where it undergoes mixed reforming, namely steam reforming (an endothermic reaction) and dry reforming (an endothermic reaction):

Steam reforming

$$CH_4 + H_2O \rightarrow CO + 3H_2, \quad \Delta H°(1,000\,°C) = +226.1\,kJ/mol$$

Dry reforming

$$CH_4 + CO_2 \rightarrow 2CO + 2H_2, \quad \Delta H(1,000\,°C) = +258.9\,kJ/mol$$

Combining these two reactions yields the following reaction:

$$CH_4 + 0.5O_2 \rightarrow CO + 2H_2, \quad \Delta H(1,000°C) = -35.67\,kJ/mol$$

Autothermal reactors are adiabatic. Since heat is provided by partial combustion of methane, no external source of heating is needed, which simplifies reactor design and operation. Combustion consumes approximately 25% of the feed methane, and is either homogeneous in the burner or catalytic at the top of the bed. The temperature of the combustion gases rises to approximately 1,000–1,500 °C (1,830–2,730 °F), and subsequently drops as the gases undergo the endothermic mixed reforming reactions in the catalyst bed, to exit as syngas at a lower temperature. If the hydrogen-to-carbon monoxide ratio in the synthesis gas is not as desired, steam may be added to the feed.

Dry reforming is less common than steam reforming, and the main use is for processes that require high proportion of carbon monoxide in the synthesis gas. The thermodynamics of dry reforming is similar to steam reforming. The main operational difference of dry reforming from steam reforming is its tendency for coking, made more severe by the lack of steam to remove carbon. In some applications, such as in mixed reforming (combination of steam and dry reforming), steam is added for effective containment of coking problems. Since coking quickly deactivates nickel (Ni) catalysts, rhodium (Rh) and ruthenium (Ru) catalysts are used in most dry reforming applications.

Even though natural gas contains mostly methane, in some instances it may contain large amounts of higher hydrocarbon derivatives (such as the C_2 to C_4 derivatives). This would require large reforming reactors and can even produce excess carbon deposits in the reactor. In such instances, a pre-reforming process step may be added to a gas reforming process to remove the higher molecular weight hydrocarbon constituents. Thus:

$$2C_xH_y + 2zH_2O \rightarrow 2(x - z)CH_4 + 2zCO + (6z + y - 4x) H_2$$

In this equation, $y, x \geq 2$.

Pre-reforming units are adiabatic with the feed entering the reactor at a temperature on the odder of 300 °C (570 °F) and subsequently reacting on a catalyst comprised of nickel (Ni) on γ-alumina (γ-Al_2O_3) substrate, to produce gas at 550 °C (1,020 °F). Methane in this gas is then converted to syngas in a downstream reforming unit that is smaller than would be required if the pre-reformed gas was fed directly into the reactor at 550 °C (1,020 °F).

The indirect conversion process for the conversion of natural gas (methane) to liquid products relies on three basic steps: (i) reforming (catalytic conversion) of natural gas to synthesis, which is a mixture of carbon monoxide and hydrogen at varying proportions, (ii) Fischer–Tropsch synthesis, which involves the catalytic conversion of synthesis gas to liquid hydrocarbons or oxygenates, and (iii) upgrading of products via a number of standard refinery processes, such as hydrocracking, isomerization, or catalytic reforming. Thus:

$$CH_4 + oxygen\,source \rightarrow CO + H_2$$

The source of the oxygen source can be steam, carbon dioxide or air. The product resulting from reforming is composed predominantly of carbon monoxide (CO) and hydrogen (H_2) (synthesis gas). The next step is Fischer–Tropsch synthesis:

$$nCO + (n + 2)H_2 \rightarrow (C_nH_{2n+2}) + H_2O \;\; and/or \, CO_2$$

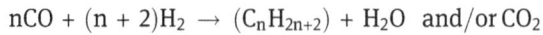

In this equation, (C_nH_{2n+2}) refers to a mixture of liquid straight-chain hydrocarbons that include alkane derivatives (paraffin derivatives) and alkene derivatives (olefin derivatives) with n ranging from 1 to more than 40, depending on process conditions, catalyst, and the composition of the synthesis as. The final step is product upgrading, and usually involves operations such as hydrocracking, isomerization, catalytic reforming, or alkylation – standard refinery technology can be used in this step (Speight, 2014, 2017). For example, waxes (C_{18+}) are converted into naphtha $(C_5H_{12}$ to $C_{11}H_{24})$ and diesel $(C_{12}H_{26}$ to $C_{18}H_{38})$ in a hydrocracking unit.

In one process, the natural gas is converted to a liquid, such as methanol, and transported as such. In the process, methane is first mixed with steam and converted to syngas (mixtures of carbon monoxide and hydrogen, $CO + H_2$) by one of a number of routes using suitable new catalyst technology.

In the steam-methane reforming process, methane reacts with steam under pressure (45–360 psi) in the presence of a catalyst to produce hydrogen, carbon monoxide, and a relatively small amount of carbon dioxide. The process is endothermic insofar as heat must be supplied to the process for the reaction to proceed. Subsequently, in what is referred to as the water gas shift reaction, the carbon monoxide and steam are reacted using a catalyst to produce carbon dioxide and more hydrogen. In a final process step (pressure-swing adsorption step) carbon dioxide and other impurities are removed from the gas stream, leaving essentially pure hydrogen. Steam reforming can also be used to produce hydrogen from other fuels, such as ethanol, propane, or even gasoline.

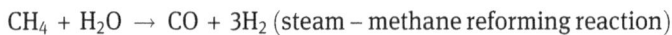

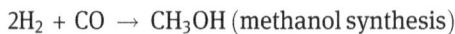

$$CH_4 + H_2O \rightarrow CO + 3H_2 \; (steam - methane\, reforming\, reaction)$$

$$CO + H_2O \rightarrow CO_2 + H_2 \, (+ \, small\, amount\, of\, heat) \;\; (water\, gas\, shift\, reaction)$$

$$2H_2 + CO \rightarrow CH_3OH \, (methanol\, synthesis)$$

Also:

$$2CH_3OH \;\; \rightarrow \; CH_3OCH_3 + H_2O$$

$$CH_3OCH_3 \rightarrow C_2H_4 + H_2O$$

The ethylene is polymerized and hydrogenated to give gasoline with hydrocarbon constituents having five or more carbon atoms that constitute approximately making up 80% w/w of the fuel.

An issue that occurs during the steam reforming reaction is the formation of carbon (coking), which can be deposited on the reactor walls, creating heat transfer problems, or on the catalyst, resulting in its deactivation. Thus:

Methane cracking:

$$CH_4 \rightarrow C + 2H_2$$

Boudouard reaction:

$$2CO \rightarrow C + CO_2$$

The methane cracking reaction is endothermic and consequently creates coking problems at higher temperatures. On the other hand, the Boudouard reaction is exothermic, and therefore favored at lower temperatures. As a result, the Boudouard reaction does not create problems at ordinary steam reforming temperatures (700–1,000 °C, 1,290–1,830 °F) except for cases where temperature is locally lower due to poor heat transfer. Coking in steam reformers using Ni catalysts results in carbon deposition on the surface of the catalyst, and subsequent polymerization until a significant part of the catalyst surface is covered and its activity decreases. Carbon atoms may also diffuse through the Ni bulk to the catalyst/support interface, where they form the so-called carbon whiskers that have detrimental effects (rupturing of catalyst pellets and plugging of the reactor).

In addition to lowering yield, catalyst deactivation creates overheating problems, because the lower rate of the endothermic reforming reaction results in lower absorption of the heat provided to the reactor (via hot gases or any other mechanism). This may lead to hot spots in the reactor wall that may eventually rupture with potentially catastrophic consequences. To prevent catalyst deactivation by coking, most commercial Ni catalysts contain promoters, such as alkalis or alkaline earth oxides (Appendix), that accelerate the removal of carbon via the reaction

$$C + H_2O \rightarrow CO + H_2$$

To promote this reaction, a large surplus of steam is usually required in the feedstock.

In the partial oxidation process, methane and other hydrocarbons in natural gas react with a limited amount of oxygen (typically from air) that is not enough to completely oxidize the hydrocarbons to carbon dioxide and water. With less than the stoichiometric amount of oxygen available, the reaction products contain primarily hydrogen and carbon monoxide (and nitrogen, if the reaction is carried out with air rather than pure oxygen), and a relatively small amount of carbon dioxide and other compounds. Subsequently, in a water gas shift reaction, the carbon monoxide reacts with water to form carbon dioxide and more hydrogen.

Partial oxidation is an exothermic process insofar as heat is evolved during the process. It is, typically, a much more rapid reaction than steam reforming and requires a smaller reactor vessel. As can be seen in chemical reactions of partial oxidation, this process initially produces less hydrogen per unit of the input fuel than is obtained by steam reforming of the same fuel.

Partial oxidation of methane

$$CH_4 + 1/2O_2 \rightarrow CO + 2H_2$$

Water gas shift reaction

$$CO + H_2O \rightarrow CO_2 + H_2$$

As stated above, the synthesis gas is then converted into a liquid using a Fischer–Tropsch process (in the presence of a catalyst) or an oxygenation method (mixing synthesis gas with oxygen in the presence of a suitable catalyst). The Fischer–Tropsch synthesis of hydrocarbon derivatives involves a vast number of reactions. For example, the following general reactions may occur:
Formation of n-paraffin derivatives

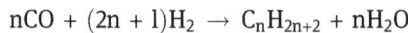

$$nCO + (2n + 1)H_2 \rightarrow C_nH_{2n+2} + nH_2O$$

Formation of olefin derivatives

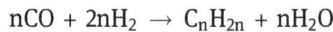

$$nCO + 2nH_2 \rightarrow C_nH_{2n} + nH_2O$$

Formation of aromatic derivatives

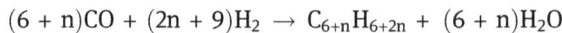

$$(6 + n)CO + (2n + 9)H_2 \rightarrow C_{6+n}H_{6+2n} + (6 + n)H_2O$$

Formation of alcohols or other oxygenates

$$nCO + 2nH_2 \rightarrow {\sim}C_nH_{2n+2}O + (n - 1)H_2O$$

Products other than the paraffin hydrocarbon derivatives are typically in small amounts and are usually neglected. As a result, the overall Fischer–Tropsch reaction is conveniently represented by the equation:

$$nCO + (n + 2)H_2 \rightarrow (C_nH_{2n+2}) + H_2O \text{ and/or } CO_2$$

Whether the products of this reaction contain alkane derivatives or alkene derivatives (with from one to 40+ carbon atoms) and water and carbon dioxide depends primarily on: (i) the hydrogen-to-carbon monoxide ratio, (ii) the catalyst, (iii) the type of reactor, and (iv) the process parameter.

The Fischer–Tropsch synthesis may be used with synthesis gas from other sources, such as coke, coal, and residue ($H_2/CO = 0.6–0.8$) and fuel oil or low Btu gas ($H_2/CO = 0.9–1.1$). Products other than alkane derivatives may be pursued, such as alkenes (desirable for subsequent production of chemicals) or oxygenate derivatives (such as alcohol derivatives, ketone derivatives, and aldehyde derivatives) but these products are minimized when Fischer–Tropsch is part of a GTL process.

The produced liquid can be a fuel, usually a clean burning motor fuel (syncrude) or lubricant, or ammonia or methanol or some precursor for plastics manufacture (e.g., urea, dimethyl ether (DME), which is also used as (i) a transportation fuel, (ii) a LPG substitute or power generation fuel as well as (iii) a chemical feedstock).

Dimethyl ether (DME, CH_3OCH_3) is a colorless gas at ambient temperature, chemically stable, with a boiling point of -25 °C (-13 °F). As its vapor pressure is approximately 0.6 mPa at 25 °C (77 °F), DME is easily liquefied. Liquid DME is colorless. The viscosity of the liquid is on the order of 0.12–0.15 kg/ms, which is almost equivalent to the viscosity of liquid propane or liquid butane. DME has a Wobbe index (ratio of calorific value and flow resistance of gaseous fuel) 52–54 that of natural gas, cooking stove for natural gas can be used for DME without any modification. The thermal efficiency and emissions with DME are almost same as with natural gas. Its physical properties are so similar to those of LPG that DME can be distributed and stored.

The reaction formulas and reaction heat concerning the exothermic synthesis of DME is

$$3CO + 3H_2 \rightarrow CH_3OCH_3 + CO_2$$

$$2CO + 4H_2 \rightarrow CH_3OCH_3 + H_2O$$

$$2CO + 4H_2 \rightarrow 2CH_3OH$$

$$2CH_3OH \rightarrow CH_3OCH_3 + H_2O$$

$$CO + H_2O \rightarrow CO_2 + H_2$$

It is more important to control the reaction temperature than in the case of methanol synthesis because the higher equilibrium conversion of DME synthesis could give higher reaction heat, and hot spot in the reactor could damage the catalyst. Synthesis gas produced by high temperature gasification of coal, oil residue and woody bio-mass has a composition of hydrogen-to-carbon monoxide ratio of approximately 0.5–1, which should be adjusted to $H_2/CO = 2$ before the DME synthesis by shift converter.

Methanol is a GTL option that has been in commission since the mid-1940s. While methanol produced from gas was originally a relatively inefficient conversion process, optimized technology has improved the efficiency. Methanol can be used in internal combustion engines as a fuel, but the current market for methanol as a fuel is limited, although the development of fuel cells for motor vehicles may change this. Methanol is best used as a basic chemical feedstock for the manufacture of plastics.

Other GTL processes are being developed to produce clean fuels, for example, syncrude, diesel, or many other products including lubricants and waxes, from gas but require complex (expensive) chemical plant with novel catalyst technology.

3.6.8 Gas-to-commodity

The gas-to-commodity (GTC) concept involves use of the components in natural gas, methane, ethane, propane, n- and iso-butanes, and pentanes that useful in their own right. The higher paraffin derivatives are particularly valuable for a wealth of chemicals and polymer precursors such as acetic acid, formaldehyde, olefin derivatives, polyethylene, polypropylene, acrylonitrile, ethylene glycol, as well as portable premium fuels, such as propane. In addition, methane can be converted via syngas to methanol, ammonia, syncrude, lubricant, or some precursor for chemicals manufacture, for example, DME, urea, and then used to make chemicals for export.

Thus, when using the GTC concept, the gas is converted to thermal or electrical power, which is then used in the production of the commodity, which is then sold, on the open market. It is the energy from the gas, heat via electricity or direct combustion, and not the components of the GTL concept that is used. The gas energy is, in essence, transported via the commodity.

In some case, the gas feedstock may be suitable for hydrogen production, which can be obtained through various thermochemical methods utilizing methane (natural gas), liquefied petroleum gas, coal gasification, or biomass (biomass gasification), from electrolysis of water, or by a thermolysis process. Hydrogen or H_2 gas is highly flammable and will burn at concentrations as low as 4% H_2 in air. For automotive applications, hydrogen is generally used in two forms: internal combustion or fuel cell conversion. In combustion, it is essentially burned as conventional gaseous fuels are, whereas a fuel cell uses the hydrogen to generate electricity that in turn is used to power electric motors on the vehicle. Hydrogen gas must be produced and is therefore an energy storage medium, not an energy source. The energy used to produce it usually comes from a more conventional source. Hydrogen holds the promise of low vehicle emissions and flexible energy storage but the technical challenges (such as the flammability of the gas) required to realize these benefits may delay the widespread implementation for several decades. However, there are disadvantages that must be given serious consideration.

For example, hydrogen poses a number of hazards to human safety, from potential detonation and fires when mixed with air to causing an asphyxia. In addition, liquid hydrogen is a cryogenic material and does present dangers (such as frostbite). Also, hydrogen can dissolve in many metals, and, in addition to leaking out, may have adverse effects such as hydrogen embrittlement, which can lead to cracks and explosions. Hydrogen gas leaking into external air may spontaneously ignite.

In fact, on occasions when the use of hydrogen as a fuel is proposed, the memory of the Hindenburg disaster always arises. At 7.25 p.m. (local time) on May 6, 1937, the German passenger airship LZ 129 Hindenburg caught fire and was destroyed during its attempt to dock with its mooring mast at the Lakehurst Naval Air Station in Manchester Township, New Jersey, United States. The airship had 97 people on board (36 passengers and 61 crewmen) of which there were 36 fatalities (13 passengers, 22 crewmen, and 1 worker on the ground).

References

Ahmed, T. 2019. Reservoir Engineering Handbook 5th Edition. Gulf Professional Publishing, Elsevier BV, Amsterdam, The Netherlands.

Arnold, K., and Stewart, M. 1999. Surface Production Operations, Volume 2: Design of Gas-Handling Systems and Facilities, 2nd Edition. Gulf Professional Publishing, Houston, TX.

ASTM. 2021. Annual Book of Standards. ASTM International, West Conshohocken, Pennsylvania.

Baker, R.O., Yarranton, H.W., and Jensen, J.L. 2015. Practical Reservoir Engineering and Characterization. Gulf Professional Publishing, Elsevier BV, Amsterdam, The Netherlands.

Basu, S., Jones, A., and Mahzari, P. 2020. Best Practices for Shale Core Handling: Transportation, Sampling and Storage for Conduction of Analyses. J. Mar. Sci. Eng., 8: 136–152.

Børrehaug, A. and Gudmundsson, J.S. 1996. Gas Transportation in Hydrate Form, EUROGAS 96, 3–5 June, Trondheim, 35–41.

Burt, S.L. 2016. Who Owns the Right to Store Gas: A Survey of Pore Space Ownership in U.S. Jurisdictions. http://www.duqlawblogs.org/joule/wp-content/uploads/2016/07/Who-Owns-the-Right-to-Store-Gas-A-Survey-of-Pore-Space-Ownership-in-U.S.-Jurisdictions-.pdf

Cretì, A. (Editor). 2009. The Economics of Natural Gas Storage. Springer-Verlag Berlin, Germany.

Faramawy, S., Zaki, T., and Sakr, A.A-E. 2016. Natural Gas Origin, Composition, and Processing: A Review. Journal of Natural Gas Science and Engineering, 34: 34–54.

Gary, J.G., Handwerk, G.E., and Kaiser, M.J. 2007. Crude oil Refining: Technology and Economics, 5th Edition. CRC Press, Taylor & Francis Group, Boca Raton, Florida.

Godec, M., Koperna, G., Petrusak, R., and Oudino, A. 2014. Enhanced Gas Recovery and CO2 Storage in Gas Shales: A Summary Review of its Status and Potential. Energy Procedia, 63: 5849–5857.

Goodway, B., Perez, M., Varsek, J., and Abaco, C. 2010. Seismic Petrophysics and Isotropic-Anisotropic AVO Methods for Unconventional Gas Exploration. The Leading Edge, 29(12): 1500–1508. Society of Exploration Geophysicists, Tulsa, Oklahoma.

Gudmundsson, J.S., Hveding, F. and Børrehaug, A. 1995. Transport of Natural Gas as Frozen Hydrate, Proc. 5th International Offshore and Polar Engineering Conf., The Hague, June 11–16, Vol. I, 282–288.

Gudmundsson, J.S. 1996. Method for Production of Gas Hydrate for Transportation and Storage, U.S. Patent No. 5,536,893.

Gudmundsson, J.S. and Børrehaug, A. 1996. Frozen Hydrate for Transport of Natural Gas, Proc. 2nd International Conf. Natural Gas Hydrates, June 2–6, Toulouse, 415–422.

Gudmundsson, J.S., Andersson, V. and Levik, O.I. 1997. Gas Storage and Transport Using Hydrates, Offshore Mediterranean Conference, Ravenna, March 19–21.

Gudmundsson, J.S., Andersson, V., Levik, O.I., and Parlaktuna, M. 1998. Hydrate Concept for Capturing Associated Gas. Proceedings. SPE European Crude oil Conference, The Hague, The Netherlands, 20–22 October 1998

Guha, R., Chowdhury, M., Singh, S., Herold, B. 2013. Application of Geomechanics and Rock Property Analysis for a Tight Oil Reservoir Development: A Case Study from Barmer Basin, India. Proceedings. 10th Biennial International Conference & Exposition – KOCHI 2013. Society of Petroleum Geophysicists (SPG) India, Kochi, Kerala, India. November 23- 25.

Guo, P., Jing, S., and Peng, C. 2014. Technologies and Countermeasures for Gas Recovery Enhancement. Natural Gas Industry B1: 96–102.

Herwanger, J.V., Bottrill, A.D., and Mildren, S.D. 2015. Uses and Abuses of the Brittleness Index with Applications to Hydraulic Stimulation. Paper No. URTeC 2172545. Proceedings. Unconventional Resources Technology Conference, San Antonio, Texas. July 20–22. Society of Petroleum Engineers, Richardson, Texas.

Hsu, C.S., and Robinson, P.R. (Editors) 2017. Handbook of Crude oil Technology, C.S. Springer International Publishing AG, Cham, Switzerland.

Jarvie, D.M., Jarvie, B., Courson, D. Garza, A., Jarvie, J., and Rocher, D. 2011. Geochemical Tools for Assessment of Tight Oil Reservoirs. Article No. 90122/2011. AAPG Hedberg Conference, Austin, Texas. December 5–10, 2010, American Association of Petroleum Geologists, Tulsa, Oklahoma.

Jones, F.O. and Owens, W.W. 1980. A Laboratory Study of Low-Permeability Gas Sands. Journal of Petroleum Technology, 32(9): 1631–1640.

Kidnay, A., McCartney, D., and Parrish, W. 2011. Fundamentals of Natural Gas Processing. CRC Press, Taylor & Francis Group, Boca Raton, Florida.

Loskutova, Yu.V., Yadrevskaya, N.N., Yudina, N.V., and Usheva, N.V. 2014. Study of Viscosity-Temperature Properties of Oil and Gas-Condensate Mixtures in Critical Temperature Ranges of Phase Transitions. Procedia Chemistry, 10: 343–348.

Lyons, W.C., and Plisga, G.J. (Editors). 2005. Standard Handbook of Petroleum & Natural Gas Engineering Second Edition, Elsevier BV, Amsterdam, The Netherlands.

Mamora, D.D., and Seo, J.G., 2002. Enhanced Gas Recovery by Carbon Dioxide Sequestration in Depleted Gas Reservoirs. SPE Paper No. 77347. Proceedings. SPE Annual Technical Conference and Exhibition, Houston, Texas. September 29-October 2. Society of Petroleum Engineers. Richardson, Texas.

Mokhatab, S., Poe, W.A., and Speight, J.G. 2006. Handbook of Natural Gas Transmission and Processing. Elsevier, Amsterdam, The Netherlands.

Parkash, S. 2003. Refining Processes Handbook. Gulf Professional Publishing, Elsevier, Amsterdam, The Netherlands.

Pooladi-Darvish, M., Hong, H., Theys, S., Stocker, R., Bachu, S., and Dashtgard, S., 2008. CO_2 injection for enhanced gas recovery and geological storage of CO_2 in the Long Coulee Glauconite F Pool, Alberta. SPE Paper No. 115789, SPE Annual Technical Conference and Exhibition, Denver, Colorado. September 21–24. Society of Petroleum Engineers, Richardson, Texas.

Riazi M., Eser, S.E., Agrawal, S., and Peña Díez, J. 2013. Crude oil Refining and Natural Gas Processing. Manual 58 MNL58. ASTM International, West Conshohocken, Pennsylvania.

Rickman, R., Mullen, M.J., Petre, J.E., and Grieser, W.V. 2008. A Practical Use of Shale Petrophysics for Stimulation Design Optimization: All Shale Plays Are Not Clones of the Barnett Shale. Paper No. SPE 115258. SPE Annual Technical Conference and Exhibition, Denver, Colorado. September 21–24. Society of Petroleum Engineers, Richardson, Texas.

Rojey, A., Jaffret, C., Cornot-Gandolph, S., Durand, B., Jullin, S., and Valais, M. 1997. Natural Gas Production, Processing, Transport. Editions Technip, Paris, France.

Satter, A., and Iqbal, G.M. 2016. Reservoir Engineering: The Fundamentals, Simulation, and Management of Conventional and Unconventional Recoveries. Elsevier BV, Amsterdam, The Netherlands.

Seo, J.G., and Mamora, D.D., 2003. Enhanced Gas Recovery by Carbon Dioxide Sequestration in Depleted Gas Reservoirs. SPE Paper No. 81200. Proceedings. SPE/EPA/DOE Exploration and Production Environmental Conference, San Antonio, Texas. March 10–12.

Sim, S.S.K., Turta, A.T., Signal, A.K., and Hawkins, B.F., 2008. Enhanced Gas Recovery: Factors Affecting Gas-Gas Displacement Efficiency. CIMPC Paper No. 2008-145. Proceedings. Canadian International Petroleum Conference/SPE Gas Technology Symposium 2008 Joint Conference, Calgary, Alberta, Canada. June 17- 19.

Soeder, D.J. and Randolph, P.L. 1987. Porosity, Permeability, and Pore Structure of the Tight Mesaverde Sandstone, Piceance Basin, Colorado. SPE Form Evaluation 2(2): 129–136. Society of Petroleum Engineers, Richardson, Texas.

Speight, J.G. 2013. The Chemistry and Technology of Coal 5th Edition. CRC Press, Taylor & Francis Group, Boca Raton, Florida.

Speight, J.G. 2014a. The Chemistry and Technology of Crude oil 5th Edition. CRC Press, Taylor & Francis Group, Boca Raton, Florida.

Speight, J.G. 2014b. Oil and Gas Corrosion Prevention. Gulf Professional Publishing, Elsevier, Oxford, United Kingdom.

Speight, J.G. 2015. Handbook of Crude oil Product Analysis 2nd Edition. John Wiley & Sons Inc., Hoboken, New Jersey.

Speight, J.G. 2016. Handbook of Hydraulic Fracturing. John Wiley & Sons Inc., Hoboken, New Jersey.

Speight, J.G. 2017. Handbook of Crude oil Refining. CRC Press, Taylor & Francis Group, Boca Raton, Florida.

Speight, J.G. 2018. Handbook of Natural Gas Analysis. John Wiley & Sons Inc., Hoboken, New Jersey.

Speight, J.G. 2019. Natural Gas: A Basic Handbook 2nd Edition. Gulf Publishing Company, Elsevier, Cambridge, Massachusetts.

Taylor, M., Dawe, R.A., Thomas, S. 2003. Fire and Ice: Gas Hydrate Transportation – A Possibility for the Caribbean Region. Paper no. SPE 81022. Proceedings. SPE Latin American and Caribbean Crude oil Engineering Conference. Port-of-Spain, Trinidad, West Indies. April 27–30. Society of Crude oil Engineers, Richardson, Texas.

Terry, R.E., and Rogers, J.B. 2015. Applied Petroleum Reservoir Engineering 3rd Edition. Prentice Hall, Pearson Education Inc. Upper Saddle River, New Jersey.

Thomas, R.D. and Ward, D.C. 1972. Effect of Overburden Pressure and Water Saturation on Gas Permeability of Tight Sandstone Cores. Journal of Petroleum Technology, 24(2): 120–124.

Ward, D.E., and Pierce, A.P. 1973 Helium. Professional Paper No. 820. In: United States Mineral Resources, US Geological Survey, Reston, Virginia. Page 285–290.

Wylie, G. 2012. Shale Gas. In: World Petroleum Council Guide: Unconventional Gas. World Petroleum Council, London, United Kingdom. Page 46–51. http://www.world-petroleum.org/docs/docs/gasbook/unconventionalgaswpc2012.pdf; accessed March 15, 2015.

Chapter 4
Production engineering

4.1 Introduction

Natural gas production engineering is a sub-division of crude oil production engineering which involves the design and use of subsurface equipment to produce the gas (and in some cases the associated crude oil). More specifically, natural gas production includes the following steps (i) evaluating inflow and outflow performance between the reservoir and the well bore, (ii) designing completion systems including tubing selection, sand control, and, in the case of tight formation, hydraulic fracturing, selecting artificial lift equipment, including sucker-rod lift gas lift, electrical submersible pumps, subsurface hydraulic pumps, progressing-cavity pumps, and plunger lift, which involves outflow that is defined as the flow of the gas from the casing perforations to the surface facilities, and (iii) selecting equipment for surface facilities that separate and measure the produced fluids (oil, natural gas, water, and impurities), prepare the oil and gas for transportation to market, and handle disposal of any water and impurities.

Thus, the production engineer is responsible for monitoring the production operations, ensuring adherence to safety protocols, and strategizing on maximizing productivity to deliver efficient results that would drive revenues and increase profitability. In fact, the choice of a production engineering teams involves engineers from other technical disciplines such as chemical engineering, mechanical engineering, mechanical engineering, geologists, and chemists. In short, few problems are solved by a single discipline and no one discipline can satisfy the needs of the production of the gas which not only the engineering aspects of gas production but also a thorough knowledge of the underground formation and the properties of the gas to ensure maximum efficiency of the production process. In some cases, the production engineer may lead the site production team and monitor the efficiency of several on-site wells while in other cases (when the underground formations differ considerably from one well to another) each individual well may require a production engineer to supervise the behavior of the well to ensure that the well is producing optimally.

Physically, natural gas exists in nature under pressure in rock reservoirs in the crust of the Earth, either in conjunction with and dissolved in crude oil (associated gas) or without crude oil (non-associated gas). In conventional natural gas reservoirs, the gas typically flows easily up through wells to the surface. However, reservoirs containing natural gas vary in size can be anything from a few hundred yards to miles across in plan, and tens to hundreds of yards thick, with the gas trapped

https://doi.org/10.1515/9783110691023-004

against an impermeable layer similar to crude oil traps (Speight, 2014a). Natural gas is a member of the class of fuels known as gaseous fuels (Chapter 1) as defined in various standard test methods (Appendix E). As a result, production engineering operations are often site specific and there is no one process pattern that is applicable to all reservoirs.

Chemically, natural gas occurs deep beneath the earth's surface and consists predominantly of methane (CH_4) but also contains small amounts of hydrocarbon gas liquids and nonhydrocarbon gases (Speight, 2014a; Faramawy et al., 2016; Speight, 2017). While natural gas from different sources may seemingly have similar behavior, the composition of the gas in not necessarily the same and other properties will vary and the gas production sequence should be analyzed with this in mind during development of a well and production of gas from that well (Tables 4.1 and 4.2) (Esteves et al., 2016). In fact, changes in composition can occur from one step in the production train to the next depending upon the composition of the gas and the processes used in the processing train (Table 4.3).

reservoir gas → produced gas → wellhead gas → transported gas → stored gas → sales gas

Table 4.1: Variation in the composition of natural gas.

Constituent	Formula	% v/v
Methane	CH_4	>85
Ethane	C_2H_6	3–8
Propane	C_3H_8	1–5
n-Butane	C_4H_{10}	1–2
iso-Butane	$C4H_{10}$	<0.3
n-Pentane	C_5H_{12}	1–5
iso-Pentane	C_5H_{12}	<0.4
Hexane, heptane, octane*	C_nH_{2n+2}	<2
Carbon dioxide	CO_2	1–2
Hydrogen sulfide	H_2S	1–2
Oxygen	O_2	<0.1
Nitrogen	N_2	1–5
Helium	He	<0.5

*Hexane (C_6H_{14}) and higher molecular weight hydrocarbon derivatives up to octane as well as benzene (C_6H_6) and toluene ($C_6H_5CH_3$).

Table 4.2: Information required for safe development of a natural gas resource.

Stage of development	Information needed	Comment
Origin	Strata mineralogy	Identify potential for gas adsorption
Exploration	Gas quality	Identify potential for gas processing
	Gas reserves	Identify economic aspects of resource development
Reservoir	Reservoir mineralogy	Tendency of gas constituents to adsorb or react with mineral
Production	Gas quality	Identify need for wellhead processing prior to transportation
Transportation	Gas composition	Identify potential for corrosion during transportation

Table 4.3: Representation of natural gas processing.

Feedstock	Processes	Examples of uses
Raw gas	Removal of carbon dioxide	Enhanced oil recovery
	Removal of hydrogen sulfide	Sulfur production
	Removal of mercaptans	Sulfur production
Product production	Fractionation (methane)	Gas to the consumer
	Fractionation (ethane)	Petrochemical feedstock
	Fractionation (propane)	Liquefied petroleum gas
	Fractionation (butane)	Liquefied petroleum gas
	Fractionation (pentane$^+$)	Gasoline blending stock

Since the different gas supplies enter the gas system at different locations, the exact composition of the natural gas at any site will vary among the different regions and over time. As an example, the heating value of the gas is dependent on the composition of the gas and can vary considerably, for example:

	Btu/ft3	Btu/lb
Natural gas (typical)*	950–1150	17,500–22,000
Natural gas (atypical)*	650–890	11,700–16,000

*Reservoir- and site-dependent.

The anticipation of differences in the composition must also consider the differences in the reservoir gas due to differences in the maturation processes from one locale to another (Speight, 2014a; Faramawy et al., 2016). Thus, in terms of natural gas definition, there are three basic approaches to defining the gas: (i) a qualitative description of the natural gas by origin, type, and constituents, (ii) classification by characteristics based upon testing procedures, and (iii) classification as a result of the concentration of specific constituents in preference to other constituents. In addition, the recent increases in domestic natural gas supplies have been made possible by two technologies (i) horizontal drilling and (ii) hydraulic fracturing which allow energy companies to develop natural gas supplies once considered to be inaccessible (Speight, 2016).

In order to understand natural gas as it exists in the reservoir or is produced at the wellhead, understanding the origin of the gas can offer pointers to the natural gas scientist or natural gas engineer as to the properties and behavior of the gas and also the means by which the gas originated. The key is to recognize that natural gas varies in composition not only because of the reservoir in which the gas exists but also because of the relative placement of wells within a reservoir.

Extracting natural gas from deposits deep underground is dependent upon several factors in the underground environment including the (i) pressure of the gas in the reservoir, (ii) the composition of the gas, (iii) the porosity of the reservoir rock, and (iv) the permeability of the reservoir rock. Primary recovery relies on underground pressure to drive the gas though the production well to the surface. When the pressure decreases due to production of the gas, the remaining gas is brought to the surface using artificial lift technologies, such as the horsehead pump at the surface or a downhole pump at the bottom of the production well.

The use of natural has in the past six decades has grown steadily in use – even replacing coal gas in many markets (Speight, 2013) – and, currently, is used extensively in residential as well as in commercial and industrial applications. While natural gas provides approximately one quarter of the energy used worldwide, and makes up nearly a quarter of electricity generation, as well as playing a crucial role as a feedstock for industry, natural gas is, in fact, a predominant source of energy in North America – it is the dominant energy used for domestic heating with slightly more than 50% of the homes in North America heated by natural gas. The use of natural gas is also rapidly increasing in electric power generation with natural gas power–generating facilities replacing coal power–generating facilities and crude oil power–generating facilities. This trend is expected to continue well into the foreseeable future. In addition, whether produced via conventional methods or renewable methods, this clean-burning alternative fuel is also being used as fuel for various vehicles but for such use, the gas must be compressed or liquefied.

Environmentally, natural gas (which is predominantly methane even in the raw unrefined form) (Burrus and Ryder, 2003, 2014; Speight, 2014a, 2019) is the cleanest burning of the fossil fuels and fossil fuel products and produces (by combustion) primarily carbon dioxide, water vapor, and small amounts of nitrogen oxides (which is produced not so much from the natural gas but from the air used as the oxidant in the combustor):

$$CH_4 + O_2 \rightarrow CO_2 + H_2O\,(+NO_x)$$

Methane is 75% w/w carbon and is not the lowest carbon fuel but, in the absence of any contaminants (after gas processing) is the cleanest burning fossil fuel (Speight, 2013, 2014a, 2019). In fact, the growth in natural gas use is linked in part to the environmental benefits (relative to other fossil fuels) of using natural gas, particularly because there is less air pollution and less greenhouse gas emissions. However, the potential to produce as much carbon dioxide as other fossil fuels and the related products is real, but whether natural gas has lower life cycle greenhouse gas emissions than crude oil or coal depends on (i) the assumed leakage rate of methane during production, transportation, and storage, (ii) the global warming potential of methane over different time frames, and (iii) the energy conversion efficiency.

Briefly, methane is a potent greenhouse gas – trapping 86 times more heat in the atmosphere than carbon dioxide – the release of methane into the air can contribute to global climate change as one of several phenomena and other naturally occurring phenomena (Speight, 2020).

Because of the fears related to the release of methane into the atmosphere, natural gas that is scheduled to be transported and stored must meet specific quality measures so that the pipeline network (or the natural gas grid system) can provide uniform quality gas. Wellhead natural gas will contain other hydrocarbon derivatives, inert gases, and contaminants which must be removed before the gas can be safely delivered to the high-pressure, long-distance pipelines that transport natural gas to consumers. Natural gas processing can be complex and usually involves several processes, or stages, to remove oil, water, hydrocarbon gas liquids, and other impurities such as sulfur, helium, nitrogen, hydrogen sulfide, and carbon dioxide (Mokhatab et al., 2006; Speight, 2019). In fact, the composition of the natural gas at the wellhead determines the number of processes (or processing stages) that are required to produce pipeline-quality natural gas. These processes or stages may be (i) integrated into one unit or operation, (ii) performed in a different order on site, (iii) or at alternative locations (lease/plant), or (iv) not be required at all; this latter option is the least likely option.

The stages of natural gas processing – also called gas treating or gas refining (Appendix D) – involve the use of (i) a gas-oil-water separator, in which pressure relief causes a natural separation of the liquids from the gases in the natural gas; (ii) a condensate separator, in which gas condensate and/or natural gasoline are removed from the natural gas stream at the wellhead with separators much like gas-oil-water separators; (iii) a dehydrator, in which water is removed to reduce the potential for corrosion in the pipeline as well as the formation of undesirable hydrates and water condensation in pipelines; (iv) a series of contaminant removal units, in which non-hydrocarbon gases such as hydrogen sulfide, carbon dioxide, water vapor, helium, nitrogen, and oxygen are removed from the natural gas stream; (v) a nitrogen rejection unit, in which nitrogen is removed and the gas stream is further dehydrated using molecular sieve beds; (vi) a methane separator, in which methane is removed from the gas stream can occur as a separate operation by cryogenic processing and absorption methods; and (vii) a fractionator, in which the gas stream is separated into component hydrocarbons (ethane, propane, butane) using the varying boiling points of the individual hydrocarbon derivatives (Mokhatab et al., 2006; Speight, 2019). Also, there is the need to determine the purity of the gas streams as well as the efficiency of each process unit by means of gas stream analysis, usually by means of online or off-line monitoring of the product streams by means of a technique such as gas chromatography (Speight, 2018, 2019).

In sour-gas streams, the primary corrosion-causing constituents of gas streams are hydrogen sulfide (H_2S) and carbon dioxide (CO_2), with contributions from other corrosive constituents. Streams containing ammonia should be dried before processing. Antifouling additives may be used in absorption oil to protect heat exchanger units. Corrosion inhibitors may be used to control corrosion in overhead systems (Speight, 2014b).

4.2 Properties of gas mixtures

Hydrocarbon fluids usually are classified on the basis of the phase behavior exhibited by the mixture and various factors play a role in the identification of the reservoir fluid (Table 4.4), with the reservoir temperature and pressure playing roles in the determination of the fluid type. Generally, natural gas streams are classified as (i) dry gas, (ii) wet gas, and (iii) gas condensate. In the case of dry gas – a low-boiling hydrocarbon mixture existing entirely in the gas phase at reservoir conditions – a decline in reservoir pressure will not usually result in the formation of a liquid phase in the reservoir. However, the gas in the reservoir typically falls into the wet gas category.

Table 4.4: Data required for the identification of reservoir fluids, including natural gas and gas condensate.

Formation volume factor (FVF)	The ratio of a phase volume (water, oil, gas, or gas plus oil) at reservoir conditions, relative to the volume of a surface phase (water, oil, or gas) at standard conditions resulting when the reservoir material is brought to the surface. Denoted mathematically as B_w (bbl/STB), B_0 (bbl/STB), B_g (ft^3/SCF), and B_t (bbl/STB).

Table 4.4 (continued)

Solution gas-oil ratio (GOR)	The amount of surface gas that can be dissolved in a stock tank oil when brought to a specific pressure and temperature. Denoted mathematically as R_s (SCF/STB).
Solution oil-gas ratio (OGR)	The amount of surface condensate that can be vaporized in a surface gas at a specific pressure and temperature; sometimes referred to as liquid content. Denoted mathematically as r_s (STB/MMSCF).
Liquid specific gravity	The ratio of density of any liquid measured at standard conditions (usually 14.7 psia and 60 °F) to the density of pure water at the same standard conditions. Denoted mathematically as γ_o (where water = 1).
API gravity	A common measure of the specific gravity of crude oil, defined by $\gamma_{API} = (141.5/\gamma_o) - 131.5$, with units in °API.
Gas specific gravity	The ratio of density of any gas at standard conditions (14.7 psia and 60 °F) to the density of air at standard conditions; based on the ideal gas law ($pV = nRT$), gas gravity is also equal to the gas molecular weight divided by air molecular weight ($M_{air} = 28.97$). Denoted mathematically as γ_g (where air = 1).
Bubblepoint pressure	At a given temperature, this condition occurs when an oil releases an infinitesimal bubble of gas from solution when pressure drops below the bubblepoint.
Retrograde pressure	At a given temperature, this condition occurs when a gas condenses a drop of oil from solution when pressure drops below the dewpoint; also called "retrograde dewpoint pressure."
Saturation pressure	An oil at its bubblepoint pressure or a gas at its dewpoint pressure.
Critical point	The pressure and temperature of a reservoir fluid where the bubblepoint pressure curve meets the retrograde dewpoint pressure curve; a unique state where all properties of the bubblepoint oil are identical to the dewpoint gas.
Composition	Quantifies the amount of each component in a reservoir mixture, usually reported in mole fraction. Typical components in crude oil reservoir mixtures include the nonhydrocarbons N_2, CO_2, and H_2S and the hydrocarbons C_1, C_2, C_3, iC_4, nC_4, iC_5, nC_5, C_6, and C_{7+} (heptanes-plus).
Saturated condition	A condition where an oil and gas are in thermodynamic equilibrium, that is, the chemical force exerted by each component in the oil phase is equal to the chemical force exerted by the same component in the gas phase, thereby eliminating mass transfer of components from one phase to the other.
Undersaturated condition	A condition when crude oil or natural gas is in a single phase but not at its saturation point (bubblepoint or dewpoint), that is, the mixture is at a pressure greater than its saturation pressure.

The presence of gas condensate occurs when the critical temperature of the reservoir is between the critical temperature and the cricondentherm, which is the highest temperature at which liquid and vapor can coexist. If the pressure is reduced to the cricondenbar pressure – which is the maximum pressure at which two phases can coexist – the liquid phase increases, but the liquid phase may reevaporate under favorable conditions. This phenomenon – the condensation of liquid upon decrease in pressure – is referred to as isothermal retrograde condensation. The liquid phase recovered from a condensate system is recovered from a phase that is vapor at reservoir conditions. This is also partly true of hydrocarbon systems where the vapor phase in equilibrium with the reservoir liquid phase is particularly rich in liquefiable constituents (C2 to C4) and a substantial proportion of stock tank liquid may derive from the vapor phase in a reservoir. Retrograde behavior does not usually occur at reservoir pressures below approximately 2,500 psi.

Gases have various properties (Appendix C) that can be observed readily, including the pressure (p), temperature (T), mass (m), and volume (V) that contains the gas. Careful, scientific observation has determined that these variables are related to one another, and the values of these properties determine the state of the gas.

The ideal gas law $PV = RT$ (for 1 mol) relates the measurable quantities pressure, P, volume, V, and temperature, T of a perfect gas at low pressures. If any two of the properties are fixed, the nature of the relationship between the other two properties can be determined. If the pressure and temperature are held constant, the volume of the gas depends directly on the mass, or amount of gas which. This allows the definition of single additional property – the gas density (ρ), which is the ratio of mass of the gas to the volume of the gas. If the mass and temperature are held constant, the product of the pressure and volume are observed to be nearly constant for a real gas (Boyle's law). If the mass and pressure are held constant, the volume is directly proportional to the temperature for an ideal gas (Charles' law and Gay-Lussac's law).

The gas laws of Boyle and Charles and Gay-Lussac can be combined into a single equation of state:

$$(P \times V)/T = n \times R$$

In this equation, P is the pressure, V is the volume, T is the temperature, n is the mass of the gas expressed in moles.

The temperature given in the equation of state must be an absolute temperature that begins at absolute zero. In the metric system of units, the temperature must be specify in degrees Kelvin (°K) and not degrees Celsius (°C), while in the non-metric (Imperial) system the absolute temperature is in degrees Rankine (°R) and not degrees Fahrenheit (°F).

There are also modified forms of the equation of state that are specialized for particular gases. For example, if both sides of the general equation are divided by the mass of the gas, the volume becomes the specific volume which is the inverse of the gas density. There is also a different gas constant (R), which is equal to the universal gas constant divided by the mass per mole of the gas. The value of the new constant depends on the type of gas as opposed to the universal gas constant, which is the same for all gases. There are in fact many different forms for the equation of state for different gases.

4.3 Flow in the reservoir

The fluid in the reservoir is defined as a single phase of gas or liquid or both. Each sort of flow results in a pressure drop. On the way from reservoir to pipeline or storage tank, the temperature and pressure of the fluid are changing and, consequently, the composition of each phase is also changing. In case of dry gas reservoir, a change in pressure and temperature does not create two-phase flow.

The flow in a reservoir is often characterized as being one of two types: transient or boundary-dominated. Transient flow takes place during the early life of a well, when the reservoir boundaries have not been felt, and the reservoir is said to be infinite-acting. During this period, the size of the reservoir has no effect on the well performance, and reservoir size cannot be determined except to deduce minimum contacted volume. Since the boundary of the reservoir has not been contacted during the transient flow period, static pressure at the boundary remains constant.

Pressure transient analysis (PTA)/well testing theory relies heavily on the assumption that the well flows at a constant rate. Several terms are often used when describing flow from a well at constant rate: which are: (i) transient flow, in which the pressure transient migrates outward from the well without encountering any boundaries; (ii) steady-state flow, in which the pressure transient has reached all of the boundaries but the static pressure at the boundary does not decline, and this is often referred to as the constant pressure boundary; (iii) pseudosteady-state flow, in which the pressure transient has reached all of the boundaries and the static pressure is declining at the boundary and uniformly throughout the reservoir; and (iv) boundary-dominated flow, in which the pressure transient has reached all of the boundaries and the static pressure is declining at the boundary, but not uniformly because the flow rate is not constant; this is also often called tank-type flow.

The constant rate and constant pressure solutions are boundaries that the actual walls approach but never really achieve. Early production and formal production tests often very closely approach the constant rate solution whereas late time production generally very closely approaches the constant pressure solution. Constant rate solutions are important for conducting PTA for the determination of key reservoir parameters such as permeability and damage, whereas the constant pressure solutions are important for determination of original in-place volumes and recoverable reserves.

Steady-state flow is defined as a flow condition under which the pressure at any point in the reservoir remains constant over time. The constant-pressure boundary might be the edge of an aquifer, or the region surrounding a water injection well. Under steady-state flow conditions due to a circular constant-pressure boundary at distance r_e from the wellbore centerline, assuming single-phase flow, the following theoretical relationship for an oil reservoir can be derived from Darcy's law which states that the discharge rate q is proportional to the gradient in hydraulic head and the hydraulic conductivity:

$$q = Q/A$$

In this equation, A is the cross-sectional area (m²) of the cylinder, Q is the flow rate (m³/s) of the fluid flowing through the area A.

On the other hand, transient flow is a fluid dynamics condition where the velocity and pressure of a fluid flow change over time due to changes in system status. Transient flow is usually characterized by a powerful pressure wave or waves that may persist for a significant time after the precipitating event has ceased.

Single-phase flow in reservoir conditions occurs if the well flowing pressure P_{wf} is higher than bubblepoint pressure P, or dewpoint pressure P_b and two-phase flow occurs by a wellbore. Reservoir depletion and production consist of two separate processes which are (i) flash liberation, that is, vaporization and (ii) differential liberation. In the flash vaporization process, at stage 1 reservoir fluid is under reservoir pressure and temperature at known volume. The pressure in the cell is covered by increasing the space available in the cell for the fluid. This procedure is repeated until a large change in the pressure-volume slope is indicated. The above procedure indicates that flash vaporization is a phase-changing process due to change pressure and temperature if the mass of reservoir fluid or total composition system remains constant and can be expressed as follows:

$$zF = XL + yV$$

In this equation, z is the mole fraction of component in a reservoir fluid mixture, F is the number of moles of sample at initial reservoir pressure and temperature, X is the mole fraction of component in liquid, L is the number of moles of equilibrium liquid, y is the mole fraction component in gas mixture, and V is the number of moles of equilibrium gas phase.

Equilibrium or flash liberation calculations may be made for reservoir fluid that divides into two phases at any temperature and pressure. The differential liberation begins in the same manner as flash vaporization. The sample is placed in a pressure higher than bubblepoint pressure. The pressure is lowered until such time that free gas is liberated in the cell. Then for pre-determined pressure or volume increments, mercury is withdrawn from the cell, gas is released from solution and the cell is agitated until the liberated gas is in the equilibrium with the oil. All the free gas is ejected from the cell at a constant pressure by injection mercury. The volume of the free gas is displaced and the oil remaining in the cells are thus measured at cell conditions. This procedure is repeated for all the pressure increments until only oil remains in the cell at reservoir temperature and atmospheric pressure.

The term "flash process" refers to the conditions where the mass of the considered system does not vary with changes in pressure and temperature. The flash process in two-phase regions (vaporization or condensation) can be defined in terms of total system composition. The total system composition can be measured at any point outside of saturation line. As a substitution the following treatment can be used; the total system composition in two-phase region flash process remains constant. The flash process may ensue for a composition that separates into two phases for the values of pressures and temperatures inside the saturation curve area. After the temperature and pressure are chosen, all the gas is in equilibrium with all the oil. In other words, a change of pressure or temperature, or both, in a flash process can change the equilibrium conditions according to the Gibbs phase rule which provides the number of independent variables that, in turn, define intensive properties. Flash vaporization may be a batch or a continuous process. Treating two-phase flow in tubing as a steady state, neglecting the gas storage effect, and gas slippage results in a flash process. In a horizontal flow, and in separators, a similar flash process is employed.

The same kind of equilibrium, but with the fluid mass decreasing differentially, is called a differential process (liberation or condensation). In reservoir conditions the hydrocarbon pore volume (HCW) remains constant if the expansion of interstitial water and rock compressibility are neglected. For such constant HCW it must be made clear that differential process occurs always as differential vaporization or differential condensation. Differential vaporization takes place when the reservoir temperature is less than critical temperature of solution ($TmS < TJ$), and also it takes place during retrograde gas reservoir depletion, but only in the region pressure and temperature where the retrograde liquid is vaporized.

In differential condensation, the oil reservoir pressure is maintained constant or almost constant-for example, by gas injection. Differential condensation can also occur just below the dewpoint in a gas-condensate reservoir. Above the bubblepoint and the dewpoint curves, the virtual (apparent) value of vaporization and/or condensation is zero, but because the mass of the fluid in a depleted reservoir is changing as a result of decreasing pressure, the process could be assumed to be differential. One important statement has to be added: there is no qualitative difference between the reservoir fluid in either differential or flash process if pressure and temperature fall into the area outside of the saturation curve.

In many cases, it has become possible to treat two-phase pipeline problems with empirical numerical techniques that yield reasonably accurate pressure drops. More sophisticated models include the temperature effect on fluid properties as well. The multicomponent **or** compositional approach is designed for gas condensate and volatile oil systems.

4.3.1 Natural flow

The most important parameters that are used to evaluate performance or behavior of petroleum fluids flowing from an upstream point (in reservoir) to a downstream point (at surface) are pressure and flowrate. According to basic fluid flow through a reservoir, the production rate is a function of flowing pressure at the bottomhole of the well for a specified reservoir pressure and the fluid and reservoir properties. The flowing bottomhole pressure required to lift the fluids up to the surface may be influenced by size of the tubing string, choke installed at downhole or surface, and pressure loss along the pipeline.

In gas fields (and oil fields) the flowing systems may be divided into at least four components, as follows: (i) the reservoir, (ii) the wellbore, (iii) the chokes and valves, and (iv) the surface flowline. Each individual component, through which reservoir fluids flow, has its own performance and, of course, affects each other. The combined performances are often used as a tool for optimizing well production and sizing equipment.

Furthermore, engineering and economic judgments can depend on good information on the well and reasonable prediction of the future performances. As has been discussed in previous sections, hydrocarbon fluids produced can be either single phase (oil or gas) or two phases. Natural flow performance of oil, gas, and the mixture will therefore be discussed separately. Some illustrative examples are given at the end of each subsection.

Inflow performance represents behavior of a reservoir in producing the oil through the well. For a heterogeneous reservoir, the inflow performance might differ from one well to another. The performance is commonly defined in terms of a plot of surface production rate (stb/d) versus flowing bottomhole pressure (P, in psi) on Cartesian coordinate. This plot is defined as inflow performance relationship curve and is useful in estimating well capacity, designing tubing string, and scheduling an artificial lift method.

As for oil wells, performance curves characterizing a gas production system are useful tools used to visualize and graphically predict the effects of declining reservoir pressure, changes in tubular size, increasing water production, or installing gas compressors.

4.3.2 Artificial lift methods

The walking beam sucker rod system for producing fluids from wells is of ancient origin, and widely used. Over 90% of artificially lifted wells use beam-type pumps. The system consists of surface and downhole components.

As the prime mover drives the gear reducer, the walking beam oscillates around the saddle bearing and imparts a reciprocating motion to the sucker rods. As the sucker rods start the upstroke, they lift the plunger and traveling valve, creating a reduced pressure below it inside the working barrel. During this part of the cycle, the traveling valve is closed and the standing valve is open. The reduced pressure within the working barrel allows fluid to flow from the formation and the casing-tubing annulus through the mud anchor, dip tube, and standing valve and into the working barrel. On the downstroke, the traveling valve opens and the standing valve closes, allowing fluid in the barrel to move up into the tubing above the traveling valve. The fluid is in the annulus between the rods and the inside of the tubing. With each complete stroke, more fluid fills this annulus, and within a short time production appears at the surface. Because of compression of the fluid and elasticity in the system, production at the surface may appear continuous, but in reality production only occurs on the upstroke of the pump.

Various types of beam pumping units are recognized. The classification is based on where the fulcrum is placed (Class I or Class III) and how they are counterbalanced (air, crank, or beam).

In a Class I lever system (also called a conventional pumping unit), the fulcrum is near the center of the walking beam and the pitman applies lifting force by pulling downward at the rear of the walking beam. In a Class II lever system, the lifting force of the pitman is applied upward near the center of the beam. Class II units are also referred to as units with front mounted geometry. A variation on Class III system that normally has a crank-type counterbalance has, instead, a piston and cylinder filled with compressed air as the counterbalance.

In selecting a pumping unit, three principal specifications should be given: (i) torque rating of the pumping unit gear reducer in in-lb; (ii) structural capacity of the unit, that is, the load-bearing capacity, in-lb; and (iii) maximum stroke length in in.

4.4 Stimulation and remedial operations

Reservoir stimulation deals with well productivity, and a successful stimulation treatment requires the accurate identification of the parameters controlling the well production. As a result, causes of impaired production must be identified. Furthermore, it must be determined whether or not a particular treatment can improve production. This is the very first step of the stimulation job design.

A low value of the permeability (tight reservoir) or a high value of the skin effect (damaged or badly completed well) would result in low well productivity. There is virtually nothing practical that can be done to the permeability, although certain investigators have erroneously suggested that hydraulic fracturing increases the effective reservoir permeability. A hydraulic fracture is a superimposed structure on a reservoir which remains largely undisturbed outside of the fracture. The fracture, however, can greatly improve the well productivity by creating a large contact surface between the well and the reservoir. The production improvement results from effectively increasing the wellbore radius.

Matrix stimulation is generally intended to reduce a large skin effect resulting from permeability damage around the wellbore during completion or production. Often, there is confusion in distinguishing matrix acidizing (a form of matrix stimulation) from acid fracturing. The latter requires that the treatment is done at formation fracturing pressure and it relies on a residual etched width of the created fracture. The two methods of stimulation are applicable to entirely different types of formations: matrix acidizing is applied to high-permeability reservoirs whereas acid fracturing is appropriate for low-permeability, acid-soluble, reservoirs such as carbonates.

Hydraulic fracturing of natural gas (and crude oil) reservoirs is a reasonably new activity (Speight, 2016). The understanding of fracture propagation, its geometry, and direction is even newer, and additions to the body of knowledge of fracturing as a reservoir stimulation treatment are an active process. The expected functions of the fracturing fluid are to initiate and propagate the fracture and to transport the proppant with minimum leak-off and minimum treating pressure. Fluid viscosity is thus critical. An ideal fracturing fluid should have low viscosity in the tubing (to avoid unnecessary friction pressure losses), and high viscosity within the fracture where a large value can provide bigger fracture width and transport the proppant more efficiently. However, what a high viscosity fracturing fluid does, inadvertently, is to plug the formation creating a highly unfavorable mobility. A mechanism to reduce the viscosity after the job to a low value is then necessary.

A typical, water-base, fracturing fluid consists of water and a thickening polymer. The polymer concentrations could vary from 20 to 80 lb/1,000 gallons depending on the required viscosity. Such polymer solutions produce viscous fluids at ambient temperatures. At reservoir conditions, these solutions thin substantially. Hence, organometallic crosslinkers (some of them "delayed" to reduce tubing friction pressure) have been used resulting in substantial viscosity increases. For low to moderate temperature applications (120 °C, 250 °F) borates have been found effective. For higher temperatures (up to 175 °C, 350 °F) titanate and zirconate-delayed-crosslinked complexes have produced excellent fracturing fluids.

In addition to water-base fluids, there are other types of fracturing fluids. Oil-base fluids were the first fluids to be used. They can be thickened via an "associative" mechanism using an aluminum-phosphate ester polymer chain. However, oil-base fluids are expensive to use and dangerous to handle. Hence, they are applied to formations that are particularly water sensitive. Multiphase fluids such as emulsions (oil and water) and foams (gas and water) have been used widely.

The most common emulsion is composed of 67% hydrocarbon (internal) phase and 33% water (external) viscosified phase. Emulsions are very viscous fluids, providing good proppant transport, but result in high friction pressures and high fluid costs. They also thin significantly and thus they cannot be used in hot wells. Foams (gas, water, and a surfactant to stabilize the mixture) are particularly popular as fracturing fluids because the contained gas provides a very rapid "cleanup" following the treatment. However, fracture face damage and speed of cleanup must be balanced with the expected fracture performance which is affected to a much greater degree by the quality of the proppant pack.

For all fracturing fluids, the injected polymer chains need to be broken after the treatment. Breakers, such as oxidative compounds (e.g., peroxydisulfate derivatives) or enzymes (such as hemicellulose) are used to reduce the length of the polymer chains and the molecular weight. The viscosity of the fluid is reduced and cleanup is accomplished. Unfortunately, as can be expected, not all chains are broken or cleaned up. Certain polymers are more resistant than others, leading to only partial decomposition and bridging of polymer residue within the proppant pack. This phenomenon, if uncontrolled, can lead to a substantial proppant pack permeability damage with devastating effects on the fractured well performance.

Encapsulated, effective, breakers are desirable because they become active when the fracturing treatment is over. Early action is detrimental since it degrades the needed viscosity. No action is particularly problematic because it may lead to permanent proppant-pack permeability impairment. Thus, appropriate amounts of encapsulated breakers are the most desirable method before fracturing applications.

During the execution of the fracture treatment, the imposed hydraulic pressure holds the fracture open. However, when the pumping stops, it is up to the injected particulates to hold the fracture propped. The propped fracture width and, thus, the amount of proppant required will be addressed in the design subsection. However, two other variables are important in the determination of the proppant pack permeability: the proppant strength and the grain size. For a given stress under which the proppant pack will be subjected, the maximum value of the fracture permeability can be estimated. Bauxite, a high-strength proppant, and an intermediate-strength proppant (ISP, a synthetic material) maintain a large portion of their permeability at high stresses. Sand, however, experiences more than an order of magnitude permeability reduction when the stress increases from 4,000 to 8,000 psi. This is important in the selection of a proppant because, while sands are less costly, they crumple readily and therefore higher-strength, but more costly, proppants are indicated at higher stresses.

The proppant size is also important aspect of the hydraulic fracturing process. Larger grain sizes result in larger fracture permeability, but the larger size grains are more susceptible to crumpling as stresses increase and the relative reduction in the pack permeability is much larger in the larger-size proppants.

Fracture permeability damage is caused by unbroken polymer residue which is by far the biggest culprit. Thus, while the proppant strength and size selection can be done using formation strength criteria, damage due to fracturing f hid residue must be controlled. Otherwise, additional damage factors, as high as 80–90%, can be experienced after the stress-induced permeability impairment is taken into consideration.

4.5 Surface production systems

Surface production systems for natural gas are assembled at or near to the wellhead which is the component at the surface of a gas (or oil) well that provides the structural and pressure-containing interface for the drilling and production equipment. The primary purpose of the wellhead system is to provide the suspension point and pressure seals for the casing strings that run from the bottom of the well sections to the surface pressure control equipment. It is at the wellhead (or near to the wellhead) that the natural gas is prepared to meet the specifications of the transportation system, particularly the specifications that allow the gas to be transported by pipeline. Major transportation pipelines usually impose restrictions on the makeup of the natural gas that is allowed into the pipeline.

These surface facilities comprise equipment and systems that are used to extract, process, and export natural gas in a safe, controlled, and efficient way which are located at or near to the wellhead.

In more general terms, the purpose of the surface production facility is to separate the well stream into three phases which are (i) gas, (ii) liquid, and (iii) water. The actual practice of processing natural gas to pipeline dry gas quality levels can be complex but typically involves processes to remove the various impurities: (i) oil and condensate removal, (ii) water removal, (iii) separation of natural gas liquids, (iv) hydrogen sulfide and carbon dioxide removal, both of which in the presence of water are extremely corrosive and can cause irreparable damage to a pipeline or storage vessel.

In addition, heaters and scrubbers are installed, usually at or near the wellhead. The scrubbers serve primarily to remove sand and other large-particle impurities. The heaters ensure that the temperature of the gas does not drop too low. With natural gas that contains even low quantities of water, natural gas hydrates (solid or semi-solid compounds, resembling ice like crystals; Chapter 1) have a tendency to form when temperatures drop. If the hydrates accumulate, they can impede the passage of natural gas through valves and gathering systems. To reduce the occurrence of hydrates, small natural gas-fired heating units are typically installed along the gathering pipe wherever it is likely that hydrates may form.

In the current context, wellhead equipment is a series of unit processes that have been installed to provide a degree of gas treatment to that the gas meets pipeline specifications. Examples of wellhead equipment are (i) the separating vessel, (ii) the flash tank, and (iii) an expansion vessel.

Separating vessels are pressure vessels used for separating well fluids produced from oil and gas wells into gaseous and liquid components. The scrubber vessel (or knockout vessel) is designed to handle streams with high gas-liquid ratios. The liquid is generally entrained as mist in the gas or is free, flowing along the pipe wall. These vessels usually have a small liquid collection section. The terms "scrubber" and "knockout" are often used interchangeably. The separator vessel is used to separate a mixed-phase stream into gas and liquid phases that are "relatively" free of each other. Other terms used are "scrubbers," "knockouts," "line drips," and "decanters." A three-phase separator vessel is used to separate gas and two immiscible liquids of different densities (such as gas, water, and oil).

The liquid–liquid separator is a vessel in which two immiscible liquid phases can be separated using the same principles as for gas and liquid separators. Liquid–liquid separators are fundamentally the same as gas-liquid separators except that they must be designed for much lower velocities. Because the difference in density between two liquids is less than between gas and liquid, separation is more difficult. The filter separator vessel is a filter separator that usually has two compartments. The first compartment contains filter-coalescing elements. As the gas flows through the elements, the liquid particles coalesce into larger droplets, and when the droplets reach sufficient size, the gas flow causes them to flow out of the filter elements into the center core. The particles are then carried into the second compartment of the vessel (containing a vane-type or knitted wire mesh mist extractor) where the larger droplets are removed. A lower barrel or boot may be used for surge or storage of the removed liquid.

The line drip vessel is typically used in a pipeline with high gas/liquid ratios to remove only free liquid from a gas stream, and not necessarily all the liquid from a gas stream. Line drips provide a place for free liquids to separate and accumulate. This type of vessel is typically installed in the lower points of flow lines and must be blown periodically to remove liquids. The slug catcher vessel is a particular separator design able to absorb sustained inflow of large liquid volumes at irregular intervals. Thus type of vessel is usually found on gas-gathering systems or other two-phase pipeline systems. A slug catcher may be a single large vessel or a manifolded system of pipes.

The flash tank (chamber, trap or vessel) is a conventional gas and oil separator operated at low pressure, with the liquid from a higher-pressure separator being flashed into it. The flash chamber is quite often the second or third stage of separation, with the liquid being discharged from the flash chamber to storage.

The expansion vessel is a first-stage separator on a low-temperature or cold-separation unit. The vessel may be equipped with a heating core to melt hydrates or a hydrate-preventive liquid (such as glycol) may be injected into the well fluid just before expansion into this vessel.

4.6 Production engineering

Production is the process of extracting the hydrocarbon derivatives followed by processing (at the wellhead or in a gas processing facility) and separating the mixture of liquid hydrocarbon derivatives, gas, water, and solids, removing the constituents that are non-saleable, and selling the liquid hydrocarbon derivatives and gas.

Production sites often handle crude oil from more than one well and, after any necessary wellhead treatment, is sent to a refinery for processing whereas natural gas may receive more assiduous processing to remove impurities either in the field (at the wellhead) or at a nearby natural gas processing plant.

During the many years when natural gas was produced as a by-product of crude oil productions its value was largely ignored. In many cases large volumes of natural gas were flared with no effort being made to conserve this valuable material. Only a slight improvement was realized when many of the first sales contracts were written for ridiculously low prices. Until recently, large volumes of natural gas were sold for less than the cost of compression to the sales pressures. Such contracts completely ignored the intrinsic value of this high-quality fuel. This poor situation was made worse when these distress prices were recognized by a federal regulatory body as proper price levels for natural gas moving in interstate commerce.

In fact, natural gas reservoirs, like crude oil reservoirs, exist in many forms such as the dome (syncline–anticline) structure, with water below, or a dome of gas with a crude a crude oil interface and water below the crude oil. When the water is in direct contact with the gas, pressure effects may dictate that a considerable portion of the gas (20% v/v or more) is dissolved in the crude oil as well as in the water. As the gas is produced the reservoir pressure will decline allowing the gas to enter the gas phase. In addition, and because of the variability of reservoir structure, gas does not always flow equally throughout the length, breadth, and depth of the reservoir and at equal pressure, recovery wells must be distributed throughout the reservoir to recovers as much of the gas as possible.

As the gas pressure in the reservoir declines, the reservoir energy declines and the natural gas requires stimulation for continued production. Furthermore, reduction in the gas pressure may allow compaction of the reservoir rock by the weight of rock above, eventually resulting in subsidence of the surface above the reservoir. This can be gradual process or a sudden catastrophic process depending on the structures of the above geological features.

A reservoir containing *wet* gas with a large amount of valuable natural gas liquids (ethane, propane, and higher molecular weight hydrocarbon derivatives) and even light crude oil and condensate has to be treated carefully. When the reservoir pressure drops below the critical point for the mixture, the liquids may condense out and remain in the reservoir. Thus, it is necessary to implement a *cycling* process in which the wet gas is produced to the surface and the natural gas liquids are condensed as a separate stream and the gas is compressed and injected back into the reservoir to maintain the pressure.

The production of natural gas from a reservoir is inherently more efficient than the production of crude oil from a reservoir. In volumetric reservoirs, at moderate depths with high formation permeabilities, recoveries of 90% v/v of the gas originally in place are common. However, in many cases, high recoveries are not attained. In a reservoir with a low-to-no recoveries, the producing rates may fall below the economic limit while large portions of the original gas are still in the reservoir. The minimum economic gas production rate is raised by the production of liquids or by any other difficulties which tend to increase operating problems and expenses. In reservoirs with strong water drives and unequal water invasion, large amounts of gas may be lost in the residual gas saturation or by premature watering out of the producing wells. In such cases, the recovery of the gas originally in place may be as low as 50% of the volume in place.

Thus, the suitability of a natural gas reservoir for the production of gas is impacted by evaluation and implementation of sound reservoir management and investment decisions during different stages of field life (delineation, primary depletion, waterflooding, infill primary depletion, waterflooding, infill drilling, and flooding). Reservoir characterization has a big impact on field development decisions and process and facilities design. The two most challenging issues in reservoir characterization are: (i) estimation and description of key reservoir properties (porosity, permeability, transmissivity, geometry, saturations and permeability, transmissivity, geometry, saturations and pressures, and rock–fluid interactions) at locations not pressures, and rock–fluid interactions at locations not sampled by well data, and (ii) representation of these properties as effective properties for reservoir simulation properties as effective properties for reservoir simulation and prediction of reservoir performance.

Nevertheless, the necessary properties are evident when the reservoir is first prepared for gas production the through following categories: (i) gas wells, (ii) well completion, (iii) the wellhead, which must be suitable for preliminary treatment of the gas when necessary, (iv) well treatment, and (v) natural gas production. Each of these necessary issues is presented in the following sections.

Natural gas is recovered (extracted) from the reservoir through a well (a drilled structure that is not a natural occurrence). However, extraction of natural gas through a well leads to decrease in pressure in the reservoir which, in turn, will lead to decreased rates of gas production from the reservoir.

Once a potential natural gas reservoir has been located, the decision of whether or not to drill a well depends on a variety of factors, not the least of which is the economic characteristics of the gas reservoir. After the decision to drill has been made, the exact placement of the drill site depends on a variety of factors, including the nature of the potential formation to be drilled, the characteristics of the subsurface geology, and the depth and size of the target deposit. During this time, it is also necessary for the drilling company to ensure that they complete all the necessary steps to ensure that they can legally drill in that area. This usually involves securing permits for the drilling operations, establishment of a legal arrangement to allow the natural gas company to extract and sell the resources under a given area of land, and a design for gathering lines that will connect the well to the pipeline.

Gas wells are technically similar to crude oil wells with casing, tubing, and a wellhead and controls at the top (Rojey et al., 1997; Arnold and Stewart, 1999; Speight,

2019). Conventional wells use casings *telescoped* inside each other that is given pressure resistance by cement. However, with *expanded tubulars* each tubular casing is expanded against the previous one by pumping a tool called a *mandrel* down the casing. Thus, the well can either be thinner and cheaper for a given size of final tubing, or, alternatively, using the conventional 20-in diameter external casing, there is space to install wider and higher-capacity tubing.

When gas is contained in deep low-permeability (*tight*) reservoirs, it is often economic to drill large-diameter wells with horizontal sections through the reservoirs to collect more gas. In addition, gas production can be increased by fracturing the reservoir rock using over-pressure and by pumping in ceramic beads that maintaining the integrity of the fractures. It is also attractive to install *smart wells* that have measurement gauges permanently installed downhole. This eliminates the need for lowering gauges down the wells during production. With the high gas flow rates, sand can often be pulled into the well with the gas and therefore wearing parts are hardened against sand erosion.

If the natural gas is to be recovered from a crude oil reservoir (as is often the case), the production methods differ somewhat because of the presence of crude oil hydrocarbons in the natural gas. Thus, once a crude oil reservoir is discovered and assessed, production engineers begin the task of maximizing the amount of oil or gas that can ultimately be recovered from it. However, before a well can produce crude oil or gas the borehole must also be stabilized with casing that is cemented in place. The casing also serves to protect any freshwater intervals that the well passes through, so that oil cannot contaminate the water. A small-diameter tubing string is centered in the wellbore and held in place with packers. This tubing will carry the crude oil and natural from the reservoir to the surface.

Reservoirs are typically at elevated pressure because of underground forces. To equalize the pressure and avoid the wasteful *gushers* of the early 1900s (and the Hollywood movies), a series of valves and equipment is installed on top of the well. This wellhead (also called a *Christmas tree*) regulates the flow of hydrocarbons out of the well.

Early in its production life, the underground pressure (often referred to as the *reservoir energy*) will push the crude oil and natural gas up the wellbore to the surface and depending on reservoir conditions, this *natural flow* may continue for many years. When the pressure differential is insufficient for the crude oil and natural gas to flow naturally, mechanical pumps must be used to bring the products to the surface (*artificial lift*).

Most wells produce crude oil and natural gas in a predictable pattern (*decline curve*) in production increases for a short period, then peaks, and follow a long, slow decline. The shape of this decline curve, how high the production peaks, and the length of the decline are all driven by reservoir conditions. There are two steps that can be taken to influence the decline curve of a well: (i) perform a periodic *work over* which cleans out the wellbore to help the crude oil or natural gas move more easily to the surface, or (ii) fracture or treat the reservoir rock with acid around the bottom of the wellbore to create better pathways for the crude oil and natural gas to move through the subsurface to the producing well.

Throughout their productive life, most crude oil wells produce crude oil, natural gas, and water. This mixture is separated at the surface. However, although natural gas wells usually do not produce crude oil, they do produce varying amounts of liquid hydrocarbons (*natural gas liquids*) that are removed in the field or at a gas processing plant (which may remove other impurities as well). Natural gas liquids often have significant value as petrochemical feedstocks. Natural gas wells also often produce water, but the volumes are much lower than is typical for oil wells. Once it is produced, onshore natural gas is usually transported through a pipeline, although alternate methods may be more appropriate depending on the location of the well and the destination of the gas.

Just as many crude oil reservoirs require enhancement to continue oil production, gas reservoirs also require application of enhanced production methods to maximize the recovery of the gas. Thus, extracting natural gas from deposits deep underground is not merely a matter of drilling and completing a well. Any number of factors in the underground environment – such as the porosity of the reservoir rock – can impede the free flow of product into the well (Sim et al., 2008; Godec et al., 2014). In the past, it was common to recover as little a minority of the gas in the reservoir.

Enhanced gas recovery (EGR) is used to describe the recovery of unconventional, deep, or otherwise difficult-to-recover natural gas (Guo et al., 2014). Unconventional natural gas is the gas contained in geologic formations which have not been exploited traditionally by the oil and gas industry and include shale formation, tight gas sands, formations with active aquifers, and coal bed methane (Chapter 3). A number of these geologic formations are known to contain sizable quantities of natural gas. However, EGR is a recent consideration because of the large storage capacity of depleted gas pools. Considering that any additional gas recovered as a result of carbon dioxide injection might offset the cost of the storage of carbon dioxide, a number of experimental and simulation studies have been undertaken to examine displacement efficiency of reservoir gas by gas injection (Mamora and Seo, 2002; Seo and Mamora, 2003; Pooladi-Darvish et al., 2008).

In order to enhance the recovery of natural gas from a gas reservoir, nitrogen is injected into the reservoirs to artificially increase pressure which results in more yield from the well (often referred to as EGR). This process can also be accomplished with other inert gases such as carbon dioxide (CO_2) which sourced from underground wells is an economical alternative.

However, nitrogen can be recovered virtually anywhere from atmospheric air – via an air separation plant. Physically, there are also differences between the two gases: (i) nitrogen requires less compression that carbon dioxide and (ii) greater amounts of carbon dioxide are needed to create high pressure in the reservoirs.

Thus, the EGR process using compressed nitrogen is more energy-efficient than using carbon dioxide. Another option to maintain the pressure in the reservoir involves injection of part of the recovered natural gas.

The principles of EGR can also be applied to the recovery of coalbed methane which, in the context of this book, is an unconventional source of methane gas (Chapter 3). Carbon dioxide has a greater adsorption affinity onto coal than methane and if carbon dioxide is pumped into a coal seam toward the end of a coalbed methane production project, it displaces any remaining methane at the adsorption sites (allowing methane recovery jointly with carbon dioxide storage). However, the low permeability of coal seams means that a large number of wells may be needed to inject sufficient amounts of carbon dioxide to make methane recovery economical. Moreover, the methane in coal represents only a small proportion of the energy value of the coal, and the remaining coal could not be mined or gasified underground without releasing the carbon dioxide to the atmosphere. Also, methane is a far more potent greenhouse gas than carbon dioxide and the necessary precautions would have to be in place to ensure no methane leakage to the atmosphere occurred.

Production is the process of extracting the hydrocarbon derivatives followed by processing (at the wellhead or in a gas processing facility) and separating the mixture of liquid hydrocarbon derivatives, gas, water, and solids, removing the constituents that are non-saleable, and selling the liquid hydrocarbon derivatives and gas. Production sites often handle crude oil from more than one well and, after any necessary wellhead treatment, is sent to a refinery for processing whereas natural gas may receive more assiduous processing to remove impurities either in the field (at the wellhead) or at a nearby natural gas processing plant.

Offshore drilling operations used to be some of the most risky and dangerous undertakings but improved offshore drilling rigs, dynamic positioning devices, and modern navigation systems allow efficient offshore drilling in waters more than 10,000 feet deep. For offshore drilling, the wells are deep beneath the surface of the ocean, and artificial platforms are constructed on the surface. The first offshore rig was built and used in 1869, but it was not until 1974 that drilling was done far out in the ocean in deep water – namely, in the Gulf of Mexico. The original rigs were designed to work solely in very shallow water, but the modern rigs have a similar four-legged design to the earliest models but are able to drill in very deep water. In either case (i.e., onshore or offshore), once the gas reservoir is found and penetrated, gas flows up through the well to the surface of the ground and into large pipelines. Some of the gases that are produced along with methane, such as ethane, propane, and butane (*gas liquids*) are separated and cleaned at a gas processing plant. The gas liquids, once removed, are employed individually or collectively for various uses.

Deep sea rigs have specific components that allow them to function efficiently and the two most important features are (i) the subsea drilling template and (ii) the blowout preventer. The subsea drilling template connects the drilling site to the platform at the surface of the water, and the blowout preventer is in place to prevent oil or gas from leaking into the water. Also, modern offshore (deep-sea) rigs fall into two main types of deep sea rigs: (i) movable rigs and (ii) unmovable rigs. As the name indicates, moveable rigs can be move from location to location and drill in multiple places, while unmovable rigs remain in one place only. There are also various other types of rigs, including drill ships and drilling barges (Speight, 2015).

4.6.1 Gas wells

Well development is an essential part of the production process and is instituted immediately after exploration has located a reservoir (or field) that can economically produce natural gas. This involves the construction of one or more wells from the beginning (often referred to as spudding) to either abandonment if no hydrocarbon derivatives are found, or to well completion if hydrocarbon derivatives are found in sufficient quantities. The well development program depends upon the quality of the natural gas and the reservoir pressure.

Gas wells are technically similar to crude oil wells with casing, tubing, and a wellhead and controls at the top (Rojey et al., 1997; Arnold and Stewart, 1999; Speight, 2019). Conventional wells use casings *telescoped* inside each other and given pressure resistance by cement. However, with *expanded tubulars* each tubular casing is expanded against the previous one by pumping a tool called a *mandrel* down the casing. Thus, the well can either be thinner and cheaper for a given size of final tubing, or alternatively, using the conventional 20-in diameter external casing, there is space to install a wider and thus higher capacity tubing.

When gas is contained in deep, relatively tight, low-permeability reservoirs, it is often economic to drill large-diameter wells with horizontal sections through the reservoirs to collect more gas. In addition, gas production can be increased by fracturing the reservoir rock using over-pressure and by pumping in ceramic beads that maintaining the integrity of the fractures. It is also attractive to install *smart wells* that have measurement gauges permanently installed downhole. This eliminates the need for lowering gauges down the wells during production. With the high gas flow rates, sand can often be pulled into the well with the gas and therefore wearing parts are hardened against sand erosion.

4.6.2 Well completion

Once a natural gas or oil well is drilled, and it has been verified that commercially viable quantities of natural gas are present for extraction, the well must be completed to allow for the flow of crude oil or natural gas out of the formation and up to the surface. This process includes strengthening the well hole with casing, evaluating the pressure and temperature of the formation, and then installing the proper equipment to ensure an efficient flow of natural gas out of the well.

Since oil is commonly associated with natural gas deposits, a certain amount of natural gas may be obtained from wells that were drilled primarily for oil production. In some cases, this associated natural gas is used to help in the production of oil, by providing pressure in the formation for the oils extraction. The associated natural gas may also exist in large enough quantities to allow its extraction along with the oil.

Condensate wells are wells that contain natural gas, as well as a liquid condensate that is predominantly hydrocarbon derivatives (Chapters 1 and 2). Thus, the gas condensate is actually a mixture of hydrocarbon derivatives that are liquid at ambient temperature and pressure mixture that is often separated from the natural gas either at the wellhead, or during the processing of the natural gas. Depending on the type of well that is being drilled, completion may differ slightly. It is important to remember that natural gas, being lighter than air and given the appropriate conditions, will naturally rise to the surface of a well. Because of this, in many natural gas and condensate wells, lifting equipment and well treatment are not necessary.

Recovery, as applied in the natural gas industry and the crude oil industry and the natural gas industry, is the production of crude oil and/or natural gas from a reservoir. Once the well has been drilled to the designated depth and the gas reservoir is evaluated to be economically attractive, the well is then ready to be completed. The completion is an important aspect of the recovery program as it is the channel to connect the wellbore and the reservoir. It is a multi-disciplinary exercise that requires the completion, drilling, reservoir, and production engineers and rock mechanics specialists to work together to make it successful.

There are several methods by which this can be achieved that range from recovery due to reservoir energy (i.e., the oil flows from the well hole without assistance) to enhanced recovery methods in which considerable energy must be added to the reservoir to produce the oil. However, the effect of the method on the oil and on the reservoir must be considered before application.

Surface casing is the next type of casing to be installed and can be anywhere from a few hundred to 2,000 feet long and is smaller in diameter than the conductor casing. When installed, the surface casing fits inside the top of the conductor casing. The primary purpose of surface casing is to protect freshwater deposits near the surface of the well from being contaminated by leaking hydrocarbon derivatives or salt water from deeper underground. It also serves as a conduit for drilling mud returning to the surface and helps protect the drill hole from being damaged during drilling. Surface casing, like conductor casing, is also cemented into place. Regulations often dictate the thickness of the cement to be used, to ensure that there is little possibility of freshwater contamination.

Intermediate casing is usually the longest section of casing found in a well. The primary purpose of intermediate casing is to minimize the hazards that come along with subsurface formations that may affect the well. These include abnormal underground pressure zones, underground shale formations, and formations that might otherwise contaminated the well, such as underground salt-water deposits. In many instances, even though there may be no evidence of an unusual underground formation, intermediate casing is run as insurance against the possibility of such a formation affecting the well. These intermediate casing areas may also be cemented into place for added protection. Liner strings are sometimes used instead of intermediate casing and are commonly run from the bottom of another type of casing to the open well area. However, liner strings are usually just attached to the previous casing with hangers, instead of being cemented into place. This type of casing is thus less permanent than intermediate casing.

Production casing (alternatively called the oil string or long string) is installed last and is the deepest section of casing in a well. This casing that provides a conduit from the surface of the well to the crude oil producing formation. The size of the production casing depends on a number of considerations, including the lifting equipment to be used, the number of completions required, and the possibility of deepening the well at a later time. For example, if it is expected that the well will be deepened at a later date, the production casing must be wide enough to allow the passage of a drill bit when required.

Once the casing has been set, and in most cases cemented into place, proper lifting equipment is installed to bring the hydrocarbon derivatives from the formation to the surface. Once the casing is installed, tubing is inserted inside the casing, from the opening well at the top, to the formation at the bottom. The hydrocarbon derivatives that are extracted run up this tubing to the surface. This tubing may also be attached to pumping systems for more efficient extraction, should that be necessary.

Thus, a wellbore usually contains several casing strings such as (i) drive pipe, (ii) conductor pipe, (iii) surface casing, and (iv) production casing as well as intermediate casing and liner(s). Typically, the casing is cemented in place to either protect freshwater (the surface pipe) or to prevent loose shale, sand, and gravel (if gravel is used in the completion step) from entering the wellbore causing wellbore damage. Inside these casing strings, the production tubing, where the reservoir fluid will be produced from the reservoir, enter through the well completion, and move to the surface.

Between the production tubing and casing, annular fluid is filled in to prevent tubing burst due to the pressure inside of the tubing. Several completion types can be chosen. An open completion (sometimes referred to as a barefoot completion) consists of a packer and tubing above the interval of interest. Slotted liners or gravel packed wells with screens often in association with cemented, cased, and perforated wells is another family of completions. Finally, fully automated completions with measurement and control systems optimize well and reservoir performance and reservoir economics without human intervention (often referred to as an intelligent completion). Choice of the correct completion type is an important issue and usually depends on (i) the reservoir rock properties to determine if sand control is needed and (ii) well life expectancy. Also, consideration must be given to the potential for turbulent flow which is critical in gas well production.

Thus, the term "well completion" refers to the process of finishing a well so that it is ready to produce oil or natural gas. In essence, completion consists of deciding on the characteristics of the intake portion of the well in the targeted hydrocarbon formation. There are a number of types of completions, including: (i) open hole completion, (ii) conventional perforated completion, (iii) sand exclusion completion, (iv) permanent completion, (v) multiple zone completion, and (vi) drain hole completion. The use of any type of completion depends on the characteristics and location of the hydrocarbon formation to be mined.

Open hole completion is the most basic type and is only used in very competent formations, which are unlikely to cave in. An open hole completion consists of simply running the casing directly down into the formation, leaving the end of the piping open, without any other protective filter. Very often, this type of completion is used on formations that have been treated with hydraulic of acid fracturing.

Conventional perforated completion consists of production casing being run through the formation. The sides of this casing are perforated, with tiny holes along the sides facing the formation, which allows for the flow of hydrocarbons into the well hole, but still provides a suitable amount of support and protection for the well hole. The process of actually perforating the casing involves the use of specialized equipment designed to make tiny holes through the casing, cementing, and any other barrier between the formation and the open well. In the past, bullet perforators were used, which were essentially small guns lowered into the well. The guns, when fired from the surface, sent off small bullets that penetrated the casing and cement. Currently, jet perforating is preferred and consists of small, electrically ignited charges, lowered into the well that, when ignited, poke tiny holes through to the formation, in the same manner as bullet perforating.

Sand exclusion completion is designed for production in an area that contains a large amount of loose sand. These completions are designed to allow for the flow of natural gas and oil into the well, but at the same time prevent sand from entering the well. Sand inside the well hole can cause many complications, including erosion of casing and other equipment. The most common method of keeping sand out of the well hole are screening or filtering systems. This includes analyzing the sand experienced in the formation and installing a screen or filter to keep sand particles out. This filter may either be a type of screen hung inside the casing, or adding a layer of specially sized gravel outside the casing to filter out the sand. Both of these types of sand barriers can be used in open hole and perforated completions.

Once the well is completed, the flow of natural gas and oil into the well is commenced. For limestone reservoir rock, acid is pumped down the well and out the perforations. The acid dissolves channels in the limestone that lead oil into the well. For sandstone reservoir rock, a specially blended fluid containing *proppants* (sand, walnut shells, aluminum pellets) is pumped down the well and out the perforations. The pressure from this fluid makes small fractures in the sandstone that allow oil to flow into the well, while the proppants hold these fractures open. Once the oil is flowing, the oil rig is removed from the site and production equipment is set up to extract the oil from the well.

Tight formations are occasionally encountered and it becomes necessary to encourage flow. Several methods are used, one of which involves setting off small explosions to fracture the rock. If the formation is mainly limestone, hydrochloric acid is sent down the hole to dissolve channels in the rock. The acid is inhibited to protect the steel casing. In sandstone, the preferred method is hydraulic fracturing. A fluid with a viscosity high enough to hold coarse sand in suspension is pumped at high pressure into the formation, fracturing the rock. The grains of sand remain, helping to hold the cracks open.

Directional drilling is also used to reach formations and targets not directly below the penetration point or drilling from shore to locations under water. A controlled deviation may also be used from a selected depth in an existing hole to attain economy in drilling costs. Various types of tools are used in directional drilling along with instruments to help orient their position and measure the degree and direction of deviation; two such tools are the whipstock and the knuckle joint. The whipstock is a gradually tapered wedge with a chisel-shaped base that prevents rotation after it has been forced into the bottom of an open hole. As the bit moves down, it is deflected by the taper approximately 5 from the alignment of the existing hole.

4.6.3 Wellhead

The wellhead is the surface termination of a wellbore that incorporates facilities for installing casing hangers during the well construction phase. The wellhead also incorporates a means of hanging the production tubing and installing the Christmas tree and surface flow-control facilities in preparation for the production phase of the well. The primary purpose of a wellhead is to provide the suspension point and pressure seals for the casing strings that run from the bottom of the well to the surface pressure control equipment. While drilling the well, surface pressure control is provided by a blowout preventer (BOP).

The wellhead equipment is the component at the surface of an oil or gas well, which offers the pressure-containing and structural interface for the drilling equipment. The wellhead equipment includes pressure seals and suspension point for the casing strings and forms an integral structure of the well. Thus, the wellhead consists of the pieces of equipment mounted at the opening of the well to regulate and monitor the extraction of hydrocarbons from the underground formation. It also prevents leaking of oil or natural gas out of the well, and prevents blowouts due to high pressure formations. The subsea wellhead system – which is located on the ocean floor, and must be installed remotely with running tools and drill pipe – is a pressure-containing vessel that provides a means to hang off and seal off casing used in drilling the well.

Whether on land or under the sea, formations that are under high pressure typically require wellheads that can withstand a great deal of upward pressure from the escaping gases and liquids. These wellheads must be able to withstand pressures of up to 20,000 psi (pounds per square inch). The wellhead consists of three components: the casing head, the tubing head, and the Christmas tree.

The casing head consists of heavy fittings that provide a seal between the casing and the surface. The casing head also serves to support the entire length of casing that is run all the way down the well. This piece of equipment typically contains a gripping mechanism that ensures a tight seal between the head and the casing itself. The tubing head is much like the casing head and provides a seal between the tubing, which is run inside the casing, and the surface. Like the casing head, the tubing head is designed to support the entire length of the casing, as well as provide connections at the surface, which allow the flow of fluids out of the well to be controlled.

The Christmas tree (so named because of its many branches that make it appear somewhat like a Christmas tree) is the piece of equipment that fits atop the casing and tubing heads, and contains tubes and valves that serve to control the flow of hydrocarbons and other fluids out of the well. The *Christmas tree* is the most visible part of a producing well, and allows for the surface monitoring and regulation of the production of hydrocarbons from a producing well.

4.6.4 Well treatment

Well treatment is another method of ensuring the efficient flow of hydrocarbon derivatives out of a formation. Essentially, this type of well stimulation consists of injecting acid, water, or gases into the well to open up the formation and allow the crude oil to flow through the formation more easily. Acidizing a well consists of injecting acid (usually hydrochloric acid) into the well. In limestone or carbonate formations, the acid dissolves portions of the rock in the formation, opening up existing spaces to allow for the flow of crude oil. Fracturing consists of injecting a fluid into the well, the pressure of which "cracks" or opens up fractures already present in the formation. In addition to the fluid being injected, *propping agents* are also used and consist of sand, glass beads, epoxy, or silica sand, and serve to prop open the newly widened fissures in the gas-bearing or oil-bearing formation (Speight, 2016).

In addition to the fluid being injected, propping agents are also used and consist of sand, glass beads, epoxy, or silica sand, and serve to prop open the newly widened fissures in the formation. Hydraulic fracturing involves the injection of water into the formation, while carbon dioxide fracturing uses gaseous carbon dioxide. Fracturing, acidizing, and lifting equipment may all be used on the same well to increase permeability.

For example, carbon dioxide–sand fracturing involves using a mixture of sand and liquid carbon dioxide to fracture formations, creating and enlarging cracks through which oil and natural gas may flow more freely. The carbon dioxide then vaporizes, leaving only sand in the formation, holding the newly enlarged cracks open. Because there are no other substances used in this type of fracturing, there are no *leftovers* from the fracturing process that must be removed. This type of fracturing effectively opens the formation and allows for increased recovery of natural gas and (i) does not damage the deposit, (ii) generates no below ground wastes, and (iii) protects groundwater resources.

Because it is a low-density gas under pressure, the completion of natural gas wells usually requires little more than the installation of casing, tubing, and the wellhead. Unlike crude oil, natural gas is much easier to extract from an underground formation. However, as deeper and less conventional natural gas wells are drilled, it is becoming more common to use stimulation techniques on gas wells.

Hydraulic fracturing involves the injection of water into the formation, while carbon dioxide fracturing uses gaseous carbon dioxide. Fracturing, acidizing, and lifting equipment may all be used on the same well to increase permeability. For example, carbon dioxide–sand fracturing involves using a mixture of sand and liquid carbon dioxide to fracture formations, creating and enlarging cracks through which oil and natural gas may flow more freely. The carbon dioxide then vaporizes, leaving only sand in the formation, holding the newly enlarged cracks open. Because there are no other substances used in this type of fracturing, there are no leftovers from the fracturing process that must be removed. This type of fracturing effectively opens the formation and allows for increased recovery of natural gas and (i) does not damage the deposit, (ii) generates no below ground wastes, and (iii) protects groundwater resources.

Because it is a low-density gas under pressure, the completion of natural gas wells usually requires little more than the installation of casing, tubing, and the wellhead. Natural gas, unlike crude oil, is much easier to extract from an underground formation. However, as deeper and less conventional natural gas wells are drilled, it is becoming more common to use stimulation techniques on gas wells.

4.6.5 Gas production

Once the well is completed, production of natural gas can commence. In some instances, the natural gas that exists in pressurized formations will naturally rise up through the well to the surface. Once a conduit to the surface is opened, the pressurized gas will rise to the surface with little or no interference. Also, once the

Christmas tree is installed, the natural gas will flow to the surface because of the reservoir energy.

In order to understand the nature of the well more fully, a potential test is typically run in the early days of production. This test allows well engineers to determine the maximum amount of natural gas that the well can produce in a 24-h period. From this, and other knowledge of the formation, the engineer may make an estimation of the most efficient recovery rate (the rate at which the greatest amount of natural gas may be extracted without harming the formation itself).

Another important aspect of producing wells is the decline rate which is the rate of change of production with respect to time and, for exponential decline, is constant for all time; the mathematical equation defining exponential decline has two constants, the initial production rate and the decline rate. To calculate the decline rate, the current year production number is subtracted from the production number of the previous year and the result is multiplied by 100. Thus, simply, the decline rate is

$$\% \text{ change } = [(\text{current year production})/(\text{previous year production})] \times 100$$

If the number is negative, this result is the percentage that the production declined during the year whereas a positive number shows an increase in production. More complex calculations are available and the formula for such calculations is dependent upon the data required.

When a well is first drilled, the formation is under pressure (the natural energy of the reservoir) and produces natural gas at a high rate. As more natural gas is extracted from the formation, the production rate of the well decreases (the decline rate). Certain techniques, including lifting equipment and well stimulation, can increase the production rate of a well. In some natural gas wells, and oil wells that have associated natural gas, it is more difficult to ensure an efficient flow of hydrocarbons up the well. The underground formation may be very tight, making the movement of petroleum and natural gas through the formation and up the well a slow and inefficient process. In these cases, lifting equipment or well treatment is required.

Lifting equipment consists of a variety of specialized equipment used to help lift petroleum out of a formation. The most common lifting method is known as rod pumping in which the pumping mechanism is powered by a surface pump that moves a cable and rod up and down in the well, providing the lifting pressure required to bring the oil to the surface. The most common type of cable rod lifting equipment is the horse head (balanced conventional beam, sucker rod) pump. These pumps are recognizable by the distinctive shape of the cable feeding fixture, which resembles the head of a horse.

In order to process and transport associated dissolved natural gas, it must be separated from the oil in which it is dissolved. This separation of natural gas from crude oil is most often done using equipment installed at or near the wellhead. The actual process used for water-crude oil-natural gas separation starts at the wellhead and the equipment used can vary considerably. Although dry pipeline quality natural gas is virtually identical across different geographic areas, raw natural gas from different regions will vary in composition and therefore separation requirements may emphasize or de-emphasize any one of several optional separation processes. Typically, it is expected that a separator vessel will cause the water-oil-gas to separate into the three individual phases is the separator. The most basic type of separator is known as a conventional separator and consists of a simple closed tank, where the force of gravity serves to separate the heavier liquids like oil, and the lighter gases, like natural gas.

However, in many instances, natural gas is dissolved in oil underground primarily due to the formation pressure and may give rise to what is known as foamy oil. In some cases, when natural gas and crude oil is produced, it is possible (but subject to pressure and temperature effects in the reservoir) that the natural gas will readily separate from the crude oil. In these cases, separation of oil and gas is relatively easy, and the two hydrocarbon derivatives are sent separate ways for further processing. In many instances, however, specialized equipment is necessary to separate oil and natural gas. An example of this type of equipment is the low-temperature separator, and this type of separator is most often used for wells producing high-pressure natural gas as well as light crude oil or gas condensate. These separators use pressure differentials to cool the wet natural gas and separate the oil and condensate.

In the separation process, the natural gas enters the separator, being cooled slightly by a heat exchanger. The gas then travels through a high-pressure liquid knockout pot, which serves to remove any liquids into a low-temperature separator. The gas then flows into this low-temperature separator through a choke mechanism, which expands the gas as it enters the separator. This rapid expansion of the gas allows for the lowering of the temperature in the separator. After liquid removal, the dry gas then travels back through the heat exchanger and is warmed by the incoming wet gas. By varying the pressure of the gas in various sections of the separator, it is possible to vary the temperature, which causes the oil and some water to be condensed out of the wet gas stream. This basic pressure–temperature relationship can work in reverse as well, to extract gas from a liquid oil stream. In addition, the potential for corrosion of the equipment and pipelines due to the presence of water and acid gases (hydrogen sulfide, H_2S, and carbon dioxide, CO_2) is high (Speight, 2014b). Hence the need to know the composition of the natural gas mixture that is produced at the wellhead.

References

Arnold, K., and Stewart, M. 1999. Surface Production Operations, Volume 2: Design of Gas-Handling Systems and Facilities, 2nd Edition. Gulf Professional Publishing, Houston, TX.

Burruss, R.C., and Ryder, R.T. 2003. Composition of Crude Oil and Natural Gas Produced from 14 Wells in the Lower Silurian "Clinton" Sandstone and Medina Group, Northeastern Ohio and Northwestern Pennsylvania. Open-File Report 03-409, United States Geological Survey, Reston, Virginia.

Burruss, R.C., and Ryder, R.T. 2014. Composition of Natural Gas and Crude Oil Produced From 10 Wells in the Lower Silurian "Clinton" Sandstone, Trumbull County, Ohio. In: Coal and Crude oil Resources in The Appalachian Basin; Distribution, Geologic Framework, and Geochemical Character. L.F. Ruppert and R.T. Ryder (Editors). Professional Paper 1708, United States Geological Survey, Reston, Virginia.

Esteves, I.A.A.C., Sousa, G.M.R.P.L., Silva, R.J.S., Ribeiro, R.P.P.L., Eusébio, M.F.J., and Mota, J.P.B. 2016. A Sensitive Method Approach for Chromatographic Analysis of Gas Streams in Separation Processes Based on Columns Packed with an Adsorbent Material. Advances in Materials Science and Engineering Volume 2016, Article ID 3216267. Hindawi Publishing Corporation: http://dx.doi.org/10.1155/2016/3216267; accessed December 1, 2017)

Faramawy, S., Zaki, T., and Sakr, A.A-E. 2016. Natural Gas Origin, Composition, and Processing: A Review. Journal of Natural Gas Science and Engineering, 34: 34–54.

Godec, M., Koperna, G., Petrusak, R., and Oudino, A. 2014. Enhanced Gas Recovery and CO2 Storage in Gas Shales: A Summary Review of its Status and Potential. Energy Procedia, 63: 5849–5857.

Guo, P., Jing, S., and Peng, C. 2014. Technologies and Countermeasures for Gas Recovery Enhancement. Natural Gas Industry B1: 96–102.

Mamora, D.D., and Seo, J.G., 2002. Enhanced Gas Recovery by Carbon Dioxide Sequestration in Depleted Gas Reservoirs. SPE Paper No. 77347. Proceedings. SPE Annual Technical Conference and Exhibition, Houston, Texas. September 29-October 2. Society of Petroleum Engineers. Richardson, Texas.

Mokhatab, S., Poe, W.A., and Speight, J.G. 2006. Handbook of Natural Gas Transmission and Processing Elsevier, Amsterdam, Netherlands.

Pooladi-Darvish, M., Hong, H., Theys, S., Stocker, R., Bachu, S., and Dashtgard, S., 2008. CO2 injection for enhanced gas recovery and geological storage of CO2 in the Long Coulee Glauconite F Pool, Alberta. SPE Paper No. 115789, SPE Annual Technical Conference and Exhibition, Denver, Colorado. September 21–24. Society of Petroleum Engineers, Richardson, Texas.

Rojey, A., Jaffret, C., Cornot-Gandolph, S., Durand, B., Jullin, S., and Valais, M. 1997. Natural Gas Production, Processing, Transport. Editions Technip, Paris, France.

Seo, J.G., and Mamora, D.D., 2003. Enhanced Gas Recovery by Carbon Dioxide Sequestration in Depleted Gas Reservoirs. SPE Paper No. 81200. Proceedings. SPE/EPA/DOE Exploration and Production Environmental Conference, San Antonio, Texas. March 10–12.

Sim, S.S.K., Turta, A.T., Signal, A.K., and Hawkins, B.F., 2008. Enhanced Gas Recovery: Factors Affecting Gas-Gas Displacement Efficiency. CIMPC Paper No. 2008-145. Proceedings. Canadian International Petroleum Conference/SPE Gas Technology Symposium 2008 Joint Conference, Calgary, Alberta, Canada. June 17- 19.

Speight, J.G. 2013. The Chemistry and Technology of Coal 5th Edition. CRC Press, Taylor & Francis Group, Boca Raton, Florida.

Speight, J.G. 2014a. The Chemistry and Technology of Crude oil 5th Edition. CRC Press, Taylor & Francis Group, Boca Raton, Florida.

Speight, J.G. 2014b. Oil and Gas Corrosion Prevention. Gulf Professional Publishing, Elsevier, Oxford, United Kingdom.

Speight, J.G. 2015. Handbook of Offshore Oil and Gas Operations. Gulf Professional Publishing, Elsevier, Oxford, United Kingdom.

Speight, J.G. 2016. Handbook of Hydraulic Fracturing. John Wiley & Sons Inc., Hoboken, New Jersey.

Speight, J.G. 2017. Handbook of Crude oil Refining. CRC Press, Taylor & Francis Group, Boca Raton, Florida.

Speight, J.G. 2018. Handbook of Natural Gas Analysis. John Wiley & Sons Inc., Hoboken, New Jersey.

Speight, J.G. 2019. Natural Gas: A Basic Handbook 2nd Edition. Gulf Publishing Company, Elsevier, Cambridge, Massachusetts.

Speight, J.G. 2020. Global Climate Change Demystified. Scrivener Publishing, Beverly, Massachusetts.

Chapter 5
Field operations

5.1 Introduction

Most natural gas fields are at a considerable distance from the refineries that convert crude oil into usable products, and therefore the oil must be transported in pipelines and tankers. Typically, natural gas needs some form of treatment near the reservoir before it can be carried to considerable distances through the pipelines or in tankers as well as railroad cars and motor vehicles.

The field operations for the production natural gas are highly variable. In some cases, the gas is sufficiently high in methane content that it can be piped directly to customers without processing. Most often, however, the gas contains unacceptable levels of higher weight hydrocarbon derivatives as well as impurities, and it is available only at low pressures. For these reasons, field operations in a natural gas often need to involve multiple stages of compression to remove the higher molecular weight hydrocarbon derivatives and the impurities as well as a reduction in the temperature of the gas stream in order to conserve the power requirements of compressor stations along the transport pipeline.

One major issue is the prediction of delivery rates from a group of wells or field to a point of sales or transfer. The gas must arrive at this point on the main pipeline at a specified pressure. The elements in the overall gas production system must then (i) include flow through the reservoir, (ii) flow through the production strings of the wells, (iii) flow through the field gathering system and processing equipment, (iv) compression of the gas, and, finally, (v) flow through an auxiliary pipeline to the point of sales.

5.2 Reserves and reservoir performance

After a natural gas field is first discovered, in order to build a better picture of the structure, several appraisal wells must be drilled and the number of wells needed depends on the quality of the seismic evaluation.

Using the seismic map, the amount of gas in place must be estimated, and this necessitates measuring the total gross volume of rock that is gas-bearing. It is also necessary to measure the porosity of the formation (Chapter 2), which is that proportion of the rock containing pores filled with reservoir fluids, and the fraction of this that contains water (i.e., the water saturation since the remainder contains gas). The porosity can be measured by coring the logging in the wells that have been drilled. Since these represent only a small proportion of the total rock volume and since the data obtained from this small area are being applied to the total probable area of the field, there is a considerable room for inaccuracies.

https://doi.org/10.1515/9783110691023-005

The total amount of gas in place can be calculated by multiplying the gross volume of rock by the porosity and the gas saturation (which equals one minus the water saturation). The product of these three values gives the gas volume at reservoir conditions, that is, at reservoir pressure and temperature. This volume must be converted to standard conditions. This, however, does not represent the amount of gas that can be economically recovered. For this purpose, the recovery factor must be known, that is, that fraction of the gas in place that is recoverable under normal economic operating conditions. To assess the value of the recovery factor it is necessary to have an understanding of the producing performance of the reservoir.

There are several factors that affect gas recovery from a reservoir. For example, if the reservoir is a closed unit (not underlain by water) it is known as a depletion reservoir and as gas is produced, the pressure will drop, as indicated on the line marked depletion. Recovery of gas is possible from such a field up to a certain abandonment pressure which is the lowest pressure at which gas can still be produced from the wells at a rate sufficiently high to cover the operating costs. The point at which the line representing the abandonment pressure crosses the depletion line indicates the ultimate economic gas recovery, which in this case is between 80% and 90% of the gas in place, which is an average figure for a depletion-type reservoir. If the gas in the reservoir is underlain by water, as the gas pressure in the reservoir begins to drop, water will start to flow and enter the gas reservoir. This effect (also called water encroachment) will then maintain the reservoir pressure to a greater or lesser extent.

Permeability (Chapter 2) is a property of a reservoir formation that allows the flow of fluids through the formation. Because of formations that have a high permeability, the flow of gas is relatively easy and occurs at low-pressure drops, while through low-permeability formations even high-pressure drops will result in low flow rates. The same applies for water insofar as low-permeability formations decrease to potential for a strong water drive.

Second, the strength of the water drive depends on the size of the reservoir. The larger the reservoir the weaker the water drive. This is because the volume of water needed to maintain pressure depends on the area of the field which, like a circle, is proportional to the radius of the field squared. The circumference of such a field through which all water fluid must pass is, however, directly proportional to the radius. Consequently, the amount of water influx in a given period of time and for a given pressure drop is roughly proportional to the radius; but the amount of water required to maintain the reservoir pressure at the given level during this period of time, expressed as a fraction of the volume of the reserves, will be proportional to the inverse of the radius squared. Combining the two effects, the relative strength of the water drive is approximately proportional to the inverse of the radius and, consequently, for comparable conditions, is relatively weaker for larger sized fields.

Third, there is the time factor which involved the time that water was flowing in the reservoir. If a high rate of production is being maintained from the field, a high amount of influx is required during a short time period; consequently, the water drive may be weak. However, the same field with a low rate of production may have a strong water drive.

Summarizing, reservoir engineers must estimate the likely depletion pattern of the field; they must know how permeable the rock is; they must assess which other factors will be of importance, what the expected reservoir drive will be, and also the amount of gas that will be left behind if there is water encroachment. The recovery factor can then be estimated which, multiplied by the gas in place, will produce an estimate of the reserves of natural gas.

The problem of actually producing the gas in the formation in the most economical way is solved by determining a drilling and production schedule. Some issues that must be given consideration in determining a drilling and production schedule are (i) the number of wells that are needed, (ii) the timing of the drilling of the wells, and (iii) the amount of gas that should be produced from any one well. To assess this, several production tests must be carried out on the first discovery well. The results of these tests will be supplemented by further tests in the appraisal wells at a later date.

5.3 Deliverability

The gas stored in the reservoir must flow through the formation to the wellbore, this process being called the inflow performance of the gas well. It must then flow upward through the well tubing to the surface. During this phase of the production two factors are important: the friction loss experienced in the well tubing and the resultant pressure drop, and the amount of suspended water present. Even for a well producing hardly any water at all, an accumulation of water in the well tubing will build up in time, depending on the production rate. This will lead to an increased overall density of the flowing gas with a consequent higher hydrostatic pressure drop. This phenomenon, liquid holdup, is particularly important at low flow rates. Finally, after leaving the wellhead, the gas will have to be dehydrated and treated to pipeline quality before delivery. Under special circumstances, when reservoir pressures will have dropped to low values, compression of the gas may also be required before delivery into the pipeline.

The deliverability of a gas well does not depend only on the capacity of the reservoir to produce. The production must also pass through the tubing, separators, dehydrators, meter run, and flow line to the pipeline. Some pressure drop is associated with each one of these pieces of equipment, and the pressure drop is a function of the flow rate. Consequently, in many cases the production rate is limited by the capacity of the equipment rather than the capacity of the reservoir to produce. When such a situation arises it may be possible to install larger diameter equipment.

The ability of the reservoir to deliver a certain quantity of gas depends both on the inflow performance relationship and the flowing-bottom-hole pressure. The flowing bottom-hole pressure on its part depends on the separator pressure and the configuration of the piping system.

In order to determine the deliverability of the total well system, it is necessary to calculate all the parameters and pressure drops. Static and flowing pressures are a major concern and methods for determining pressure drops in tubing for single-phase gas flow and for multiphase gas–liquid flow. Inflow and outflow performance curves must be determined. The static and flowing pressure at the formation must be known in order to predict the productivity or absolute open flow potential of gas wells. The preferred method is to measure the pressure with a bottom-hole pressure gauge. It is often impractical or too expensive to measure static or flowing bottom-hole pressures with bottom-hole gauges. However, for many problems, a sufficiently pressure and temperature, formation temperature, and well depth is the remedy. The static pressure is equal to the weight of the column of gas. In the case of flowing wells, the gas column weight and friction effects must be evaluated and summed up. There are several ways to determine the pressure either static or flowing and these are (i) the static bottom-hole pressure (SBHP), (ii) flowing bottom-hole pressure, and (iii) the annular flow.

The estimation of SBHP from surface measurement only involves calculating the additive pressure exerted by the weight of the static fluid column. A well-known technique to calculate the pressure is (i) average temperature and deviation factor method: This is the simplest method most people use, and the only thing to consider is that pressure and temperature must be the average.

The flowing bottom-hole pressure of a gas well is the sum of the flowing wellhead pressure, the pressure exerted by the weight of the gas column, the kinetic energy change, and the energy losses resulting from friction. This leaves the general mechanical energy equating in a simple form, for the situation of no heat loss from gas to surroundings and no work performed by the system.

5.3.1 Annular flow

With respect to the annular flow, in most instances, gas wells are produced through tubing. Occasionally, however, a well may be produced through the casing-tubing annulus. The tubing flow equations may be used for annular flow, provided proper account is taken of the flow diameter variable.

For a particular set of equipment, pipeline pressure, and state of reservoir depletion there is some maximum rate that can be produced; this is represented by the intersection of the two capacity curves. At this point the reservoir flow results in a bottom-hole pressure that just matches the pressure drop needed for flow through the production equipment at this rate. At any other rate, the capacity of the well to produce is limited by either the reservoir or the equipment.

5.3.2 Predicting reservoir performance

To predict the production history of a reservoir properly, it is necessary to consider the capacity of the reservoir to produce, the capacity of the equipment, and the state of depletion of the reservoir as predicted by material balance. The producing system capacity curve must be considered along with material balance and deliverability curves to predict how a reservoir will perform under any given set of conditions. For example, it may be necessary to know the number of wells to be drilled in a particular reservoir to fulfill some stated flow rate contract from a gas reservoir for some stated period of time. From a rate standpoint, the critical time will be the time at the end of the contract when the reservoir pressure has declined to a minimum under this contract. At this particular time, it is necessary to know one must be certain that a sufficient number of wells has been drilled to provide the required producing rate. As the number of wells is increased, the required rate per well will be reduced. Also remember that as the number of wells increases, the production capacity of each well at any fixed state of depletion will be slightly increased because each well will be draining a lesser volume of the reservoir.

Consequently, this problem must be solved by trial and error. With the state of depletion fixed by the contract length and total reservoir rate, the average pressure or the pressure at the external drainage boundary of each well can be determined. Now, if a number of wells is assumed, one can use the basic deliverability curve and determine the deliverability curve for the resulting spacing.

With the number of wells assumed, the rate per well will be assumed since the contract fixes the total reservoir rate. Based on the per-well rate, the bottom-hole pressure necessary to supply that rate from the deliverability curve can be found. Then, this bottom-hole pressure can be used with the equipment capacity curve to determine if the equipment capacity can supply the required per-well rate at the subject bottom-hole pressure. If the equipment will not supply the rate, a greater number of wells is considered. The number of wells must be bracketed before the engineer can be certain that he has determined the most economical solution to this problem.

5.3.3 Optimum development patterns

The production schedule of a gas field should be such that the market can absorb the gas produced. This will normally lead to a restriction of the rate at which production can be built up. On the other hand, the rate of production buildup may be limited by drilling schedules and processing and transportation facilities. Economic considerations may also play a part in determining the production schedule of a gas field.

The production pattern consists of three parts: (i) a period of production buildup; (ii) a period of constant rate production; and (iii) a period of production decline. The procedure is as follows: (i) At any point in time, determine the total amount of gas produced since the beginning of production, (ii) determine the corresponding well production rate at a given tubing head pressure, and (iii) divide the field production rate by the well production rate to obtain the number of wells required at any point in time.

The field development pattern indicates a period of drilling during production buildup, followed by a period of production at constant rate without further drilling. To maintain total field output at the same level, additional wells are drilled while still producing at high tubing head pressure. To avoid drilling too many wells, field potential may be maintained by lowering tubing head pressures and installing gas compressors. The compressor phase will continue until the tubing head pressure falls below an efficient and economic compressor intake pressure. The field production rate will then begin to'decline.

Instead of maintaining the tubing head pressure at a high value and drilling additional wells to maintain constant field production rate, tubing head pressures are first lowered with subsequent installation of compressor. Then additional wells are drilled to maintain field potential until the start of the production decline period.

5.3.4 Well performance

A gas well is a flowing well producing predominantly gaseous hydrocarbons. The gas may contain subordinate amounts of liquid hydrocarbons and water. Condensate, that is, the hydrocarbons produced in gas form but liquid under surface conditions, is a colorless or pale liquid composed of low-molecular-weight hydrocarbons. Gas-oil ratio (*GOR*) is in the order of ten thousand at least. Gas wells may therefore be regarded also as oil wells with a high (sometimes infinite) GOR.

The first subject relates to a productivity analysis of gas wells; the compressibility of gas being much greater than that of most well fluids composed of oil and gas, the flow of gas in the reservoir is governed by relationships other than those in connection with the performance of crude oil wells. The often high pressure, high temperature and possible corrosivity of gas raise the need of completing wells with a view to these features.

The productivity index (PI) is an important indicator of the production capacity of a well. For conventional reservoirs, well productivity is usually calculated using the pressure response of the reservoir in its pseudosteady-state period. There are numerous studies for different well completion schemes which developed correlations for pseudosteady-state PI for specific cases, such as horizontal wells and fractured wells. Most of the developed models for complex well completion schemes use some approximations for calculation of the PI calculation and they have some limitations in use. Furthermore, as the industry goes toward producing lower-quality reservoirs like low- and ultralow-permeability reservoirs, the period of transient flow covers a larger part of the lifetime of the well and these pseudosteady-state productivity calculations become less applicable in prediction of the production behavior of the reservoir.

In order to determine the deliverability of the total well system, it is necessary to calculate all of the parameters and pressure drops. Static and flowing pressures are of major concern and methods for determining pressure drops in tubing for single-phase gas flow and for multiphase gas–liquid flow. Inflow and outflow performance curves must be determined. The static and flowing pressure at the formation must be known in order to predict the productivity or absolute open flow potential of gas wells. The preferred method is to measure the pressure with a bottom-hole pressure gauge. The static pressure is equal to the weight of the column of gas. In the case of flowing wells, the gas column weight and friction effects must be evaluated and summed up.

There are several methods by which the pressure (either static or flowing) can be determined and these are (i) the SBHP method, (ii) the average temperature and deviation factor method, (iii) the Sukkar and Cornell method, (iv) the Cullender and Smith method, (v) the flowing bottom-hole pressure, and (vi) the annular flow method.

In the SBHP method, the estimation of SBHP from surface measurement involves calculation of the additive pressure exerted by the weight of the static fluid column. The average temperature and deviation factor method – this method is the most convenient method to use and the only parameter to consider is that pressure and temperature must be the average. The Sukkar and Cornell method is also one of the easiest methods to use for estimating bottom-hole pressure and has the advantage of improved accuracy and the method also permits calculations of bottom-hole pressures without a trial-and-error procedure. The Cullender and Smith method uses a more definitive approach insofar as the gas deviation factor is a function of both temperature and pressure. The flowing bottom-hole pressure acknowledges that the flowing bottom-hole pressure of a gas well is the sum of (i) the flowing wellhead pressure, (ii) the pressure exerted by the weight of the gas column, (iii) the kinetic energy change, and (iv) the energy losses resulting from friction which leaves the general mechanical energy equation in a relatively simple form, for the situation of no heat loss from gas to surroundings and no work performed by the system. Finally, the annular flow method acknowledges that, in most instances, gas wells are produced through tubing but there are instances when a well may be produced through the casing-tubing annulus; thus, the tubing flow equations may be used for annular flow, provided proper account is taken of the variability of the flow diameter.

Overall, there is also the PI which is the ratio of the total liquid surface flowrate to the pressure drawdown at the midpoint of the producing interval. The units typically are in field units, STB/D/psi. Thus:

$$\text{Productivity index} = J = Q/(P_e - P_{wf})$$

In this equation, J is the PI, STB/day/psi, Q is the surface flowrate at standard conditions, STB/D, P_e is the external boundary radius pressure, psi, and P_{wf} is the well sand-face mid-perf pressure, ps.

5.3.5 Material balance

Material balance (and energy balance) (Chapter 3) is based on a conservation law which is stated generally in the form:

$$input + source = output + sink + accumulation.$$

The individual terms can be plural and can be rates as well as absolute quantities.

The material balance equation, when combined with reliable relative permeability data, can be used to predict future reservoir performance. Many times, reservoirs do not conform to the assumptions made in the material balance equation. Few reservoirs are homogeneous, and no reservoirs respond instantaneously to changes in pressure. The precision with which reserves can be calculated or predicted with the material balance equation is affected by the quality of data available and the degree of agreement between the assumptions made in the equation and the actual reservoir conditions.

If a field development program has been well planned and executed, enough information should be available to calculate reserves by the material balance equation. The material balance equation is derived on the assumption that the reservoir is a homogeneous vessel with uniform porosity, permeability, and fluid properties. The equation accounts for all quantities of materials that enter or leave the vessel. The simplest form of the equation is that initial volume is equal to the volume remaining plus the volume removed. As material is withdrawn from a constant-volume reservoir the pressure declines and remaining material expands to fill the reservoir.

Balances of particular entities are made around a hounded region called a system. Input and output quantities of an entity cross the boundaries. A source is an increase in the amount of the entity that occurs without a crossing of the boundary – for example, an increase in the sensible enthalpy or in the amount of a substance as a consequence of chemical reaction. Analogously, sinks are decreases without a boundary crossing, as the disappearance of water from a fluid stream by adsorption onto a solid phase within the boundary.

Accumulations are time rates of change of the amount of the entities within the boundary. For example, in the absence of sources and sinks, an accumulation occurs when the input and output rates are different. In the steady state, the accumulation is zero. Although the principle of balancing is simple, its application requires knowledge of the performance of all the kinds of equipment comprising the system and of the phase relations and physical properties of all mixtures that participate in the process. As a consequence of trying to cover a variety of equipment and processes, the books devoted to the subject of material and energy balances always run to several hundred pages. Throughout this book, material and energy balances are utilized in connection with the design of individual kinds of equipment and some processes.

Knowing the amount of fluid withdrawn from the reservoir and the drop in pressure one can calculate the corresponding volume of fluid at the original reservoir pressure. The calculated reservoir size should remain constant as fluid is withdrawn and pressure drops. If the calculated reservoir size changes constantly in one direction as the field is produced the assumed production mechanism is probably wrong. Calculations should be repeated assuming different mechanisms until one is found that yields a constant reservoir size.

After the development of the original material balance equation in 1936 it has been rearranged to solve almost any unknown. The most frequently used forms of the equation are for these types of recovery mechanisms: (i) oil reservoir with gas cap and active water drive, (ii) oil reservoir with gas cap and no active water drive, (iii) initially undersaturated oil reservoir with active water drive (a) above bubble point and (b) below bubble point, (iv) initially undersaturated oil reservoir with no active water drive (a) above bubble point and (b) below bubble point, (v) gas reservoir with active water drive, and (vi) gas reservoir with no active water drive.

The material balance equation, when combined with reliable relative permeability data, can be used to predict future reservoir performance. Many times, reservoirs do not conform to the assumptions made in the material balance equation. Few reservoirs are homogeneous and no reservoirs respond instantaneously to changes in pressure. The precision with which reserves can be calculated or predicted with the material balance equation is affected by the quality of data available and the degree of agreement between the assumptions made in the equation and the actual reservoir conditions.

5.4 Wellhead operations

A wellhead is the component at the surface of a crude oil or a natural gas well that provides the structural and pressure-containing interface for the drilling and production equipment. Wellheads are typically welded onto the first string of casing, which has been cemented in place during drilling operations, to form an integral structure of the well. In exploration wells that are later abandoned, the wellhead may be recovered for refurbishment and re-use. Offshore, where a wellhead is located on the production platform it is referred to as a *surface wellhead* and if the wellhead is located beneath the water it is referred to as a subsea wellhead or a mudline wellhead.

The primary purpose of a wellhead is to provide the suspension point and pressure seals for the casing strings that run from the bottom of the drill-hole sections to the surface pressure control equipment. While drilling the oil well, surface pressure control is provided by a blowout preventer.

Some accumulations furnish natural gas of high purity (almost methane, CH_4). Natural gases of this composition do not require any processing; they only require a dehydration treatment before being conveyed to the transportation pipeline. Hydrocarbon streams as produced from other reservoirs are complex mixtures of hundreds of different compounds. A typical well-stream is a high velocity, turbulent, constantly expanding mixture of gases and hydrocarbon liquids, intimately mixed with water vapor, free water, solids, and other contaminants. As it flows from the hot high-pressure petroleum reservoir, the well-stream undergoes continuous pressure and temperature reduction. Gases evolve from the liquids, water vapor condenses, and some of the gas stream from the well changes in character from liquid to bubbles, mist, and free gas. The high-velocity gas carries liquid droplets, and the liquid carries gas bubbles.

In most cases, the gas, liquid hydrocarbon derivatives, and free water should be separated as soon as possible after bringing them to the surface; these phases should be handled and transported separately. This separation of liquids from the gas phase is accomplished by passing the well-stream through an oil-gas or oil-gas-water separator.

Field processing simply removes undesirable components and separates the well-stream into salable gas and hydrocarbon liquids, recovering the maximum amounts of each at the lowest possible overall cost. Field processing of natural gas consists of four basic processes: (i) separation of the gas from free liquids such as crude oil, hydrocarbon condensate, water, and entrained solids, (ii) processing the gas to remove condensable and recoverable hydrocarbon vapors, (iii) processing the gas to remove condensable water vapor which, under certain conditions, might cause hydrate formation, and (iv) processing the gas to remove other undesirable compounds, such as hydrogen sulfide or carbon dioxide.

Natural gas, as it is used by consumers (which is composed almost entirely of methane) is much different from the natural gas that is brought from underground up to the wellhead. Although the processing of natural gas is in many respects less complicated than the processing and refining of crude oil, it is equally as necessary before its use by end users. However, natural gas found at the wellhead, although still composed primarily of methane, is by no means as pure. A wellhead natural gas processing plant is a facility designed to clean raw natural gas by separating impurities and various non-methane hydrocarbons and fluids to produce what is known as pipeline quality dry natural gas. A gas processing plant is also used to recover natural gas liquids (condensate, natural gasoline and liquefied petroleum gas) and sometimes other substances such as sulfur.

Raw natural gas comes from three types of wells: oil wells, gas wells, and condensate wells. Natural gas that comes from oil wells is typically termed "associated gas." This gas can exist separate from oil in the formation (free gas), or dissolved in the crude oil (dissolved gas). Natural gas from gas and condensate wells, in which there is little or no crude oil, is termed "nonassociated gas." Gas wells typically produce raw natural gas by itself, while condensate wells produce free natural gas along with a semi-liquid hydrocarbon condensate. Whatever the source of the natural gas, once separated from crude oil (if present) it commonly exists in mixtures with other hydrocarbon derivatives – principally ethane, propane, butane, and pentanes.

In addition, raw natural gas contains impurities (other than the low molecular weight hydrocarbon derivative enumerated above) which are (i) water, (ii) sulfur species, (iii) mercury, (iv) naturally occurring radioactive materials (NORM), (v) diluents, such as carbon dioxide, and (vi) oxygen that must be removed before the gas is sent to the consumers (Appendix D).

Most gas produced contains water – concentrations range from trace amounts to saturation – which must be removed. If the hydrogen sulfide (H_2S) concentration is greater than 2–3% v/v, carbonyl sulfide (COS), carbon disulfide (CS_2), elemental sulfur, and mercaptan derivatives (RSH) may be present. Also, trace quantities of mercury may be present in some gases; levels reported vary from 0.01 to 180 $\mu g/Nm^3$. Because mercury can damage the brazed aluminum heat exchangers used in cryogenic applications, conservative design requires mercury removal to a level of 0.01 $\mu g/Nm^3$.

NORM may also present problems in gas processing. The radioactive gas radon can occur in wellhead gas at levels from 1 to 1,450 pCi/l (pico-curie per liter) have extreme amounts of undesirable components. For some wells produce gas streams that contain diluents such as carbon dioxide, hydrogen sulfide, and nitrogen have been observed. Finally, some gas-gathering systems operate below atmospheric pressure and, as a result of leaking pipelines, open valves, and other system compromises, oxygen is an important impurity to monitor. A significant amount of corrosion in gas processing is related to oxygen ingress into the gas stream.

By way of clarification, the concentration of an air pollutant is given in micrograms (one millionth of a gram) per cubic meter air or $\mu g/m^3$. The designation Nm^3 (normal cubic meter) or Sm^3 (standard cubic meter) are complete definitions in themselves and is essential to know the standard reference conditions of temperature and pressure to define the gas volume since there are various debates about what normal and standard should be. On the other hand, 1 pCi is one trillionth (1×10^{-12}) of a Curie, 0.037 disintegrations per second, or 2.22 disintegrations per minute. Therefore, at 4 pCi/L (picocuries per liter, the EPA's recommended action level), there will be approximately 12,672 radioactive disintegrations in one liter of air during a 24-h period.

Processing of wellhead natural gas into pipeline-quality natural gas (e.g., 99.9% methane) can be quite complex and usually involves several processes. The actual practice of processing natural gas to pipeline dry gas quality levels can be quite complex, but usually involves four main processes to remove the various impurities: (i) oil and condensate removal, (ii) water removal, (iii) separation of natural gas liquids, and (iv) sulfur and carbon dioxide removal. In addition to these four processes above, heaters and scrubbers are installed, usually at or near the wellhead. The scrubbers serve primarily to remove sand and other large-particle impurities. The heaters ensure that the temperature of the gas does not drop too low. With natural gas that contains even low quantities of water, natural gas hydrates have a tendency to form when temperatures drop. These hydrates are solid or semi-solid compounds, resembling ice like crystals. Should these hydrates accumulate, they can impede the passage of natural gas through valves and gathering systems. To reduce the occurrence of hydrates, small natural gas–fired heating units are typically installed along the gathering pipe wherever it is likely that hydrates may form.

It is often necessary to install scrubbers and heaters at or near the wellhead which primarily serve to remove sand and other large particle impurities. The heaters ensure that the temperature of the natural gas does not drop too low to form a hydrate with the water vapor content of the gas stream. The natural gas hydrates are crystalline solids that block the passage of natural gas through valves and pipes.

Natural gas processing consists of separating all of the various hydrocarbons and fluids from the pure natural gas, to produce what is known as "pipeline quality" dry natural gas. Major transportation pipelines usually impose restrictions on the make-up of the natural gas that is allowed into the pipeline. That means that before the natural gas can be transported it must be purified. While the ethane, propane, butane, and pentanes must be removed from natural gas, this does not mean that they are all waste products.

While some of the needed processing can be accomplished at or near the wellhead (field processing), the complete processing of natural gas takes place at a processing plant, usually located in a natural gas producing region. The extracted natural gas is transported to these processing plants through a network of gathering pipelines, which are small-diameter, low pressure pipes. A complex gathering system can consist of thousands of miles of pipes, interconnecting the processing plant to upward of 100 wells in the area.

In addition to processing done at the wellhead and at centralized processing plants, some final processing is also sometimes accomplished at "straddle extraction plants." These plants are located on major pipeline systems. Although the natural gas that arrives at these straddle extraction plants is already of pipeline quality, in certain instances there still exist small quantities of NGLs, which are extracted at the straddle plants.

5.4.1 Gas and liquid separation

If the gas is being stripped from oil, separators knock out both oil and water. For high-pressure wells, the oil passes through up to three separators to recover the light ends. Because the last-stage separator is near ambient conditions, the gas may be compressed before it flows into a gathering line to the gas plant.

Separator vessel orientation can be vertical or horizontal. Vertical separators are most commonly used when the liquid-to-gas ratio is low or gas flow rates are low and are preferred offshore because they occupy less platform area. However, gas flow is upward and opposes the flow of liquid droplets. Therefore, vertical separators can be bigger and, thus, more costly than horizontal separators. Inlet suction scrubbers at compressor stations are usually vertical. Horizontal separators are favored for large liquid volumes or if the liquid-to-gas ratio is high. Lower gas flow rates and increased residence times offer better liquid dropout.

Separation of well-stream gas from free liquids is by far the most common of all field-processing operations and also one of the most critical of the processes. The composition of the fluid mixture determines the design criteria for sizing and selecting a separator for a hydrocarbon stream. In the case of low-pressure oil wells, the liquid phase will be large in volume compared to the gas phase. In the case of high-pressure gas-distillate wells, the gas volume will be higher compared to the liquid volume. The liquid produced with high-pressure gas is generally a high API gravity hydrocarbon, usually referred to as distillate or condensate. However, both low-pressure oil wells and high-pressure gas-distillate wells may contain free water.

The liquid collection section of the separator acts as a holder for the liquids removed from the gas in the above three separation sections. This section also provides for degassing of the liquid and for water and solids separation from the hydrocarbon phase. The most common solid is iron sulfide from corrosion, which can interfere with the liquid–liquid separation. If a large amount of water is present, separators often have an area (often referred to as the boot) at the bottom of the separator for the water to collect.

As an example of separator action, the well fluid is introduced into the primary separator where most of the produced gas is separated from the produce oil and water. The liquid from the primary separator is then flashed into a stock tank and any gas from the stock tank is added to the primary gas to arrive at the total produced surface gas. At this point, the produced amounts of oil and gas are measured, samples are taken, and these data are used to evaluate and forecast the performance of the well.

Separators are also used in many locations other than at wellhead production batteries, such as gasoline plants, upstream and downstream of compressors, liquid traps in gas transmission lines, inlet scrubbers to dehydration units, and gas sweetening units. At some of these other locations, separators are referred to as scrubbers, knockouts, and free liquid knockouts. All these vessels serve the same primary purpose: to separate free liquids from the gas stream.

A correctly designed well-stream separator must perform the following functions: (i) cause a primary-phase separation of the mostly liquid hydrocarbons from those that are mostly gas, (ii) refine the primary separation by removing most of the entrained liquid mist from the gas, (iii) further refine the separation by removing the entrained gas from the liquid, and (iv) discharge the separated gas and liquid from the vessel and ensure that no re-entrainment of one into the other takes place.

5.4.2 Separator internals

The principal items of construction that should be present in a good liquid–gas separator are the same, regardless of the overall shape or configuration of the vessel. Some of these features are as follows: (i) a centrifugal inlet device where the primary separation of the liquid and gas is made, (ii) a large settling section of sufficient length or height to allow liquid droplets to settle out of the gas stream with adequate surge room for slugs of liquid, (iii) a mist extractor or eliminator near the gas outlet to coalesce small particles of liquid that will not settle out by gravity, and (iv) adequate controls consisting of level control, liquid dump valve, gas back pressure valve, safety relief valve, pressure gauge, gauge glass, instrument gas regulator, and piping.

The bulk of the gas–liquid separation occurs in the inlet centrifugal separating section. Here, the incoming stream is spun around the walls of a small cylinder, which would be the wails of the vessel in the case of a vertical or spherical separator. This subjects the fluids to a centrifugal force up to 500 times the force of gravity. This action stops the horizontal motion of the free liquid entrained in the gas stream and forces the liquid droplets together where they will fall to the bottom of the separator into the settling section.

The settling section lets the turbulence of the fluid stream subside and allows liquid droplets to fail to the bottom of the vessel because of the difference in the gravity between the liquid and gas phases. A large open space in the vessel is adequate for this purpose. Introducing special quieting plates or baffles with narrow openings only complicates the internal construction of the separator and provides places for sand, sludge, paraffin, and other materials to collect and eventually plug the vessel and stop the flow. The separation of liquid and gas using the centrifugal inlet feature and a large, open settling section produces a more stable liquid product, which can be contained in atmospheric or low-pressure storage tanks. Minute scrubbing of the gas phase by use of internal baffling or plates may produce more liquid to be discharged from the separator, but the product will not be stable since light ends will be entrained in it, incurring more vapor losses from the storage system.

Sufficient surge room should be allowed in the settling section to handle slugs of liquid without carryover to the gas outlet. This can be accomplished to some extent by placement of the liquid-level control in the separator, which ln turn determines the liquid level. The amount of surge room required is often difficult, if not impossible, to determine from well test or flowing data. In most cases, the separator size used for a particular application is a compromise between initial cost and possible surging requirements.

Another major item required for good and complete liquid–gas separation is a mist eliminator or extractor near the gas outlet. Small liquid droplets that will not settle out of the gas stream, due to little or no gravity difference between them and the gas phase, will be entrained and pass out of the separator with the gas. This can be almost eliminated by passing the gas through a mist eliminator near the gas outlet, which has a large surface impingement area. The small liquid droplets will hit the surfaces, coalesce, and collect to form larger droplets that will then drain back to the liquid section in the bottom of the vessel. A stainless steel woven-wire mesh mist eliminator is probably the most efficient type since it removes up to 99.9% or more of the entrained liquids from the gas stream. This type offers more surface area for collecting liquid droplets per unit volume than vane types, ceramic packing, or other configurations. The vane mist eliminators apply in areas where there is entrained solid material in the gas phase that may collect and plug a wire mesh mist eliminator.

5.4.3 Types of separators

There are four major types or basic configurations of separators, generally available from manufacturers: vertical, horizontal single tube, horizontal double tube, and spherical. Each type has specific advantages, and selection is usually based on which one will accomplish the desired results at the lowest cost.

5.4.3.1 Vertical separators
A vertical separator (Figure 5.1) is often used on low to intermediate GOR well streams and where relatively large slugs of liquid are expected. This type of separator will handle larger slugs of liquid without carryover to the gas outlet, and the action of the liquid-level control is not as critical. A vertical separator occupies less floor space, an important consideration where space might be expensive as on an offshore platform. Due to the greater vertical distance between the liquid level and the gas outlet, there is less tendency to revaporize the liquid into the gas phase. However, because the natural upward flow of gas in a vertical vessel opposes the falling droplets of liquid, it takes a larger diameter separator for given gas capacity than a horizontal vessel. Also, vertical vessels are more expensive to fabricate and ship in skid-mounted assemblies.

5.4.3.2 Horizontal single-tube separators
The horizontal separator (Figure 5.2) may be the best separator for the money. The horizontal separator has a much greater gas–liquid interface area consisting of a large, Jong, baffled gas-separation section that permits much higher gas velocities. This type of separator is easier to skid-mount and service and requires less piping for field connections and a smaller diameter for a given gas capacity. Several separators can be stacked easily into stage-separation assemblies, minimizing space requirements.

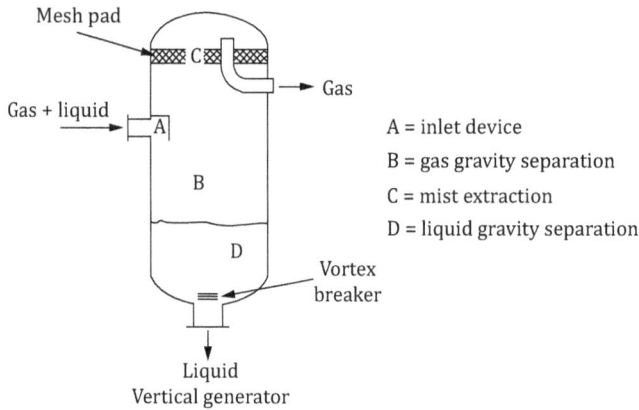

A = inlet device
B = gas gravity separation
C = mist extraction
D = liquid gravity separation

Figure 5.1: A vertical separator.

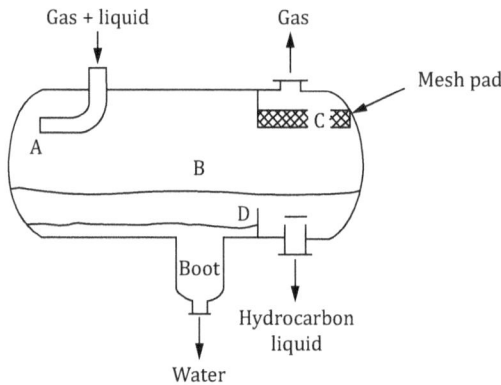

Figure 5.2: A horizontal separator.

In operation, gas flows horizontally and, at the same time, falls toward the liquid surface. The gas flows in the baffle surfaces and forms a liquid film that is drained away to the liquid section of the separator. The baffles need only be longer than the distance of liquid trajectory travel at the design gas velocity. The liquid-level control placement is more critical than in a vertical separator, and surge space is somewhat limited. Horizontal separators are almost always used for high GOR well streams, for foaming well streams, or for liquid-from-liquid separation.

5.4.3.3 Horizontal double-tube separators
A horizontal double-tube or double-barrel separator has all the advantages of a normal horizontal separator plus a much higher liquid capacity. Incoming free liquid is

immediately drained away from the upper section into the lower section. The upper section is filled with baffles, and gas flow is straight through and at higher velocities.

5.4.3.4 Spherical separators

Spherical separators offer an inexpensive and compact vessel arrangement. However, these types of vessels have a limited surge space and liquids settling section. The placement and action of the liquid-level control in this type of vessel are critical.

Three-phase or oil-gas-water separation can be easily accomplished in any type of separator by installing either special internal baffling to construct a water leg or water siphon arrangement, or by using an interface liquid-level control. A three-phase feature is difficult to install in a spherical separator because of the limited internal space available.

With three-phase operation, two liquid-level controls and two liquid dump valves are required. Three-phase separators are used commonly for well testing and in instances where free water readily separates from the oil or condensate. From an evaluation of the advantages and disadvantages of the various types of separators, the horizontal single-tube separator has emerged as the one that gives the most efficient operation for initial investment cost for high-pressure gas-distillate wells with high GORs. For high liquid loadings, vertical separators should be considered.

5.4.4 Factors affecting separation

Separator operating pressure, separator operating temperature, and fluid stream composition affect the operation and separation between the liquid and gas phases in a separator. Changes in any one of these factors on a given fluid well stream will change the amount of gas and liquid leaving the separator. Generally, an increase in operating pressure or a decrease in operating temperature will increase the liquid covered in a separator. However, there are optimum points in both cases beyond which further changes will not aid in liquid recovery. In fact, storage system vapor losses may become too great before these points are reached.

In the case of wellhead separation equipment, an operator generally wants to determine the optimum conditions for a separator to effect the maximum income. Again, generally speaking, the liquid recovered is worth more than the gas. So, high liquid recovery is a desirable feature, providing it can be held in the available storage system. Also, pipeline requirements for the Btu content of the gas may be another factor in separator operation. Without the addition of expensive mechanical refrigeration equipment, it is often unfeasible to try to lower the operating temperature of a separator. However, on most high-pressure wells, an indirect heater heats the gas prior to pressure reduction to pipeline pressure in a choke. By careful operation of this indirect heater, the operator can prevent overheating the gas stream ahead of the choke, which will adversely affect the temperature of the separator downstream from the indirect heater.

The operator can also control operating pressure to some extent by use of back-pressure valves within the limitation of the flowing characteristics of the well against a set pressure head and the transmission line pressure requirements. As previously mentioned, higher operating pressure will generally result in higher liquid recovery.

An analysis can be made using the well stream composition to find the optimum temperature and pressure at which a separator should operate to give maximum liquid or gas phase recovery. These calculations, known as flash vaporization calculations, require a trial-and-error solution and are more generally adapted to solution by a programmed computer. An operator can, however, make trial settings within the limitations of the equipment to find the operating conditions that result in the maximum amount of gas or liquids. In the case where separators are used as scrubbers or knockouts ahead of other treating equipment or compressors, it is generally best to remove the maximum amount of liquid from the gas stream to prevent operational damage to the equipment downstream from the scrubber.

5.4.4.1 Low-temperature separation

A low-temperature separation unit is another type of equipment employed for gas–liquid separation that consists primarily of a high-pressure separator, pressure-reducing chokes, and various pieces of heat exchange equipment. As described previously, lowering the operating temperature of a separator increases the liquid recovery. When the pressure is reduced on a high-pressure gas condensate stream by use of a pressure-reducing choke, the fluid temperature also decreases. This is known as the Joule–Thomson or throttling effect, which is an irreversible adiabatic process where the heat content of the gas remains the same across the choke but the pressure and temperature of the gas stream is reduced.

Low-temperature separation is probably the most efficient means yet devised for handling high-pressure gas and condensate at the wellhead. The process separates water and hydrocarbon liquids from the inlet well-stream, recovers more liquids from the gas than can be recovered with normal-temperature separators, and dehydrates gas, usually to pipeline specifications. By special construction of low-temperature separation units, the pressure-reducing choke is mounted directly on the inlet of the high-pressure separator.

Hydrates formed downstream of the choke due to the low gas temperature after the pressure reduction are blown into the separator and fall to the bottom settling section where they are heated and melted by liquid heating coils located in the bottom of the vessel. The low-temperature effect is used in low-temperature units to increase the liquid recovery. The lower the operating temperature of the separator, the higher the liquid recovery will be. However, the maximum flowing pressure from the well at a given flow rate and the transmission line pressure will indicate the maximum amount of pressure drop available across the choke.

Enthalpy curves on natural gas can be used to determine the temperature drop expected based on the available pressure drop. For example, a 1,500-psi reduction from 3,000 psi at 120 °F will provide a final temperature of only 78 °F, while a 1,500-psi reduction from 2,000 psi at 120 °F will give a final temperature of 49 °F. The actual temperature drop per pounds per square inch of pressure drop will depend on composition of gas stream, gas and liquid flow rates, bath temperature, and ambient temperature.

In general, at least a 2,500-psi and 3,000-psi pressure reduction should be available from wellhead flowing pressure to pipeline pressure before a low-temperature separation unit will pay out in increased liquid recovery. The lowest operating temperature recommended for low-temperature units is usually on the order of 20 °F. Carbon steel embrittlement occurs below this temperature and high-alloy steels for lower temperatures are usually not economical for most oil field installations. Generally, low-temperature separation units are operated in the range of 0–20 °F.

5.4.4.2 High-temperature separation

The high-temperature gas separation unit efficiently removes liquids from natural gas streams while also introducing heat to protect production equipment, valves and pipelines from clogging due to the buildup of paraffin. Our standard unit is a horizontal three-phase separator equipped with an internal fire tube and is rated to handle higher pressures than a traditional heater treater, up to 500 psi.

5.5 Dehydration of natural gas

Natural gas destined for transport by pipeline must meet certain specifications. Items usually included in the United States and Canada are maximum water content (water dew point), maximum condensable hydrocarbon content (hydrocarbon dew point), allowable concentrations of contaminants such as H_2S, CO_2, mercaptans, minimum heating value, and cleanliness (allowable solids content). In addition there will be specifications regarding delivery pressure, rate, and possibly temperature.

The reservoir fluids flowing to the surface during production operations normally contain water regardless of which zone the well has been completed. The gas-producing zone of a reservoir tends to contain more water vapor and less liquid or free water than the oil-producing zone. Water vapor is probably the most common undesirable impurity found in untreated natural gas. The principal reasons for the removal of water vapor from natural gas for long-distance transmission include the following: (i) liquid water and natural gas can form solid, ice-like hydrates that plug equipment, (ii) natural gas containing liquid water is corrosive, particularly if it also contains CO_2 or H_2S, (iii) water vapor in natural gas may condense in pipelines potentially causing slugging flow conditions, (iv) water vapor increases the volume and decreases the heating value of natural gas; this leads to reduced line capacity, and (v) the most rigorous approach uses physical chemistry techniques to account for the effect of differences in gas composition on water vapor content.

In almost all cases, the inlet natural gas to a dehydration system will be saturated with water vapor at process temperature and pressure, but the upstream process configuration should be examined to confirm this. Note, however, that the assumption of saturation is a conservative approach.

5.5.1 Gas hydrates

The concept of natural gas occurring as a hydrate (called methane hydrate (also called gas hydrate, methane clathrate, natural has hydrate, methane ice, hydromethane, methane ice, fire ice) is relatively new but does offer the potential to recover hitherto unknown reserves of methane that can be expected to extend the availability of natural gas (Giavarini et al., 2003; Giavarini and Maccioni, 2004; Giavarini et al., 2005; Makogon et al., 2007; Makogon, 2010; Wang and Economides, 2012; Yang and Qin, 2012). In terms of gas availability from this resource, one liter of solid methane hydrate can contain up to 168 L of methane gas.

The methane in gas hydrates is predominantly generated by bacterial degradation of organic matter in low oxygen environments. Organic matter in the uppermost few centimeters of sediments is first attacked by aerobic bacteria, generating carbon dioxide, which escapes from the sediments into the water column. In this region of aerobic bacterial activity sulfate derivatives ($-SO_4$) are reduced to sulfide derivatives ($=S$). If the sedimentation rate is low (<1 cm per 1,000 years), the organic carbon content is low (<1%), and oxygen is abundant, and the aerobic bacteria use up all the organic matter in the sediments. However, when sedimentation rate is high, and the organic carbon content of the sediment is high, the pore waters in the sediments are anoxic at depths of less than one foot or so and methane is produced by anaerobic bacteria.

The two major conditions that promote hydrate formation are thus: (i) high gas pressure and low gas temperature and (ii) the gas at or below its water dew point with free water present (Sloan, 1998b; Collett et al., 2009). The hydrates are believed to form by migration of gas from depth along geological faults, followed by precipitation, or crystallization, on contact of the rising gas stream with cold sea water. At high pressures methane hydrates remain stable at temperatures up to 18 °C (64 °F) and the typical methane hydrate contains one molecule of methane for every six molecules of water that forms the ice cage. However, the methane (hydrocarbon)-water ratio is dependent on the number of methane molecules that fit into the cage structure of the water lattice.

Chemically, gas hydrates are non-stoichiometric compounds formed by a lattice of hydrogen bonded molecules (host) which engage low molecular weight gases or volatile liquids (guest) with specific properties that differentiate them from ice (Chapter 1) (Bishnoi and Clarke, 2006). No actual chemical bond exists between guest and host molecules. Hydrate formation is favored by low temperature and high pressure (Makogon, 1997; Sloan, 1998a; Lorenson and Collett, 2000; Carrol, 2003a, 2003b; Seo et al., 2009). Most methane hydrate deposits also contain small amounts of other hydrocarbon hydrates; these include ethane hydrate and propane hydrate. In fact, gas hydrates of current interest are composed of water and the following molecules: methane, ethane, propane, isobutane, normal butane, nitrogen, carbon dioxide, and hydrogen sulfide. However, other nonpolar components such as argon (Ar) and ethyl cyclohexane ($C_6H_{11}C_2H_5$) can also form hydrates. Typically, gas hydrates form at temperatures on the order of 0 °C (32 °F) and elevated pressures (Sloan, 1998a).

In the hydrate structure, methane is trapped within a cage-like crystal structure composed of water molecules in a structure that resembles packed snow or ice (Lorenson and Collett, 2000). Under the appropriate pressure, gas hydrates can exist at temperatures significantly above the freezing point of water, but the stability of the hydrate derivatives depends on pressure and gas composition and is also sensitive to temperature changes (Stern et al., 2000; Stoll and Bryan, 1979; Collett, 2001; Belosludov et al., 2007; Collett, 2010). For example, methane plus water at 600 psia forms hydrate at 5 °C (41 °F), while at the same pressure, methane with 1% v/v propane forms a gas hydrate at 9.4 °C (49 °F). Hydrate stability can also be influence by other factors, such as salinity.

Methane hydrates are restricted to the shallow lithosphere (i.e., at depths less than 6,000 feet below the surface). The necessary conditions for the formation of hydrates are found only either in polar continental sedimentary rocks where surface temperatures are less than 0 °C (32 °F) or in oceanic sediment at water depths greater than 1,000 feet where the bottom water temperature is on the order of 2 °C (35 °F).

Caution is advised when drawing generalities related to the formation and the stability of gas hydrates. Methane hydrates are also formed during natural gas production operations, when liquid water is condensed in the presence of methane at high pressure. Higher molecular weight hydrocarbons such as ethane and propane can also form hydrates, although larger molecules (butane hydrocarbons and pentane hydrocarbons) cannot fit into the water cage structure and, therefore, tend to destabilize the formation of hydrates (Belosludov et al., 2007). However, for this text, the emphasis is focused on methane hydrates.

5.5.1.1 Formation

Natural gas hydrates are solid crystalline compounds formed by the chemical combination of natural gas and water under pressure at temperatures considerably above the freezing point of water. In the presence of free water, hydrates will form when the temperature is below a certain degree (hydrate temperature). The chemical formulas for natural gas hydrates are:

Methane, $CH_4 \cdot 7H_2O$
Ethane, $C_2H_6 \cdot 8H_2O$
Propane, $C_3H_8 \cdot 18H_2O$
Carbon dioxide, $CO_2 \cdot 7H_2O$

Gas hydrate crystals resemble ice or wet snow in appearance but do not have ice's solid structure, are much less dense, and exhibit properties that are generally associated with chemical compounds. The main framework of their structure is water; the hydrocarbon molecule occupies the void space in a crystalline network held together by chemically weak bonds with the water. The water framework is icelike; unlike ice, however, it has void space and a network structure.

Hydrate formation is often confused with condensation, and the distinction between the two must be clearly understood. Condensation of water from a natural gas under pressure occurs when the temperature is at or below the dew point at that pressure. Free water obtained under such conditions is essential to formation of hydrates, which will occur at or below the hydrate temperature at the same pressure. Hence, the hydrate temperature would be less than or equal to the dew point temperature.

During the flow of natural gas, it becomes necessary to define, and thereby avoid, conditions that promote the formation of hydrates. This is essential since hydrates may choke the flow string, surface lines, and other equipment. Hydrate formation in the flow string results in a lower value for measured wellhead pressures. In a flow rate measuring device, hydrate formation results in lower flow rates. Excessive hydrate formation may also completely block flow lines and surface equipment.

The conditions that tend to promote the formation of natural gas hydrates are: (i) natural gas at or below its water dew point with liquid water present, (ii) temperatures below the "hydrate formation" temperature for the pressure and gas composition considered, (iii) high operating pressures that increase the "hydrate formation" temperature, (iv) high velocity or agitation through piping or equipment, (v) the presence of a small "seed" crystal of hydrate, and (vi) the presence of H_2S or CO_2 is conducive to hydrate formation since these acid gases are more soluble in water than hydrocarbons.

5.5.1.2 Gas production

Gas production of from gas hydrates can be achieved by methods such as (1) thermal stimulation, (ii) depressurization, or (iii) chemical inhibition.

The gas content of gas hydrates typically involves the controlled decomposition of the hydrate in a dissociation chamber and allowed to decompose at room temperature. The gas pressure of the chamber will increase and reach a stable value over a period of time after which the hydrate gas is ready for analysis. The volume of water that remains in the dissociation chamber is then measured and stored in a glass ampoule for determination of the chemical composition and to ensure that no gases remain dissolved. Gas analysis will show that the gas released from the hydrate is predominantly methane with trace-to small-(typically <5% v/v) of ethane, propane, isobutane, normal butane, nitrogen, carbon dioxide, and hydrogen sulfide (Gabitto and Tsouris, 2010).

To this point, the heat of dissociation (ΔH_d) is the enthalpy change to dissociate the hydrate phase to a vapor and aqueous liquid, with values given at temperatures just above the ice point. Also, the heat pf dissociation is a function of the number of crystal hydrogen bonds (generally assumed to be the same as the hydration number). However, the value of the heat of dissociation is relatively constant for molecules which occupy the same cavity, within a wide range of component sizes.

The enthalpy of dissociation may be determined via the univariant slopes of phase equilibrium lines (ln P versus 1/T) using the Clausius–Clapeyron relationship (Sloan, 2006):

$$[(\Delta H_d) = - zR_d(\ln P)/d(1/T)]$$

Gas hydrates may formally be referred to as chemical compounds because they have a fixed composition at a certain pressure and temperature. However, hydrates are the compounds of a molecular type. They form because of the Van der Waals attraction forces between the molecules. Covalent bonding is absent in the gas hydrates because during their formation there is no pairing of valence electrons and no spatial redistribution of electron cloud density.

Natural gas hydrates are metastable minerals, where the formation and dissociation depend on the pressure and temperature, composition of gas, salinity of the reservoir water, and the characteristics of the porous medium in which they were formed. Hydrate crystals in reservoir rocks can be dispersed in the pore space without the destruction of pores; however, in some cases, the rock is affected. Hydrates can be in the form of small nodules (from 5 to 12 cm in size), in the form of small lenses, or in the form of layers that can be several meters thick.

The composition of natural gas hydrates is determined by the composition of the gas and water, and the pressure and temperature which existed at the time of their formation. Over geologic time, there will be changes in the thermodynamic conditions and the vertical and lateral migration of gas and water; therefore, the composition of hydrate can change both due to the absorption of free gas and the recrystallization of already-formed hydrate. Based on the cores taken while drilling in gas hydrate deposits, the hydrate usually consists of methane with small ad mixtures of heavier components. However, in a number of cases the hydrate contains a significant volumes of higher molecular weight gases.

The recovery of methane gas from natural has hydrates can be achieved by dissociating the solid gas hydrate structure into gas and water. The conventional transportation method then can be used to transport this dissociated gas. The methods that have been proposed for the extraction of natural gas from natural has hydrates include: (i) depressurization, (ii) thermal stimulation, (iii) chemical inhibitor injections, and (iv) carbon dioxide replacement (Castaldi et al., 2007; Makogon, 2010; Lee et al., 2013). The method for reducing pressure (the depressurization method) for obtaining gas from natural has hydrates is reported to be more effective and economical than the thermal stimulation method (Demirbaş, 2010).

The depressurization method operates by lowering the pressure inside the well with embedded and adjacent zones of gas hydrates those results in dissociation of methane gas (Fan et al., 2017; Wang et al., 2017). The target is to lower the pressure below the dissociation pressure, which results in decomposition of hydrates into gas and water. Ahmadi et al. (2007) studied the production of natural gas from dissociation of natural has hydrates in a confined and pressurized reservoir utilizing the depressurizing method. The depressurization method was proved to be a successful method with both well pressure and reservoir temperature being the sensitive parameters for gas production and overall well output (Ahmadi et al., 2007).

In thermal stimulation method, heat is applied to natural gas hydrate system at constant pressure to increase the temperature above the dissociation temperature, to decompose the hydrates to produce methane gas. Different heat sources can be used to apply direct heat such as steam injection, hot water or other liquids, or indirect heat by electric current or sonication. The methane gas, mixed with hot water returned to the surface, is then separated from hot water to be used as a pure gas (Goel et al., 2001).

The third method for dissociation of national gas hydrates is chemical inhibition, in which a chemical inhibitor is injected adjacent to the hydrate to displace the natural gas hydrate equilibrium conditions beyond the thermodynamic conditions of hydrate stability zone. A chemical inhibitor, for example methanol, is used and injected into the methane hydrate–bearing layers that results in separation of methane gas from the solid hydrate structure.

The choice of method for methane extraction from natural gas hydrates not only depends on the geological locations and thermodynamic conditions of natural gas hydrate, but also on amounts of hydrate deposits, capital and maintenance cost, environmental impact and simplicity of the selected method. The thermal stimulation and chemical inhibition methods are relatively more expensive. In comparison, depressurization method for production of methane gas from natural has hydrates is considered to be most economical and efficient method. Methane gas is an attractive fuel for its least greenhouse effect because it produces less carbon dioxide during its combustion as compared to all other petroleum-derivate fuels. The real positive contribution of methane gas fuel for minimization of environmental pollution must also be considered in more details together with its extraction from natural gas hydrates and potential as an unconventional future energy resource (Davies 2001).

Thus, it is necessary to use much of the energy contained in the gas hydrate deposits for heating the rock layers near the gas hydrate deposits. Preliminary estimates show that the coefficient of extraction of the gas hydrate can be as high as 50 to 70%. However, from total world potential resource it has been estimated that the coefficient of extraction should average from 17% to 20%.

For offshore conditions, with the depths of water ranging from 0.7 to 2.5 km, effective production of gas from gas hydrate deposits in the majority of the cases may occur when hydrate saturation of porous media exceeds 30% to 40%. However, each geologic region will have to be studied in detail to establish the minimal hydrate saturation that is required. To change a gas hydrate deposit to natural gas, it is necessary to (i) decrease reservoir pressure to lower than equilibrium one, (ii) increase the temperature to higher than the equilibrium temperature, (iii) inject active reagents, which facilitate the decomposition of hydrate, and (iv) use some new technology. The easiest method is to lower the reservoir pressure in the gas hydrate deposit – this method is feasible only when free gas is found below the gas hydrate deposit.

Several properties of gas hydrates are unique. For example, 1 m^3 of water may accommodate up 207 m^3 of methane to form 1.26 m^3 of solid hydrate, whereas without gas, 1 m^3 of water freezes to form 1.09 m^3 of ice. One volume of methane hydrate at a pressure of 3,800 psi and temperature of 0 °C (32 °F) contains 164 volumes of gas. In a hydrate, 80% v/v is occupied by water and 20% v/v by gas. Thus, 164 m^3 of gas are contained in a volume of 0.2 m^3 of hydrate.

The dissociation of methane hydrate by increasing the temperature in a constant volume will be accompanied by a substantial increase in pressure. For methane hydrate formed at a pressure of 3,800 psi and a temperature of 0 °C (32 °F), it is possible to obtain a pressure increase of up to 23,500 psi. Hydrate density depends on its composition, pressure, and temperature. Depending on the composition of gas, pressure and temperature, the density of the hydrates range be on the order of 0.9 g/cm^3 – typically falling in the range 0.8–1.2 g/cm^3 – and, thus, methane hydrate will float to the surface of the sea or of a lake unless it is bound in place by being formed in or anchored to sediment.

In the thermal stimulation method of decomposition and gas release, the temperature of the formation is raised by the injection of heated fluid or potentially direct heating of the formation. Thermal stimulation is energy intensive and will lead to relatively slow, conduction limited dissociation of gas hydrates unless warmer pore fluids become mobilized and increase the volume of the formation exposed to higher temperatures. The endothermic nature of gas hydrate dissociation also presents a challenge to thermal stimulation – the cooling associated with dissociation (and, in some cases, gas expansion) will partially offset artificial warming of the formation, meaning that more heat must be introduced to drive continued dissociation and prevent formation of new gas hydrate. In terrestrial settings thermal stimulation must be carefully controlled to minimize permafrost thawing, which might lead to unintended environmental consequences and alter the permeability seal for the underlying gas hydrate deposits.

On the other hand, the depressurization method of decomposition and gas release does not require large energy expenditure and can be used to drive dissociation of a significant volume of gas hydrate relatively rapidly (Lorenson and Collett, 2000). On the other hand, the chemical inhibition exploits the fact that gas hydrate stability is inhibited in the presence of certain organic compounds (such as glycol) or ionic compounds (such as seawater or brine). Seawater or other inhibitors might be needed during some stages of production of gas from methane hydrate deposits but would not be the primary means of dissociating gas hydrate nor used for an extended period or on a large scale.

5.5.1.3 Prediction of hydrate formation

Of the common components in natural gas, hydrogen sulfide is known to form hydrates at the lowest pressure and they persist to the highest temperatures (Carroll, 2003a, 2003b). In addition, hydrogen sulfide has a significant effect on hydrate formation of a mixture. Based on experience, it is known that sour gas more readily forms a hydrate than does sweet gas.

To prevent hydrate formation, it is important to predict the temperature and pressure under which gas hydrates will form.

In the absence of inhibitors and at the same temperature, a gas with lighter specific gravity forms hydrates at higher pressure, but the presence of components such as propane, isobutane, and nitrogen should be considered. The presence of propane and isobutane in a gas mixture decreases the hydrate-formation pressure and increases the hydrate-formation temperature, while the presence of nitrogen in a gas mixture increases the hydrate-formation pressure and decreases the hydrate formation temperature.

The compositions of a gas system play a very important role in determining the hydrate-formation temperature or pressure which can be interpreted that two gas systems with equal specific gravity may form hydrates at very different conditions. For example, a binary mixture of methane and isobutane forms hydrates at lower pressure and higher temperature than a mixture of methane and butane with the same composition and specific gravity.

5.5.2 Hydrate control in gas production

Gas producers, processors, and pipeline operators have often been plagued by operating difficulties that cause undesirable and costly interruptions in gas production, field processing, and distribution. While many operating difficulties are mechanical in nature, two constantly recurring operating problems, gas hydrate formation and freeze-up in separator water dump lines, are more costly and more serious than others.

These difficulties most often occur in cold weather when seasonal demand for gas is at its highest level, thus making shutdowns or interruptions in production or gas service most expensive. These two operating problems are somewhat related, and their solutions in some cases are quite similar.

5.5.3 Preventing hydrate formation

Hydrate formation can be prevented by heating the cold, unprocessed gas stream. If the pressure and water content of the gas remain constant during heating, the gas will become undersaturated, thereby eliminating one of the conditions for hydrate formation.

A variety of methods can prevent the formation of gas hydrates. The permanent solution is to extract water prior to transport by pipeline, such as using an offshore dehydration or subsea separation facility, which is not always the most cost-effective solution. Another way to avoid hydrate plugs is to maintain pressure and temperature conditions outside the hydrate-forming zone. But all in all, there are methods to prevent hydration which are (i) remove the free and dissolved water from the system with separators, glycol dehydrators, molecular sieves, or other methods, (ii) maintain high temperatures so that hydrates do not form, (iii) maintain low pressures to keep all phases fluid, and (iv) inject chemical inhibitors which mitigate the formation of hydrates.

Two major classes of chemical inhibitors are available: (i) thermodynamic inhibitors and (ii) synthetic inhibitors. The addition of thermodynamic inhibitors such as methanol (CH_3OH) or diethylene glycol (DEG, $HOCH_2CH_2OCH_2CH_2OH$) is a common method for preventing hydrates from forming. The addition of a synthetic inhibitor typically prevents or delays hydrate crystal nucleation and disrupts crystal growth thereby preventing the chemical connection between molecules of gas and water.

5.5.4 Preventing wax deposition

A popular wax mitigation technique known as pigging is used to regularly clean the pipeline by removing any deposited wax from the internal walls of the pipeline. Wax deposition is a complex problem involving various aspects of study such as fluid dynamics, mass and heat transfers, and during steady-state flows. For some subsea pipelines, especially export lines where hydrate is not a concern, pigging would normally be the main wax management strategy.

The pig (pipeline inspection gauge) is a synthetic body, which lies tightly against the wall and is sent through the pipeline with the medium water – there are numerous different sizes of pigs, depending on the requirements of the pipeline. Due to the space-filling closure up to the pipe wall, any air present is also pushed through the pipe and can escape at the other end of the pipe at which point the pig is removed from the pipeline. The pipeline is then completely filled with water and almost free of air. Only very small residual amounts of air, for example, at pipe connections, can remain.

The pig is sent down from a pig launcher into the pipeline and is pushed by petroleum extraction or all other liquids, such as dead oil or coal. The pig scrapes the wax from the pipe wall manually and re-deposits it in front of the pig back into the crude. For the effectiveness of pigging activities, a regularly scheduled pigging program is very important. If the pigging process is not planned often enough, there could be too much wax accumulated on the wall of the vessel. A pig may get trapped inside the pipe during the pigging process because of the excess volume of wax in front of it. Based on wax deposition models, the pigging schedule software will be built and will be fine-tuned when the wax deposition rate is best understood in field operations.

5.5.5 Dehydration systems

There are essentially three methods in current use for dehydrating gas streams: (i) adsorption, (ii) absorption, and (iii) cooling, which includes compression followed by cooling. Generally, the first two methods result in sufficiently low water dew points to permit injection into a pipeline and additional dehydration is not required.

5.5.5.1 Adsorption of water vapor by solid desiccants

Adsorption is defined as the ability of a substance to hold gases or liquids on its surface and dehydration plants using solid desiccants can remove practically all water from natural gas (as low as 1 ppm). Because of their great drying ability, solid desiccants may be employed where higher efficiencies are required.

In the adsorption dehydration process, water vapor from the gas is concentrated and held at the surface of the solid desiccant by forces thought to be caused by residual valency. Solid desiccants have large surface areas per unit weight to take advantage of these surface forces. The most common solid adsorbents include silica, alumina, and certain silicates known as molecular sieves.

In a solid desiccant dehydration plant, the incoming wet gas must be carefully cleaned, preferably by a filter separator, to remove virtually all solid and liquid contaminants in the gas. The filtered liquid-free gas generally flows downward during dehydration through one adsorber containing a desiccant bed. The downflow arrangement lessens disturbance of the bed due to high gas velocity during the dehydration or adsorption step. While one adsorber is dehydrating, the other adsorber is being regenerated by a hot slip stream of inlet gas from the regeneration gas heater.

A direct-fired heater, hot oil, steam, or an indirect heater can supply the necessary regeneration heat. The regeneration gas normally flows upward through the bed to ensure thorough regeneration of the bottom of the bed, which is the last area contacted by the gas being dehydrated. The hot regenerated bed is cooled by shutting off or bypassing the heater. The cooling gas flows downward through the bed so that any water adsorbed from the cooling gas will be at the top of the bed and will not be desorbed into the gas during the dehydration step. The still hot regeneration gas and the cooling gas flow through the regeneration gas cooler to condense the desorbed water. Power-operated valves activated by a timing device switch the adsorbers between the dehydration, regeneration, and cooling steps.

All desiccants become less effective in normal use through loss of effective surface area. The loss of effective surface area is rapid at first, then becomes more gradual as the desiccant ages. Abnormally fast degradation occurs through blockage of the small pores and capillary openings, which contain most of the effective surface area. Lubricating oils, amines, glycols, corrosion inhibitors, and other contaminants, which cannot be removed during regeneration cycle, will eventually ruin the bed. Hydrogen sulfide may poison the desiccant and reduce its capacity. Activated alumina has good resistance to liquids, but it tends to powder due to mechanical agitation of the flowing gas.

There are several advantages of solid-desiccant dehydration. First, lower dew points are obtainable over a wide range of operating conditions. Second, essentially dry gas (moisture content <1.0 lb/MMcf) can be produced. Third, higher contact temperatures can be tolerated with some adsorbents. Fourth, greater adaptability is offered to sudden load changes, especially on starting up. Fifth, the plant may be put in operation more quickly after a shutdown.

Unloading towers and recharging them with new desiccant should be completed well ahead of the operating season. In the interest of maintaining continuous operation when most needed, this may require discarding desiccant before the end of its normal operating life. To conserve material, the inlet part of the tower may be recharged and the remainder of the desiccant retained since it may still possess some useful life. Additional service life may be obtained if the direction of gas flow is reversed at a time when the tower would normally be recharged.

Sudden pressure surges should be avoided. They may upset the desiccant bed and channel the gas stream with poor dehydration. If a plant is operated above its rated capacity, pressure Joss will increase and some attrition may occur. Attrition causes fines, which may in turn cause excessive pressure loss with resulting Joss in capacity. Finally, if water cooling is used during the reactivating cycle, provision must be made against freezing in water coolers. This problem generally occurs only during cooler shutdown.

5.5.5.2 Absorption of water vapor by liquid desiccants

In the absorption dehydration process, the water vapor is removed from the gas by intimate contact with a hygroscopic liquid desiccant which is usually contained in trayed or packed towers. Glycol derivatives are used frequently and have been proved to be the most effective liquid desiccants in current use and the types of glycol that have been successfully used to dehydrate natural gas are (i) ethylene glycol (EG, $HOCH_2CH_2OH$), (ii) DEG ($HOCH_2CH_2OCH_2CH_2OH$), (iii) triethylene glycol (TEG, $HOCH_2CH_2OCH_2CH_2OCH_2CH_2OH$), and (iv) tetraethylene glycol (sometimes referred to as T4EG, $HOCH_2CH_2OCH_2CH_2OCH_2CH_2OCH_2CH_2OH$).

In almost all cases a single type of pure glycol is used; however, under certain circumstances, a glycol blend may be economically attractive. In over 40 years of use in the gas industry, TEG has gained nearly universal acceptance as the most cost effective of the glycols due to superior dew point depression, operating cost, and operation reliability.

TEG has been successfully used to dehydrate sweet and sour natural gases over the following range of operating conditions: (i) dew point depression: 40–140 °F, (ii) gas pressure: 25–2,500 psig, and (iii) gas temperature: 40–160 °F. The dew point depression obtained depends on the equilibrium dew point temperature for a given concentration of TEG and contact temperature. Increased glycol viscosity may be a problem at the lower end of the contact temperature range; consequently, heating of the natural gas may be desirable. Hot gas streams contain more water, vaporize more TEG, have lower equilibrium dew points, and are therefore often cooled prior to dehydration.

In the process, the west inlet gas stream first enters the unit through an inlet gas scrubber where any liquid accumulations are removed. If any liquid water is in the gas stream, a three-phase scrubber may be used to discharge the distillate and water from the vessel separately. The mist eliminator aids in removing any entrained liquid particles from the wet gas stream leaving the top of the inlet scrubber.

The wet gas then enters the bottom of the glycol contactor and flows upward through the trays as illustrated countercurrent to the glycol flowing downward through the column. The gas contacts the glycol on each tray and the glycol absorbs the water vapor from the gas stream. The dry gas leaves the top of the contactor vessel through another mist eliminator which aids in removing any entrained glycol droplets from the gas steam. The gas then flows down through a vertical glycol cooler, usually fabricated in the form of a concentric pipe heat exchanger, where the outlet dry gas aids in cooling the hot regenerated glycol before it enters the contactor. The dry gas then leaves the unit from the bottom of the glycol cooler.

The dry glycol enters the top of the glycol-gas contactor from the glycol cooler and is injected onto the top tray. The glycol flows across each tray and down through a downcomer pipe onto the next tray. The bottom tray downcomer is fitted with a seal pot to hold a liquid seal on the trays. The wet glycol, which has now absorbed the water vapor from the gas stream, leaves the bottom of the glycol-gas contactor column, passes through a high pressure glycol filter, which removes any foreign solid particles that may have been picked up from the gas stream, and enters the power side of the glycol pump.

The wet glycol stream leaves the heat exchange coil in the surge tank and enters the stripping still mounted on top of the reboiler at the feed point in the still. The stripping still is packed with a ceramic saddle-type packing, and the glycol flows downward through the column and enters the reboiler. The wet glycol passing downward through the still is contacted by hot rising glycol and water vapors passing upward through the column. The water vapors released in the reboiler and stripped from the glycol in the stripping still pass upward through the still column through an atmospheric reflux condenser that provides a partial reflux for the column. The water vapor then leaves the top of the stripping still column and is released to the atmosphere.

The glycol flows through the reboiler in essentially a horizontal path from the stripping still column to the opposite end. In the reboiler, the glycol is heated to approximately 350–400 °F to remove enough water vapor to reconcentrate it to 99.5% or more. In field dehydration units, the reboiler is generally equipped with a direct-fired firebox, using a portion of the natural gas stream for fuel. In plant-type units, the reboiler may be fitted with a hot oil-heated coil or steam coil. A temperature control in the reboiler operates a fuel gas motor valve to maintain the proper temperature in the glycol. The reboiler is also generally equipped with a high-temperature safety overriding temperature controller to shut down the fuel gas system in case the primary temperature control should malfunction.

In order to provide extra-dry glycol, 99% plus (i.e., <1% v/v water), it is usually necessary to add some stripping gas to the reboiler. A valve and small pressure regulator are generally provided to take a small amount of gas from the fuel gas system and inject it into the bottom of the reboiler through a spreader system. This stripping gas will roll the glycol in the reboiler to allow any pockets of water vapor to escape that might otherwise remain in the glycol due to its normal high viscosity. This gas will also sweep the water vapor out of the reboiler and stripping still. By lowering the partial pressure of the water vapor in the reboiler and still column, the glycol can be reconcentrated to a higher percentage.

The reconcentrated glycol leaves the reboiler through an overflow pipe and passes into the shell side of the heat exchanger-surge tank. In the surge tank the hot reconcentrated glycol is cooled by exchanging heat with the wet glycol stream passing through the coil. The surge tank also acts as a liquid accumulator for feed for the glycol pump. The reconcentrated glycol flows from the surge tank through a strainer and into the glycol pump. From the pump it passes into the shell side of the glycol cooler mounted on the glycol-gas contactor. It then flows upward through the glycol cooler where it is further cooled and enters the column on the top tray.

Glycol dehydrators have several advantages. Initial equipment costs are lower, and the low-pressure drop across absorption towers saves power. Operation is continuous; this is generally preferred to batch operation. Also, makeup requirements may be added readily; recharging of towers presents no problems. Finally, the plant may be used satisfactorily in the presence of materials that would cause fouling of some solid adsorbents. There are also several operating problems. First, suspended foreign matter, such as dirt, scale, and iron oxide, may contaminate glycol solutions. Also, overheating of solution may produce both low and high boiling decomposition products. The resultant sludge may collect on heating surfaces, causing some loss in efficiency, or, in severe cases, complete flow stoppage. Placing a bypass mechanical filter ahead of the solution pump usually prevents such troubles.

When both oxygen and hydrogen sulfide are present in the gas stream, corrosion may become a problem because of the formation of acid material in glycol solution. Also, liquids (e.g., water, low-boiling hydrocarbon derivatives) in the inlet gas may require installation of an efficient separator ahead of the absorber. Highly mineralized water entering the system with inlet gas may, over long periods, crystallize and fill the reboiler with solid salts. Foaming of the solution may occur with a resultant carry-over of liquid. The addition of a small quantity of antifoam compound usually remedies this trouble.

5.5.5.3 Dehydration by cooling

The ability of natural gas to contain water vapor decreases as the temperature is lowered at constant pressure. Cooling for the specific purpose of gas dehydration is sometimes economical if the gas temperature is unusually high. Cooling is often used in conjunction with other dehydration processes. This method is limited in that gases dehydrated by cooling are still at their water dew point unless the temperature is raised again or the pressure is decreased.

Gas processing facilities that cool water may use different methods. In one type, some water is condensed and removed from gas at compressor stations by the compressor discharge coolers. Note that the saturation water content of gases decreases at higher pressure.

Low-temperature separation systems use the expansion-refrigeration cooling of the Joule–Thomson effect to condense water and hydrocarbons from the gas. In this method, hydrates are intentionally allowed to form by expanding the gas from wellhead pressure to pipeline pressure. The resulting dry gas, hydrates, and gas condensate enter a separator containing coils through which the warm well-stream flows to melt the hydrates. EG or DEG may be injected if the available wellhead pressure is insufficient to produce the required dehydration by expansion alone. Low-temperature separation can be an attractive technique since it can dehydrate gas as well as recover valuable liquid hydrocarbons.

All recent lean oil absorption gas plants use mechanical refrigeration to chill the inlet gas stream. EG is usually injected into the gas chilling section of the plant, which simultaneously dehydrates the gas and recovers liquid hydrocarbons, in a manner similar to the low-temperature separators.

References

Ahmadi, G., Ji, C., Smith, D.H. 2007. Production of Natural Gas from Methane Hydrate by a Constant Downhole Pressure Well. Energy Convers. Management, 48: 2053–2068.

Belosludov, V.R., Subbotin, O.S., Krupskii, D.S., Belosludov, R.V., Kawazoe, Y., and Kudoh, J. 2007. Physical and Chemical Properties of Gas Hydrates: Theoretical Aspects of Energy Storage Application. Materials Transactions, 48(4): 704–710.

Bishnoi, P.R., and Clarke, M.A. 2006. Natural Gas Hydrates. In: Encyclopedia of Chemical Processing. 2006 by Taylor & Francis. Taylor & Francis Publishers, Philadelphia, Pennsylvania.

Carroll, J.J. 2003a. Natural Gas Hydrates. Gulf Professional Publishing: Burlington, Vermont.

Carroll, J.J. 2003b. Natural Gas Hydrates: A Guide for Engineers, Gulf Professional Publishing, Amsterdam, Netherlands.

Castaldi, M.J., Zhou, Y., and Yegulalp, T.M. 2007. Down-hole Combustion Method for Gas Production From Methane Hydrates. Journal of petroleum Science and Engineering, 56: 176–185.

Collett, T.S. 2001. Natural-Gas Hydrates; Resource of The Twenty-First Century? Journal of the American Association of Petroleum Geologists, 74: 85–108.

Collett, T.S., Johnson, A.H., Knapp, C.C., and Boswell, R. 2009, Natural Gas Hydrates: A Review. In: Natural Gas Hydrates – Energy Resource Potential and Associated Geologic Hazards. T.S. Collett, A.H. Johnson, C.C. Knapp, and R. Boswell (Editors). AAPG Memoir No. 89, p. 146–219. American Association of Petroleum Geologists, Tulsa, Oklahoma.

Collett, T.S. 2010. Physical Properties of Gas Hydrates: A Review. Journal of Thermodynamics Volume 2010, Article ID 271291; doi:10.1155/2010/271291; https://www.hindawi.com/journals/jther/2010/271291/ accessed November 1, 2017.

Davies, P. 2001. The New Challenge of Natural Gas. Proceedings. OPEC and the Global Energy Balance: Towards A Sustainable Future. Vienna, September 28.

Demirbaş, A. 2010. Methane Hydrates as a Potential Energy Resource: Part 2 – Methane Production Processes from Gas Hydrates. Energy Convers. Manage. 51: 1562–1571.

Fan, Z., Sun, C., Kuang, Y., Wang, B., Zhao, J., and Song, Y. 2017. MRI Analysis for Methane Hydrate Dissociation by Depressurization and the Concomitant Ice Generation. Energy Procedia, 105: 4763–4768.

Gabitto, J., and Tsouris, C. 2010. Physical Properties of Gas Hydrates: A Review. Journal of Thermodynamics, Volume 2010, Article ID 271291. https://www.hindawi.com/journals/jther/2010/271291/citations/

Giavarini, C., Maccioni, F., and Santarelli, M.L. 2003. Formation Kinetics of Propane Hydrate Ind. Eng. Chem. Res., 42: 1517–1521.

Giavarini, C., and Maccioni, F. 2004. Self-Preservation at Low Pressure of Methane Hydrates with Various Gas Contents. Ind. Eng. Chem. Res., 43: 6616–6621.

Giavarini, C., Maccioni, F., and Santarelli, M.L. 2005. Characterization of Gas Hydrates by Modulated Differential Scanning Calorimetry. Pet. Sci. and Technol., 23: 327–335.

Goel, N., Wiggins., M., and Shah, S. 2001. Analytical Modeling of Gas Recovery from In Situ Hydrates Dissociation. J. Pet. Sci. Eng., 29: 115–127.

Lee, S., Lee, Y., Lee, J., Lee, H., and Seo Y. 2013. Experimental Verification of Methane-Carbon Dioxide Replacement in Natural Gas Hydrates Using a Differential Scanning Calorimeter. Environ. Sci. Technol., 47: 13184–13190.

Lorenson, T.D., and Collett, T.S. 2000. Gas Content and Composition of Gas Hydrate from Sediments of the Southeastern North American Continental Margin. In: Proceedings of the Ocean Drilling Program, Scientific Results. C.K., Paull, R. Matsumoto, P.J. Wallace, and W.P. Dillon, W.P. (Editors). Volume 164: 37–46.

Makogon, Y.F. 1997. Hydrates of Hydrocarbons. PennWell Books, Tulsa, Oklahoma.

Makogon, Y.F., Holditch, S.A., Makogon, T.Y. 2007. Natural Gas Hydrates – A Potential Energy Source for the twenty-first Century, Journal of Petroleum Science and Engineering 56 (1–3): pp. 14–31

Makogon, Y.F. 2010, Natural Gas Hydrates – A Promising Source of Energy. Journal of Natural Gas Science and Engineering, 2(1): pp. 49–59.

Seo, Y., Kang, S.P., and Jang, W. 2009. Structure and Composition Analysis of Natural Gas Hydrates: 13C NMR Spectroscopic and Gas Uptake Measurements of Mixed Gas Hydrates. J. Phys. Chem., 113(35): 9641–9649.

Sloan, E.D. Jr. 1998a. Gas Hydrates: Review of Physical/Chemical Properties. *Energy & Fuels*, 12(2): 191–196.

Sloan, E.D., Jr. 1998b. Clathrate Hydrates of Natural Gases 2nd Edition. Marcel Dekker Inc., New York.

Sloan, E.D. Jr. 2006. Clathrate Hydrates of Natural Gases 3rd Edition. Marcel Dekker Inc., New York.

Stern, L., Kirby, S., Durham, W., Circone, S., and Waite, W.F. 2000. Laboratory Synthesis of Pure Methane Hydrate Suitable for Measurement of Physical Properties and Decomposition Behavior. Proceedings. Natural Gas Hydrate in Oceanic and Permafrost Environments. M.D. Max (Editor). Kluwer Academic Publishers, Dordrecht, Netherlands. Page 323–348.

Stoll, R.G., and Bryan, G.M. 1979. Physical Properties of Sediments Containing Gas Hydrates. Journal of Geophysical Research, 84(B4): 1629–1634.

Wang, X., and Economides, M.J. 2012. Natural Gas Hydrates as an Energy Source – Revisited 2012. Proceedings. SPE International Petroleum Technology Conference 2012. 1: 176–186. Society of Petroleum Engineers, Richardson, Texas.

Wang, B., Dong, H., Fan, Z., Zhao, J., and Song, Y. 2017. Gas Production from Methane Hydrate Deposits Induced by Depressurization in Conjunction with Thermal Stimulation. Energy Procedia, 105: 4713–4717.

Yang, X., Qin, M. 2012. Natural Gas Hydrate as Potential Energy Resources in the Future, Advanced Materials Research, 462: 221–224.

Chapter 6
Production

6.1 Introduction

In conventional natural gas reservoirs, the gas generally flows from the reservoir, through wells, and to the surface facilities. Typically, the wells are formed by drilling vertically from the surface of the Earth into the gas-bearing formation. However, natural gas is also produced from tight (low-to-no permeability) formations (such as shale formations) and carbonate formations by forcing water, chemicals, and sand down a well and into the formation under high pressure. In these formations, techniques such as horizontal drilling and hydraulic fracturing are processes that are used to increase the productivity of the reservoir.

The design of an optimum development plan for the recovery of the natural gas depends on the characteristics of the producing formation. Thus, a knowledge of the field parameters, such as the total natural gas reserves, the potential productivity of a well and the dependence of production rates on the depletion of the gas reserves, is required prior to designing the development scheme for the reservoir.

One major issue for a gas recovery project is the prediction of delivery rates from a group of wells or field. The elements in the overall gas production system must then include (i) flow through the reservoir, (ii) flow out of the reservoir and through the production strings of the wells, (iii) flow through the field gathering system and processing equipment, (iv) transmission to the gas processing plant, and (v) then to the point of sales.

6.2 Darcy and non-Darcy flow in porous media

In the phenomenon known as Darcy flow, the velocity (v) of a fluid traveling through a porous medium is directly proportional to the pressure gradient, $\Delta P/\Delta r$ (a difference in pressure ΔP over some finite distance Δr), and inversely proportional to the viscosity of the fluid or gas. On the other hand, the flow of a reservoir fluid that deviates from Darcy's law assumes laminar flow in the formation. Non-Darcy flow is typically observed in high-rate gas wells when the flow converging to the wellbore reaches flow a velocity that exceeds the Reynolds number for laminar flow or for Darcy flow and results in turbulent flow.

The radial flow of gas in the reservoir drainage volume of a well is based on Darcy's law for viscous flow and on high-velocity effects that may occur near the wellbore. Thus, to perform natural gas well deliverability calculations, it is essential to understand the fundamentals of gas flow in porous media. Fluid flow is affected by the competing inertial and viscous effects, combined by the well-known Reynolds

https://doi.org/10.1515/9783110691023-006

number, whose value delineates laminar from turbulent flow. In porous media, the limiting Reynolds number is equal to 1 based on the average grain diameter.

The Reynolds number (R_e, which is an important dimensionless quantity in fluid mechanics) assists in the prediction of flow patterns in different fluid flow situations. Thus:

$$R_e = \rho \upsilon L / \mu$$

In this equation, ρ is the density of the fluid (kg/m^3), υ is the flow speed (meters per second), L is a characteristic linear dimension (m), and μ is the dynamic viscosity of the fluid [Pa·s or N·s/m^2 or kg/(m·s)].

If the Reynolds numbers are low, the flow tends to be dominated by laminar (sheet-like) flow, while at a high Reynolds number, the flow tends to be turbulent. The turbulence results from differences in the fluid's speed and direction, which may sometimes intersect or even move counter to the overall direction of the flow (eddy currents). These eddy currents begin to churn the flow, using up energy in the process, which for liquids increase the chances of cavitation.

The Reynolds number has wide applications (such in the current context, as liquid flow in a pipe) and is used to predict the transition from laminar flow to turbulent flow, and is used in the scaling of similar but different-sized flow situations. The predictions of the onset of turbulence and the ability to calculate scaling effects can be used to help predict fluid behavior on a larger scale.

When a gas well is first produced after being shut in for a period of time, the gas flow in the reservoir follows an unsteady-state behavior until the pressure drops at the drainage boundary of the well. Then the flow behavior passes through a transition period, after which it attains a steady-state or semi-steady-state condition. More generally, gas well flow in a reservoir is influenced by damage or improvement to the permeability near the wellbore. Also, if the flow rate is sufficiently high, high-velocity flow phenomena will occur around the wellbore. These effects are incorporated into a generalized radial flow equation. Also, pseudo radial flow is a type of radial flow, which appears at a later time. For fractured wells, if the fracture half-length is short, pseudo radial flow will appear and for those horizontal wells whose horizontal sections are relatively short, pseudo radial flow may also appear.

Fluid flow that deviates from Darcy's law, which assumes laminar flow in the formation and non-Darcy flow is typically observed in high-rate gas wells when the flow converging to the wellbore reaches a flow velocity that exceeds the Reynolds number for laminar or Darcy flow, and results in turbulent flow.

Non-Darcy flow occurs in the near-wellbore region of high-capacity gas and condensate reservoirs. As the flow area is reduced substantially, the velocity increases, inertial effects become important, and the gas flow becomes non-Darcy. Thus, Darcy's law is inadequate for representing high-velocity gas flow in porous media, such as near the wellbore.

6.2.1 Gas well inflow under Darcy flow and non-Darcy flow

Well inflow is a term that describes the fluid flow from the reservoir into the sand face, takes into account the reservoir characteristics, the well geometry (vertical, horizontal, complex architecture), the near-wellbore zone or other features such as hydraulic or natural fractures and the pressure drawdown. Different flow regimes that take into account boundary effects such as steady-state, pseudosteady-state, and transient behavior are considered.

For low permeability reservoirs in mature environments such as the United States and continental Europe, it is sufficient to assume that gas flow in the reservoir obeys Darcy's law. Newly found reservoirs are primarily offshore, in developing nations, and are of moderate to high permeability, that is, 1–100 md. As well deliverability increases, turbulence becomes increasingly dominant in the production of gas wells. For reservoirs, whose permeability is more than 5 md, turbulence effects may account for a 20–60% reduction in the production rate of an open-hole well (when laminar flow is assumed). Turbulence in such cases practically overwhelms all other factors, including damage (Wang and Economides, 2009).

6.2.2 Steady-state and pseudosteady-state flow

The term "steady-state flow" describes the behavior of a reservoir fluid when the pressure (at the wellhead or at the bottomhole) and flow rates are constant. This behavior usually happens when there is pressure support, either naturally through an aquifer, or through water injection. The well performance under steady-state flow can be derived from Darcy's law (Wang and Economides, 2009).

On the other hand, pseudosteady-state flow is a flow regime that occurs in closed reservoirs, after the pressure transient has reached all the boundaries of the reservoir. This includes not only the case of physically bounded reservoirs, but also the case of a well surrounded by other producing wells.

Drainage areas can either be described by natural limits such as faults, and pinch-outs (no-flow boundary), or can be artificially induced by the production of adjoining wells, which is often referred to as "pseudosteady state." The pressure at the outer boundary is not constant but instead declines at a constant rate with time, that is, = const. Therefore, a more useful expression for the pseudosteady-state equation would be one using the average reservoir pressure, p. It is defined as a volumetrically weighted pressure and in practice can be obtained from periodic pressure buildup tests (Wang and Economides, 2009).

6.2.3 Transient flow

Transient flow is a fluid dynamics condition where the velocity and pressure of a fluid (gas or liquid) flow change over time due to changes in the system. These changes may be caused by (i) the starting or shutting down of a pump, (ii) the opening or closing of valves, or (iii) the fluctuations in supply pressure from reservoirs or tanks.

At early time, the flowing bottomhole pressure of a producing well is a function of time if the rate is held largely constant. This type of flow condition is called transient flow and is used deliberately during a pressure transient test. In practice, the well is usually operated under the same wellhead pressure (which is imposed by the well hardware such as chokes), the resulting flowing bottomhole pressure is also largely constant, and the flow rate will vary with time.

6.2.4 Turbulent flow

Turbulence is a flow regime where random fluctuations occur with time and is due to high flow rate due, for example, to an increase in the pressure difference. Under these conditions, conventional flow becomes very difficult to obtain/maintain.

Thus, turbulent flow is due to high flow rate due to increase in pressure difference, gas molecule continuously change its direction and pore cross section area due to pressure difference close to the wellbore. Under these conditions, conventional flow becomes very difficult to obtain.

Reservoir rock properties and flow rate are affected by gas flow through the formation at high velocity due to turbulent effects. These effects are due to the fluid deceleration at pore body and acceleration at pore throat and tortuosity is also present there. Turbulence occurs especially when the permeability is high and the well deliverability increases and plays a considerable role in well performance, showing that the production rate is affected by itself; the larger the potential rate, the larger the relative detrimental impact (Wang and Economides, 2009).

6.3 Hydraulic fracturing

Extraction of natural gas from conventional reservoirs involves drilling through impervious rock that traps concentrated underground reservoirs of natural gas (Speight, 2014a, 2016a).

However, not all of the natural gas is conveniently located in conventional and accessible reservoirs. Many natural gas resources are trapped in the pore spaces and cracks within impermeable sedimentary rock formations – shale formations, tight sandstone formations, and tight carbonate formations are examples of such reservoirs. These reservoirs can vary in thickness – the shale formations are rela-tively thin layers (albeit deep under the ground) but cover extensive horizontal areas and a vertically drilled well will only access a small area of the reservoir and, by inference due to the impermeable nature of the formation, a minimal part of the resource. However, when the drilling operation can deviate from the conventional vertical plane and move in the horizontal plane much more of the reservoir resource becomes accessible (Ely, 1985; Gidley et al., 1990).

In order to improve recovery of natural gas from reservoirs, hydraulic fracturing has become an essential part of natural gas production, especially, production of natural gas that is trapped in low-permeability (shale, sandstone, and carbonate) formations (Figures 6.1 and 6.2) (Agarwal et al., 1979). The procedure significantly improves the recovery from the reservoir by stimulating the movement of natural gas.

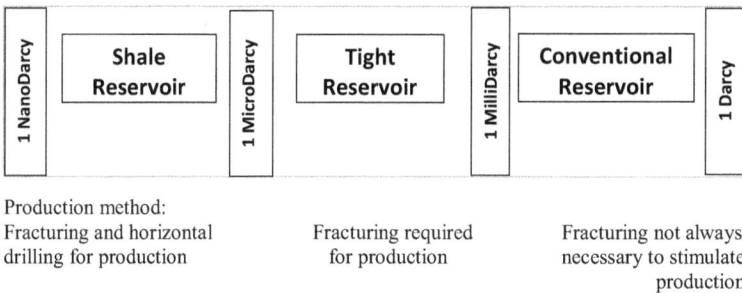

Production method:
Fracturing and horizontal drilling for production

Fracturing required for production

Fracturing not always necessary to stimulate production

Figure 6.1: Illustration of reservoir types based on permeability and production methods.

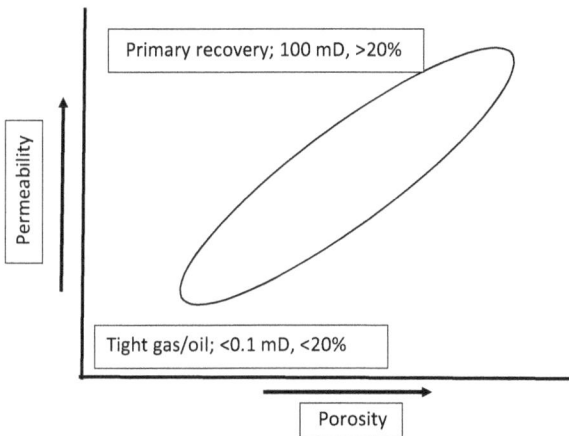

Figure 6.2: General trends in the relationship between porosity and permeability.

   Application of hydraulic fracturing (hydraulic fracture stimulation, fraccing, frack-ing) has evolved as one of the key techniques that allow access to natural gas from tight formations. Also key to shale gas development is the presence of natural fractures and planes of weakness that can result in complex fracture geometries during stimula-tion (Reddy and Nair, 2012). Furthermore, the presence and ability to open and main-tain flow in both primary and secondary natural fracture systems are critical to shale gas production (King, 2010). The technology involves pumping water, a proppant such as sand (Tables 6.1 and 6.2) to keep the fractures open, and a small amount of one or more chemical additives into the well to assist the fracture process, by which the natu-ral gas is enabled to flow to the wellbore.

Table 6.1: Highlights in the development of hydraulic fracturing.

Date	Comment
Early 1900s	Natural gas extracted from shale wells
	Vertical wells fractured with foam
1947	Klepper gas unit no. 1; first well to be fractured to increase productivity
1949	Stephens County, Oklahoma; first commercial fracturing treatment
1950	Fracturing with cement pumpers
1950s	Evolution of fracture geometry
	Increasing well productivity
1960s	Fracturing pumpers and blenders
1970s	Massive hydraulic fracturing
	Increase recoverable reserves
	Hydraulic fracturing in Europe
1983	First gas well drilled in Barnett Shale in Texas
1980s	Evolution of proppant transport
	Fracture conductivity testing
	Cross-linked gel fracturing fluids developed; used in vertical wells
1990s	First horizontal well drilled in Barnett Shale
	Orientation of induced fractures identified
	Foam fracturing
1996	Slickwater fracturing fluids introduced
1996	Microseismic post-fracturing mapping developed

Table 6.1 (continued)

Date	Comment
1997	Hydraulic fracturing in Barnett shale
	Slickwater fracturing developed
1998	Slickwater refracturing of originally gel-fractured wells
2002	Multi-stage slickwater fracturing of horizontal wells
2003	First hydraulic fracturing of Marcellus Shale
2004	Horizontal wells become dominant
2005	Increased emphasis on improving the recovery factor
2007	Use of multi-well pads and cluster drilling

Table 6.2: Proppant-type definition.

Sand	Includes all raw sand types.
Resin-coated sand	Includes only resin-coated proppants for which the substrate is sand; does not include any double-counting with the "sand category" described above.
Ceramic	Any proppant for which the substrate is a ceramic or otherwise manufactured proppant, resin-coated ceramic proppant is included in this category.

In fact, the use of horizontal drilling in conjunction with hydraulic fracturing has greatly expanded the ability of producers to profitably recover natural gas from low-to-no permeability geologic plays – particularly, shale plays and other tight reservoirs (Speight, 2016b). In the process, the pressure exceeds the rock strength and the fluid opens or enlarges fractures in the rock. As the formation is fractured, a propping agent, such as sand or ceramic beads, is pumped into the fractures to keep them from closing as the pumping pressure is released. The fracturing fluids (water and chemical additives) are then returned back to the surface. The natural gas will then flow from pores and fractures in the rock into the well for subsequent extraction to the surface.

Moreover, advances in directional drilling technology have allowed wells to deviate from nearly vertical to extend horizontally into the reservoir formation, which further increases contact of a well with the reservoir. Directional drilling technology (sometimes referred to as *deviated drilling technology*) also enables drilling a number of wells from a single well pad and combined with hydraulic fracturing with directional drilling has opened up the production of tight (low-to-no permeability) natural gas reservoirs, particularly unconventional gas shales.

Recovery, at any point during the well development/production can be defined as the percentage of original natural gas that has been produced. Recovery from any natural gas reservoir can be expressed using an equation of the type:

$$G_{pD} = G_p/G = (G - G_a)/G$$

In this equation, G_{pD} is the recovery factor, which is a dimensionless number, Gp is the cumulative volume of gas produced, MSCF, G is the original volume of gas in place, MSCF, and Ga is the volume of gas in place, MSCF, at the time of abandonment of the well.

If the recovery of the gas in the number of moles of gas, n, is preferred instead of the volume, the above equation can be rewritten in the in terms of the real gas law and in the form:

$$G_{pD} = 1 - (P_a V_a Z_i)/(P_i V_i Z_a)$$

In this equation, P_i is the initial reservoir pressure, psia, V_i is the initial hydrocarbon pore volume (HCPV), P_a is the reservoir pressure at abandonment, psia, V_a is the hydrocarbon pore volume (HCPV) at abandonment, and Z is the real gas compressibility factor, which is a dimensionless number. Furthermore, since P_i and V_i are fixed, the only way that recovery can be increased is to decrease P_a or V_a.

Typically, in volumetric reservoirs, P_a is minimized while V_a remains nearly unchanged. The opposite is true of strong water drive reservoirs, which there is often little decrease in pressure while nearly the entire reservoir is swept by the aquifer. In most reservoirs recovery can be increased by minimizing either the pressure or initial hydrocarbon pore volume and neglecting the other factor.

Finally, naturally fractured reservoirs contain secondary porosity or induced porosity in addition to their original primary porosity. Induced porosity is formed by tension or shear stresses causing fractures in a competent or brittle formation. Fracture porosity is small. Values between 0.0001 and 0.001 of rock volume are typical (0.01–0.1%). Fracture-related porosity, such as solution porosity in granite or carbonate reservoirs, may attain much larger values, but the porosity in the actual fracture is still small. There are, of course, exceptions to all rules of thumb. In rare cases, such as the cooling of intrusive minerals or surface lava flows, in which natural fracture porosity may exceed 10%. When buried and later filled with hydrocarbons, they form interesting reservoirs.

6.3.1 The process

Hydraulic fracturing is applied after well completion to facilitate movement of the reservoir fluids to the well and thence to the surface. This process creates access to more natural gas but requires the use of large quantities of water and fracturing fluids, which are injected underground at high volumes and pressure. Thus, the sequence of fracturing a particular formation typically consists of: (i) an acid stage, (ii) a pad stage, (iii) a prop sequence stage, and (iv) a flushing stage.

Briefly, the process involves use of a *perforating gun,* which is lowered into a newly drilled well and lined up precisely within the target formation (a tight shale, sandstone, or carbonate formation) using seismic images, well logs, global positioning systems, and other indicators to target the spots from which natural gas are most likely to occur. When fired, the gun punches small holes into the well casing, cement, and rock after which the fracturing fluid is forced out of the perforations under high pressure. This creates fractures (small cracks) in the formation that allow the natural gas to flow from the reservoir into the wellbore. The fracturing fluid contains proppants such as sand or other similarly sized materials (in order to maintain the fractures) created by the pressure treatment in the open position, thus preventing closure when the pressure treatment is terminated. Although the fracturing fluid (*slickwater*) is predominantly water, it does contain chemicals (in addition to the proppant), which can pose an environmental risk (Green, 2014, 2015).

The acid stage consists of several thousand gallons of water mixed with a dilute acid, such as hydrochloric or muriatic acid, which serves to clear cement debris in the wellbore by dissolving carbonate minerals and opening fractures near the wellbore. The *pad stage* consists of use of approximately 100,000 gallons of water (or more), 100,000 gallons of slickwater without proppant material. In this slickwater pad stage, the slickwater solution fills the wellbore and opens the formation, which helps to facilitate the flow and placement of proppant material.

The *prop sequence stage* may consist of several sub-stages of the use of water combined with proppant material, which consists of a fine mesh sand or ceramic material, and is intended to keep open (prop) the fractures created and/or enhanced during the fracturing operation after the pressure is reduced; this stage may collectively use several hundred thousand gallons of water. The proppant material may vary from a finer particle size to a coarser particle size throughout this sequence. The *flushing stage* consists of a volume of freshwater sufficient to flush the excess proppant from the wellbore; the amount of water used is dependent upon the site characteristics (including the character of the subterranean formations).

Most of the fluids used in hydraulic fracturing are water and chemicals (typically 1% v/v of the water). The formulas for fracturing fluids vary, partly depending on the composition of the gas-bearing formations, remembering that all gas-bearing formations are not the same even when the formations are composed of the same minerals (shale, sandstone, or carbonate). In addition, some of the chemical additives can be hazardous if not handled carefully (Table 6.3) and caution is advised since the amount of the chemical(s) must not exceed the amount specified in regulatory requirements related to handling hazardous materials. Even if the chemical is one that is indigenous to the sub-surface (and supposedly benign because it is found naturally), the amount used must not exceed the indigenous amount – in some case, exceeding the indigenous amount of a chemical can cause environmental problems.

Table 6.3: Additives used in the hydraulic fracturing process.

Water and sand (ca. 98% v/v)			
Water	Expands the fracture and delivers sand	Some stays in the formation while the remainder returns with natural formation water as produced water (actual amounts returned vary from well to well)	Landscaping and manufacturing
Sand (proppant)	Allows the fractures to remain open so that the oil and natural gas can escape	Stays in the formation, embedded in the fractures (used to "prop" fractures open)	Drinking water filtration, play sand, concrete and brick mortar
Other additives (ca. 2%)			
Acid	Helps dissolve minerals and initiate cracks in the rock	Reacts with the minerals present in the formation to create salts, water, and carbon dioxide (neutralized)	Swimming pool chemicals and cleaners
Antibacterial agent	Eliminates bacteria in the water that produces corrosive by-products	Reacts with microorganisms that may be present in the treatment fluid and formation; these micro-organisms break down the product with a small amount returning to the surface in the produced water	Disinfectant; sterilizer for medical and dental equipment
Breaker	Allows a delayed breakdown of the gel	Reacts with the crosslinker and gel in the formation making it easier for the fluid to flow to the borehole; this reaction produces ammonia and sulfate salts, which are returned to the surface in the produced water	Hair colorings, as a disinfectant and in the manufacture of common household plastics
Clay stabilizer	Prevents clay minerals in the formation from swelling	Reacts with clay minerals in the formation through a sodium – potassium ion exchange; this reaction results in sodium chlorine (table salt), which is returned to the surface in the produced water	Low-sodium table salt substitutes, medicines, and IV fluids

Table 6.3 (continued)

Other additives (ca. 2%)			
Corrosion inhibitor	Prevents corrosion of the pipe	Bonds to the metal surfaces, such as pipe, downhole; any remaining product that is not bonded is broken down by micro-organisms and consumed or returned to the surface in the produced water	Pharmaceuticals, acrylic fibers, and plastics
Crosslinker	Maintains fluid viscosity as temperature increases	Combines with the "breaker" in the formation to create salts that are returned to the surface in produced water	Laundry detergents, hand soaps, and cosmetics
Friction reducer	Minimizes friction	Remains in the formation where temperature and exposure to the breaker allows it to be broken down and consumed by naturally occurring micro-organisms; a small amount returns to the surface with the produced water	Cosmetics including hair, make-up, nail and skin products
Gelling agent	Thickens the water to suspend the sand	Combines with the breaker in the formation making it easier for the fluid to flow to the borehole and return to the surface in the produced water	Cosmetics, baked goods, ice cream, toothpastes, sauces, and salad dressings
Iron control	Prevents precipitation of metal in pipe	Reacts with minerals in the formation to create simple salts, carbon dioxide, and water, all of which are returned to the surface in the produced water	Food additives; food and beverages; lemon juice
Non-emulsifier	Breaks or separates oil/water mixtures (emulsions)	Generally, returns to the surface with produced water, but in some formations it may enter the gas stream and return to the surface in the produced oil and natural gas	Food and beverage processing, pharmaceuticals, and wastewater treatment

Table 6.3 (continued)

Other additives (ca. 2%)			
pH adjusting agent	Maintains the effectiveness of other components, such as crosslinkers	Reacts with acidic agents in the treatment fluid to maintain a neutral (non-acidic, non-alkaline) pH; this reaction results in mineral salts, water and carbon dioxide – a portion of each is returned to the surface in the produced water	Laundry detergents, soap, water softeners, and dish washer detergents

Safe handling of all water and other fluids on the site, including any added chemicals, must be a high priority and compliance with all regulations regarding containment, transport, and spill handling is essential. When it comes to disposal of the fracturing fluid, there are options. For example, the fluid, when possible without causing adverse effects to the environment, can be reused for additional wells in a single field – this reduces the overall use of freshwater and reduces the amount of recovered water and chemicals that must be sent for disposal. However, in such cases, recognition of the geological or mineralogical similarities or difference within a site must has been determined to assure minimal environmental damage. In addition, tanks (or *lined* storage pits) for the storage of recovered water are also a necessity until the water can be sent for disposal of in a permitted saltwater injection disposal well or taken to a treatment plant for processing. The linings of such pits must be in accordance with local environmental regulations.

In addition, production zones should have that have multiple confining layers above the zone to keep the injected fluids within the target gas-bearing formation. In addition, multiple layers of well casing and cement (similar to production wells) should be used with periodic mechanical integrity tests to verify that the casing and cement are holding the liquids. The amount and pressure of the injected fluid (specified in each well permit) should be monitored to maintain the fluids in the target zone and the pressure in the injection well and the spaces between the casing layers (also called the annuluses) should also be monitored check and verify the integrity of the injection well.

In reasonable non-fractured reservoirs, it is usually possible to estimate permeability, and hence productivity (Speight, 2014a, 2016b), but this is not always possible in fractured reservoirs. Although both the presence of fractures and the presence of a reservoir can be determined from logs, a production test will be needed to determine whether economic production is possible. The test must be analyzed carefully to avoid over optimistic predictions based on the flush production rates associated with the fracture system. Local correlations between fracture intensity observed on logs and production rate are also used to predict well quality.

6.3.2 Equipment

Conventional cement-pumping and acid-pumping equipment was used initially to execute fracturing treatments. One to three units equipped with one pressure pump delivering 75–125 hhp were adequate for the small volumes injected at the low rates. However, as treating volumes increased, accompanied by a demand for greater injection rates as well as special pumping and blending equipment was developed and development continues.

Initially, sand was added to the fracturing fluid by pouring it into a tank of fracturing fluid over the suction. More recently, with less-viscous fluid(s), a ribbon or paddle type of batch blender was used after which a continuous proportioning blender utilizing a screw to lift the sand into the blending tub was developed. As the procedure evolved, blending equipment has also had to evolve to meet the need for proportioning a large number of dry and liquid additives, then uniformly blending them into the base fluid and adding the various concentrations of sand or alternate propping agents. In fact, the hydraulic fracturing treatment follows the actual drilling and completion of the well (Hubbert and Willis, 1957; Hibbeler and Rae, 2005; Arthur et al., 2009).

In the initial stages of a fracturing project, the drilling may be the same as drilling a conventional reservoir. Thus, in the process, a borehole is drilled vertically, then a casing is put in place after which cement and mud are pumped into the annulus to form a barrier between the borehole and adjacent formations. Drilling is then continued, to an adequate depth adjacent (sometimes within) within the producing reservoir. From this point (the *kick-off point*), the wellbore is deviated gradually until it curves into a horizontal plane and drilled a distance on the order of 1,000 feet to more than 5,000 feet (Arthur et al., 2009).

The hydraulic fracturing procedure is then initiated. The process involves fracturing at isolated intervals along the horizontal well since it is difficult (if not impossible) to apply pressure along the entire length of the wellbore because of loss of pressure efficiency over the distance involved (1,000–5,000 feet). The fracturing areas are isolated using packers and perforations are created in the wellbore within the interval bounded by packers (Arthur et al., 2009). In some fracture treatments, acid is pushed through the perforated interval to help breakdown any barrier that might be due to the characteristics of the formation(s) surrounding the wellbore.

In order to develop a formation containing natural gas, the wellbore is drilled in successive sections through the rock layers. Once the desired length of each wellbore section has been drilled, the drilling assembly is removed, and steel casing is inserted and cemented in place. As the well is constructed, concentric layers of steel casing and cement form the barrier to protect groundwater resources from the fluids that will later flow inside the well. In the next step, a section of casing within the formation is perforated at the desired location for gas production.

At this stage, the well is ready for hydraulic fracturing process, which involves pumping fluid through the perforations. The fracturing fluid exerts pressure against the rock, creating tiny cracks, or fractures, in the reservoir deep underground. The fluid is predominantly water, proppant (grains of sand or ceramic particles), and a small amount (on the order of 1% v/v) of chemical additives. Once fluid injection stops, pressure begins to dissipate, unless the necessary steps are taken, the fractures previously held open by the fluid pressure begin to close. The necessary steps include the injection of proppants, which act as wedges to hold open the narrow fractures, thereby creating pathways for the natural gas and the fracturing fluids to flow more easily to the well. A plug is set inside the casing to isolate the stimulated section of the well and the perforate-inject-plug cycle is repeated at regular intervals along the targeted section of the reservoir. Finally, the plugs are drilled out, allowing the natural gas to flow into the well casing and up to the surface. The fracturing fluid mixture is separated at the surface, and the fracturing fluid (also known as flowback water) is captured in tanks or lined pits. The fracturing fluids are then disposed of according to the regulatory-approved methods.

Briefly, several definitions exist for the terms *produced water* and *flowback water* and indicate the confusion in the use of the terminology. The common definitions are (i) produced water is any of the many types of water produced from gas wells and (ii) flowback water is the hydraulic fracturing fluid that returns to the surface after a hydraulic fracture is completed.

Fractures from both horizontal and vertical wells can propagate vertically out of the intended zone thereby (i) reducing stimulation effectiveness, (ii) wasting proppant and fluids, and potentially connecting up with other hydraulic fracturing stages or unwanted water or gas intervals, which can also lead to a variety of environmental issues. The direction of lateral propagation is largely dictated by the horizontal stress regime, but in areas where there is low horizontal stress anisotropy or in reservoirs that are naturally fractured, fracture growth is not always easy to predict (Hammack et al., 2014). In shallow zones, horizontal hydraulic fractures can develop because the weight of the overburden – the vertical stress component – is smallest. A horizontal hydraulic fracture reduces the effectiveness of the stimulation treatment because it most likely forms along horizontal areas of weakness – such as the areas between the formation strata – and is aligned preferentially to formation vertical permeability, which is typically much lower than horizontal permeability.

More specifically, after a hydraulic fracture is initiated, the degree to which the fracture grows laterally or vertically depends on numerous factors, such as confining stress, fluid leak from the fracture, fluid viscosity, fracture toughness, and the number of natural fractures in the reservoir. Thus, prediction of the precise behavior of the fracture is difficult and, in many cases, may even be impossible because of incorrect information and assumptions used in planning the fracturing project.

The extent of a hydraulic fracture is a complex relationship between the strength of the rock and the pressure difference between the rock and the fracturing pressure.

The extent is defined by the fracture dimensions – height, depth of penetration (wing length or fracture length), and aperture (width or opening). One measure of the strength of the rock is the Poisson ratio. Thus, when a material is compressed in one direction, it tends to expand in the other two directions perpendicular to the direction of compression (the Poisson effect) and the Poisson ratio (υ, the fraction or percent) of expansion divided by the fraction (or percent) of compression is a measure of this effect. The Poisson ratio is low (0.10–0.30) for most sandstone formations and carbonates – rocks that fracture relatively easily. On the other hand, the Poisson ratio is high (0.35–0.45) for shale, sandstone, and coal – rocks that are more elastic and are harder to fracture (Sone and Zoback, 2013a, 2013b). Shale is often the upper and lower barrier to the height of a fracture in conventional sandstone.

One other aspect of equipment consideration relates to the inducement of fracturing by pneumatic methods. Pneumatic fractures can be generated in geologic formations if air or any other gas is injected at a pressure that exceeds the natural strength as well as the in-situ stresses present in the formation. As noted earlier, pneumatic fracture propagation will be predominantly horizontal at over-consolidated formations. However, in shallow recent fills, some upward inclination of the fractures has been observed, the reason for which is attributed to the lack of stratification and consolidation in these formations. The amount of pressure required to initiate pneumatic fractures is dependent on the cohesive or tensile strength of the formation, as well as the overburden pressure (dependent upon the depth and density of the formation). The most important system parameter for efficient pneumatic fracturing is injection flow rate, as it largely determines the dimensions of a pneumatic fracture. Once a fracture has been initiated, it is the high volume airflow, which propagates the fracture and supports the formation. The design goal of a pneumatic fracturing system therefore becomes one of providing the highest possible flow rate. Field observations indicate that pneumatic fractures reach their maximum dimension in less than 20 s, after which continued injection simply maintains the fracture network in a dilated state (in essence, the formation is "floating" on a cushion of injected air). Pneumatically induced fractures continue to propagate until they intersect a sufficient number of pores and existing discontinuities, so that leak-off (fluid loss) rate into the formation exactly equals the injection flow rate.

An individual pneumatic fracture is accomplished by (i) advancing a borehole to the desired depth of exploration and withdrawing the auger, (ii) positioning the injector at the desired fracture elevation, (iii) sealing off a discrete 1 or 2 feet interval by inflating the flexible packers on the injector with nitrogen gas, (iv) applying pressurized air for approximately 30 s, and (v) repositioning the injector to the next elevation and repeating the procedure. A typical fracture cycle takes approximately 15 min, and a production rate of 15–20 fractures per day is attainable with one rig.

The pneumatic fracturing procedure typically does not include the intentional deposition of foreign propping agents to maintain fracture stability. The created fractures are thought to be self-propping, which is attributed to both the asperities present along the fracture plane and the block shifting, which takes place during injection. The aperture or thickness of a typical pneumatically induced fracture is approximately 0.5–1 mm. Testing to date has confirmed fracture viability in excess of 2 years, although the longevity is expected to be highly site-specific.

Without the carrier fluids used in hydraulic fracturing, there are no concerns with fluid breakdown characteristics for pneumatic fracturing. There is also the potential for higher permeability within the fractures formed pneumatically, in comparison to hydraulic fractures, as these are essentially air space and are devoid of propping agents. The open, self-propped fractures resulting from pneumatic fracturing are capable of transmitting significant amounts of fluid flow.

6.3.3 Well development and completion

Well development is an integral part of the hydraulic fracturing process and is typically divided into two stages: (i) the drilling stage and (ii) the completion stage. For a successful fracturing operation, it is important that drilling equipment is properly maintained and that their rated capacity is not exceeded. Completions involve the final stages of the well development process, which include casing and cementing design.

The rotary drilling process for a vertical well or for a directional well involves: (i) application of a force downward on a drill bit, (ii) rotation of the drill bit, and (iii) circulation of the drilling fluid from the surface through the tubular (drill string), and back to the surface through the annular space, which is the area between drill string and borehole wall or casing (Azar and Samuel, 2007). On the other hand, horizontal drilling involves directing the drill bit to follow a horizontal path, oriented at approximately 90° from the vertical, through the reservoir rock (Azar and Samuel, 2007). Over the years, hydraulic fracturing has been performed on vertical, deviated, and horizontal wells. However, the coupling of horizontal wells and hydraulic fracturing have been proven to improve well performance in gas reservoirs (Britt et al., 2010; Devereux, 2012) and enhances the recovery of natural gas by reducing the number of vertical wells to develop fields of interest. In the current context, horizontal wells have found ready application in the Barnett shale, the Marcellus shale as well as in other shale plays.

Horizontal wellbores allow a far greater exposure to a formation than a conventional vertical wellbore, which is particularly useful in tight formations that do not have a sufficiently high permeability to produce natural gas economically from a vertically drilled well. Furthermore, the type of wellbore completion used will influence the number of times that a formation is fractured and the locations along the horizontal section of the wellbore that fracturing is necessary. The method by which the fractures are placed along the wellbore is most commonly achieved by one of two methods: (i) as the *plug and perf* method and (ii) the *sliding sleeve* method.

The wellbore for the plug and perf method is generally composed of standard joints of steel casing, either cemented or uncemented, which is set in place at the conclusion of the drilling process. Once the drilling rig has been removed, a perforation is created near the end of the well, following which a fracturing stage is initiated. Once the fracturing stage is completed, a plug is set in the well to temporarily to seal off that section of the wellbore. Another stage is then pumped, and the process is repeated as necessary along the entire length of the horizontal part of the wellbore.

On the other hand, the wellbore for the sliding sleeve method is different insofar as the sliding sleeves are included at set distances (spacing) in the steel casing at the time the casing is set in place. The sliding sleeves are usually all closed at this time and when the well is ready for application of the fracturing process, the bottom sliding sleeve is opened and the first stage is pumped. Once finished, the next sleeve is opened, which concurrently isolates the first stage, and the process is repeated. These completion techniques may allow more than 30 stages to be pumped into the horizontal section of a single well if required, which is many more stages than would typically be pumped into a vertical well.

Finally, to optimize the completion, it is necessary to understand the mechanical properties of all the layers above, within, and below the gas pay intervals. Basic rock properties such as in-situ stress, Young's modulus and Poisson ratio are needed to design a fracture treatment. The in-situ stress of each rock layer affects how much pressure is required to create and propagate a fracture within the layer. The values of Young's modulus relate to the stiffness of the rock and help determine the width of the hydraulic fracture. The values of Poisson ratio relate to the lateral deformation of the rock when stressed. Poisson ratio is a parameter required in several fracture design formulas.

The most important mechanical property is in-situ stress, often called the minimum compressive stress or the fracture closure pressure. When the pressure inside the fracture is greater than the in-situ stress, the fracture is open. When the pressure inside the fracture is less than the in-situ stress, the fracture is closed. To optimize the completion, it is important to know the values of in-situ stress in every rock layer.

6.3.4 Fracturing fluids

Initially, the fluid is injected that does not contain any propping agent and is injected to create a fracture that is multi-directional and spreads in several directions. This creates a fracture that is sufficiently open for insertion of the proppant, which is injected as a slurry – a mix of the proppant and the carrier fluid and proppant material. In shallow reservoirs, sand is often used can be used and remains the most common proppant but in deep reservoirs, ceramic beads (in place of the usual sand proppant) may be used to prop open the fractures. Once the fracture has initiated, fluid is continually pumped into the wellbore to extend the created fracture and develop a fracture network.

However, each formation has different properties and, therefore, diffident in-situ stress forces are operational so that each hydraulic fracture project is unique and must be designed accordingly by identification of the properties of the target formation including estimating fracture treating pressure, amount of material and the desired length for optimal economics. Furthermore, the fracturing fluid should have a number of properties that are suited to the properties of the formation, such as: (i) compatibility with the formation rock, (ii) compatibility with the formation fluid, (iii) suitability to generate sufficient pressure drop down the fracture to create a wide enough fracture, and (iv) a sufficiently low viscosity to allow clean-up after the treatment.

Water-based fluids are commonly used – *slickwater* is the most common fluid used for shale gas fracturing, where the major chemical added is a surfactant polymer to reduce the surface tension or friction, so that water can be pumped at lower treating pressures. Other fluids that have been considered are oil-based fluids, foams, and emulsions but caution is advised when using non-aqueous fluids since these fluids must be allowable for injection (by regulation) and must not have any detrimental effect on the environment.

Additives for fracturing fluids are chosen according to the properties of the reservoir and include (Table 6.3): (i) polymers, which allow for an increase in the viscosity of the fluid, together with crosslinkers, (ii) crosslinkers, which increase the viscosity of the linear polymer base gel, (iii) breakers, which are used to break the polymers and crosslink sites at formation temperature, for better cleanup, (iv) biocides, which are used to kill bacteria in the mix water, (v) buffers, which are used to control the pH, (vi) fluid loss additives, which are used to control excessive fluid leak-off into the formation, and (vii) stabilizers, which are used to keep the fluid viscous at higher temperature. However, it must be emphasized that additives are used for every site and in general as few additives as possible are added to avoid potential environmental contamination (use of the additives must be controlled) and production problems with the reservoir.

6.3.5 Fracture design

An important concept in hydraulic fracturing design is the number of holes per stage. Designing the number of holes per stage in a conventional reservoir is completely different from in an unconventional shale reservoir. Limited entry is a term of art used in the industry and is referred to as the practice of limiting the number of perforations (holes) in a completion stage to help the development of perforation friction pressure during a frac stimulation treatment.

Limited entry can be achieved using the following steps: (i) determine the friction pressure of a single perforation for the limited entry design – a value of at least 200–300 psi is recommended since a value of this magnitude should be noticeable and (ii) once the friction pressure is chosen, solve for rate per perforation to determine the rate per perforation. This will provide the rate per perforation needed to develop the friction pressure of a single perforation.

The choking effect produces back pressure in the casing, which allows simultaneous entry of fracturing fluid into multiple zones of varying in-situ stress states. Treatment distribution among zones can be controlled to a degree.

6.3.5.1 Fracture patterns

Hydraulic fracturing may be performed on a single reservoir interval in a vertical well. Horizontal wells, however, by virtue of their significant wellbore length in the target formation, are generally isolated into several discrete intervals along the horizontal wellbore – there may be four-to-twenty intervals for each horizontal well, with each interval requiring its own fracturing stage. This is due to the difficulty in maintaining pressures sufficient to induce fractures over the complete length of the lateral leg.

The most important data for designing a fracture treatment and the resulting fracture patterns are (i) the in-situ stress profile, (ii) formation permeability, (iii) fluid-loss characteristics, (iv) total fluid volume pumped, (v) propping agent – type and amount, (vi) viscosity of the fracture fluid, (vii) injection rate, and (viii) formation modulus. The in-situ stress profile and the permeability profile of the zone to be stimulated must be quantified and identification of the layers of rock above and below the target zone must be identified since these formations will influence fracture height growth. In order to design the optimum treatment, the effect of fracture length and fracture conductivity on the productivity and the ultimate recovery from the well must be determined.

The selection of the fracture fluid for the treatment is a critical decision and selection of fracture fluid on the basis of factors such as (Economides and Nolte, 2000): (i) reservoir temperature, (ii) reservoir pressure, (iii) the expected value of fracture half-length, and (iv) any water sensitivity. The definition of what comprises a water-sensitive reservoir and what causes the damage is not always evident. Most reservoirs contain water, and most natural gas reservoirs can be successfully waterflooded. Thus, most fracture treatments should be pumped with suitable water-base fracture fluids. Acid-base fluids can be used in carbonates but many deep carbonate reservoirs have been stimulated successfully with water-base fluids containing propping agents.

When selecting a propping agent, it is necessary to determine the maximum effective stress on the agent must be determined. The maximum effective stress depends on the minimum value of flowing bottomhole pressure expected during the life of the well. To confirm exactly which type of propping agent should be used during a specific fracture treatment, the designer should factor in the estimated values of formation permeability and optimum fracture half-length (Cinco-Ley, et al., 1978). The treatment must be designed to create a fracture wide enough, and pump proppants at concentrations high enough, to achieve the conductivity required to optimize the treatment. There is a tendency to compromise fracture length and conductivity in an often unsuccessful attempt to prevent damage to the formation around the fracture and substantial damage to the formation around the fracture can be tolerated as long as the optimum fracture length and conductivity are achieved (Holditch, 1979). However, damage to the fracture or the propping agents can be very detrimental to the productivity of the fractured well. Ideally, the optimum fracture length and conductivity can be created while minimizing the damage to the formation.

Finally, in horizontal wells, transverse fractures are relatively more difficult to achieve than longitudinal fractures. However, for gas-bearing formations that are typically characterized by low permeability, transverse fractures in horizontal wells have greater production benefits. Transverse vertical fractures move along the path of least resistance, which is normal to the minimum horizontal stress. In horizontal wells or deviated wells, there are effects in the immediate vicinity around the wellbore that lead to the transverse fractures taking unpredicted paths before eventually aligning normal to the horizontal stress. These effects are increased by the presence of natural fractures in the formation and the deviation of the horizontal well at an angle from the minimum horizontal stress.

In the design of a hydraulic fracturing procedure, most procedures to optimize well productivity begin with the fracture size. Limitations in the different hydraulic fracture design methods are inherent in their assumptions of (i) fracture geometry, (ii) dependence on fracture fluid properties, (iii) dependence on reservoir properties, (iv) dependence on whether or not the formations are layered, and (v) a variety of other factors, such as stress intensity. Challenges in fracture geometry when fracturing unconventional reservoirs include fracture azimuth and dip, not creating expected length, brittle and ductile rocks – complex and simple networks, wellbore axis (vertical or horizontal drilling) (Kennedy et al., 2012). In all cases however, knowledge of existing in-situ stress is essential to developing a fracture propagation model, which describes the methods of obtaining a desired hydraulic fracture geometry definitely including the fracture (half) length, width, height, and fracture complexity.

The ideal formation evaluation would be one where the value of the in-situ stress obtained from injection tests and those calculated from logs and core analysis all result in a consistent stress profile (Holditch et al., 1987). The hydraulic methods are also the most reliable for determining in-situ stress in deep (>160 feet) formations (Amadei and Stephansson, 1997).

Hydraulic fractures are formed in the direction perpendicular to the least stress. Typically, horizontal fractures will occur at depths less than approximately 2,000 feet because the overburden at these depths provides the least principal stress. If pressure is applied to the center of a formation under these relatively shallow conditions, the fracture is most likely to occur in the horizontal plane, because it will be easier to part the rock in this direction than in any other. In general, therefore, these fractures are typically parallel to the bedding plane of the formation.

As depth increases beyond approximately 2,000 feet, overburden stress increases by approximately 1 psi/foot, making the overburden stress the dominant stress. This means the horizontal confining stress is now the least principal stress and, since hydraulically induced fractures are formed in the direction perpendicular to the least stress, the resulting fractures at depths greater than approximately 2,000 feet will be oriented in the vertical direction.

In the case where a fracture might cross over a boundary where the principal stress direction changes, the fracture would attempt to reorient itself perpendicular to the direction of least stress. Therefore, if a fracture propagated from deeper to shallower formations it would reorient itself from a vertical to a horizontal pathway and spread sideways along the bedding planes of the rock strata.

The extent that a created fracture will propagate is controlled by the upper confining zone or formation, and the volume, rate, and pressure of the fluid that is pumped. The confining zone will limit the vertical growth of a fracture because it either possesses sufficient strength or elasticity to contain the pressure of the injected fluids or an insufficient volume of fluid has been pumped. This is important because the greater the distance between the fractured formation and the underground source of drinking water, the more likely it will be that multiple formations possessing the qualities necessary to impede the fracture will occur. However, while it should be noted that the length of a fracture can also be influenced by natural fractures or faults, natural attenuation of the fracture will occur over relatively short distances due to the limited volume of fluid being pumped and dispersion of the pumping pressure regardless of intersecting migratory pathways.

6.3.5.2 Fracture optimization

Hydraulic communication is a key factor for determining hydrocarbon or thermal energy recovery sweep efficiency in an underground reservoir. Sweep efficiency is a measure of the effectiveness of heat, gas recovery process that depends on the volume of the reservoir contacted by an injected fluid (Britt, 2012). Artificial (stimulated) hydraulic fractures are usually initiated by injecting fluids into the borehole to increase the pressure to the point where the minimal principal stress in the rock becomes tensile.

Continued pumping at an elevated pressure causes tensile failure in the rock, forcing it to split and generate a fracture that grows in the direction normal to the least principal stress in the formation. Hydraulic fracturing activities often involve injection of a fracturing fluid with proppants in order to better propagate fractures and to keep them open (Britt, 2012). The design of fracturing treatment should involve the optimization of operational parameters, such as the viscosity of the fracturing fluid, injection rate and duration, as well as proppant concentration, so that a fracture geometry is created that favors increased sweep efficiency. The net present value is the economic criterion is usually used as an objective for optimal fracturing treatment design. Some studies have been reported to use a sensitivity-based optimization procedure coupled with a fracture propagation model and an economic model to optimize design parameters leading to maximum net present value (Hareland et al., 1993; Rueda et al., 1994; Aggour and Economides, 1998; Mohaghegh et al., 1999; Chen et al., 2013).

Optimization of the fracture geometry involves defining the desired fracture half-length, width, and conductivity for maximized production. While there are several optimization methods, all involve a relative comparison of the flow potential of the fracture to that of the reservoir. Thus, a fracture is often considered (or defined) as a high permeability path in a low permeability rock formation but if the fracture is filled with a cementing material, such as calcite (calcium carbonate, $CaCO_3$), resulting in a fracture with little or no permeability. Thus, in any evaluation of the reservoir, it is important to distinguish between open fractures and healed (plugged or filled) fractures. The total volume of fractures is often small compared to the total pore volume of the reservoir. Thus, natural fractures in reservoir rocks (especially in tight reservoir formations) contribute significantly to natural gas production. Therefore, it is important to glean every scrap of information from open hole logs to locate the presence and intensity of fracturing. Even though some modern logs, such as the formation micro-scanner and televiewer, are the tools of choice for fracture indicators, many wells lack this data.

Most natural fractures are vertical – a horizontal fracture may exist for a short distance, propped open by bridging of the irregular surfaces. Most horizontal fractures, however, are sealed by overburden pressure. Both horizontal and semi-vertical fractures can be detected by various logging tools. The vertical extent of fractures is often controlled by thin layers of plastic material, such as shale beds or laminations, or by weak layers of rock, such as stylolites in carbonate sequences. The thickness of these beds may be too small to be seen on logs, so fractures may seem to start and stop for no apparent reason.

The nucleation and propagation of hydraulic rock fractures are chiefly controlled by the local in-situ stress field, the strength of the rock (stress level needed to induce failure), and the pore fluid pressure. Temperature, elastic properties, pore water chemistry, and the loading rate also have an influence (Secor, 1965; Phillips, 1972; Sone and Zoback, 2013a, 2013b). Fractures in rock can be classified as tensile, shear or hybrid (a mixture of tensile and shear). If the dominant displacement of the wall rocks on either side of the fracture is perpendicular to the fracture surface, then the fracture is deemed tensile. New tensile fractures form when the pore fluid pressure in the rock exceeds the sum of the stress acting in a direction perpendicular to the fracture wall and the tensile strength of the rock. Note that any preexisting fractures that are uncemented (i.e., have zero cohesion) can be opened at a lower value of pore fluid pressure, when it exceeds the stress acting in a direction perpendicular to the fracture wall.

The formation or reactivation of shear fractures depends on the shear stress, the normal stress, the pore fluid pressure, and the coefficient of friction for the specific rock type. It is important to recognize that the hydraulic fracturing process of pumping large volumes of water into a borehole at a certain depth cannot control the type of fractures that are created or reactivated. The array of fractures created and/or reactivated or reopened depends on a complex interplay of the *in-situ* stress, the physical properties of the local rock volume and any pre-existing fractures, and the pore fluid pressure (Phillips, 1972). This could have implications for the risk of groundwater contamination by hydraulic fracturing operations, as the fracture network generated by the hydraulic fracturing fluid could be complex and difficult to predict in detail. The orientations, sizes, and apertures of permeable rock fractures created by a hydraulic fracturing operation ultimately control the fate of the hydraulic fracturing fluid and the released shale gas, at least in the deep subsurface. Geomechanical models used to predict these fracture pattern attributes therefore need thorough testing/benchmarking, together with ongoing and future developments.

A common issue encountered in hydraulic fracturing operations in tight formations (especially in shale formations) is the variability and unpredictability of the outcome of the fracturing process. The injection pressure required to fracture the formation (fracture gradient) often varies significantly along a well, and there can be intervals where the formation cannot be fractured successfully by the injection of the fluid. The use of real-time fracture mapping allows real-time observation of changes in fracture design and also allows changes in restimulation design to maximize the *effective stimulation volume* (the reservoir volume that has been effectively contacted by the stimulation treatment). A correlation of microseismic activity with log data allows estimation of fracture geometry to be made after which the data can be used to design a stimulation that has the greatest chance of maximizing production (Fisher et al., 2004; Baihly et al., 2006; Daniels et al., 2007). Shale gas reservoirs also respond to fluid injection in a variety of modes although distribution of activated seismicity can be confined along a macroscopic fracture plane, but most time they are dispersed throughout a wide region in the reservoir reflecting the development of a complex fracture network (Waters et al., 2006; Cipolla et al., 2009; Das and Zoback, 2011; Maxwell, 2011).

In recent years, various attempts have been made to optimize the design of transverse fractures of horizontal wells for shale gas reservoirs. However, these optimization methods may not provide the optimal design. Hence, the optimization of hydraulic fracturing treatment design for shale gas production remains a challenge.

Finally, earlier literature on fracture analysis suggested that fractures might contribute as much as a few to several percent porosity but more recent work using the fracture aperture (calculated from resistivity micro-scanner logs) indicates a much lower contribution to the porosity of the formation. The term *secondary porosity* also includes rock-volume shrinkage due to dolomitization, porosity increase due to solution or recrystallization, and other geological processes. *Secondary porosity* should not be confused with *fracture porosity*, which can be determined by processing the formation micro-scanner curves for fracture aperture and fracture frequency (*fracture intensity*). The effect of fracture porosity on reservoir performance, however, is large due to the substantial contribution to permeability. As a result, naturally fractured reservoirs behave differently than non-fractured reservoirs with similar porosity due to the relatively high flow rate and capacity of the secondary porosity system. This provides high initial production rates, which can lead to extremely optimistic production forecasts and sometimes, economic failures when the small reservoir volume is not properly taken into account.

6.3.6 Fracture monitoring

During the hydraulic fracturing process, fluid leak-off that is loss of fracturing fluid from the fracture channel into the surrounding permeable rock can (and often does) occur. If not controlled properly, the fluid loss can exceed 70% v/v of the injected volume, which may result in formation matrix damage, adverse formation fluid interactions, or altered fracture geometry and thereby decreased production efficiency. Thus, fracture geometry and fracture monitoring are important aspects of the hydraulic fracturing process.

Monitoring technologies are used to map where fracturing occurs during a stimulation treatment and includes such techniques as microseismic fracture mapping, and tilt meter measurements (Arthur et al., 2008). These technologies can be used to define the success and orientation of the fractures created during a stimulation process. Measurements of the pressure and rate during the growth of a hydraulic fracture, as well as knowing the properties of the fluid and proppant being injected into the well provide the most common and simplest method of monitoring a hydraulic fracture treatment. This data, along with knowledge of the underground geology, can be used to model information such as length, width, and conductivity of a propped fracture.

Microseismic monitoring is the process by which the seismic waves generated during the fracturing of a rock formation are monitored and used to map the locations of the fractures generated. Monitoring is done using a similar technology to that used to monitor larger naturally occurring seismic events associated with earthquakes and other natural processes. Microseismic monitoring is an active monitoring process performed during a hydraulic fracture treatment. As an active monitoring process microseismic monitoring can be used to develop real-time changes to a fracture program. Microseismic monitoring provides engineers the ability to manage the resource through intelligent placement of additional wells to take advantage of the natural conditions of the reservoir and expected fracture results in new wells.

Microseismic theory and mapping are based on earthquake seismology. Similar to earthquakes, but at a much higher frequency (200–2,000 Hz), microseismic events emit elastic P waves (compressional) and S waves (shear waves) (Jones and Britt, 2009). During hydraulic fracture, there is an increase in formation stress proportional to the net fracturing pressure, and an increase in pore pressure due to fracturing fluid leak-off. The increase in stresses at the fracture tip and pore pressure increments causes shear slippages to occur. Microseismic technology thus uses earthquake seismology methodologies to detect and locate these hydraulic fracturing induced shear slippages, which resemble micro-earthquakes. Micro-seismic events or micro-earthquakes occur with fracture initiation and are observed with receivers placed on an offset wellbore like with the downhole tiltmeters.

Microseismic mapping technology involves installing an array of tri-axial geophone or accelerometer receivers into an offset well at approximately the depth of the fracture (like in downhole tiltmeters), orienting the receivers (geophones), recording seismic data, finding micro-earthquakes in the data and locating them. Locating the earthquake events requires the determination of compressional (P) and shear (S) wave arrivals and consequent acoustic interpretation of the velocity of the P–S waves (Davis et al., 2008; Jones and Britt, 2009). Standard microseismic mapping use P–S arrival time separation for distance location. Horizontal and vertical plane holograms are used to determine azimuth and inclination (Warpinski et al., 2005).

Tiltmeters are passive monitoring technologies, which record the deformation of rocks that are induced by the hydraulic fracture process. Tiltmeters can be placed at the ground surface away from a well or downhole in a nearby wellbore tightly into the rock. Tiltmeters measure changes in inclination in two orthogonal directions, which can then be translated into the strain rotation that results from hydraulic fracturing. Engineers can then determine based on the strain rotation the location of the hydraulic fracturing event that caused the strain rotation.

Downhole tiltmeter mapping technology was developed to circumvent the limitations of the surface tiltmeter by giving estimates of the fracture dimensions. The downhole tiltmeters have the same operational principle as the surface tiltmeters, but instead of being at the surface, the tiltmeters are positioned by wireline in one or multiple offset wellbores at the depth of the hydraulic fracture. Typically, the array consists of 7–12 tiltmeters coupled to the borehole with standard centralizer springs (Wright et al., 1999). Downhole tiltmeters provide a map of the deformation of the Earth adjacent to the hydraulic fracture. Thus, what is obtained is an estimate of an ellipsoid that best approximates the fracture dimensions. The tiltmeters are located closer to the fracture than the surface tiltmeter and hence more sensitive to fracture dimensions (Cipolla and Wright, 2002). The closer the downhole tiltmeter to the fracture, the better the quality of data obtained to determine fracture height (Jones and Britt, 2009), which may be limited by the volume of the hydraulic fracturing fluid volume regardless of whether the fluid interacts with faults (Flewelling et al., 2013). The downhole array tilts in a continuous fashion, similar to surface tiltmeter records but the arrays span the same depth interval as the zone being fractured. The total interval covered by a downhole tilt array ranges from 300 feet to more than 1,000 feet, depending on the design conditions. Conventionally, surface and downhole tiltmeter analysis are done separately but techniques have been proposed to combine them for evaluating fracture geometry during drill cuttings disposal.

The greatest advantage of both surface and downhole tiltmeter fracture mapping is that for a given fracture geometry, the induced deformation field is almost completely independent of formation properties. Also, the required degree of formation description is lower in tiltmeter mapping than microseismic mapping (such as velocity profiles and attenuation thresholds) as will be described in a later section. Complex fracture growth would yield independent fractures at different orientations or depths but in tiltmeter mapping a simpler analysis is required.

At the completion of the stimulation process, approximately 20–30% v/v of the water flows back up the wellbore, where it is collected and then recycled in a subsequent well completion operation. Over the productive life of the well, additional "produced" water slowly comes to the surface, where it is collected in on-site storage tanks and transported to permitted treatment facilities.

6.3.7 Fracture diagnostics

As part of the fracturing process, it is necessary that the fracture be analyzed through the various procedures known as *fracture diagnostics*, which are the techniques used to analyze the original state of the formation (pre-fracture analysis) and the created fractures and after hydraulic fracture treatment (post-fracture analysis) (Barree et al., 2002; Vulgamore et al., 2007). The data will allow the determination of the dimensions of the created fractures and also whether or not the fractures are effectively maintained in an open mode (*propped*). Thus, the fracture diagnostic techniques can be conveniently sub-divided into three groups: (i) direct far-field techniques, (ii) direct near-wellbore techniques, and (iii) indirect fracture techniques.

6.3.7.1 Direct far-field techniques
The *direct far-field* techniques require the use of a tiltmeter – an instrument designed to measure small changes from the horizontal level, either on the ground or in subterranean structures – and microseismic fracture mapping techniques, which

require that the instrumentation is placed in boreholes surrounding, and near to, the well through which fracturing will occur. The microseismic fracture mapping technique relies on the use of a downhole receiver array of geophones, to locate any *micro-seismic events* that are triggered by movement in any natural fractures surrounding the hydraulic fracture.

6.3.7.2 Direct near-wellbore techniques

Direct near-wellbore techniques are used in the well that is being fractured to locate the portion of fracture that is near wellbore. These techniques consist of tracer logs, temperature logging, production logging, borehole image logging, downhole video logging, and caliper logging. The caliper log provides a continuous measurement of the size and shape of a borehole along its depth and is commonly used when drilling wells to detect natural gas formations. The measurements that are recorded can be an important indicator of cave-ins or swelling in the borehole, which can affect the results of other well logs.

6.3.7.3 Indirect fracture techniques

Indirect fracture techniques consist of modeling the hydraulic fracturing process followed by matching of the net surface treating pressures, together with subsequent pressure transient test analyses and production data analyses. As fracture treatment data and the post-fracture production data are normally available on every well, the indirect fracture diagnostic techniques are widely used methods to determine (or estimate) the shape and dimensions of the created fractures and the propped hydraulic fracture.

6.3.8 Performance of a hydraulically fractured well

For successful development of a shale gas reservoir, it is necessary to maximize productivity by analyzing the flow characteristic and reservoir property and effective fracture treatment design, which influence on well performance. When a low-permeability reservoir is hydraulically fractured, a fracture network system is formed comprising primary and secondary fractures and it should be considered to understand the production performance of a shale gas reservoir.

Horizontal wells with multistage hydraulic fracturing technique are the stimulation method of choice and have been successful in shale gas reservoirs. However, there are several factors affecting the hydraulic fracture treatment and the post-fracture gas production in shale gas reservoirs.

When a shale gas reservoir is hydraulically fractured, the fracture network systems are formed comprising primary and secondary fractures. In the case of low permeability reservoir, especially, the more complex fracture network systems is, the more the productivity is improved. The productivity is improved when the secondary fracture conductivity increases or spacing decreases with various conditions of reservoir.

6.3.9 Fracturing horizontal gas wells

Hydraulically fractured horizontal wells have become the predominant method of new natural gas development. The combination of horizontal drilling and hydraulically fracturing has contributed to increases in natural gas production. Horizontal drilling is a process in which the drill path is turned horizontally at the target depth. The procedure is superior to vertical drilling, which limits the contact to rock and drastically reduces the potential gas recovery. Because they are longer, and the drilling process is more complex, a horizontal well is expected to produce more natural gas.

Horizontal drilling allows more of the wellbore to remain in contact with the producing formation, increasing the amount of natural gas that can be recovered. This method also results in horizontal wells having more drilled footage than vertical wells – hence total footage drilled using horizontal drilling techniques surpassed vertical footage before the actual number of horizontal wells surpassed the number of vertical wells.

6.4 Fluids management

For the purposes of this section, the term *fluids* refer to any fluid material that is either introduced into the formation to aid in the recovery of natural gas as well as any fluid material that is recovered from the fractured formation.

Water is essential to the development of the resources of natural gas in these tight formations but as developers escalate the exploration for suitable tight formations, the limited availability of water will (if it has not already) become a reality. Extracting these resources requires large amounts of water for not only for the drilling operations but particularly for the hydraulic fracturing operations. In most cases, these demands are met by local freshwater – the financial aspects of shipping water to a remote site are not favorable. Nevertheless, the developers of tight resources are significant users and managers of water at local and regional levels and, as a result, water and energy have become interdependent. For example, improving the efficiency of water use reduces the need to develop, transport, pump, treat, and distribute additional water resources thereby reducing the amount of power or energy required for recovery of the resources. Alternatively, improving energy efficiency reduces the demand on electricity generation and fuel consumption for water transportation, which, in turn, reduces the need for water resources for power generation cooling and fuel processing, along with reducing the water resources needed to extract the natural gas.

Thus, the development of natural gas resources in deep tight, low-to-no permeability shale, sandstone, or carbonate formations, which are typically found thousands of feet below the surface of the Earth, requires copious amount of water than can floral and faunal communities, including human communities (Stillwell et al., 2010).

Finally, hydrological conditions vary spatially and seasonally across the regions where subterranean tight formations occur, with variation among not only between adjacent formation but within a single formation. This variability causes different demands on the water sources and makes the ability of the developers to meet the water demands for horizontal drilling and hydraulic fracturing highly unpredictable. Thus, estimates based on previous seasonal demands as well as demands by previously developed formations are not always transferable to new tight formations that are under investigation for the near-future development. Each formation becomes a development that must be considered to be site specific and subject not only to the depth and properties of the formation as well as the formation fluids but also to the seasonal availability of water. Furthermore, public concern over increased competition and impacts on freshwater availability for domestic purposes (with the guarantee of untarnished drinking water sources) can influence the awarding of a development license and ability of a developer to operate, which, in turn, can lead to changes in government regulations that could impact both the short-term and long-term financial aspects of the development. Therefore, responsible water management practice is critical to a developer obtaining a license to operate at the site, which is particularly true when a resource is located within a water-poor region, such as may be the case in certain regions Texas and North Dakota.

The quality of water used to in the hydraulic fracturing fluid and the effect and impact of the fluid on well production is a critical factor in developing a water management strategy. Furthermore, regulatory requirements (mandated federal, state and local regulatory authorities) often dictate water management options. Also within the United States, multi-state and regional water permitting agencies may also be responsible for maintaining water quality and water supply. In fact, all authorities may dictate water withdrawal and/or disposal options that are available for consideration and use and injection wells that may be used for disposal of flowback water and other produced waters require state or federal permits, which are subject to suitable analysis by standardized text methods (Blauch, 2009). The primary objective of any such program, whether administered at the state or federal level, is protection of underground sources of drinking water but, in many cases, the responsible authority is a function of the acquisition or disposal option chosen. For example, surface water discharge may be regulated by a different agency than subsurface injection. Therefore, regardless of the regulatory agency with program authority over subsurface injection, new injection wells will require a permit that meets the appropriate state or federal regulatory requirements.

Thus, it is the purpose of this section to introduce the reader to the various aspects of tight resource development and the necessity for water as part of the development of natural gas resources in tight formations as well as the effects of this development on water use, water production, and water quality.

6.4.1 Water requirements, use, and sources

A significant part of the development of natural gas resources in tight formations is the hydraulic fracturing operation, a necessary part of which involves (i) securing access to reliable sources of water, (ii) the timing associated with this accessibility, and (iii) the requirements for obtaining permission to secure these supplies (API, 2009, 2010).

Thus, it is not surprising that hydraulic fracturing, the technology central to the success development of unconventional resources – the extraction of natural gas from tight formations – has been the subject of intense scrutiny come under fire because of concerns related to the chemicals used and any other real or unproven effects. The industry is working to improve transparency and collaboration with landowners, mineral rights holders, regulators, and communities.

Thus, when considering a project involving horizontal drilling and a hydraulic fracturing, the location of rivers and the amount of streamflow in rivers must be given serious consideration – the key concept is to determine the watershed of the river and identify the area of land where all of the water that falls in it and drains off of it goes to a common outlet, remembering that a watershed can be as small as a small pond or sufficiently large enough to encompass all the land that drains water into a major bay.

In order to commence a water requirements project, after pre-project communications have commenced, the operator(s) should conduct a detailed, documented review of the identified water sources available in an area that could be used to support hydraulic fracturing operations. Considerations factoring in this review should include: (i) evaluating source water requirements, (ii) fluid handling and storage, and (iii) transportation considerations.

6.4.1.1 Requirements
Although water is used in several stages of the natural gas recovery from tight formations, the majority of water is typically consumed during the production stage. This is primarily due to the large volumes of water (2.3–5.5 million gallons) required to hydraulically fracture a tight formation (Clark et al. 2011). In addition, water in amounts on the order of 190,000–310,000 gallons is also used to drill and cement a

natural gas well during the drilling phase of the project (Clark et al. 2011). Once the natural gas is produced, it is processed, transported and distributed to customers, and ultimately used. Water consumption occurs in each of these stages as well, with the most significant non-production consumption potentially occurring during end use.

A rate-limiting factor in further development of the tight formation resources, particularly in certain regions of the western United States is not only the effect of water quality but also the availability of water, which may not be a problem in other regions and countries. Water from a nearby river may be made available for the drilling process and for the hydraulic fracturing process but an important factor that must be taken into consideration in any water-use plan is the potential salt loading of the river water. With tight formation development near the river, the average annual salinity is anticipated to increase, unless some prevention or treatment is implemented. The economic damages associated with these higher salinity levels could be significant and have been the subject of extensive economic studies.

Water requirements for hydraulically fracturing projects may vary widely, but on average required several million gallons of water for deep unconventional tight reservoirs. While these water volumes may seem large, they generally represent a relatively small percentage of total water use in the areas where hydraulic fracturing is operative (Satterfield et al., 2008). Water used for hydraulic fracturing operations can come from a variety of sources, including (i) surface water bodies, (ii) municipal water supplies, (iii) groundwater, (iv) wastewater sources, or (v) water that has been recycled from other sources including previous hydraulic fracturing operations.

Typically, water for hydraulic fracturing is withdrawn from one location or watershed over a period of several days (Veil, 2010). Additionally, in some cases, the water may be acquired from remote, often environmentally sensitive headwater areas where even small withdrawals can have a significant impact on the flow regime. As a result, while hydraulic fracturing may account for a small fraction of the water supply, there can be more severe local impacts. In addition, much of the water injected underground is either not recovered or is unfit for further use once it is returned to the surface and often usually requires disposal in an underground injection well. This water use represents a consumptive use if it is not available for subsequent use within the basin from which it was extracted. Alternatively, the water may be treated and re-used for subsequent hydraulic fracturing projects.

In terms of local impact, especially in areas of water-scarcity, the extraction of water for drilling and hydraulic fracturing (or even the production of water, as is the case in the production of coalbed methane) can have broad and serious environmental effects. The extraction of the water can (more likely, will) lower the water table thereby affecting biodiversity (i.e., the variety of the floral and faunal populations) and harm the local ecosystem as well as reducing the availability of water for use by local communities and in other productive activities, such as agriculture. Thus, the limited availability of water for hydraulic fracturing could become a significant constraint on the development of natural gas in tight formations in water-stressed areas. In fact, water stress (water scarcity) is increasing around the world and all domestic and industrial users are becoming subject to intense scrutiny in terms of water consumption. Thus, water management and groundwater protection are vital issues for the natural gas industry.

When the time comes (and this should be sooner rather than later) to evaluate the water requirements for a hydraulic fracturing project, it is necessary to conduct a pre-project comprehensive evaluation of the cumulative water demand on a stage-by-stage basis as well as the timing of these needs at an individual well site. This should include consideration of the water requirements for (i) drilling operations, (ii) dust suppression, (iii) emergency response, as well as (iv) the water requirements for hydraulic fracturing operations. A decision must be made as to whether or not the sources of water are adequate to support the total operation, with water of the desired quality, and can be accessed when needed for the planned development program.

6.4.1.2 Use
Hydraulic fracturing uses between 1.2 and 3.5 million US gallons of water per well, with large projects using up to 5 million US gallons per well. Typically, a well requires 3 to 8 million US gallons of water over its lifetime. Water consumption for hydraulic fracturing occurs during: (i) drilling, (ii) extraction and processing of proppant sands, (iii) testing natural gas transportation pipelines, and (iv) gas processing plants. Typically, for most shale basins, water is acquired from local water supplies, including: (i) surface water bodies, such as rivers, lakes, and ponds, (ii) groundwater aquifers, (iii) municipal water supplies, (iv) treated wastewater from municipal and industrial treatment facilities, and (v) produced and/or flowback water that is recovered, treated, and re-used. In regions where hydraulic fracturing occurs, the sources of water should be well documented.

Thus, although water is used in several stages of the tight formation development, the majority of water is typically consumed during the production stage. This is primarily due to the large volumes of water (on the order of several million gallons depending upon the depth of the formation and the formation characteristics) required to hydraulically fracture a well (Clark et al., 2011). Water in amounts of several hundred thousand gallons is also used to drill and cement a shale gas well during construction (Clark et al., 2011). After fracturing a well, anywhere from 5% to 20% of the original volume of the fluid will return to the surface within the first two weeks as flowback water. An additional volume of water, equivalent to anywhere from 10% to almost 300% of the injected volume, will return to the surface as produced water over the life of the well. It should be noted that there is no clear distinction between so-called flowback water and produced water, with the terms typically being defined by operators based upon the timing, flow rate, or sometimes composition of the water produced.

Deep shale gas projects use water primarily during drilling and stimulation, but produces a tremendous amount of energy over the approximate 20-year lifespan of the natural gas well (Mantell, 2009). Thus, water is an essential component of deep shale natural gas development. Operators use water for drilling, where a mixture of clay and water is used to carry rock cuttings to the surface as well as to cool and lubricate the drill bit. Drilling a typical deep shale well requires between 65,000 and 1 million gallons of water. Water is also used in hydraulic fracturing, where a mixture of water and sand is injected into the deep shale at high pressure to create small cracks in the rock and allows gas to freely flow to the surface. Hydraulically fracturing a typical deep shale well requires an average of 3.5 million gallons of water.

The water supply requirements of deep shale natural gas development are isolated in that the water needs for each well are limited to drilling and development, and the placement of shale gas wells is spread out over the entire shale gas play. In other words, these shale gas wells are not drawing water from one single source. Subsequent hydrofracturing treatments of wells to re-stimulate production may be applied, though their use is dependent upon the particular characteristics of the producing formation and the spacing of wells within the field.

The majority of water use occurs during the early exploration and drilling phase of gas production from tight formations. Water is a necessary input and, although only a relatively small amount of water is needed for drilling, more significant quantities of water are used during completions, or hydraulic fracturing operations. Thus, in terms of water use, the issue is the application of hydraulic fracturing to release the gas form the shale formation – high-volume hydraulic fracturing to create fissures in the rock to release gas trapped inside. Thus, water-related issues in shale drilling are leading to growing and complex policy and regulatory challenges and environmental compliance hurdles that could potentially challenge shale gas production expansion and increase operational costs.

Hydraulic fracturing is a technique used in natural gas production to stimulate the production of hydrocarbons. Recall, after a well is drilled into reservoir rock that contains natural gas, and water, every effort is made to maximize the production of the gas (Speight, 2014a). In hydraulic fracturing, a fluid (usually water containing special high-viscosity fluid additives) is injected under high pressure. The pressure exceeds the rock strength and the fluid opens or enlarges fractures in the rock. These larger, man-made fractures start at the well and extend deep into the reservoir rock. After the formation is fractured, a *propping agent* (usually sand carried by the high-viscosity additives) is pumped into the fractures to keep them from closing when the pumping pressure is released. This allows the natural gas to move more freely from the rock pores to a production well and thence to the surface.

Even though it is often looked upon as a temporary process, hydraulic fracturing (and the accompanying horizontal drilling process) requires considerable amount of water. For example, the drilling and completion (including hydraulic fracturing) phases of a well (leading to initiation of the production of natural gas) is a short time period – typically on the order of two to eight weeks – compared to an expected production lifetime of a well on the order of 20–40 years. In terms of the total amount of water, up to 170,000 barrels of water per well may be used for drilling and fracturing and after the fracturing process has been completed, 30–70% v/v of the injected water injected flows out of the well, but this amount is dependent upon the properties (such as size and mineralogy) of the reservoir and the returned water must be treated before disposal or before re-use.

Such a level of water consumption can put a heavy strain on local water sources and, if projections of natural gas production are to be met, technology development to reduce consumption will be necessary. Another aspect of water use is that 3,000,000 to 5,00,000 gallons of water are typically necessary to hydraulically fracture a multi-stage well. The volume and quality of water used is also characteristic of the reservoir properties (which vary by region and by play) and also by design of the well.

6.4.1.3 Sources

The types of water sources that are used by the natural gas industry (for hydraulic fracturing) typically fall into three categories: (i) potable water (freshwater), (ii) natural gas generated water (produced and flowback water), and (iii) the alternative water sources that are generally not usable by the public (brackish or non-potable water). The choice among sources depends on a variety of factors such as volume, availability, source-water quality, competing water uses, economics, and regulatory requirements.

Potable water sources typically include freshwater from a groundwater source, surface water (from a lake or river) and municipal water supplies. Water is drawn from these sources over time in accordance with agreements and regulatory requirements and stored for use when a greater amount of water is necessary. A regular and slow withdrawal rate can minimize impacts on water sources that are also used by communities.

Produced water (re-used and recycled water) used in natural gas operations is typically flowback and produced water from the field operations. There are few instances of operators using other produced water sources, such as municipal and industrial. In an effort to reduce the use of freshwater sources, the industry is increasing the amount of flowback and produced and brackish water in hydraulic fracturing. There are several key factors that operators must consider concerning the re-use of produced water, including: the quantity of produced water; duration and consistency of produced water; produced water quality; target formation characteristics; scale of the operation, and cost to re-use the water. Operational logistics such as fluid-handling capabilities, transportation considerations, storage capabilities and access to treatment locations or onsite treatment options, will influence the ability to re-use produced water. The availability of flowback and produced water is dependent on the amount of water that returns from the fracturing process and formation.

Fluids other than water may be used in hydraulic fracturing process, including carbon dioxide, nitrogen, or propane, although their use is currently much less widespread than water. The use of flowback and produced water is also dependent on the type of fracturing fluids being used for that play. Hydraulic fracturing fluids (fracking fluids) are a complex mixture of many ingredients that are designed to perform a diverse set of functions and accommodate a variety of factors, including local geology, well depth, and length of the horizontal segment of the well. Although the precise recipe is unique to the formation, the fluid is typically composed of proppants (typically sand) to hold open the fractures and allow the natural gas to flow into the well, and chemicals that serve as friction reducers, gelling agents, breakers, biocides, corrosion inhibitors, and scale inhibitors. Industry representatives point out that chemicals represent a small percentage of the fracturing fluid; on average, fracking fluid for shale gas consists of more than 99% water and sand. Given the large volume of fluid that is injected underground, however, a small percentage can represent a large quantity of chemicals.

The sources for the supply of water for hydraulic fracturing project will depend on the cumulative amount of water that will be required for the long-term project that is planned. Water sources will need to be appropriate for the forecasted pace and level of development anticipated. Thus, the selection of a water source (or, more likely, because of the amount of water needed, water sources) ultimately depends upon (i) the volume required, (ii) the water quality requirements, (iii) regulatory issues, (iv) physical availability, (v) competing uses for the water, and (vi) the characteristics of the formation to be fractured, including water quality and compatibility considerations. If possible, wastewater from an industrial facility (or industrial facilities) should be considered as a water source, followed by groundwater sources and surface water sources (with the preference over non-potable sources over potable sources), with the least desirable (at least for long-term, large scale development) being any municipal water supplies.

However, in regard to the choice of the water source(s) will depend on local conditions and the availability of groundwater and surface water resources in proximity to planned operations. Most important, not all options may be available for all situations, and the order of preferences can vary from area to area. Furthermore, for water sources such as industrial wastewater, power plant cooling water, or recycled flow back water and/or produced water, additional treatment may be required prior to use for hydraulic fracturing. Contaminants in the water which are not removable to the desired levels for re-injection may render the water as unsatisfactory because it may deliver subterranean contamination in and around the drilling site and the reservoir and open up the real potential for groundwater contamination as well as contamination of aquifers that provide municipal water supplies and such results cannot assure project success.

Thus, water sources for hydraulic fracturing projects consists of: (i) surface water, (ii) groundwater, (iii) municipal water suppliers, (iv) wastewater and power plant cooling water, and (v) reservoir water and recycled flowback water.

6.4.1.3.1 Surface water

Many municipalities draw their principal water supplies from surface water sources, so the large-scale use of this source for hydraulic fracturing operations will undoubtedly impact not only municipal usage but also other competing uses and will, therefore, be of concern to local water management authorities and other public officials. In some circumstances there will be a need to identify water supply sources capable of meeting the needs of water requirements for vertical and horizontal drilling as well as for hydraulic fracturing and, in addition, it must be made clear to the relevant authorities that the water needs of the drilling/fracturing project will not compete or interfere with community needs as well as with currently existing uses.

Thus, necessary considerations in evaluating water supply requirements from surface water sources include not only the volume of water supplies required but also the sequence and scheduling of acquiring these supplies of water. Withdrawal from surface water bodies, such as rivers, streams, lakes, natural ponds, and private stock ponds will likely require permits from state agencies or from multi-state regulatory agencies, as well as permission from the relevant landowners – in some regions, water rights are also a key consideration. In addition, the water quality standards and regulations that have been established by the various regulatory authorities may prohibit any alteration in water flow (such as pertains, for example, to river flow and stream flow as well as flow into lakes) that would impair a high priority use of the fresh (surface) water, which is often defined by the local water management authorities. Also consideration should be given to ensure that water withdrawal from rivers and streams during periods of low river and/or stream flow does not affect fish and other aquatic life, fishing and other recreational activities, municipal water supplies, and the operational demands for water by other existing industrial facilities, such as power plants.

When the options for application of permits to use water are being considered, the applicant must know that water withdrawal permits can require compliance with specific metering, monitoring, reporting, record keeping, and other consumptive use requirements. In addition, compliance could also include specifications for the minimum measured quantity of water that must pass a specific point downstream of the water intake in order for a withdrawal to occur. Furthermore, in the case where stream flow is less than the prescribed minimum quantity, the withdrawal of water may be required to be reduced or even to cease. Thus it is necessary to consider the various issues that can arise with the timing and location of water withdrawals since impacted watersheds may be sensitive, especially in drought years, when the periods of flow are low or diminished during the years of the project or during periods of the year when activities such as agricultural irrigation place additional demands on the surface supply of water.

To be fully aware of the potential for water use, requests for (or proposals for the use of) surface water withdrawal should be a consideration for the following potential impacts that could control the timing and volume available: (i) ownership, allocation, or appropriation of existing water resources, (ii) water volume available for other needs, including public water supply, (iii) degradation of a stream's designated best use, (iv) impacts to downstream habitats and users, (v) impacts to fish and wildlife, (vi) aquifer volume diminishment, and (vii) mitigation measures to prevent transfer of invasive species from one surface water body to another (as a result of water withdrawal and subsequent discharge into another surface water body).

With the continued development of resources in tight formations, additional regulatory requirements are likely to be associated with water use and requirements. For example, water could be sent to storage impoundments during periods of high flow to allow water withdrawal at a time of peak water availability. However, this approach would typically require the development of water storage capabilities to meet the overall demands of drilling and hydraulic fracturing projects over the course of the project, which may even be a multi-year period to accommodate possible periods of drought.

Another alternative for ensuring adequate water supply is to use abandoned surface coal mining pits for the storage of water, which provide more permanent facilities for the installation of a comprehensive water distribution system. However, the water quality in such storage areas must meet the operational requirements and with all regulatory requirements which would depend upon the nature of the exposed overburden that could allow subject leaching of undesirable chemical species during periods of heavy rain or winter run-off. In keeping with the concept of using surface coal-mining pits, another option is to excavate low lying areas to allow for collection of rain water. Again, such an option must meet with approval from state, regional, or local water management authorities to ensure compliance with storm water runoff program elements.

6.4.1.3.2 Groundwater

In order to use groundwater for drilling and hydraulic fracturing projects, many of the same types of considerations for groundwater as for surface water will need to be addressed. The primary concern regarding groundwater withdrawal is volume diminishment and, in some areas, the availability of fresh groundwater is limited, so withdrawal limitations could be imposed. To overcome such issues and where possible, consideration should be given to the use of non-potable water for drilling and hydraulic fracturing projects. Another that applies to the protection of groundwater sources is to locate water source wells for gas operations at an appropriate distance from municipal wells, public wells, or private water supply wells. Furthermore, public wells or private water supply wells and freshwater springs within a defined distance of any proposed drilling/fracturing project for a water supply well should be identified and the characteristics of the wells evaluated for production capacity and water quality. As part of the well evaluation, there may also be the need to test the water currently available from these sources. This will require locating the public and private water wells and gathering obtaining information related to (i) well depth, (ii) completed interval and use, including whether the well is public or private, community or non-community, and (iii) the type of facility or establishment if it is not a private residence.

Guidance for groundwater protection related to well drilling and hydraulic fracturing operations is available (API, 2009, 2010) and maintaining well integrity is featured is a key design principle of all natural gas wells, which is essential for two primary reasons: (i) to isolate the internal conduit of the well from the surface and subsurface environment, and (ii) to isolate and contain the well's produced fluid to a production conduit within the well.

6.4.1.3.3 Municipal water supplies

Obtaining water supplies from municipal water suppliers can be considered, but again, the water needs for fracturing would need to be balanced with other uses and community needs. This option might be limited, since some areas may be suffering from current water supply constraints, especially during periods of drought, so the long-term reliability of supplies from municipal water suppliers needs to be carefully evaluated.

6.4.1.3.4 Wastewater and power plant cooling water

Other possible options for source water to support hydraulic fracturing operations that could be considered are municipal wastewater, industrial wastewater, and/or power plant cooling water. However, the properties or specifications of this water source need to be compatible with the target formation and the plan for fracturing as well as whether treating is technically possible and whether treatment can deliver an overall successful project. In some cases, required water specification could be achieved with the proper mixing of supplies from these sources with supplies from surface water or groundwater sources.

6.4.1.3.5 Reservoir water and recycled flow back water

Produced reservoir water and recycled flow back water can be treated and re-used for fracturing, depending on the quality of the water. Natural formation water has been in contact with the reservoir formation for millions of years and thus contains minerals native to the reservoir rock. Some of this formation water is recovered with the flowback water after hydraulic fracturing, so that both contribute to the characteristics of the flow back water. However, the salinity, total dissolved solids (TDS, sometime referred to as *total dissolved salts*), and the overall quality of this formation/flow back water mixture can vary by geologic basin and specific rock strata.

Finally, whenever water is recycled and/or re-used, or additional sources of industrial wastewater are used to supply water for drilling and hydraulic fracturing operations, additional make up water may be required. In such cases, water management alternatives to be considered will depend on the volume and quality of both the recycled water and the make-up water, to ensure compatibility with each other and with the formation being fractured.

Much of the focus of recent policy decisions has been on the identification and the disclosure of the chemical constituents used in hydraulic fracturing fluid. In addition, if toxic chemical additives are used in the fracturing fluid, many regulatory authorities require open disclosure of the necessary information and in many cases disclosure of the chemicals employed in a fracturing project is mandated. However, some authorities do require the complete disclosure of the chemical constituents of fracturing fluid along with the concentration and volume of these chemicals. In addition, some regulations require disclosure to the state regulatory authorities rather than to the public, and most authorities allow companies to apply for trade secret exemptions. In regard to the concentration and volume of the chemicals used, it is worth noting at this point that many chemicals that have been classed as non-toxic below a specified dosage. For example, table salt (sodium chloride) is used by many persons at meal time but it is advisable not to try to consume a pound (454 g) at one sitting or it could be lethal. Thus, the laws of the use of toxic chemicals in hydraulic fracturing fluids need tweaking to specify the amount of chemicals used. This also applies to the use of chemicals that are indigenous to the environments since a concentration of the chemicals related to the indigenous amount may become toxic to the environment.

The primary hydraulic fracturing fluid systems are slickwater insofar as chemicals are added to the water to increase the fluid flow and increase the speed at which the pressurized fluid can be pumped into the wellbore. Such chemicals include crosslink or crosslinked gels. However, the procedure is not (and cannot be) standardized since the selection of the fracturing fluids is based on a variety of factors and the fluid systems are designed on the basis of several factors such as: (i) the characteristics of the target formation, (ii) the formation fluids, and (iii) the source of the make-up water source. In addition, the use of recycled flowback water and produced water will be more likely in a slickwater hydraulic fracture.

6.4.1.3.6 Other sources

Brackish or non-potable water is considered publicly unusable for hydraulic fracturing projects without significant treatment. Typically, brackish water is high in salinity, but has a lower salinity that than high-saline brine, and has TDS levels are on the order of 1,000 parts per million (ppm) or even higher. The type and dose of friction reducer can be adjusted to accommodate for higher TDS water.

In the early exploration phase, water with a relatively low TDS content will be used for hydraulic fracturing. However, for tight formations (especially shale plays) that are beyond the phase of early exploration, the use of brackish water and recycled produced water is becoming an increasingly feasible option due to the advancements in the use of chemical additives and water-treatment technologies. Other wastewater sources are also given consideration for hydraulic fracturing projects. For example, acid mine drainage water is often given consideration for use in hydraulic fracturing projects. However, acid mine drainage is one of most serious threats to water since a mine draining acid can devastate rivers, streams, and aquatic life for prolonged periods measuring in decades and additional caution is advised when considering the use of such water. Pre-treated seawater is also being explored as an alternative water source for hydraulic fracturing. However, the viability of any alternative water source is dependent upon factors such as the fracturing fluid type considerations, water treatment and water transportation costs (e.g., proximity to an ocean).

6.4.2 Water contamination

Water contamination, especially groundwater contamination from the retrieval of natural gas from tight formations, can (and often does) occur through a variety of pathways. While the formations may be located at varying depths, often (but not always) far below underground, caution must be taken to assure that sources of drinking water are not contaminated and there must be, in the early stages of project planning, an accurate geological assessment of the formations – the project must be recognized as a multi-disciplinary project and not merely a recovery engineering project. However, if a well from the surface to the formation level must be drilled through any drinking water sources in order to access the natural gas, assurance of well integrity has to be an essential part of the drilling operation. Vibrations and pressure pulses associated with drilling can cause (at least) short-term impacts to groundwater quality, including changes in color, turbidity, and odor. Chemicals (from the drilling fluid) can escape the wellbore if it is not properly sealed and cased and even though there are, in most cases, regulatory requirements for well casing and well integrity, accidents and failures can still occur. In addition, older and abandoned wells that have not been sealed correctly can also potentially serve as migration pathways for contaminants to enter groundwater systems. Natural fractures in subterranean formations as well as the fractures created during the fracturing process could also serve as pathways to contamination of groundwater. Finally, coalbed methane is generally found at shallower depths and in closer proximity to underground sources of drinking water and therefore accessing the natural gas from this source might pose a greater risk of contamination.

Bacteria in the fracturing fluid can cause formation biofouling, reducing permeability and gas production. The presence of sulfate-reducing bacteria can form hydrogen sulfide, making the well sour, creating safety issues and increasing costs. Permeability and gas production can be reduced by metals in water, specifically iron, which can oxidize and form deposits, and by suspended solids in the frac fluid such as sand, silt, clay minerals, and scale particles. Since some formations are a composite of shale carbonate shales, and salt crystals, the use of low a fracturing fluid low in TDS will increase dissolution of formation salts, potentially increasing reservoir permeability and gas production. Understanding the relationship between the fracturing fluid water quality and long-term well production is crucial when considering water treatment technologies for a good water management plan.

Another effect, the scaling tendency of source water, usually caused by poor compatibility of source water with formation water and poor compatibility of re-use water, is another consideration. Scaling can occur within the formation, potentially creating reduced permeability and ultimately reduced gas production. Scaling can also damage equipment casing, reducing functionality. Multivalent ions and chlorides in the water can limit friction reducer effectiveness and drive up horsepower costs for pumping in a hydraulic fracturing project. As brine and gas proceed from the formation to the surface, pressure and temperature change and certain dissolved salts can precipitate (referred to as *self-scaling*). If a brine is injected into the formation to maintain pressure and sweep the gas to the producing wells, there will eventually be a commingling with the formation water. Additional salts may precipitate in the formation or in the wellbore (scale from *incompatible waters*) and other factors (Table 6.1). Many of these scaling processes can and do occur simultaneously. Thus, wells producing water are likely to develop deposits of inorganic scales. Scales can and do coat perforations, casing, production tubulars, valves, pumps, and downhole completion equipment, such as safety equipment and gas lift mandrels. If allowed to proceed, this scaling will limit production, eventually requiring abandonment of the well. However, technology is available for removing scale from tubing, flowline, valves, and surface equipment, restoring at least some of the lost production level. Technology also exists for preventing the occurrence or reoccurrence of the scale, at least on a temporary basis.

Scale remediation techniques must be quick and non-damaging to the wellbore, tubing, and the reservoir. If the scale is in the wellbore, it can be removed mechanically or dissolved chemically. Selecting the best scale-removal technique for a particular well depends on knowing the type and quantity of scale, its physical composition, and its texture. As a simple example of scale reduction, seawater can reduce the likelihood of scaling caused by a formation brine that has a high calcium carbonate ($CaCO_3$) scaling potential. Seawater is often injected into reservoirs for pressure maintenance and enhancing gas recovery. When comingled with formation brine, the seawater can markedly decrease the calcium carbonate scaling potential of the mixture. As the percentage of seawater in the mixture increases, scaling potential decreases and may even become negative. Once the scaling potential becomes negative, the produced water can dissolve the calcium carbonate scale that may have formed in the flow lines.

Considerations related to water disposal should also be given suitable acknowledgment during the early stages of field development with the recognition that the need for water disposal will increases as activity increases through evolution of the project – moreover, demand for water disposal in terms of the volume of water can be surprising. Thus, as project evolves and water sources for long-term supply are considered, equal consideration should be given to disposal of the water respect to the collective volumes of all water use. In fact, water disposal issues are challenging in most project areas. Facilities accepting flowback water may be able to be designed to accept other water disposal streams thereby reducing overall treatment costs and creating more make-up water for re-use.

6.4.3 Water disposal

Water that originates from a hydraulic fracturing project often contains chemical additives to help carry the proppant and may become enriched in salts after being injected into tight formations. Therefore, the water that is recovered during natural gas production from tight formations must be either treated or disposed of in a safe manner – a suggestion for disposal of such water is typically to inject the water into deep, highly saline formations through one or more wells drilled specifically for that purpose but the disposal method must follow clearly defined regulations. Flowback water is infrequently re-used in hydraulic fracturing because of the potential for corrosion or scaling, where the dissolved salts may precipitate out of the water and block parts of the well or the formation thereby interfering with and influencing flow of the fluids.

In addition to chemicals that are added to the fracturing fluid, wastewater from natural gas extraction may contain high levels of TDS, which can be a complex collection of inorganic compounds. Furthermore, the amount of saline water produced from tight formations can vary widely – from zero to (at least, depending upon the formation characteristics) several hundred barrels per day. The water may originate from the tight reservoir formation or from any adjacent formations that are connected to the reservoir formation through a natural fracture network or, more likely, through the process-induced fracture network. Typically, this water, like flowback water, is highly saline and must be treated or sent for disposal by injection into deep saline formations, which is also subject to clearly defined regulations. In fact, in some regulatory authorities that oversee the production of natural gas has implemented regulations regarding the disclosure of chemicals used in the process of hydraulic fracturing to ensure that the subterranean environment is protected when re-injection of these chemicals is the method of disposal.

In terms of water use management, one alternative that states and natural gas production projects are pursuing is to make use of seasonal changes in river flow to capture water when surface water flows are greatest. Utilizing seasonal flow differences allows planning of withdrawals to avoid potential impacts to municipal drinking water supplies or to aquatic or riparian communities. Also included is monitoring of stream water quality as well as game and nongame fish species in the reach of river surrounding the intake. In addition, new treatment technologies have made it possible to recycle the water recovered from hydraulic fracturing. The re-use of treated flowback fluids from hydraulic fracturing is being used (or at least, considered for use) for various projects.

Briefly, and by way of explanation, a riparian community is a community that exits in a riparian zone (riparian area), which is the interface between land and a river or between land and a stream. Plant habitats and communities along the river margins and banks, stream margins and banks (*riparian vegetation*) are characterized by collections of hydrophilic plants. Riparian zones are important in ecology and in environmental management because of their role in soil conservation, their habitat biodiversity, and the influence they have on fauna and aquatic ecosystems, including grasslands, woodlands, wetlands, or even non-vegetative growing areas. In some regions the terms riparian woodland, riparian forest, riparian buffer zone, and riparian strip are used to characterize a riparian zone.

Once the natural gas is produced, it is processed, and transported for further processing (gas cleaning, gas refining). Water consumption occurs in each of these stages as well, with the most significant non-production consumption potentially occurring during end use. On the other hand, although natural gas can be combusted directly with no additional water consumption, if the end use of the gas is in the form of a transportation fuel, storage in a vehicle tank is required and the natural gas will most likely have to be compressed by means of an electric compressor (King and Webber, 2008; Wu et al., 2011). It then remains to select an option (or options) for treatment of the water either for re-use or for disposal.

Put simply, water treatment is any process that removes contaminants from the water or reduces the concentration of the contaminants so that the water is more acceptable for a specific end use. In the present context, the end use may be industrial water supply, irrigation, river flow maintenance, water recreation or many other uses including being safely returned to the environment. In some cases, exceptional and assiduous treatment of the water can produce water suitable for use as drinking water. Thus, management of flowback water can generally by realized by any one (or more) of the management strategies: (i) disposal, (ii) re-use, and (iii) recycling (Halldorson and Horner, 2012).

In regard to the *disposal* option for the water, the flowback is transported to an injection well for disposal. This option for disposal is often chosen if there is a ready and relatively inexpensive, abundant supply of freshwater nearby and, most important, nearby injection wells can handle the flowback disposal volumes according to the relevant regulatory guidelines or laws. However, as freshwater availability decreases, costs increase and/or distance to injection wells for disposal increases, making the disposal scenario less appealing.

The *re-use* option (which may be the least expensive strategy) involves careful treatment of the flowback water to remove suspended solids and any soluble organic constituents (Speight, 2014b) followed by blending the treated water with freshwater to generate a fluid that is suitable for use in a hydraulic fracturing project for a new well. This option reduces the amount of freshwater required for the project and eliminates the need for the disposal option if all flowback water can be treated and re-used.

Finally, the *recycling* option treatment of the flowback water to produce a product that is typically of freshwater quality. In this option, the recycled water is blended with make-up water from freshwater sources to generate a hydraulic fracturing fluid that is low in TDS. Recycling is used when (i) freshwater costs are high, a high quality, (ii) a fracturing fluid with a low TDSs content is required, or (iii) other logistics such as hydraulic fracturing schedules do not permit re-use of the water. If the fracturing fluid is transferred by means of temporary above-ground pipelines (*fastlines*) may recycle the water to minimize potential environmental liability from spills or rupture of the pipeline.

The issue of the TDS in water is a continuous issue and during an active fracking project involving natural gas production is continually being subject to inconsistent flowback water containing TDS – ranging from 5,000 to 200,000 ppm – and/or total suspended solids ranging from 100 to 3,000 ppm. In order to combat such impurities, a clarifier can pump can be used to pump flowback water from the source (such as a such as a *frac tank* or pit) into a unit which uses a solids settling system which enables suspended solids settle to the bottom of the unit where they are collected and dewatered. A major benefit of re-using water is that it reduces the financial, social and environmental costs associated with water transportation.

In addition, evaporation technology can be used to recycle contaminated water – the process involves boiling a solution so that contaminants remain in the liquid phase, while pure water vapor evaporates and can be condensed into distilled water. Also, mechanical vapor recompression evaporation is a leading means of treating wastewater – the process differs from conventional evaporation insofar as a compressor is used to generate steam rather than a heat source such as a boiler. High energy efficiency is achieved by utilizing the latent heat of the condensing steam as the primary energy source for boiling the wastewater.

6.4.4 Waste fluids

A variety of waste fluids are generated on site at natural gas wells and one of the biggest challenges for protecting water resources from tight formation natural gas activities is the wastewater generated during production. Wastes, such as drill cut-tings, and wastewater generated during exploration, development, and production of natural gas are categorized by the US Environmental Protection Agency as *special wastes* that are exempted from federal hazardous waste regulations under Subtitle C of the Resource Conservation and Recovery Act.

Fluids management involves the environmentally friendly disposal or re-use of excess fluids and is an integral and essential phase of project involved in the devel-opment of and recovery of natural gas from tight formations because pollutants in the natural waterways (rivers, stream, and some case, he oceans) result in the re-duction of the oxygen content of water leading to serious effects on (even elimina-tion of) the aquatic life. This gives rise to the need, not only for water purification but also for water conservation.

Furthermore, an important challenge is fluids management, which involves: (i) treatment, (ii) recycling, (iii) re-use, and (iv) disposal of the flowback water and produced water. This water mix often contains residual fracturing fluid and may also contain substances found in the reservoir formations, such as trace elements of heavy metals and even naturally occurring radioactive materials. Also, the im-pact of the recovery operations on land and air must also be minimized. Produc-tion wells require roadways to connect drilling pads and there will be the need for either pipelines or trucks or both to transport natural gas, or process waste as well as the need for storage units, and water treatment facilities. With the potential for several hundred truck trips per well site, pipeline use has the potential to mini-mize surface disturbance.

Wastewater from natural gas exploration is generally classified into flowback and produced waters. Flowback water is the fluid that returns to the surface after the step of hydraulic fracturing and before natural gas production begins, primarily during the days to weeks of well completion. This fluid can consist of 10–40% v/v of the injected fracturing fluids and chemicals pumped underground that return to the surface mixed with an increasing proportion of natural brines from the tight for-mation formations through time.

Produced water is the fluid that flows to the surface during extended natural gas production and primarily reflects the chemistry and composition of deep forma-tion waters and capillary-bound fluids. These naturally occurring brines are often saline to hypersaline and contain potentially toxic levels of elements such as barium, arsenic, and radioactive radium. Wastewater from hydraulic fracturing operations is disposed of in several ways. Deep underground injection of wastewater comprises >95% of disposal in the United States. In contrast, deep injection of wastewater is not permitted in Europe unless the water is used to enhance natural gas recovery. Waste-water in the United States is also sent to private treatment facilities or, increasingly, is recycled or re-used (see above). More recently, wastewater is increasingly being sent to facilities with advanced treatment technologies such as desalination at rates approaching 90% reuse.

One of the first steps in water management involves proper well construction to isolate the wellbore by use of cement and several different sets of steel casing, is a critical step taken by the natural gas industry to protect groundwater sources. The casings are individually cemented into the wellbore to provide barrier, which isolate wellbore fluids from the rock formations. In the process, cement is pumped into the center of casing so that it circulates back to the surface in the space outside of the casing (the *annulus*), following the installation of each length (*string*) of casing in the well. After these steps are complete, the cement must be allowed to set prior to the continuation of drilling – a geophysical log is run to determine the integrity of the cement that surrounds the casing. This is an aid to ensure that the wellbore is adequately cemented and capable of withstanding the pressure associated with hydraulic fracturing. Prior to stimulation, the well is pressure tested to ensure the integrity of the casing system that has been installed underground.

Thus, under the various regulations, comprehensive rules are in place to ensure that wells are constructed in a manner (well integrity) that protects freshwater supplies. Specific guidelines vary between the various regulatory authorities but, in all cases, steel casing and cement are used to isolate and protect groundwater zones from deeper natural gas and saline water zones.

6.4.4.1 Fracturing fluid requirements

The primary factor influencing water management associated with hydraulic fracturing is related to the fluid requirements for a successful hydraulic fracturing operation. All phases of water management ultimately depend on the requirements of the hydraulic fracturing properties needed for conducting a successful fracturing project. These requirements are the result of the geology of the reservoir formation and the formations above and below the reservoir, the operating environment, the design of the hydraulic fracturing process, the scale of the development process, and the results required for total project success.

The primary issue in understanding water management for a hydraulic fracturing project involves knowledge of the reservoir rock need and what the rock will produce after completion of the fracturing process. The choice of the fracturing fluid dictates the fracturing design as well as the types of fracturing fluids and the additives that are required. Furthermore, the choice of the fracturing fluid dictates the transport and ultimate fate of the fracturing fluid used in the fracturing operations as well as the means by which the recovered fluids will need to be managed and disposed.

The typical hydraulic fracturing practice (if there is such a process that can be described as *typical*) is designed to create single fractures or multiple fractures in specific rock formations. These hydraulic fracture treatments are controlled and monitored processes designed for the site-specific conditions of the reservoir. Moreover, the process conditions are guided by (i) extraction of the target product, that is, natural gas, (ii) the respective properties of the target product, (iii) the properties – including the mineralogy – of the target formation, (iv) the rock fracturing charac- teristics of the formation, (v) the properties of the formation water, (vi) the antici- pated water production, that is, formation water versus fracturing flow back water, and (vii) the type of well drilled (horizontal or vertical) into the formation.

Thus, understanding the in-situ reservoir conditions is critical to successful stim- ulations, and in the design of the fracture treatment and fluid used as well as to water management. Hydraulic fracturing designs are continually evolving both during the fracture stimulation itself and over time as more related to fracturing the target formation is understood and this understanding evolves. Thus, while the concepts and general practices are similar, the details of a specific fracture operation can vary substantially from resource to resource, from area to area, and even from well to well.

6.4.4.2 Fracturing fluid composition

There is a wide variety of additives that could be included in the fracturing fluid mix to achieve successful fracturing. These include proppants, gel and foaming agents, salts, acids, and other fluid additives and there is a movement (and a need) to maximize the utilization of environmentally benign additives and minimize the amount of additives required.

The characteristics of the resource target determine the required composition of the hydraulic fracturing fluid composition, which, in turn can affect water manage- ment. For example, the tight formation may contain various naturally occurring trace metals and compounds that become available because of the induced frac- tures and are the leached from the reservoir rock by acidic water, by conversion to soluble species as a result of oxidation, and by the action of ionic species that occur in brines. Numerous inorganic and organic compounds have been formed naturally and occur in the tight formation and a stimulation fluid pumped into a well may require various chemicals to counteract any negative effects these compounds may have in the well or the reservoir. For example, iron compounds require an iron se- questering agent so that the compounds of iron will not precipitate out of the frac- turing fluid and be deposited within the pore spaces of the reservoir, reducing the permeability of the reservoir and adding further complications to water management.

One of the major aspects of water management is, when developing plans for a hydraulic fracturing project in addition to considerations associated with success- fully fracturing the target formations, the fluid management and disposal implications of the project and the fracture fluid formulations should receive major consideration. The best water management practice is to use additives that pose minimal risk of possi- ble adverse environmental effects to the extent possible in delivering the needed effec- tiveness of the fracturing operation. While this is a highly desirable option, product substitution may not be possible in all situations because effective alternatives may not be available for all additives.

6.4.4.3 Fracturing fluid handling and storage

Fracturing fluid handling and storage is perhaps that major issue that arises when considering water management options. Fracturing fluid requirements and fracturing fluid composition are all aspects of fluid handling and storage that contribute to water management. Fluids handled at the well site both before and after hydraulic fracturing often must be stored on site, and must be transported from the source of supply to the point of ultimate treatment and/or disposal. Fluids used for hydraulic fracturing will generally be stored onsite in tanks or lined surface impoundments. Returned fluids, or flowback, may also be directed to tanks or lined pits. Furthermore, the volume of initial flowback water recovered during the first month or so following the completion of hydraulic fracturing operations may account for less than 10% v/v to more than 70% v/v of the original volume of the fracturing fluid. The vast majority of fracturing fluid injected is recovered in a short period of time, usually to a maximum time period of several months.

Thus all of the components of the fracturing fluids, including and especially water, additives and proppants, should be managed properly on site before, during, and after the hydraulic fracturing process. If possible, to assist in water management, the components of the fracturing fluid should all be blended into the composite fracturing fluid on an as-needed basis. In addition, any unused products should be removed from the fracturing location as soon as is appropriate. Furthermore, the project planning process should take into consideration the possibility of unexpected delays in the fracturing operation and ensure that water and the additive materials are managed correctly.

If lined impoundments or pits are used for storage of fracturing fluids or flowback water, the pits must comply with applicable rules, regulations, good industry practice, and liner specifications. Thus, these impoundments must be designed and constructed in such a manner as to provide structural integrity for the life of their operation – correct design is imperative to the objective of preventing a failure or unintended discharge. If the fluids are to be stored in tanks, these tanks must meet appropriate state and federal standards, which may be specific to the use of the tank – for example, if the tank is used as a tank for flowback water or for more permanent production tank batteries.

6.4.4.4 Transportation

Before the onset of fracturing, water, sand, and any other additives are generally delivered separately to the well site. Water is generally delivered in tanker trucks that may arrive over a period of days or weeks, or via pipelines from a supply source or treatment/recycling facility. Thus, water supply and management approaches should take into consideration the requirements and constraints associated with fluid transport.

Transportation of water to and from a well site can be a major expense and major activity. While trucking costs can be the biggest part of the water management expense, one option to consider as an alternative to trucking is the use of temporary or permanent surface pipelines but the transport of fluids associated with hydraulic fracturing by surface pipeline may not be practical, cost effective, or even feasible. Moreover, when fracture fluids are to be transported by truck, the project will need development of a basin-wide trucking plan that includes (i) the estimated amount of trucking required, (ii) the hours of operations, (iii) the appropriate off road parking/staging areas, and (iv) the trucking routes. Furthermore, considerations for the trucking plan for large volumes of fracture fluid include the following: (i) public input on route selection to maximize efficient driving and public safety, (ii) avoidance of peak traffic hours, school bus hours, community events, and overnight quiet periods, (iii) coordination with local emergency management agencies and highway departments, (iv) upgrades and improvements to roads that will be traveled frequently to and from many different well sites, (v) advance public notice of any necessary detours or road/lane closures, and (vi) adequate off-road parking and delivery areas at the site.

The use of multi-well pads makes the use of central water storage easier, reduces truck traffic, and allows for easier and centralized management of flow back water. In some cases, it can enhance the option of pipeline transport of water. Furthermore, in order to make truck transportation more efficient and have lees impact on the surrounding environment, it is worth considering the construction of storage ponds and drilling source wells in cooperation with any non-project (but interested) private property owners. The opportunity to construct or improve an existing pond, drilling a water well, and/or improving the roads on their property can be an extremely helpful (perhaps even a *win-win*) situation for the operator and the landowner by providing close access to a water source for the project as well as adding improvements to the nearby property that could benefit the property owner.

During drilling, used mud and saturated cuttings are produced and must be managed. The volume of mud approximately correlates with the size of the well drilled, so a horizontal well may generate twice as much drilling waste as a single vertical well however the horizontal well may replace as many as four vertical wells (Arthur et al., 2008). Drilling wastes can be managed onsite either in pits or in steel tanks. Each pit is designed to keep liquids from infiltrating vulnerable water resources. Onsite pits are a standard in the natural gas industry but are not appropriate everywhere; they can be large and they disturb the land for an extended period of time. Steel tanks may be required to store drilling mud in some environments to minimize the size of the well site *footprint* or to provide extra protection for a sensitive environment. Steel tanks are not, of course, appropriate in every setting either but in rural areas or pits or ponds, where space is available at the well site, steel tanks are usually not needed (Arthur et al., 2008).

The drill cuttings are regarded as controlled or hazardous waste and can be disposed of in the following ways: (i) decontamination treatment, (ii) injection of the cuttings into the well, or (iii) transfer to a controlled hazardous-waste landfill. The lowest environmental effect for solids treatment, especially for offshore operation, is decontamination treatment followed by discharge.

Horizontal drilling development has the power to reduce the number of well sites and to group them so that management facilities such as storage ponds can be used for several wells. Make-up water is used throughout the development process to drill the well and to form the basis of the hydraulic fracturing fluid. Large volumes of water may be needed and are often stored at the well site in pits or tanks. For example, surface water can be piped into the pit during high-water runoff periods and used during the year for drilling and fracture treatments in nearby wells. Storage ponds are not suitable everywhere in the area of a natural gas resource – just as steel tanks are appropriate in some locations but not in others. Finally, it may be opportune for any hydraulic fracturing project to include consideration of utilizing agricultural techniques to transport the water used near the water sources. Large diameter, aluminum agricultural pipe is sometimes used to move the freshwater from the source to locations within a few miles where drilling and hydraulic fracturing activities are occurring. Water use by the natural gas industry when working to recover these resources from tight formations has spurred the formation or expansion business involved in the supply the temporary pipe, pumps, installation, and after-project removal of these amenities.

References

Agarwal, R.G., Carter, R.D., and Pollock, C.B. 1979. Evaluation and Performance Prediction of Low-Permeability Gas Wells Stimulated by Massive Hydraulic Fracturing. J. Pet. Technol., 31(3): 362–372. SPE-6838-PA. Society of Petroleum Engineers, Richardson, Texas.

Aggour, T.M., and Economides, M.J. 1998. Optimization of the Performance of High-Permeability Fractured Wells. Paper No. 39474. Proceedings. SPE International Symposium on Formation Damage Control. Lafayette, Louisiana. Society of Petroleum Engineers, Richardson, Texas.

Amadei, B., and Stephansson, O. 1997. Rock Stress and Its Measurement. Cambridge University Press, Cambridge, United Kingdom.

API. 2009. Hydraulic Fracturing Operations – Well Construction and Integrity Guidelines First Edition. Guidance Document HF1. American Petroleum Institute, Washington, DC.

API. 2010. Water Management Associated with Hydraulic Fracturing. API Guidance Document HF2. American Petroleum Institute, Washington, DC.

Arthur, J.D., Bohm, B., and Layne, M. 2008. Hydraulic Fracturing Considerations for Natural Gas Wells of the Marcellus Shale. ALL Consulting. Presented at the GWPC Annual Forum in Cincinnati, OH. September. Groundwater Protection Council, Oklahoma City, Oklahoma.

Arthur, J.D., Bohm, B., Coughlin, B.J., and Layne, M. 2009. Evaluating Implications of Hydraulic Fracturing in Shale Gas Reservoirs. Paper No. SPE 121038. Proceedings., 2009 SPE Americas

Environmental and Safety Conference, San Antonio, Texas. March 23–25. Society of Petroleum Engineers, Richardson, Texas.

Azar, J.J., and Samuel, G.R. 2007. Drilling Engineering. PennWell Corporation, Tulsa, Oklahoma.

ASTM. 2021. Annual Book of Standards. ASTM International, West Conshohocken, Pennsylvania.

Baihly, J., Laursen, P, Ogrin, J., Le Calvez, J.H., Villarreal, R., Tanner, K., Bennett, L. 2006. Using Microseismic Monitoring and Advanced Stimulation Technology to Understand Fracture Geometry and Eliminate Screenout Problems in the Bossier Sand of East Texas. Paper No. SPE 102493. Proceedings. SPE Annual -Conference and Exhibition, San Antonio, Texas. September 24–27.

Barree, R.D., Fisher, M.K., Woodrood, R.A. 2002. A Practical Guide to Hydraulic Fracture Diagnostics Technologies. Paper No. SPE 77442. Proceedings. SPE Annual Technical Conference and Exhibition, San Antonio, TX, USA, 29 September 29-October 2.

Blauch, M.E., Myers, R.R., Moore, T.R., and Houston, N.A. 2009. Marcellus Shale Post-Frac Flowback Waters – Where is All the Salt Coming from and What are the Implications? Paper No. SPE 125740. Proceedings. SPE Regional Meeting, Charleston, West Virginia. September 23–25.

Britt L.K, Jones J.R and Miller, W.K. 2010. Defining Horizontal Well Objectives in Tight and Unconventional Gas Reservoirs. Paper No. CSUG/SPE 137839. Proceedings. Canadian Unconventional Resources & International Petroleum Conference, Calgary Alberta, Canada. October 19–21. Society of Petroleum Engineers, Richardson, Texas.

Britt, L.K. 2012. Fracture Stimulation Fundamentals. Journal of Natural Gas Science and Engineering, 8: 34–51.

Chen, M., Sun, Y., Fu, P., Carrigan, C.R., Lu, Z., Tong, C.H., and Buscheck, T.A. 2013. Surrogate-Based Optimization of Hydraulic Fracturing in Pre-Existing Fracture Networks. Computers & Geosciences, 58: 69–79.

Cinco-Ley, H., Samaniego, V.F., and Dominguez-A., N. 1978. Transient Pressure Behavior for a Well with a Finite-Conductivity Vertical Fracture. SPE Journal, 18(4): 253–264.

Cipolla, C., and Wright, C.A. 2002. Diagnostic Techniques to Understand Hydraulic Fracturing: What? Why and How? Proceedings. 2002 SPE/CERI Gas technology symposium, Calgary, Canada. April 3–5. Society of Petroleum Engineers, Richardson, Texas.

Cipolla, C.L., Lolon, E.P., Mayerhofer, M.J., and Warpinski, N.R. 2009. Fracture Design Considerations in Horizontal Wells Drilled in Unconventional Gas Reservoirs. Paper No. SPE 119366. Proceedings. SPE Hydraulic Fracturing Technology Conference, The Woodlands, Texas. January 19–21.

Clark, C., Han, J., Burnham, A., Dunn, J., and Wang, M. 2011, Life-Cycle Analysis of Shale Gas and Natural Gas. Report No. ANL/ESD/11-11, Argonne National Laboratory, Argonne, Illinois.

Daniels, J., Waters, G., LeCalvez, J., Lassek, J., and Bentley, D. 2007. Contacting More of the Barnett Shale Through an Integration of Real-Time Microseismic Monitoring, Petrophysics and Hydraulic Fracture Design. Paper No. 110562. Proceedings. SPE Annual Technical Conference and Exhibition, Anaheim, California.

Das, I, and Zoback, M.D. 2011. Long-Period, Long-Duration Seismic Events during Hydraulic Fracture Stimulation of a Shale Gas Reservoir. Leading Edge, 30: 778–786.

Davis, J., Warpinski, N.R., Davis, E.J., Griffin, and Malone, S. L.G. 2008. Joint Inversion of Downhole Tiltmeter and Microseismic Data and its Application to Hydraulic Fracture Mapping in Tight Gas Sand Formation. Paper No. ARMA 08-344. Proceedings. 42nd US Rock Mechanics Symposium and 2nd US-Canada Rock Mechanics Symposium. San Francisco, June 29-July 2.

Devereux, S. 2012. Drilling Technology in Non-Technical Language 2nd Edition. PennWell Publishing Corporation, Tulsa, Oklahoma.

Economides, M.J., and Nolte, K.G. 2000. Reservoir Stimulation 3rd Edition. John Wiley & Sons Inc., Hoboken, New Jersey.

Ely, J.W. 1985. Handbook of Stimulation Engineering. PennWell Publishing, Tulsa, Oklahoma.

Fisher, M.K., Heinze, J.R., Harris, C.D., McDavidson, B.M., Wright, C.A., Dunn, K.P. 2004. Optimizing Horizontal Completion Techniques in the Barnett Shale Using Microseismic Fracture Mapping. Paper No. SPE 90051. Proceedings. SPE Annual Technical Conference and Exhibition, Houston, Texas. September 26–29.

Flewelling, S.A., Tymchak, M.P., and Warpinski, N. 2013. Hydraulic Fracture Height Limits and Fault Interactions in Tight Oil and Gas Formations. Geophysical Research Letters, 40: 3602–3606.

Gidley, J.L., Holditch, S.A., Nierode, D.E., and Veatch, R.W. 1990. Hydraulic Fracturing to Improve Production. Monograph Series SPE 12, Society of Petroleum Engineers, Richardson, Texas.

Green, K.P. 2014. Managing the Risks of Hydraulic Fracturing. Fraser Institute, Vancouver, British Columbia, Canada. http://catskillcitizens.org/learnmore/managing-the-risks-of-hydraulic-fracturing.pdf; accessed January 15, 2016.

Green, K.P. 2015. Managing the Risks of Hydraulic Fracturing: An Update. Fraser Institute, Vancouver, British Columbia, Canada. https://www.fraserinstitute.org/studies/managing-the-risks-of-hydraulic-fracturing-an-update; accessed April 4, 2016.

Halldorson, B., and Horner, P. 2012. Shale Gas Water Management. In: World Petroleum Council Guide: Unconventional Gas. World Petroleum Council, London, United Kingdom. Page 58–63. http://www.world-petroleum.org/docs/docs/gasbook/unconventionalgaswpc2012.pdf; accessed March 15, 2015.

Hammack, R., Harbert, W., Sharma, S., Stewart, B., Capo, R., Wall, A., Wells, A., Diehl, R., Blaushild, D., Sams, J., and Veloski, G. 2014. An Evaluation of Fracture Growth and Gas/Fluid Migration as Horizontal Marcellus Shale Gas Wells Are Hydraulically Fractured in Greene County, Pennsylvania. Report No. NETL-TRS-3-2014. EPAct Technical Report Series. National Energy Technology, Laboratory, Pittsburgh, Pennsylvania. US Department of Energy, Washington, DC.

Hareland, G.I., Rampersad, P., Dharaphop, J., and Sasnanand S., 1993. Hydraulic Fracturing Design Optimization. Paper No. 26950. Proceedings. SPE Eastern Regional Conference and Exhibition, Pittsburgh, Pennsylvania. Page 493–500. Society of Petroleum Engineers, Richardson, Texas.

Hibbeler, J., and Rae, P. 2005. Simplifying Hydraulic Fracturing: Theory and Practice. Paper No. SPE 97311. Proceedings. 2005 SPE Technical Conference and Exhibition, Dallas, Texas. October 9–12. Society of Petroleum Engineers, Richardson, Texas.

Holditch, S.A., Robinson, B.M., and Whitehead, W.S. 1987. Prefracture and Postfracture Formation Evaluation Necessary to Characterize the Three Dimensional Shape of the Hydraulic Fracture. Proceedings. SPE Formation Evaluation. December 1987. Society of Petroleum Engineers, Richardson, Texas.

Holditch, S.A., Jennings, J.W., and Neuse, S.H. 1978. The Optimization of Well Spacing and Fracture Length in Low Permeability Gas Reservoirs. Paper No. SPE-7496. Proceedings. SPE Annual Fall Technical Conference and Exhibition, Houston, Texas, October 1–3. Society of Petroleum Engineers, Richardson, Texas.

Holditch, S.A. 1979. Factors Affecting Water Blocking and Gas Flow from Hydraulically Fractured Gas Wells. J. Pet. Technol., 31(12): 1515–1524.

Hubbert, M.K., and Willis, D.G., 1957. Mechanics of Hydraulic Fracturing. Petroleum Transactions AIME, 210: 153.

IEA. 2013. Resources to Reserves 2013: Oil, Gas and Coal Technologies for the Energy Markets of the Future. OECD Publishing. International Energy Agency, Paris, France.

Jones, J.R., and Britt, L.K. 2009 Design and Appraisal of Hydraulic Fractures. Society of Petroleum Engineers, Richardson, Texas.

Kennedy R.L., Gupta, R., Kotov, S.V., Burton, W.A., Knecht, W.N., and Ahmed, U. 2012. Optimized Shale Resource Development: Proper Placement of Wells and Hydraulic Fracture Stages. Paper No. 162534. Proceedings. Abu Dhabi International Petroleum Conference and Exhibition. Abu

Dhabi, United Arab Emirates. November 11–14. Society of Petroleum Engineers, Richardson, Texas.

King, C.W., and Webber, M.E. 2008. Water Intensity of Transportation. *Environmental Science and Technology*, 42(21): 7866–7872.

King, G.E. 2010. Thirty Years of Gas Shale Fracturing: What Have We Learned? Paper No. SPE 133456. Proceedings. SPE Annual Technical Conference and Exhibition Florence, Italy. September.

Mantell, M.E. 2009. Deep Shale Natural Gas: Abundant, Affordable, and Surprisingly Water Efficient. Proceedings. 2009 GWPC Water/Energy Sustainability Symposium. Salt Lake City, Utah. September 13–16.

Maxwell, S. 2011. Microseismic Hydraulic Fracture Imaging: The Path Toward Optimizing Shale Gas Production. The Leading Edge, 30: 340–346.

Mohaghegh, S., Balanb, B., Platon, V., and Ameri, S. 1999. Hydraulic Fracture Design and Optimization of Gas Storage Wells. Journal of Petroleum Science and Engineering, 23: 161–171.

Phillips, W.J. 1972. Hydraulic Fracturing and Mineralization. Journal of the Geological Society of London, 128: 337–359.

Reddy, T.R., and Nair, R.R. 2012. Fracture Characterization of Shale Gas Reservoir Using Connected – Cluster DFN Simulation. Proceedings. 2nd International Conference on Drilling Technology 2012 (ICDT-2012) and 1st National Symposium on Petroleum Science and Engineering 2012 (NSPSE-2012). R. Sharma, R. Sundaravadivelu, S. K. Bhattacharyya and SP. Subramanian (Editors). Page 133–136. December 6–8.

Rueda, J.I., Rahim, Z., Holditch, S.A., 1994. Using a mixed integer linear programming technique to optimize a fracture treatment design. Paper No. 29184. Proceedings. SPE Eastern Regional Meeting, Charleston, South Carolina. Page 233–244. Society of Petroleum Engineers, Richardson, Texas.

Satterfield, J.M., Mantell, D., Kathol, F., Hiebert, K., Patterson, and Lee, R. 2008. Managing Water Resource's Challenges in Select Natural Gas Shale Plays. Proceedings. GWPC Annual Meeting. September. Groundwater Protection Council, Oklahoma City, Oklahoma.

Secor, D.T. 1965. Role of Fluid Pressure in Jointing. American Journal of Science, 263: 633–646.

Sone, H., and Zoback, M.D. 2013a. Mechanical Properties of Shale-Gas Reservoir Rocks – Part 1: Static and Dynamic Elastic Properties and Anisotropy. Geophysics, 78(5): D381–D392.

Sone, H., and Zoback, M.D. 2013b. Mechanical Properties of Shale-Gas Reservoir Rocks – Part 2: Ductile Creep, Brittle Strength, and Their Relation to The Elastic Modulus. Geophysics, 78(5): D393–D402.

Speight, J.G. 2014a. The Chemistry and Technology of Petroleum 5th Edition. CRC Press, Taylor & Francis Group, Boca Raton, Florida.

Speight, J.G. 2014b. High Acid Crudes. Gulf Professional Publishing, Elsevier, Oxford, United Kingdom, 2014.

Speight, J.G. 2016a. Introduction to Enhanced Recovery Methods for Heavy Oil and Tar Sands 2nd Edition. Gulf Publishing Company, Taylor & Francis Group, Waltham Massachusetts.

Speight, J.G. 2016b, Handbook of Hydraulic Fracturing. John Wiley & Sons Inc., Hoboken, New Jersey.

Stillwell, A.S., King, C.W., Webber, M.E., Duncan, I.J., and Herzberger A. 2010. The energy-water nexus in Texas. Ecology and Society, 16(1): 2.

Veil, J.A. 2010. Water Management Technologies Used by Marcellus Shale Gas Producers. Report No. ANL/EVR/R-10/3. Argonne National Laboratory, Argonne, Illinois. United states Department of Energy, Washington, DC. July.

Vulgamore, T., Clawson, T., Pope, C., Wolhart, S, Mayerhofer, M., Machovoe, S., and Waltman, C. 2007. Applying Hydraulic Fracture Diagnostics to Optimize Stimulations in the Woodford

Shale. Paper No. SPE 110029. Proceedings. SPE Annual Technical Conference and Exhibition, Anaheim, California. November 11–14.

Wang, X., and Economides, M. 2009. Advanced Natural Gas Engineering. Gulf Publishing Company, Elsevier BV, Amsterdam, Netherlands.

Warpinski, N.R., Engler, B.P., Young, C.J., Peterson, R., Branagan, P.T., and Fix, J.E. 2005. Microseismic Mapping of Hydraulic Fractures using Multi-Level Wireline Receivers. Paper No. SPE 30507. Proceedings. 2005 SPE Annual Technical Conference and Exhibition, Dallas, Texas. October 22–25. Society of Petroleum Engineers, Richardson, Texas.

Waters, G., Heinze, J., Jackson, R., Ketter, A., Daniels, J., and Bentley, D. 2006. Use of Horizontal Well Image Tools to Optimize Barnett Shale Reservoir Exploitation. SPE Paper No. 103202. Proceedings. SPE Annual Technical Conference and Exhibition, San Antonio, Texas.

Wright, C.A., Davis, E.J., Wang, G., and Weijers, L. 1999. Downhole Tiltmeter Fracture Mapping: A New Tool for Direct Measurement of Hydraulic Fracturing Growth. In: Rock Mechanics for Industry. B. Amadei, R.L. Krantz, G.A. Scott, and P.H. Smeallie (Editors). Balkema Publishers, Rotterdam, Netherlands.

Wu, M., Mintz, M., Wang, M., Arora, S., and Chiu, Y. 2011. Consumptive Water Use in the Production of Ethanol and Petroleum Gasoline – 2011 Update. Report No. ANL/ESD/09-1, Argonne National Laboratory, Argonne, Illinois.

Chapter 7
The environment

7.1 Introduction

All fossil fuels – natural gas, crude oil, and coal – release pollutants into the atmosphere when burned to provide energy. However, natural gas – being composed predominantly of methane, which combusts to carbon dioxide and water – is considered the most environmentally friendly fossil fuel. It is cleaner burning than coal or crude oil because it contains less carbon than some of its fossil fuel cousins.

The use of natural gas in the past seven decades has grown steadily in – even replacing coal gas in many markets and, currently, is used extensively in residential as well as in commercial and industrial applications (Speight, 2013, 2019a), while natural gas use represents approximately 25% of the energy used worldwide for electricity generation as well as a feedstock for industry. Natural gas is the dominant source of energy for domestic heating as well as the fuel for various vehicles. The use of natural gas is also rapidly increasing in electric power generation with natural gas power-generating facilities replacing coal power-generating facilities and crude oil power-generating facilities. This trend is expected to continue well into the foreseeable future.

As the gas is recovered and collected, the typical natural gas well requires minor human support on a daily basis. The pumper or switcher may visit the site for a short time to confirm normal operations and then move on to the next location. Occasionally, the well may be worked over for repair or stimulation and the activity around the wellsite increased during this time. Because of the lessened activity involved in production operations, the environmental impacts associated with it are also decreased.

Environmentally, natural gas (which is predominantly methane even in the raw unrefined form) (Table 7.1) is the cleanest burning of the fossil fuels and fossil fuel products and produces (by combustion) primarily carbon dioxide, water vapor, and small amounts of nitrogen oxides (Burrus and Ryder, 2003, 2014; Speight, 2019a):

$$CH_4 + O_2 \rightarrow CO_2 + H_2O$$

After removal of the non-methane constituents, methane, natural gas, is considered to be a versatile fuel and its growth is linked in part to its environmental benefits (even though there are environmental effects; Table 7.2) relative to other fossil fuels, particularly for air quality as well as greenhouse gas emissions.

Thus, natural gas is an extremely important source of energy for reducing pollution and maintaining a clean and healthy environment (Mokhatab et al., 2006; Speight, 2013, 2014, 2017a, 2018, 2019a). In addition, the use of natural gas also offers a number of environmental benefits over other sources of energy, particularly other fossil fuels. Furthermore, there are environmental concerns with the use of any fuel. As with other fossil fuels, burning natural gas produces carbon dioxide, which is an important and

https://doi.org/10.1515/9783110691023-007

Table 7.1: Constituents of natural gas.

Constituent	Formula	% v/v
Methane	CH_4	>85
Ethane	C_2H_6	3–8
Propane	C_3H_8	1–5
n-Butane	C_4H_{10}	1–2
Isobutane	$C4H_{10}$	<0.3
n-Pentane	C_5H_{12}	1–5
Isopentane	C_5H_{12}	<0.4
Hexane, heptane, octane*	C_nH_{2n+2}	<2
Carbon dioxide	CO_2	1–2
Hydrogen sulfide	H_2S	1–2
Oxygen	O_2	<0.1
Nitrogen	N_2	1–5
Helium	He	<0.5

*Hexane (C_6H_{14}) and higher molecular weight hydrocarbon derivatives up to octane as well as benzene (C_6H_6) and toluene ($C_6H_5CH_3$).

Table 7.2: Constituents of natural gas and the potential effects.

Constituent	Comment
Hydrocarbons	Provide the calorific value of natural gas when it is combusted Hydrocarbon-rich liquid phase can form via condensation from the gas phase Liquid phases in the network can cause blockage Control of hydrocarbon dewpoint is the preferred method to control liquid formation
Diluents or inert gases	Typical inert gases are carbon dioxide, nitrogen, helium Has very low Wobbe index if a large amount of inert gas is present Affects combustion efficiency
Water	Can cause corrosion of the pipeline Can lead to formation of hydrates Hydrates can block valves
Oxygen	Can promote pipeline corrosion in the presence of water and sulfur In underground storage, promotes bacterial activity which can lead to formation of hydrogen sulfide
Hydrogen	Associated with stress corrosion cracking of steel pipelines
Hydrogen sulfide	Hydrogen sulfide reacts with copper piping to form copper sulfide flakes, which can cause blockage
Solids or liquids	Can cause corrosion, stress, or abrasion damage Can cause blockage

effective greenhouse gas. Many scientists believe that increasing levels of carbon dioxide and other greenhouse gases in the atmosphere of the Earth are changing the global climate. However, there are issues that have been raised that are related to the means by which the carbon dioxide in ice cores (the leading evidence for increases on carbon dioxide in the atmosphere) have been measured that throw doubt upon the contribution to the carbon dioxide in the atmosphere by anthropogenic activities (Speight and Islam, 2016; Speight, 2020).

Thus, as with other fuels, natural gas also affects the environment when it is produced, stored, and transported. Because natural gas is mostly made up of methane (another greenhouse gas), there is the potential for leaks of methane into the atmosphere from wells, storage tanks, and pipelines. In addition, exploring and drilling for natural gas will always have some impact on land and marine habitats, but new technologies have greatly reduced the number and size of areas disturbed by drilling (often referred to as *environmental footprints*). Satellites, global positioning systems, remote sensing devices, and 3-D and 4-D seismic technologies make it possible to discover natural gas reserves while drilling fewer wells. Plus, use of horizontal drilling and directional drilling make it possible for a single well to produce gas from much bigger areas of the reservoirs (Speight, 2016).

On a unit basis, natural gas emits lower quantities of greenhouse gases and criteria pollutants than other fossil fuels. This occurs in part because natural gas is more easily fully combusted, and in part because processed (cleaned) natural gas contains fewer impurities than any other fossil fuel. For example, coal mined in the United States typically contains 1.6% w/w sulfur (a consumption-weighted national average). Crude oil burned at electric utility power plants ranges from 0.5% to 1.4% sulfur. Diesel fuel has less than 0.05%, while the current national average for motor gasoline is 0.034% sulfur. Comparatively, natural gas when used for power (electricity) generation has less than 0.005% sulfur compounds.

Global warming (now referred to as global climate change), or the greenhouse effect, is an environmental issue that deals with the potential for global climate change due to increased levels of atmospheric greenhouse gases. These are the gases in the atmosphere that serve to regulate the amount of heat that is kept close to the surface of the Earth. It is speculated that an increase in these greenhouse gases will translate into increased temperatures around the globe, which would result in many disastrous changes to the various environmental patterns.

The principle greenhouse gases include water vapor, carbon dioxide, methane, nitrogen oxides, and some manufactured chemicals such as chlorofluorocarbons. While most of these gases occur in the atmosphere naturally, levels have been increasing due to the widespread burning of fossil fuels by growing human populations. The reduction of greenhouse gas emissions has become a primary focus of environmental programs in many (but not all) countries around the world.

One of the principle greenhouse gases is carbon dioxide and although it does not trap heat as effectively as other greenhouse gases (making it a less potent greenhouse gas), the sheer volume of carbon dioxide emissions into the atmosphere is high, particularly from the burning of fossil fuels. The majority of greenhouse gas emissions come from carbon dioxide directly attributable to the combustion of fossil fuels. Therefore, reducing carbon dioxide emissions can play a huge role in combating the greenhouse effect and global warming. The combustion of natural gas emits almost 30% less carbon dioxide than oil, and just under 45% less carbon dioxide than coal.

One issue that has arisen with respect to natural gas and the greenhouse effect is the fact that methane, the principle component of natural gas, is itself a very potent greenhouse gas. In fact, methane has an ability to trap heat almost 21 times more effectively than carbon dioxide.

Sources of methane emissions include the waste management and operations industry, the agricultural industry, as well as leaks and emissions from the crude oil natural and gas industry itself. It is felt that the reduction in carbon dioxide emissions from increased natural gas use would strongly outweigh the detrimental effects of increased methane emissions. Thus, the increased use of natural gas in the place of other dirtier fossil fuels can serve to lessen the emission of greenhouse gases. Before describing the polluting nature of natural gas, it is worth reviewing the composition of the gas as a means of understanding the nature of the pollutants.

Briefly, natural gas is obtained principally from conventional crude oil and non-associated gas reservoirs, and secondarily from coal beds, tight sandstones, and Devonian shale. Some is also produced from minor sources such as landfills. In the not too distant future, natural gas may also be obtained from natural gas hydrate deposits located beneath the sea floor in deep water on the continental shelves or associated with thick subsurface permafrost zones in the Arctic.

The primary source of emissions is the combustion of a fossil fuel in the preheat furnace and in boilers that produce steam for process heat and stripping. However, when operating in an optimum condition and burning cleaner fuels (such as natural gas), these heating units create relatively low emissions of sulfur oxides (SO_x), nitrogen oxides (NO_x), carbon monoxide (CO), hydrogen sulfide (H_2S), particulate matter (PM), and volatile hydrocarbon derivatives (often referred to as volatile organic compounds, VOCs).

However, the potential to produce as much carbon dioxide as other fossil fuels and the related products is real but whether natural gas has lower life cycle greenhouse gas emissions than coal and oil depends on the assumed leakage rate, the global warming potential of methane over different time frames, the energy conversion efficiency, and other factors. In addtion, methane itself is a greenhouse gas and, in addtion to the production of carbon dioxide during combustion, can have adverse effect on the environment (Speight, 2020).

Moreover, natural gas that is scheduled to be transported and stored must meet specific quality measures so that the pipeline network (or grid) can provide uniform quality gas. Wellhead natural gas will contain other hydrocarbon derivatives, inert gases, and contaminants which must be removed before the natural gas can be safely delivered to the high-pressure, long-distance pipelines that transport natural gas to consumers (Mokhatab et al., 2006; Burrus and Ryder, 2003; Burruss and Ryder, 2014; Speight, 2014, 2017, 2019a). The composition of the wellhead natural gas determines the number of stages and the process required to produce pipeline-quality dry natural gas. Natural gas processing can be complex and usually involves several processes, or stages, to remove oil, water, hydrocarbon gas liquids, and other impurities such as sulfur, helium, nitrogen, hydrogen sulfide, and carbon dioxide (Mokhatab et al., 2006; Speight, 2014, 2019a). The processing options or stages may be (i) integrated into one unit or operation, (ii) performed in a different order, (iii) performed at alternative locations (lease/plant), or (iv) not be required at all.

The stages of natural gas processing/treatment are (i) a gas-oil-water separators, in which pressure relief causes a natural separation of the liquids from the gases in the natural gas; (ii) a condensate separator, in which gas condensate and/or natural gasoline is removed from the natural gas stream at the wellhead with separators much like gas-oil-water separators, (iii) a dehydrator, in which water is removed to reduce the potential for corrosion in the pipeline as well as the formation of undesirable hydrates and water condensation in pipelines, (iv) a series of contaminant removal units, in which non-hydrocarbon gases such as hydrogen sulfide, carbon dioxide, water vapor, helium, nitrogen, and oxygen are removed from the natural gas stream, (v) a nitrogen rejection unit, in which nitrogen is removed and the gas stream is further dehydrated using molecular sieve beds, (vi) a methane separator, in which the removal of methane from the gas stream can occur as a separate operation by cryogenic processing and absorption methods, and (vii) a fractionator, in which the gas stream is separated into component hydrocarbon derivatives (ethane, propane, butane) using the varying boiling points of the individual hydrocarbon derivatives. Also, there is the need to determine the purity of the gas streams as well as the efficiency of each process unit by means of gas stream analysis, usually by means of online or off-line monitoring of the product streams by means of, for example, a technique such as gas chromatography (Speight, 2018, 2019a).

This chapter addresses the many environmental aspects related to the use of natural gas, including the environmental impact of natural gas relative to other fossil fuels and some of the potential applications for increased use of natural gas. These issues include: (i) greenhouse gas emissions as well as (ii) air quality, including smog and acid rain.

7.2 Environmental issues

Once brought from underground, the natural gas receives preliminary treatment, usually to remove water, hydrogen sulfide (if required by the pipeline company), and inorganic materials (such as reservoir sand), to prepare the gas stream for transportation after which the treatment is completed at a gas-processing plant to prepare the gas for sale to the industrial and domestic markets. During the processing, some hydrocarbon derivatives (such as ethane, C_2H_6, propane, C_3H_8, and butane, C_4H_{10}) are removed and sold separately. Other impurities such as hydrogen sulfide (H_2S) are also removed and used to produce sulfur, which is then also sold separately.

When burned to provide energy, all of the fossil fuels (natural gas, crude oil, and coal) release pollutants into the atmosphere – the common pollutant is carbon dioxide which is a greenhouse gas (Speight, 2020). On the other hand, natural gas also has a lower (often low-to-none) sulfur content and nitrogen content and, after combustion, emits less ash (as PM) into the air than crude oil, crude oil-derived fuels, or coal. Thus, natural gas is an important source of energy for reducing pollution and maintaining a clean and healthy environment (Mokhatab et al., 2006; Speight, 2019a). In addition to being an abundant and secure source of energy in the United States, the use of natural gas also offers a number of environmental benefits over other sources of energy, particularly other fossil fuels. However, it must not be forgotten that methane is a potent greenhouse gas, and, on a weight basis, methane has 21 times the global warming potential of carbon dioxide, thereby making a large contribution to global warming and global climate change.

On a relative basis, natural gas is the cleanest of all the fossil fuels. Composed primarily of methane, the main products of the combustion of natural gas are carbon dioxide and water vapor:

$$2CH_4 + 3O_2 \rightarrow 2CO_2 + 2H_2O$$

Crude oil and coal are composed of much more complex molecules than are found in natural gas and which have a lower hydrgoen-to-carbon atomic ratio (a higher carbon-hydrogen atomic ratio) as well as a higher content of nitrogen and sulfur. Thus, when combusted, crude oil and coal release higher levels of harmful emissions – this includes higher amounts of nitrogen oxides (NO_x) and sulfur dioxide (SO_2). There is also the release of PM (soot and ash – products of the combustion process) into the environment, which is carried into the atmosphere and contributes to pollution. On the other hand, the combustion of natural gas releases smaller amounts (on a per-carbon basis than crude oil and coal) of sulfur dioxide and nitrogen oxides, virtually no soot or PM, and lower levels of carbon dioxide, carbon monoxide, and other reactive hydrocarbon derivatives.

Thus, on a carbon-unit basis, natural gas emits lower quantities of greenhouse gases and lower amounts of criteria air pollutants than other fossil fuels – in the current context, criteria air pollutants include (i) PM, (ii) photochemical oxidants, (iii) carbon monoxide, (iv) sulfur oxides, and (v) nitrogen oxides. This is due, in part, because natural gas is more easily fully combusted, and in part because natural gas contains fewer impurities than any other fossil fuel. For example, the crude oil used as combustion fuel at a electric utility power plant typically ranges from 0.5

to 1.4% w/w sulfur while a typical coal might contain 0.6–2.0% w/w sulfur. Comparatively, natural gas at the burner tip has less than 0.1% w/w sulfur.

The amount of carbon dioxide produced for an equivalent amount of heat production varies substantially among the fossil fuels, with natural gas producing the least. For the major fossil fuels, the amounts of carbon dioxide produced for each billion Btu of heat energy extracted are on the order of (i) 117,000 pounds for natural gas, (ii) 164,000 pounds for crude oil and cured oil products, and (iii) 208,000 pounds for coal. Thus, as long as many countries have fossil fuel-based economies, fossil fuel combustion will lead to environmental problems. In addition, the venting or leaking of natural gas into the atmosphere can have a significant effect with respect to greenhouse gases because methane, the principal component of natural gas, is much more effective in trapping these gases than carbon dioxide. The exploration, production, and transmission of natural gas, as well, can have adverse effects on the environment.

7.2.1 Hydrocarbon contamination

Spills and leaks of hydrocarbon derivatives are the most common accidents that are encountered at production sites. The spill of condensate or produced water onto the lease and the surrounding areas exhibits the greatest liability to the producer. Chemicals associated with gas-producing operations may also leak into the environment and create an untenable environmental situation. For the most part, such leaks and spills are unpreventable and are caused by equipment failure or an unpredictable event during production. The construction of a berm – a rounded mound of soil and fill that is constructed upon an otherwise level patch of land – to enable containment of a spill is one preventative measure. The berm design is based on the produced flowrate multiplied by frequency of inspection.

Soil contaminated by a spill or leak may be treated as an exempt waste, provided the leaking substance is listed as an exempt waste – a waste that has been generated from a material or process uniquely associated with the exploration, development, and production of natural gas and crude oil. In warmer climates, the appearance of a sheen during runoff conditions may be apparent, thus increasing the likelihood of hydrocarbon contamination to the immediate area. To decrease the likelihood of hydrocarbon derivatives migrating from the wellsite during periods of heavy rain, the perimeter of the location should have a berm allowing for drainage away from the location. Water collecting on location should be diverted to an onsite catchment.

As a result of a spill, a hydrocarbon may impose a hazard to groundwater, flora, and fauna (including humans). While many wastes produced at the wellsite are exempt, the operator is responsible for the degradation of the surroundings due to imprudence, neglect, and accidental release. Hydrocarbon derivatives in the soil in excess of the saturation limit (the point at which a substance will receive no more of another substance in solution or in chemical combination) may be transmitted to the surrounding environment through runoff and gravity. These hydrocarbon derivatives are loosely categorized as dense nonaqueous phase liquids (DNAPL) or light nonaqueous phase liquids (LNAPL) with the DNAPL being heavier (more dense) than water and the LNAPL being lighter (less dense) than water. In the field, both of these species are present and the LNAPLs are more mobile, which translates to a greater relative conductivity in the soil and lesser adsorption onto the soil constituents. However, LNAPLs possess a lower vapor pressure allowing the liquid to dissipate into the gaseous phase (another potential environmental issue) leaving only the less-volatile traces of the product.

7.2.2 Soluble contaminants

Contaminants that are highly soluble, such as salts (e.g., sodium chloride, NaCl) move readily from surface soils to saturated materials below the water table, which often occurs during and after rainfall events. The contaminants that are not highly soluble may have considerably longer residence times in the soil zone.

Because of the dissolved mineral content in the water and residual hydrocarbon derivatives, any produced water should be treated before use or before disposal. Safe disposal of this water is secured through (i) injection into a secure geologic formation, (ii) evaporation in lined pits, or (iii) other approved means. The use of lined pits as the sole means for disposal has been eliminated in all but a few cases where the evaporation rates are competitive with the production rate from one or more wells. Unlined pits are not considered a sound environmental approach because although the percolation in such pits has allowed for a cost-effective means of disposal but in some cases, however, this practice has led to local groundwater contamination.

Guidelines have been made for the production water pit construction, including the following: (i) a minimum freeboard of 2 feet must be maintained in the pits – by way of clarification, the freeboard is the height of the watertight portion of a construction above a given level of water in the pit, (ii) life expectancy and pertinent characteristics of the liner material, (iii) monthly evaporation/precipitation rates, and (iv) method and schedule for removal of residual solids and saturated brine.

7.2.3 Other sources of contamination

In addition to hydrogen sulfide and carbon dioxide, natural gas may (on rare occasions) contain other contaminants, such as mercaptans (RSH) and carbonyl sulfide (COS). The presence of these impurities may eliminate some of the sweetening processes since some processes remove large amounts of acid gas but not to a sufficiently low concentration. On the other hand, there are those processes that are not designed to remove (or are incapable of removing) large amounts of acid gases. However, these processes are also capable of removing the acid gas impurities to low levels when the acid gases are present in low to medium concentrations in the gas.

Glycol derivatives and other liquids used for dehydration in a gas-processing plant (Mokhatab et al., 2006; Speight, 2019a) may occasionally leak or be spilled on the ground. Due to the inherent physical properties of dehydration liquids (e.g., triethylene glycol, also known as EG), such as the viscosity as well as the complete miscibility with water and toxicity, significant losses are not realized to soil seepage or evaporation.

Unless treated, ethylene glycol (which is colorless and odorless) has been known to be mistaken for water upon casual analysis. The greatest potential danger is then contamination of surface runoff. Any spillage should be retrieved and sent for correct disposal, or waste liquid desiccants should be recycled when possible. Solid desiccants, such as alumina (Al_2O_3), silica (SiO_2), and molecular sieves become uneconomically regenerable with time, and when exhausted, these materials may be sent to a disposal site or buried on-site (to remove any adsorbed contaminants). In the sweetening processes (i.e., sulfur removal processes), amine treatment is the most prevalent. The amines, such as monoethanolamine (MEA, $HOCH_2CH_2NH_2$) and diethanolamine [DEA, $(HOCH_2CH_2)NH$] combine with carbon dioxide and hydrogen sulfide at moderate temperatures and are removed from the gas stream; the carbon dioxide and hydrogen sulfide are released when the combined solution temperature is raised.

These gases (i.e., carbon dioxide and hydrogen sulfide) may also be trapped in caustic solutions, and the pertinent chemical reactions involved are

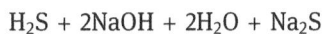

$$CO_2(g) + NaOH + NaHCO_3$$

$$H_2S + 2NaOH + 2H_2O + Na_2S$$

In these equations, the pH of the solution is lowered and a salt is formed, and the salt solutions, as with most liquid wastes, may be disposed of in Class II injection wells when recycling is not feasible.

Class II wells are used only to inject fluids associated with oil and natural gas production. Class II fluids are primarily brines (salt water) that are brought to the surface while producing oil and gas.

7.2.4 Air pollutants

The emissions from natural gas-fired boilers and furnaces include nitrogen oxides (NO_x), carbon monoxide (CO), carbon dioxide (CO_2), methane (CH_4), nitrous oxide (N_2O), VOCs, trace amounts of sulfur dioxide (SO_2), and PM.

7.2.4.1 Nitrogen oxides

The formation of nitrogen oxides occurs by three fundamentally different mechanisms and the processes are known as (i) thermal NO_x, (ii) prompt NO_x, and (iii) fuel NO_x.

The principal mechanism of formation of nitrogen oxide(s) during the combustion of natural gas is the thermal NO_x route, which occurs through the thermal dissociation and subsequent reaction of nitrogen (N_2) and oxygen (O_2) molecules in the combustion air. Most nitrogen oxide(s) formed through the thermal NO_x mechanism occurs in the high temperature flame zone near the burners. The formation of thermal NO_x is affected by three furnace-zone factors which are (i) the oxygen concentration, (ii) the peak temperature, and (iii) the time of exposure at peak temperature. As these three factors increase, NO_x emission levels increase, and the emission trends due to changes in these factors are fairly consistent for all types of natural gas-fired boilers and furnaces. However, emission levels do vary considerably with the type and size of combustor and with operating conditions (e.g., combustion air temperature, volumetric heat release rate, load, and excess oxygen level).

The second mechanism of the formation of nitrogen oxide(s) (prompt NO_x) occurs through early reactions of nitrogen in the combustion air and hydrocarbon radicals from the fuel. Prompt NO_x reactions occur within the flame and are usually negligible when compared to the amount of nitrogen oxide(s) formed through the thermal NO_x mechanism. However, prompt NO_x levels may become significant with ultra-low-NO_x burners.

The third mechanism of the formation of nitrogen oxide(s) (fuel NO_x) arises from the evolution and reaction of fuel-bound nitrogen compounds with oxygen. Due to the characteristically low fuel nitrogen content of natural gas, the formation of nitrogen oxide(s) through the fuel NO_x mechanism is insignificant.

7.2.4.2 Carbon monoxide

The rate of the emissions of carbon monoxide from gas-fired boilers depends on the efficiency of the combustion of the natural gas. Improperly tuned boilers and boilers operating at off-design levels decrease combustion efficiency, resulting in increased emissions of carbon monoxide. In some cases, the addition of control systems for nitrogen oxide(s) – such as low-NO_x burners and flue gas recirculation (FGR) – may also reduce the efficiency of the combustion process, thereby resulting in higher emissions of carbon monoxide relative to uncontrolled boilers.

7.2.4.3 Volatile organic compounds

The rate of emissions of VOCs from boilers and furnaces also depends on combustion efficiency. The emissions of VOCs can be minimized by combustion practices that promote (i) high combustion temperatures, (ii) long residence times at those temperatures, and (iii) turbulent mixing of fuel and combustion air. Trace amounts of VOCs in the natural gas fuel (such as formaldehyde, HCHO, and benzene, C_6H_6)

may also contribute to the emissions of VOCs if the gas stream is not completely combusted in the boiler.

7.2.4.4 Sulfur oxides

Emissions of sulfur dioxide from natural gas-fired boilers are low because pipeline quality natural gas typically has sulfur levels of 2,000 grains per million cubic feet. However, sulfur-containing odorants are added to natural gas for detecting leaks, leading to small amounts of emissions of sulfur dioxide. Boilers combusting unprocessed natural gas may have higher emissions of sulfur dioxide due to higher levels of sulfur in the natural gas. For these units, a sulfur mass balance should be used to determine the sulfur dioxide emissions.

7.2.4.5 Particulate matter

Because natural gas is a gaseous fuel, the emissions of filterable PM are typically low. The micrometer, also commonly known as the micron, is 1×10^{-6} m which is, one millionth of a meter (or one thousandth of a millimeter, 0.001 mm, or approximately 1 inch $\times$ 0.000039).

PM from natural gas combustion has been estimated to be less than 1 µm in size and has filterable and condensable fractions. The PM is usually composed of higher molecular weight hydrocarbon derivatives (such as polynuclear hydrocarbon derivatives, PNAs, or PHAs) that are not fully combusted. Increased emissions of PM may result from poor air/fuel mixing or maintenance problems.

7.2.4.6 Greenhouse gases

Emissions of carbon dioxide (CO_2), methane (CH_4), and nitrous oxide (N_2O) are produced during natural gas combustion. In correctly tuned boilers, nearly all of the fuel carbon (99.9%) in natural gas is converted to carbon dioxide during the combustion process, but the conversion is relatively independent of the type of boiler or the type of combustor. Any fuel carbon not converted to carbon dioxide results in the formation and emission of methane, carbon monoxide, and/or VOCs that are formed because of incomplete combustion. Even in boilers operating with poor combustion efficiency, the amount of methane, carbon monoxide, and/or VOCs is insignificant compared to the level of the carbon dioxide emissions.

Formation of nitrous oxide during the combustion process is affected by two furnace-zone factors. The emissions of this gas are minimized when the combustion temperature is high (>800 °C, >1,475 °F) and excess oxygen is kept to a minimum (<1%).

The emissions of methane are highest during low-temperature combustion or incomplete combustion, such as the start-up or shut-down cycle of boilers. Typically, conditions that favor the formation of nitrous oxide also favor emissions of methane.

7.2.5 Emissions during recovery and processing

The extraction and production of natural gas, as well as other natural gas operations, do have environmental consequences and are subject to numerous laws and regulations. In some areas, development is completely prohibited so as to protect natural habitats, wetlands, and designated wilderness areas.

7.2.5.1 Emissions during exploration, production, and delivery

The environmental side effects of natural gas production start in what is called the upstream portion of the natural gas industry, beginning with selection of a geologically promising area for possible future natural gas production. An upstream firm will collect all available existing information on the geology and natural gas potential of the proposed area and may decide to conduct new geologic and geophysical studies.

Following analysis of the geologic and geophysical data, permission to drill and produce natural gas from owners of the land and relevant government permitting authorities is necessary. In making leasing and permitting decisions, the potential environmental impacts of future development are often considered. Such considerations include the projected numbers and extent of wells and related facilities, such as pipelines, compressor stations, water disposal facilities as well as roads and power lines.

Drilling a gas well involves preparing the wellsite by constructing a road to it, if necessary, clearing the site, and flooring it with wood or gravel. The soil under the road and the site may be so compacted by the heavy equipment used in drilling as to require compaction relief for subsequent farming. In wetland areas, drilling is often accomplished using a barge-mounted rig that is floated to the site after a temporary slot is cut through the levee bordering the nearest navigable stream. However, the primary environmental concern directly associated with drilling is not the surface site but the disposal of drilling waste (spent drilling mud and rock cuttings, etc.).

Drilling of a typical natural gas well (6,000 feet deep) results in the production of approximately 150,000 pounds of rock cuttings and at least 470 barrels of spent mud. Early industry practice was to dump spent drilling fluid and rock cuttings into pits dug alongside the well and just plow them over after drilling was completed, or dump them directly into the ocean if offshore. Currently, the operator must not discharge drilling fluids and solids without permission, and it has to be determined whether such waste can be discharged or shipped to a special disposal facility.

If the drilling fluids are saltwater- or oil-based, they can cause damage to soils and groundwater, and on-site disposal is often not permitted, so operators must dispose of such wastes at an off-site disposal facility. The disposal methods include underground injection, burial in pits or landfills, land spreading, evaporation, incineration, and reuse/recycling. In areas with subsurface salt formations, disposal in man-made salt caverns may be permitted; this form of disposal poses low risks to plant and animal life because the formations where the caverns are constructed are very stable and are located beneath any subsurface fresh water supplies.

In recent years, new drilling technologies such as slimhole drilling (slim hole drilling), horizontal drilling, multilateral drilling, coiled tubing drilling, and improved drill bits have helped to reduce the generated quantity of drilling wastes. Another advanced drilling technology that provides pollution-prevention benefits is the use of synthetic drilling fluids that have a less severe environmental impact and their use results in a much cleaner well bore and less sidewall collapse.

From an environmental perspective, the most obvious problem is leaking pipelines. However, if hydrate inhibitors are used, berms are required around the storage tanks to prevent soil contamination; in the case of methanol, the potential for fire exists as well. A major issue is control of exhausts from compressors. Carbon monoxide and nitrogen oxides are the major concerns for the internal-combustion engines that drive reciprocating compressors, and nitrogen oxide(s) is (are) the main pollutant from sulfur removal or sulfur oxide emissions may be required.

7.2.5.2 Emissions during processing

Gas processing usually poses low environmental risk, primarily because natural gas has a simple and comparatively pure composition (Kidnay and Parrish, 2006; Mokhatab et al., 2006; Kerry, 2007; Speight, 2019a). Typical processes performed by a gas plant are separation of the heavier-than-methane hydrocarbon derivatives as liquefied petroleum gas (Table 7.3), stabilization of condensate by removal of lower-boiling hydrocarbon derivatives from the condensate stream, gas sweetening, and consequent sulfur production and dehydration sufficient to avoid formation of methane hydrates in the downstream pipeline.

The identified hazardous air pollutant (HAP) emission points at natural gas-processing plants are the glycol dehydration unit reboiler vent, storage tanks, and equipment leaks from components handling hydrocarbon streams that contain HAPs. Other potential HAP emission points are the tail gas streams from amine-treating processes and sulfur recovery units.

Methods vary for removing natural gas contaminants, such as hydrogen sulfide gas, carbon dioxide gas, nitrogen, and water. Commonly, hydrogen sulfide is converted to solid sulfur for sale. Likewise the carbons and nitrogen are separated for sale to the extent economically possible but otherwise the gases are vented, while the water is treated before release. Compressor operation at gas plants has a similar impact to that of compressors installed at other locations.

It is sometimes necessary either to vent produced gas into the atmosphere or to flare (burn) it. Worldwide, most venting and flaring occurs when the cost of transporting and marketing gas co-produced from crude oil reservoirs exceeds the net-back price received for the gas.

Table 7.3: Properties of liquefied petroleum gas.

Constituent	Propane	Butane
Formula	C_3H_8	C_4H_{10}
Boiling point	−44 °F	32 °F
Specific gravity of the gas (air = 1.00)	1.53	2.00
Specific gravity of the liquid (water = 1.00)	0.51	0.58
Lbs./gallon: liquid @ 60 °F	4.24	4.81
BTU/gallon: gas @ 60 °F	91,690	102,032
BTU/lb: gas	21,591	21,221
BTU/ft^3: gas @ 60 °F	2,516	3,280
Ft^3 of vapor @ 60 °F./gal. of liquid, 60 °F	36.39	31.26
Ft^3 of vapor @ 60 °F./lb. of liquid, 60 °F	8.547	6.506
Latent heat of vaporization @ boiling point BTU/gal	785.0	808.0
Combustion data:		
Flash point, °F	−155	−76
Auto-ignition temperature, °F	878	761
Maximum flame temperature in air, °F	3,595	3,615
Flammability limits, % v/v of gas in air mixture: Lower limit – % Upper limit – %	2.4 9.6	1.9 8.6
Octane number (iso-octane = 100)	100+	92

Emissions will result from gas sweetening plants only if the acid waste gas from the amine process is flared or incinerated. Most often, the acid waste gas is used as a feedstock in nearby sulfur recovery or sulfuric acid plants.

When flaring or incineration is practiced, the major pollutant of concern is sulfur dioxide. Most plants employ elevated smokeless flares or tail gas incinerators for complete combustion of all waste gas constituents, including virtually 100% conversion of hydrogen sulfide to sulfur dioxide. Little particulate, smoke, or hydrocarbon derivatives result from these devices, and because gas temperatures do not usually exceed 650 °C (1,200 °F), significant quantities of nitrogen oxides are not formed. Some plants still use older, less-efficient waste gas flares. Because these flares usually burn at temperatures lower than necessary for complete combustion, larger emissions of hydrocarbon derivatives and particulate, as well as hydrogen sulfide, can occur.

This practice of venting is now by no means as common as it was a few decades ago when oil was the primary valuable product, and there was no market for much of the co-produced natural gas. The venting and flaring that does occur now is regulated and may happen at several locations: the well gas separator, the lease tank battery gas separator, or a downstream natural gas plant.

7.2.5.3 Emissions during combustion

The majority of natural gases are mixtures of saturated hydrocarbon derivatives where methane prevails; they come from underground accumulations of gases alone or gases associated with oil. There are thus as many compositions of natural gases as exploited hydrocarbon layers. Apart from the methane, which is the prevailing element, the crude natural gas usually contains decreasing volumetric percentages of ethane, propane, butane, pentane, and so on.

Combustion is an exothermic chemical reaction, specifically the oxidation of a fuel, which liberates energy that can be used for diverse purposes. The combustion of natural gas is one of the major sources of energy, and a detailed understanding of its combustion behavior is of considerable practical importance. However, the composition of commercial natural gas can vary widely with concentration extremes of 75–98% v/v for methane, 0.5–13% v/v for ethane, and 0–2.6% v/v for propane. Therefore, it is important to understand the chemistry of each of these individual fuels and then consider how varying levels of these fuels in natural gas affect their performance. Simply put, in order to understand the formation of pollutants in combustion systems, it is essential to understand (i) the nature of the fuel being burned and (ii) the thermodynamics of the combustion process, and some aspects of flame structure, as well as the fundamental aspects of hydrocarbon fuel combustion that relate to the formation of pollutants and/or to the control of emissions (El-Mahallawy and Habik, 2002; Glassman and Yetter, 2008).

In theory, and often but not always in practice, natural gas burns more cleaner than other fossil fuels. It has fewer emissions of sulfur, carbon, and nitrogen than coal or oil, and it has almost no ash particles left after burning. Being a clean fuel is one reason that the use of natural gas, especially for electricity generation, has grown so much and is expected to grow even more in the future.

Natural gas is less chemically complex than other fuels, has fewer impurities, and its combustion accordingly results in less pollution. In the simplest case, complete combustive reaction of a molecule of pure methane (CH_4) with two molecules of pure oxygen produces a molecule of carbon dioxide gas, two molecules of water in vapor form, and heat:

$$CH_4 + 2O_2 \rightarrow CO_2 + 2H_2O + heat$$

More specifically, the complete oxidation of simple hydrocarbon fuels forms carbon dioxide (CO_2) from all of the carbon and water (H_2O) from the hydrogen, that is, for a hydrocarbon fuel with the general composition C_nH_m:

$$C_n H_m + (n + m/4)O_2 \rightarrow nCO_2 + m/2H_2O$$

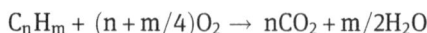

For methane, the equation is

$$CH_4 + 2O_2 \rightarrow CO_2 + 2H_2O$$

Even in the idealized case of complete combustion, the accounting of all species present in combustion exhaust involves more than simply measuring the carbon dioxide and the water. Since fuels are burned in air rather than in pure oxygen, the nitrogen in the air may participate in the combustion process to produce nitrogen oxides. Also, many fuels contain elements other than carbon, and these elements may be transformed during combustion.

Finally, combustion is not always complete, and the effluent gases contain unburned and partially burned products in addition to carbon dioxide and water.

Air is composed of oxygen, nitrogen, and small amounts of carbon dioxide, argon, and other trace species. Since the vast majority of the diluent in air is nitrogen, for general purposes it is perfectly reasonable to consider air as a mixture of 20.9% (mole basis) oxygen and 79.1% (mole basis) nitrogen. Thus, for every mole of oxygen required for combustion, 3.78 mol of nitrogen must be introduced as well. Although nitrogen may not significantly alter the oxygen balance, it does have a major impact on the thermodynamics, chemical kinetics, and formation of pollutants in combustion systems. For this reason, it is useful to carry the "inert" species along in the combustion calculations. The stoichiometric relation for complete oxidation of a hydrocarbon fuel, $CnHm$, becomes

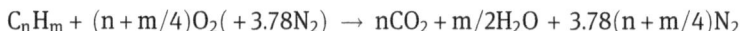

$$C_n H_m + (n + m/4)O_2(+3.78N_2) \rightarrow nCO_2 + m/2H_2O + 3.78(n + m/4)N_2$$

Another version of the equation is (with $a = x + y/4$ for stoichiometric combustion):

$$C_x H_y + a(O_2 + 0.79/0.21N_2) \rightarrow xCO_2 + y/2H_2O + a0.79/(0.21N_2)$$

Thus for every mole of fuel burned, 4.78 $(n + m/4)$ mol of air are required and 4.78 $(n + m/4) + m/4$ mol of combustion products are generated. The molar fuel/air ratio for stoichiometric combustion is $1/[4.78\,((n + m)/4)]$.

Gas compositions are generally reported in terms of mole fractions since the mole fraction does not vary with temperature or pressure as does the concentration (moles/unit volume). The product mole fractions for complete combustion of this hydrocarbon fuel are

$$yCO_2 = n/[4.78(n + M/4) + m/4]$$

$$yH_2O = (m/2)[4.78(n + m/4) + m/4]$$

$$yN_2 = 3.78(n + m/4)/[4.78(n + m/4) + m/4]$$

The large quantity of nitrogen diluent substantially reduces the mole fractions of the combustion products from the values they would have in its absence.

Minor components and impurities in the fuel somewhat complicate our analysis of combustion products. Fuel sulfur is usually oxidized to form sulfur dioxide (SO_2) (even though there are cases where sulfur compounds involving higher oxidation states of sulfur or reduced sulfur compounds are produced, it is a reasonable first approximation to assume that all of the fuel sulfur forms SO_2). Upon combustion, organically bound fuel nitrogen is converted to both N_2 and NO with molecular nitrogen generally dominating, and in such calculations it can be assumed that all of the nitrogen is the fuel molecular nitrogen (N_2).

The mineral matter in the fuel (which is manifested as combustion ash), the noncombustible inorganic (mineral) impurities in the fuel, undergoes a number of transformations at combustion temperatures, which will also be neglected for the time being, so that the ash will be assumed to be inert. For most common gaseous fuel stream, the only chemical information available is its elemental composition on a mass basis, as determined in the ultimate analysis.

Few combustion systems are operated precisely at the stoichiometric condition because of the difficulty of achieving such intimate mixing between fuel and air that perfect conversion is attained. More commonly, combustors are operated with a margin for error using more than the stoichiometric amount of air. The *fuel/air ratio* is used to define the operating conditions of a combustor. Comparison of the two examples presented above shows that the fuel/air ratio required for complete combustion varies with fuel composition. Because the mass ratios vary widely with gas stream composition, they are not a convenient base for comparison of systems burning different fuels.

Thus, in practice, the combustion process is not always perfect, and when the air supply is inadequate, carbon monoxide and PM (soot) are also produced. In fact, since natural gas is never pure methane and small amounts of additional impurities are present, pollutants are also generated during combustion. Thus, the combustion of natural gas also produces undesirable compounds but in significantly lower quantities compared to the combustion of crude oil, crude oil-derived products, and coal. The particulates produced by natural gas combustion are usually less than 1 micron in diameter and are composed of low-molecular-weight hydrocarbon derivatives that are not fully combusted.

Natural gas is one of the major combustion fuels used throughout the country. It is mainly used to generate industrial and utility electric power, produce industrial process steam and heat, and heat residential and commercial spaces. Natural gas consists of a high percentage of methane (generally >85%) and varying amounts of ethane, propane, butane, and inerts (typically nitrogen, carbon dioxide, and helium). The average gross heating value of natural gas is approximately 1,020 British thermal units per standard cubic foot (Btu/scf), usually varying from 950 to 1,050 Btu/scf.

e major types of boilers used for natural gas combustion in com-

d utility applications: (i) water-tube boiler, (ii) fire-tube boiler,

Water-tube boilers are designed to pass water through the inside of heat transfer tubes while the outside of the tubes is heated by direct contact with the hot combustion gases and through radiant heat transfer. The water-tube design is the most common in utility and large industrial boilers. Water-tube boilers are used for a variety of applications ranging from providing large amounts of process steam to providing hot water or steam for space heating to generating high-temperature, high-pressure steam for producing electricity. Furthermore, water-tube boilers can be distinguished either as field erected units or packaged units.

Field erected boilers are boilers that are constructed on-site and comprise the larger sized water-tube boilers. Generally, boilers with heat input levels greater than 100 MMBtu/h are field erected. Field erected units usually have multiple burners and given the customized nature of their construction, also have greater operational flexibility and NO_x control options. Field erected units can also be further categorized as wall-fired or tangential-fired units. Wall-fired units are characterized by multiple individual burners located on a single wall or on opposing walls of the furnace, while tangential units have several rows of air and fuel nozzles located in each of the four corners of the boiler.

Package units are constructed off-site and shipped to the location where they are needed. While the heat input levels of packaged units may range up to 250 MMBtu/h, the physical size of these units are constrained by shipping considerations and generally have heat input levels less than 100 MMBtu/h. Packaged units are always wall-fired units with one or more individual burners. Given the size limitations imposed on packaged boilers, they have limited operational flexibility and cannot feasibly incorporate some NO_x control options.

Fire-tube boilers are designed such that the hot combustion gases flow through tubes, which heat the water circulating outside of the tubes. These boilers are used primarily for space heating systems, industrial process steam, and portable power boilers. Fire-tube boilers are almost exclusively packaged units. The two major types of fire-tube units are Scotch Marine boilers and the older firebox boilers.

In the cast iron boiler, as in a fire-tube boiler, the hot gases are contained inside the tubes and the water being heated circulates outside the tubes. However, the units are constructed of cast iron rather than steel. Virtually all cast iron boilers are constructed as package boilers. These boilers are used to produce either low-pressure steam or hot water and are most commonly used in small commercial applications.

Natural gas is also combusted in residential boilers and furnaces. Residential boilers and furnaces generally resemble fire-tube boilers with flue gas traveling through several channels or tubes with water or air circulated outside the channels or tubes.

7.2.6 Smog and acid rain

Smog and acid rain might be considered to be the end point of emissions from the use of natural gas.

The use of natural gas does not contribute significantly to smog formation, as it emits low levels of nitrogen oxides and virtually no PM. For this reason, it can be used to help combat smog formation in those areas where ground level air quality is poor. The main sources of nitrogen oxides are electric utilities, motor vehicles, and industrial plants. Increased natural gas use in the electric generation sector, a shift to cleaner natural gas vehicles, or increased industrial natural gas use could serve to combat smog production, especially in urban centers where it is needed the most. Particularly in the summertime, when natural gas demand is lowest and smog problems are the greatest, industrial plants and electric generators could use natural gas to fuel their operations instead of other, more-polluting fossil fuels. This would effectively reduce the emissions of smog-causing chemicals and result in clearer, healthier air around urban centers.

Particulate emissions also cause the degradation of air quality in the United States. These particulates can include soot, ash, metals, and other airborne particles.

In fact, the Intergovernmental Panel on Climate Change (IPCC) predicts in its "Third Assessment Report" released in February 2001 that over the next 100 years, global average temperatures will rise by 2.4–10.4 °F.

The principle greenhouse gases include water vapor, carbon dioxide, methane, nitrogen oxides, and some engineered chemicals such as chlorofluorocarbons. While most of these gases occur in the atmosphere naturally, levels have been increasing due to the widespread burning of fossil fuels by growing human populations. The reduction of greenhouse gas emissions has become a primary focus of environmental programs in countries around the world.

According to the Energy Information Administration, although methane emissions account for only 1.1% of total US greenhouse gas emissions, they account for 8.5% of the greenhouse gas emissions based on global warming potential. A major study performed by the Environmental Protection Agency (EPA) and the Gas Research Institute (GRI) in 1997 sought to discover whether the reduction in carbon dioxide emissions from increased natural gas use would be offset by a possible increased level of methane emissions.

Smog and poor air quality are a pressing environmental problem, particularly for large metropolitan cities. Smog, the primary constituent of which is ground level ozone, is formed by a chemical reaction of carbon monoxide, nitrogen oxides, VOCs, and heat from sunlight. As well as creating that familiar smoggy haze commonly found surrounding large cities, particularly in the summer time, smog and ground level ozone can contribute to respiratory problems ranging from temporary discomfort to long-lasting, permanent lung damage. Pollutants contributing to smog come from a variety of sources, including vehicle emissions, smokestack emissions, paints, and solvents. Because the reaction to create smog requires heat, smog problems are the worst in the summertime.

Particulate emissions also cause the degradation of air quality in the United States. These particulates can include soot, ash, metals, and other airborne particles. A study (Union of Concerned Scientists, 1998, "Cars and Trucks and Air Pollution") showed that the risk of premature death for residents in areas with high airborne PM was 26% greater than for those in areas with low particulate levels. Natural gas emits virtually no particulates into the atmosphere: in fact, emissions of particulates from natural gas combustion are 90% lower than from the combustion of oil, and 99% lower than burning coal.

Acid rain is another environmental problem that affects much of the Eastern United States, damaging crops, forests, wildlife populations, and causing respiratory and other illnesses in humans. Acid rain is formed when sulfur dioxide and nitrogen oxides react with water vapor and other chemicals in the presence of sunlight to form various acidic compounds in the air. Acid rain occurs when the oxides of nitrogen and sulfur that are released to the atmosphere during the combustion of fossil fuels are deposited (as soluble acids) with rainfall, usually at some location remote from the source of the emissions.

The principle source of acid rain-causing pollutants, sulfur dioxide (SO_2) and nitrogen oxides (NO_x) are various fossil fuel-fired power plants. Since natural gas emits virtually no sulfur dioxide, and up to 80% less nitrogen oxides than the combustion of coal, increased use of natural gas could provide for fewer acid rain-causing emissions.

It is generally believed (the chemical thermodynamics are favorable) that acidic compounds are formed when sulfur dioxide and nitrogen oxide emissions are released from tall industrial stacks. Gases such as sulfur oxides (usually sulfur dioxide, SO_2) as well as the nitrogen oxides (NO_x) react with the water in the atmosphere to form acids:

$$SO_2 + H_2O \rightarrow H_2SO_3$$

$$2SO_2 + O_2 \rightarrow 2SO_3$$

$$SO_3 + H_2O \rightarrow H_2SO_4$$

$$2NO + H_2O \rightarrow 2HNO_2$$

$$2NO + O_2 \rightarrow 2NO_2$$

$$NO_2 + H_2O \rightarrow HNO_3$$

Acid rain has a pH less than 5.0 and predominantly consists of sulfuric acid (H_2SO_4) and nitric acid (HNO_3). As a point of reference, in the absence of anthropogenic pollution sources, the average pH of rain is 6.0 (slightly acidic; neutral pH = 7.0). In summary, the sulfur dioxide that is produced during a variety of processes will react with oxygen and water in the atmosphere to yield environmentally detrimental sulfuric acid. Similarly, nitrogen oxides will also react to produce nitric acid.

Another acid gas, hydrogen chloride (HCl), although not usually considered to be a major emission, is produced from mineral matter and the brines that often accompany petroleum during production and is gaining increasing recognition as a contributor to acid rain. However, hydrogen chloride may exert severe local effects because it does not need to participate in any further chemical reaction to become an acid. Under atmospheric conditions that favor a buildup of stack emissions in the areas where hydrogen chloride is produced, the amount of hydrochloric acid in rainwater could be quite high.

On a regional level the emission of sulfur oxides (SO_x) and nitrogen oxides (NO_x) can also cause the formation of acid species at high altitudes, which eventually precipitate in the form of *acid rain*, damaging plants, wildlife, and property. Most petroleum products are low in sulfur or are desulfurized, and while natural gas sometimes includes sulfur as a contaminant, it is typically removed at the production site.

At the global level, there is concern that the increased use of hydrocarbon-based fuels will ultimately raise the temperature of the planet (*global warming*), as carbon dioxide reflects the infrared or thermal emissions from the earth, preventing them from escaping into space (*greenhouse effect*). Whether or not the potential for global warming becomes real will depend upon how emissions into the atmosphere are handled. There is considerable discussion about the merits and de-merits of the global warming theory and the discussion is likely to continue for some time. Be that as it may, the atmosphere can only tolerate pollutants up to a limiting value. And that value needs to be determined. In the meantime, efforts must be made to curtail the use of noxious and foreign (non-indigenous) materials into the air.

There are a variety of processes which are designed for sulfur dioxide removal from gas streams (Mokhatab et al., 2006; Speight, 2019a), but scrubbing process utilizing limestone ($CaCO_3$) or lime [$Ca(OH)_2$] slurries have received more attention than other gas scrubbing processes. The majority of the gas scrubbing processes is designed to remove sulfur dioxide from the gas streams; some processes show the potential for removal of nitrogen oxide(s).

7.2.7 Emissions control

Natural gas is becoming an increasingly important fuel in the generation of electricity. As well as providing an efficient, competitively priced fuel for the generation of electricity, the increased use of natural gas allows for the improvement in the emissions profile of the electric generation industry. Power plants in the United States account for 67% of sulfur dioxide emissions, 40% of carbon dioxide emissions, 25% of nitrogen oxide emissions, and 34% of mercury emissions (Nationa   ʼmental

Trust, 2002, "Cleaning up Air Pollution from America's Power Plants"). Coal-fired power plants are the greatest contributors to these types of emissions. In fact, only 3% of sulfur dioxide emissions, 5% of carbon dioxide emissions, 2% of nitrogen oxide emissions, and 1% of mercury emissions come from non-coal–fired power plants.

Natural gas-fired electric generation, and natural gas-powered industrial applications offer a variety of environmental benefits and environmentally friendly uses, including:

1. Fewer emissions: combustion of natural gas, used in the generation of electricity, industrial boilers, and other applications, emits lower levels of NO_x, CO_2, and particulate emissions, and virtually no SO_2 and mercury emissions. Natural gas can be used in place of, or in addition to, other fossil fuels, including coal, oil, or crude oil coke, which emit significantly higher levels of these pollutants.

2. Reduced sludge: coal-fired power plants and industrial boilers that use scrubbers to reduce SO_2 emission levels generate thousands of tons of harmful sludge. Combustion of natural gas emits low levels of SO_2, eliminating the need for scrubbers, and reducing the amounts of sludge associated with power plants and industrial processes.

3. Reburning: this process involves injecting natural gas into coal- or oil-fired boilers. The addition of natural gas to the fuel mix can result in NO_x emission reductions of 50–70% and SO_2 emission reductions of 20– 25%.

4. Cogeneration: the production and use of both heat and electricity can increase the energy efficiency of electric generation systems and industrial boilers, which translates to requiring the combustion of less fuel and the emission of fewer pollutants. Natural gas is the preferred choice for new cogeneration applications.

5. Combined cycle generation: combined cycle generation units generate electricity and capture normally wasted heat energy, using it to generate more electricity. Like cogeneration applications, this increases energy efficiency, uses less fuel, and thus produces fewer emissions. Natural gas-fired combined cycle generation units can be up to 60% energy efficient, whereas coal and oil generation units are typically only 30–35% efficient.

6. Fuel cells: natural gas fuel cell technologies are in development for the generation of electricity. Fuel cells are sophisticated devices that use hydrogen to generate electricity, much like a battery. No emissions are involved in the generation of electricity from fuel cells, and natural gas, being a hydrogen-rich source of fuel, can be used. Although still under development, widespread use of fuel cells could significantly reduce the emissions associated with the generation of electricity in the future.

Essentially, electric generation and industrial applications that require energy, particularly for heating, use the combustion of fossil fuels for that energy. Because of its clean burning nature, the use of natural gas wherever possible, either in conjunction with other fossil fuels, or instead of them, can help to reduce the emission of harmful pollutants.

Currently, the two most prevalent combustion control techniques used to reduce NO_x emissions from natural gas-fired boilers are FGR and low NO_x burners. In an FGR system, a portion of the flue gas is recycled from the stack to the burner wind box. Upon entering the winebox, the recirculated gas is mixed with combustion air prior to being fed to the burner. The recycled flue gas consists of combustion products which act as inerts during combustion of the fuel/air mixture. The FGR system reduces NO_x emissions by two mechanisms. Primarily, the recirculated gas acts as a diluent to reduce combustion temperatures, thus suppressing the thermal NO_x mechanism. To a lesser extent, FGR also reduces NO_x formation by lowering the oxygen concentration in the primary flame zone. The amount of recirculated flue gas is a key operating parameter influencing NO_x emission rates for these systems. An FGR system is normally used in combination with specially designed low NO_x burners capable of sustaining a stable flame with the increased inert gas flow resulting from the use of FGR. When low NO_x burners and FGR are used in combination, these techniques are capable of reducing NO_x emissions by 60–90%.

Low NO_x burners reduce NO_x by accomplishing the combustion process in stages. Staging partially delays the combustion process, resulting in a cooler flame which suppresses thermal NO_x formation. The two most common types of low NO_x burners being applied to natural gas-fired boilers are staged air burners and staged fuel burners. NO_x emission reductions of 40–85% (relative to uncontrolled emission levels) have been observed with low NO_x burners.

Other combustion control techniques used to reduce NO_x emissions include staged combustion and gas reburning. In staged combustion (e.g., burners-out-of-service and overfire air), the degree of staging is a key operating parameter influencing NO_x emission rates. Gas reburning is similar to the use of overfire in the use of combustion staging. However, gas reburning injects additional amounts of natural gas in the upper furnace, just before the overfire air ports, to provide increased reduction of NO_x to NO_2.

Two post-combustion technologies that may be applied to natural gas-fired boilers to reduce NO_x emissions are selective noncatalytic reduction (SNCR) and selective catalytic reduction (SCR). The SNCR system injects ammonia (NH_3) or urea into combustion flue gases (in a specific temperature zone) to reduce NO_x emission. The alternative control techniques (ACT) document for NO_x emissions from utility boilers; maximum SNCR performance was estimated to range from 25 to 40% for natural gas-fired boilers. Performance data available from several natural gas-fired utility boilers with SNCR show a 24% reduction in NO_x for applications on wall-fired boilers and a 13% reduction in NO_x for applications on tangential-fired boilers. In many situations, a boiler may have an SNCR system installed to trim NO_x emissions to meet permitted levels. In these cases, the SNCR system may not be operated to achieve maximum NO_x reduction. The SCR system involves injecting NH3 into the flue gas in the presence of a catalyst to reduce NO_x emissions. No data were available on SCR performance on natural gas-fired boilers at the time of this publication. However, the ACT document for utility boilers estimates NO_x reduction efficiencies for SCR control ranging from 80 to 90%.

7.3 Natural gas and global climate change

While the primary constituent of natural gas is methane (CH_4), it may contain smaller amounts of other hydrocarbon derivatives, such as ethane (C_2H_6) and various isomers of propane (C_3H_8), butane (C_4H_{10}), and pentane (C_5H_{12}), as well as trace amounts of higher boiling hydrocarbon derivatives up to octane (C_8H_{18}). Non-hydrocarbon gases, such as carbon dioxide (CO_2), helium (He), hydrogen sulfide (H_2S), nitrogen (N_2), and water vapor (H_2O) may also be present. At the pressure and temperature conditions of the source reservoir, natural gas may occur as free gas (bubbles) or be dissolved in either crude oil or brine.

Pipeline-quality natural gas contains at least 80% methane and has a minimum heat content of 870 Btu/ft^3. Most pipeline natural gas significantly exceeds both minimum specifications. Since natural gas has by far the lowest energy density of the common hydrocarbon fuels, by volume (not weight), much more of it must be used to provide a given amount of energy. Purified natural gas (specifically methane and *not* the higher boiling constituents) is also much less physically dense, weighing approximately half as much (55%) as the same volume of dry air at the same pressure. It is consequently buoyant in air, in which it is also combustible at concentrations ranging from 5% to 15% by volume.

7.4 The future

The natural gas industry includes integrated process operations that are engaged in processing natural gas into saleable products, especially gas for domestic and industrial consumers as well as products that can serve as raw materials for the petrochemical industry (Speight, 2019b).

Over the past four decades, the industry has experienced significant changes in market dynamics, resource availability, and technological advancements. Advancements made in exploration, production, and processing technologies allow utilization of resources such as gas in tight formations that was considered economically and technically unsuitable in the middle decades of the twentieth century. Along with the many challenges, it is imperative for gas processors to raise their operations to new levels of performance. Merely extending current performance standards on an incremental basis will fail to meet most performance goals.

Natural gas recovery and processing in the twenty-first century will continue to be shaped by the factors such as (i) dramatic changes in market demand, (ii) customization of products and uses, as well as (iii) changes in the properties of the recovered and processed gas. In fact, in addition to a plentiful supply of natural gas, the future of the industry will be based on the following factors such as (i) increased operating costs or investments due to stringent environmental requirements for facilities and products and (ii) accelerating globalization resulting in stronger international price scenarios. The effect of these factors is likely to reduce profit margins further and the industry will need to make significant changes in their operation and structure to be competitive on a global basis.

As global consumption of natural gas increases and resources are depleted, the production of fuels from non-fossil fuel sources (such as biomass and landfill waste) will increase significantly (Speight, 2021). In fact, over the next decade, gas-processing operations and refineries will need to adapt to receiving a variety of gas streams, especially gas from a range of bio-feedstocks and bio-waste materials (Appendix A).

As feedstocks to gas-processing plants change, there must be an accompanying change in processing technology as the feedstocks are becoming highly variable. At the same time, more stringent anti-pollution regulations are forcing greater restrictions on fuel specifications. There are fundamental limitations on how far current processes can go in achieving proper control over feedstock behavior. Thus, the need for the development of processes continues in order to fulfill the product market demand as well as to satisfy environmental regulations.

Technological advances are on the horizon for alternate sources of transportation fuels. For example, gas-to-liquids and biomass-to-liquids are just two of the concepts currently under development (Speight, 2019b, 2021). However, the state of many of these technologies, coupled with the associated infrastructure required to implement them, leave traditional gas-processing options as the modus operandi for the foreseeable future. The near future challenge for gas-processing operations will be harnessing new technologies to remain alive in a changing global marketplace. Merely extending current process performance incrementally will fail to meet most future performance goals, and it will be necessary to reshape gas-processing technology to be more adaptive to changing feedstocks and product demand.

Furthermore, environmental regulations could either preclude unconventional production or, more likely, raise the cost significantly. If future laws limited and/or taxed greenhouse gas emissions, these laws would lead to substantial increase in the costs of production. In addition to increase in the volumes of carbon dioxide, restrictions on access to water also could prove costly, especially in the arid or semi-arid regions of the world, particularly in the western states.

The general prognosis for reduction of emissions or emission cleanup is optimistic, and it is considered likely that most of the environmental impact of natural gas recovery and processing can be substantially abated. A considerable investment in retrofitting or replacing existing facilities and equipment will be needed although a conscious goal must be to improve the efficiency with which natural gas is transformed to marketable products and consumed.

References

Burruss, R.C., and Ryder, R.T. 2003. Composition of Crude Oil and Natural Gas Produced from 14 Wells in the Lower Silurian "Clinton" Sandstone and Medina Group, Northeastern Ohio and Northwestern Pennsylvania. Open-File Report 03-409, United States Geological Survey, Reston, Virginia.

Burruss, R.C., and Ryder, R.T. 2014. Composition of Natural Gas and Crude Oil Produced From 10 Wells in the Lower Silurian "Clinton" Sandstone, Trumbull County, Ohio. In: Coal and Crude oil Resources in The Appalachian Basin; Distribution, Geologic Framework, And Geochemical Character. L.F. Ruppert and R.T. Ryder (Editors). Professional Paper 1708, United States Geological Survey, Reston, Virginia.

El-Mahallawy, F., and Habik, S. 2002. Fundamentals and Technology of Combustion, Elsevier Science, Oxford, United Kingdom.

Glassman, I., and Yetter, R.A. 2008. Combustion 4th Edition. Elsevier BV, Amsterdam, Netherlands.

Kerry, F.G. 2007. Industrial Gas Handbook Gas Separation and Purification. CRC Press, Taylor & Francis Group, Boca Raton, Florida.

Kidnay, A.J., and Parrish, W.R. 2006. Fundamentals of Natural Gas Processing. CRC Press, Taylor & Francis Group, Boca Raton, Florida.

Mokhatab, S., Poe, W.A., and Speight, J.G. 2006. Handbook of Natural Gas Transmission and Processing. Elsevier, Amsterdam, Netherlands.

Speight, J.G. 2013. The Chemistry and Technology of Coal 5th Edition. CRC Press, Taylor & Francis Group, Boca Raton, Florida.

Speight, J.G. 2014. The Chemistry and Technology of Crude oil 5th Edition. CRC Press, Taylor & Francis Group, Boca Raton, Florida.

Speight, J.G. 2016. Introduction to Enhanced Recovery Methods for Heavy Oil and Tar Sands 2nd Edition. Gulf Publishing Company, Taylor & Francis Group, Waltham Massachusetts.

Speight, J.G., and Islam, M.R. 2016. Peak Energy – Myth or Reality. Scrivener Publishing, Beverly, Massachusetts.

Speight, J.G. 2017. Handbook of Crude oil Refining. CRC Press, Taylor & Francis Group, Boca Raton, Florida.

Speight, J.G. 2018. Handbook of Natural Gas Analysis. John Wiley & Sons Inc., Hoboken, New Jersey.

Speight, J.G. 2019a. Natural Gas: A Basic Handbook 2nd Edition. Gulf Publishing Company, Elsevier, Cambridge, Massachusetts.

Speight, J.G. 2019b. Handbook of Petrochemical Processes. CRC Press, Taylor & Francis Group, Boca Raton, Florida.

Speight, J.G. 2020. Global Climate Change Demystified. Scrivener Publishing, Beverly, Massachusetts.

Speight, J.G. 2021. Chemistry and Technology of Alternate Fuels. World Scientific Publishing Co. Pte. Ltd., Singapore and Hackensack, New Jersey.

Conversion factors and constants

1 Area

1 square centimeter $(1\ cm^2)$ = 0.1550 square inches
1 square meter 1 (m^2) = 1.1960 square yards
1 square kilometer $(1\ km^2)$ = 0.3861 square miles
1 square inch $(1\ inch^2)$ = 6.4516 square centimeters = 645.2 square millimeters
1 square foot $(1\ ft^2)$ = 0.0929 square meters
1 square yard $(1\ yd^2)$ = 0.8361 square meters
1 acre = 4,046.9 square meters
1 hectare = 2.4711 acres
1 square mile $(1\ mi^2)$ = 2.59 square kilometers

2 Compressibility factor

The compressibility factor (Z) corrects for deviation from the ideal gas law and is calculated using an equation of state or industry correlation based on the following relationship:

$$Z = {\sim}f\,(\text{composition},\ P,\ T)$$

Z is the compressibility factor at P, T for a given composition, P is the absolute pressure, and T is the temperature.

The compressibility factor (Z) is used to calculate the mass density (ρ) of a gas or dense phase fluid, thus:

$$\rho = [P\,MW_{gas}]/[R \times Z \times T]$$

In this equation, ρ is the mass density of fluid, MW_{gas} is the molecular weight of gas stream, R is the universal gas constant, Z is the compressibility factor at P, T for a given composition, P is the absolute pressure, and T is the temperature.

In the USC system, the universal gas constant (R) is expressed in [psia ft^3]/[lb$_m$ mol °R] units and equals 10.73164. The pressure and temperature must be in psia and R to be consistent with the units of the universal gas constant (R). The mass density of the fluid (ρ) is in lb$_m$/ft^3 units.

In the SI system, the universal gas constant (R) is expressed in [kPaa m^3]/[kg$_m$ mol °K] units and equals 8.314510. The pressure and temperature must be in kPaa and °K to be consistent with the units of the universal gas constant (R). The fluid's mass density (ρ) is in kgm/m^3 units.

Added note:

The compressibility factor of natural gas (which corrects for the ratio of actual volume to ideal volume) is on the order of a 0.5% correction in volume per 100 psi

https://doi.org/10.1515/9783110691023-008

of pressure for an orifice meter under normal pressure and temperature conditions. However, if the gas is near to the critical point, correction factors of on the order of up to 225% may be required, and small errors in measured variables (temperature and pressure) are reflected as large errors in volume.

3 Concentration conversion

1 part per million (1 ppm) = 1 microgram per liter (1 μg/L)
1 microgram per liter (1 μg/L) = 1 milligram per kilogram (1 mg/kg)
1 microgram per liter (μg/L) × 6.243 × 10^8 = 1 lb per cubic foot (1 lb/ft³)
1 microgram per liter (1 μg/L) × 10^{-3} = 1 milligram per liter (1 mg/L)
1 milligram per liter (1 mg/L) × 6.243 × 10^5 = 1 pound per cubic foot (1 lb/ft³)
I gram mole per cubic meter (1 g mol/m³) × 6.243 × 10^5 = 1 pound per cubic foot (1 lb/ft³)
10,000 ppm = 1% w/w
1 ppm hydrocarbon in soil × 0.002 = 1 lb of hydrocarbons per ton of contaminated soil

4 Density

Density is the mass of a unit volume of a substance. Thus:

$$d = M/V$$

d is the density, M is the mass, and V is the volume; often ρ (Greek rho) is the symbol for density.

Density is commonly expressed in units of grams per cubic centimeter.

Density conversion:

	lb/inch³	lb/ft³	lb/gal	g/cm³	g/liter
1 lb/inch³	1	1,728	231	27.68	27,680
1 lb/ft³	—	1	0.1337	0.016	16.019
1 lb/gal	4.33 (10^{-3})	7.481	1	0.1198	119.83
1 g/cm³	0.03613	62.43	8.345	1	103

5 Conversion

Flow

From	To	Multiply by
LPM	m^3/h	0.06
SCFM	m^3/hr	1.7
IGPM	Uh	273
USGPM	Uh	227
IGPH	Umin	0.076
USGPH	Umin	0.06
SCFH	L/min	0.472
LPH	cc/min	16.67
PPD	kQ/h	0.0189

Mass

	Kilograms	Grams	Pounds	Ounces
Kilograms	1	1,000	2.20462	35.2740
Grams	0.001	1	0.00220462	0.0352740
Pounds	0.453592	453.952	1	16
Ounces	0.0283	28.3495	0.0625000	1

Pressure

From	to	Multiply by
Psi	Millibar	68.95
Psi	Atm	0.068
Psi	Inches H$_2$0	27.68
kPa	Millibar	10
Inches H$_2$0	Millibar	2.40

Volume

From	to		Multiply by
US galls	Liters		3.79
Imp galls	Liters		4.55
SCH	Liters	I	28.32
Pound	Kilogramg		0.454
Imp Pint	Liters		0.568

6 Gas constant

The gas constant R is the Avogadro constant N_A multiplied by the Boltzmann constant (k, sometimes represented as k_B):

$$R = N_A k$$

Since the 2019 redefinition of the SI base units, both N_A and k are defined with exact numerical values when expressed in SI units. As a consequence, the value of the gas constant is also exactly defined, as precisely 8.31446261815324 J/K · mol.

Values of R	Units
SI units	
8.31446261815324	$J/K \cdot mol$
8.31446261815324	$m^3 \cdot Pa/K \cdot mol$
8.31446261815324	$kg \cdot m^2/K \cdot mol \, s^2$
$8.31446261815324 \times 10^3$	$L \cdot Pa/K \cdot mol$
$8.31446261815324 \times 10^{-2}$	$L \cdot bar/K \cdot mol$
US customary units	
0.730240507295273	$atm \cdot ft^3 \cdot lb/mol°R$
10.731557089016	$psi \cdot ft^3 \cdot lb/mol°R$
1.985875279009	$BTU \cdot lb/mol°R$
Other common units	
297.049031214	$in. \, H_2O \cdot ft^3 \cdot lb/mol°R$
554.984319180	$torr \cdot ft^3 \cdot lb/mol°R$
0.082057366080960	$L \cdot atm/K \cdot mol$
62.363598221529	$L \cdot Torr/K \cdot mol$

(continued)

Values of R	Units
$1.98720425864083 \ldots \times 10^{-3}$	$\text{kcal/K} \cdot \text{mol}$
$8.20573660809596 \ldots \times 10^{-5}$	$\text{m}^3 \cdot \text{atm/K} \cdot \text{mol}$
$8.31446261815324 \times 10^7$	$\text{erg/K} \cdot \text{mol}$

From the ideal gas law $PV = nRT$:

$$R = (PV)/nT$$

P is the pressure, V is the volume, n is number of moles of a given substance, and T is the temperature. Since pressure is the force per unit area, the gas equation can also be written as follows:

$$R = (\text{force/area}) \times \text{volume}/(\text{amount} + \text{temperature})$$

Area is $(\text{length})^2$ and volume is $(\text{length})^3$, thus:

$$R = ([\text{force/length}^2] \times \text{length}^3)/(\text{amount} \times \text{temperature})$$
$$= (\text{force} \times \text{length})/(\text{amount} \times \text{temperature})$$

Since force $\times$ length = work:

$$R = \text{work}/(\text{amount} \times \text{temperature})$$

The physical significance of R is work per degree per mole and may be expressed in any set of units representing work or energy (such as joules), units representing degrees of temperature on an absolute scale (such as degrees Kelvin, °K, or degrees Rankine, °R), and any system of units designating a mole or a similar pure number that allows an equation of macroscopic mass and fundamental particle numbers in a system, such as an ideal gas. Instead of a mole, the constant can be expressed by considering the normal cubic meter. Thus:

$$\text{force} = (\text{mass} \times \text{length})/\text{time}^2$$

Thus:

$$R = (\text{mass} \times \text{length}^2)/(\text{amount} \times \text{temperature} \times \text{time}^2)$$
$$\text{In SI units, } R = 8.314462618 \text{ kg} \cdot \text{m}^2/\text{s}^2 \cdot \text{K} \cdot \text{mol}$$

7 Length conversion

1 inch = 2.54 centimeter = 25.4 millimeter
1 foot = 30.48 centimeter = 0.3048 meter
1 yard = 0.9111 meter
1 mile = 1.609 kilometer

8 Periodic table of the elements

1 IA	2 IIA	3 IIIB	4 IVB	5 VB	6 VIB	7 VIIB	8 VIIIB	9 VIIIB	10 VIIIB	11 IB	12 IIB	13 IIIA	14 IVA	15 VA	16 VIA	17 VIIA	18 VIIIA
1 **H** Hydrogen 1.008																	2 **He** Helium 4.002602
3 **Li** Lithium 6.94	4 **Be** Beryllium 9.0121831											5 **B** Boron 10.81	6 **C** Carbon 12.011	7 **N** Nitrogen 14.007	8 **O** Oxygen 15.999	9 **F** Fluorine 18.998403163	10 **Ne** Neon 20.1797
11 **Na** Sodium 22.98976928	12 **Mg** Magnesium 24.305											13 **Al** Aluminium 26.9815385	14 **Si** Silicon 28.085	15 **P** Phosphorus 30.973761998	16 **S** Sulfur 32.06	17 **Cl** Chlorine 35.45	18 **Ar** Argon 39.948
19 **K** Potassium 39.0983	20 **Ca** Calcium 40.078	21 **Sc** Scandium 44.955908	22 **Ti** Titanium 47.867	23 **V** Vanadium 50.9415	24 **Cr** Chromium 51.9961	25 **Mn** Manganese 54.938044	26 **Fe** Iron 55.845	27 **Co** Cobalt 58.933194	28 **Ni** Nickel 58.6934	29 **Cu** Copper 63.546	30 **Zn** Zinc 65.38	31 **Ga** Gallium 69.723	32 **Ge** Germanium 72.630	33 **As** Arsenic 74.921595	34 **Se** Selenium 78.971	35 **Br** Bromine 79.904	36 **Kr** Krypton 83.798
37 **Rb** Rubidium 85.4678	38 **Sr** Strontium 87.62	39 **Y** Yttrium 88.90584	40 **Zr** Zirconium 91.224	41 **Nb** Niobium 92.90637	42 **Mo** Molybdenum 95.95	43 **Tc** Technetium (98)	44 **Ru** Ruthenium 101.07	45 **Rh** Rhodium 102.90550	46 **Pd** Palladium 106.42	47 **Ag** Silver 107.8682	48 **Cd** Cadmium 112.414	49 **In** Indium 114.818	50 **Sn** Tin 118.710	51 **Sb** Antimony 121.760	52 **Te** Tellurium 127.60	53 **I** Iodine 126.90447	54 **Xe** Xenon 131.293
55 **Cs** Caesium 132.90545196	56 **Ba** Barium 137.327	57 - 71 Lanthanoids	72 **Hf** Hafnium 178.49	73 **Ta** Tantalum 180.94788	74 **W** Tungsten 183.84	75 **Re** Rhenium 186.207	76 **Os** Osmium 190.23	77 **Ir** Iridium 192.217	78 **Pt** Platinum 195.084	79 **Au** Gold 196.966569	80 **Hg** Mercury 200.592	81 **Tl** Thallium 204.38	82 **Pb** Lead 207.2	83 **Bi** Bismuth 208.98040	84 **Po** Polonium (209)	85 **At** Astatine (210)	86 **Rn** Radon (222)
87 **Fr** Francium (223)	88 **Ra** Radium (226)	89 - 103 Actinoids	104 **Rf** Rutherfordium (267)	105 **Db** Dubnium (268)	106 **Sg** Seaborgium (269)	107 **Bh** Bohrium (270)	108 **Hs** Hassium (269)	109 **Mt** Meitnerium (278)	110 **Ds** Darmstadtium (281)	111 **Rg** Roentgenium (282)	112 **Cn** Copernicium (285)	113 **Nh** Nihonium (286)	114 **Fl** Flerovium (289)	115 **Mc** Moscovium (289)	116 **Lv** Livermorium (293)	117 **Ts** Tennessine (294)	118 **Og** Oganesson (294)

57 **La** Lanthanum 138.90547	58 **Ce** Cerium 140.116	59 **Pr** Praseodymium 140.90766	60 **Nd** Neodymium 144.242	61 **Pm** Promethium (145)	62 **Sm** Samarium 150.36	63 **Eu** Europium 151.964	64 **Gd** Gadolinium 157.25	65 **Tb** Terbium 158.92535	66 **Dy** Dysprosium 162.500	67 **Ho** Holmium 164.93033	68 **Er** Erbium 167.259	69 **Tm** Thulium 168.93422	70 **Yb** Ytterbium 173.045	71 **Lu** Lutetium 174.9668
89 **Ac** Actinium (227)	90 **Th** Thorium 232.0377	91 **Pa** Protactinium 231.03588	92 **U** Uranium 238.02891	93 **Np** Neptunium (237)	94 **Pu** Plutonium (244)	95 **Am** Americium (243)	96 **Cm** Curium (247)	97 **Bk** Berkelium (247)	98 **Cf** Californium (251)	99 **Es** Einsteinium (252)	100 **Fm** Fermium (257)	101 **Md** Mendelevium (258)	102 **No** Nobelium (259)	103 **Lr** Lawrencium (266)

9 Sludge conversion

1,700 lbs wet sludge = 1 yd^3 wet sludge
1 yd^3 sludge = wet tons/0.85
Wet tons sludge × 240 = gallons sludge
1 wet ton sludge × % dry solids/100 = 1 dry ton of sludge

10 Standard gas conditions

Standard temperature and pressure are the standard conditions for experimental measurements to be established to allow comparisons to be made between different sets of data. The most used standards are those of the International Union of Pure and Applied Chemistry (IUPAC) and the National Institute of Standards and Technology (NIST); other organizations have established a variety of alternative definitions for their standard reference conditions.

Until 1982, STP was defined as a temperature of 273.15 K (0 °C, 32 °F) and an absolute pressure of exactly 1 atmosphere (14.696 psi, 101.325 kPa). Since 1982, STP is defined as a temperature of 273.15 °K (0 °C, 32 °F) and an absolute pressure of exactly 10^5 Pa (100 kPa, 1 bar, 14.696 psi). NIST uses a temperature of 20 °C (293.15 °K, 20 °C, 68 °F) and an absolute pressure of 1 atmosphere (14.696 psi, 101.325 kPa). Whatever standard is used, it must be acknowledged and defined in the publication.

11 Temperature conversion

°F = (°C × 1.8) + 32
°C = (°F − 32)/1.8
(°F − 32) × 0.555 = °C
Absolute zero = −273.15 °C
Absolute zero = −459.67 °F

12 Thermodynamic data at 25 °C for selected constituents of natural gas

Component	Enthalpy combustion ΔH^{o}_{c}, kJ/mol	Enthalpy formation ΔH^{o}_{f}, kJ/mol	Free energy Formation ΔG^{o}_{f}, kJ/mol	Entropy S^{o}, J/(K-mol)
CH_4 (g), methane	−890	−74.81	−50.72	186.26
C_2H_5 (g), ethane	−1,560	−84.68	−32.82	229.60
C_3H_8 (g), propane	−2,220	−103.85	−23.49	270.2
C_4H_{10} (g), butane	−2,878	−126.15	−17.03	310.1
C_5H_{12} (g), pentane	−3,537	−146.44	−8.20	349
C_6H_6 (l), benzene	−3,268	49.0	124.3	173.3
C_6H_6 (g), benzene	−3,302	−	−	−
C_7H_8 (l), toluene	−3,910	12.0	113.8	221.0
C_7H_8 (g), toluene	−3,953	−	−	−
C_8H_{18} (l), octane	−5,471	−249.9	6.4	358

13 Weight conversion

1 ounce (1 oz) = 28.3495 grams (18.2495 g)
1 pound (1 lb) = 0.454 kilogram
1 pound (1 lb) = 454 grams (454 g)
1 kilogram (1 kg) = 2.20462 pounds (2.20462 lb)
1 stone (English, 1 st) = 14 pounds (14 lb)
1 ton (US; 1 short ton) = 2,000 lbs
1 ton (English; 1 long ton) = 2,240 lbs
1 metric ton = 2,204.62262 pounds
1 ton = 2,204.62262 pounds

14 Volume conversion

Volume percent characterizes the composition of gas mixtures which obey the ideal gas law that sets the relation between the gas volume, temperature, and pressure. According to this law, the volume is proportional to the number of moles of a gas, and therefore, the mole percentage is the same as the volume percent for gas mixtures.

Weight percent refers to mass of gasses in the mixtures and is required for stoichiometry calculations in chemistry.

1 cubic inch = 16.39 cubic centimeters

1 cubic foot = 0.028 cubic meters = 28.32 liters

1 cubic yard = 0.7646 cubic meters = 764.6 liters

I UK gallon = 1 Imperial gallon = 4.546 liters

1 US gallon – 3.785 liters

1 cubic foot (cf or scf, standard cubic foot)	=	1,027 Btu
100 cubic feet (scf)	=	1 Therm (approximate)
1,000 cubic feet (Mcf)	=	1,027,000 Btu (1 MMBtu)
1,000 cubic feet (Mcf)	=	1 dekatherm (10 therms)
1 million (1,000,000) cubic feet (MMcf)	=	1,027,000,000 Btu
1 billion (1,000,000,000) cubic feet (Bcf)	=	1.027 trillion Btu
1 trillion (1,000,000,000,000) cubic feet (Tcf)	=	1.027 quadrillion Btu

Multiply flow of	By	To obtain flow of
Natural gas	0.625	Propane
	0.547	Butane
	0.775	Air
Propane	1.598	Natural gas
	0.874	Butane
	1.237	Air
Butane	1.826	Natural gas
	1.140	Propane
	1.414	Air
Air	1.290	Natural gas
	0.808	Propane
	0.707	Butane

15 Other conversions

14.7 pounds per square inch (14.7 psi) – 1 atmosphere (1 atm)

1 kiloPascal (kPa) × 9.8692 × 10^{-3} = 14.7 pounds per square inch (14.7 psi)

1 yd^3 = 27 ft^3

1 US gallon of water = 8.34 lbs

1 Imperial gallon of water = 10 lbs

1 yd^3 = 0.765 m^3

1 acre-inch of liquid = 27,150 gallons = 3.630 ft^3

1 ft depth in 1 acre (in situ) = 1,613 × (20–25% excavation factor) = ~2,000 yd^3

1 yd^3 (clayey soils excavated) = 1.1–1.2 tons (US)

1 yd^3 (sandy soils excavated) = 1.2–1.3 tons (US)

Glossary

Abandoned well	A well not in use because it was a dry hole originally, or because it has ceased to produce. Statutes and regulations in many states require the plugging of abandoned wells to prevent the seepage of oil, gas, or water from one stratum to another.
Abiogenic gas	Gas formed by inorganic chemical reactions.
Abiotic	Produced from non-organism materials.
Absolute pressure	Gauge pressure plus barometric pressure; the absolute pressure can be zero only in a perfect vacuum.
Absolute zero	The zero point on the absolute temperature scale that is equal to −273.16 °C, or −459.69 °F, or 0 °K (degrees Kelvin), or 0 °R (degrees Rankine).
Absorbed gas	Natural gas that has been dissolved into the rock and requires hydraulic fracturing to be released.
Absorbent	A material which, due to an affinity for certain substances, extracts one or more such substances from a liquid or gaseous medium with which it contacts, and which changes physically, or both, during the process.
Absorption	The process by which the gas is distributed throughout an absorbent (liquid); depends only on physical solubility and may include chemical reactions in the liquid phase (*chemisorption*).
Absorption plant	A device or unit that removes hydrocarbon compounds from natural gas, especially casinghead gas; in the plant the gas is run through absorption oil which absorbs the liquid constituents, which are then recovered by distillation.
Abyssal	Of or relating to the bottom waters of the ocean.
Accumulate	To amass or collect; when oil and gas migrate into porous formations, the quantity collected is called an accumulation.
Accumulation	Pressure increase over the maximum allowable working pressure of the vessel during discharge through the pressure relief valve (expressed as a percent of that pressure) is called accumulation.
Accuracy	The closeness of the data of an analytical method to the true value.
Acid deposition (acid rain)	Occurs when sulfur dioxide (SO_2) and, to a lesser extent, NO_x emissions are transformed in the atmosphere and return to the earth as dry deposition or in rain, fog, or snow.
Acid gas	Impurities in a gas stream usually consisting of carbon dioxide (CO_2), hydrogen sulfide (H2S), carbonyl sulfide (COS), thiols/mercaptans (RSH), and sulfur dioxide (SO_2) – the most common in natural gas are carbon dioxide, hydrogen sulfide, and carbonyl sulfide. See also: Sour gas.
Acid gas loading	The amount of acid gas, on a molar or volumetric basis, which will be picked up by a solvent.
Acid Rain	Abnormally acidic rainfall, most often containing dilute concentrations of sulfuric acid or nitric acid.
Adsorption	The process by which the gas is concentrated on the surface of a solid or liquid to remove impurities; carbon is a common adsorbing medium which can be regenerated upon *desorption*.
Aerobic bacteria	Bacteria which can grow in the presence of oxygen.

https://doi.org/10.1515/9783110691023-009

Air gun	A chamber filled with compressed air, often used offshore in seismic exploration. As the gun is trailed behind a boat, air is released, making a low-frequency popping noise, which penetrates the subsurface rock layers and is reflected by the layers. Sensitive hydrophones receive the reflections and transmit them to recording equipment on the boat.
Alkazid process	A process for removal of hydrogen sulfide and carbon dioxide from natural gas using concentrated aqueous solutions of amino acids.
Alluvial fan	A large, sloping sedimentary deposit at the mouth of a canyon, laid down by intermittently flowing water, especially in arid climates, and composed of gravel and sand. The deposit tends to be coarse and unworked, with angular, poorly sorted grains in thin, overlapping sheets. A line of fans may eventually coalesce into an apron that grows broader and higher as the slopes above are eroded away.
Amphibole	Any group of common rock-forming silicate minerals.
Anaerobic bacteria	Bacteria which can grow in the absence of oxygen.
Angle of deflection	In directional drilling, the angle at which a well diverts from vertical; usually expressed in degrees, with vertical being 0°.
Angle of dip	The angle at which a formation dips downward from the horizontal.
Analytical batch	Consists of samples which are analyzed together with the same method sequence and the same lots of reagents and with the manipulations common to each sample within the same time period or in continuous sequential time periods.
Anticlinal trap	A hydrocarbon trap in which petroleum accumulates in the top of an anticline. See: Anticline, Syncline.
Anticline	An area of the earth's crust where folding has made a dome-like shape in the once flat rock layers. Anticlines often provide an environment where natural gas can become trapped beneath the surface of the Earth and extracted. See also: Traps, Faults, Permeability, and Porosity.
Antifoam	A substance, usually a silicone or long-chain alcohol, added to the treating system to reduce the tendency to foam.
Aquifer	An underground porous, permeable rock formation that acts as a natural water reservoir.
Aquifer storage field	A sub-surface facility for storing natural gas consisting of water-bearing sands topped by an impermeable caprock.
Associated gas	Natural gas that over-lies and contacts crude oil in a reservoir. Where reservoir conditions are such that the production of associated gas does not substantially affect the recovery of crude oil in the reservoir, such gas may also be reclassified as non-associated gas by a regulatory agency. Also called associated free gas. See: Gas cap.
Associated natural gas	Gas that occurs as free gas in a petroleum reservoir. See: Dissolved natural gas, Non-associated natural gas, Gas cap.
Atmospheric discharge	The release of vapors and gases from pressure-relieving and depressurizing devices to the atmosphere.
Authigenic minerals	Minerals formed in their present position.
Authigenic sediment	A deep-sea sediment that has been formed in place on the seafloor. The most significant authigenic sediments in modern ocean basins are

metal-rich sediments which include those enriched by iron, manganese, copper, chromium, and lead.

Back pressure — Pressure on the discharge side of safety-relief valves is back pressure; the pressure that exists at the outlet of a pressure relief device as a result of pressure in the discharge system.

Balancing item — Represents differences between the sum of the components of natural gas supply and the sum of the components of natural gas disposition

Barrel (bbl) — A measure of volume for petroleum products. One barrel is the equivalent of 35 imperial gallons or 42 U.S. gallons or 0.15899 cubic meters (9,702 cubic inches). One cubic meter equals 6.2897 barrels.

Base gas — The quantity of gas needed to maintain adequate reservoir pressures and deliverability rates throughout the withdrawal season; base gas usually is not withdrawn and remains in the reservoir; all gas native to a depleted reservoir is included in the base gas volume.

Base load requirements (base load storage) — Gas that is used to meet seasonal demand increases and the facilities are capable of holding enough natural gas to satisfy long-term seasonal demand requirements.

Basement rock — The impervious geological stratum that underlays the reservoir rock and retains gas or oil in a reservoir.

Basin — A local depression in the earth's crust in which sediments can accumulate to form thick sequences of sedimentary rock.

Bcf (billion cubic feet) — Gas measurement approximately equal to one trillion (1,000,000,000,000) Btu's. See also: Mcf, Tcf, Quad.

Bed — A specific layer of earth or rock that presents a contrast to other layers of different material lying above, below, or adjacent to it.

Bedrock — Solid rock just beneath the soil.

Benthic — Relating to the seabed.

Biogenic coalbed methane — Methane formed in coal seams by naturally occurring bacteria that are associated with meteoric water recharge at outcrop or sub-crop. See: Coal bed methane.

Biogenic gas — Gas formed by organic chemical reactions.

Biomass — Organic non-fossil material of biological origin constituting a renewable energy source.

Biotic — Produced by living organisms.

Bit — The cutting or boring element used in drilling oil and gas wells. The bit consists of a cutting element and a circulating element. The cutting element is steel teeth, tungsten carbide buttons, industrial diamonds, or polycrystalline diamonds (PDCs). These teeth, buttons, or diamonds penetrate and gouge or scrape the formation to remove it. The circulating element permits the passage of drilling fluid and utilizes the hydraulic force of the fluid stream to improve drilling rates. In rotary drilling, several drill collars are joined to the bottom end of the drill pipe column, and the bit is attached to the end of the drill collars. Drill collars provide weight on the bit to keep it in firm contact with the bottom of the hole. Most bits used in rotary drilling are roller cone bits, but diamond bits are also used extensively.

Bitumen — A hydrocarbonaceous substance of dark to black color consisting almost entirely of carbon and hydrogen with little oxygen, nitrogen, or

	sulfur; bitumen occurs naturally in tar sand (oil sand) formations; This is also the term used in some countries for *asphalt*.
Black shale	A thinly bedded shale that is rich in carbon, sulfide, and organic material, formed by anaerobic (lacking oxygen) decay of organic matter; black shale formations occur in thin beds in many areas at various depths and are of interest both historically and economically.
Blanket sand reservoir	A reservoir that has good areal extension with uniform thickness and which has been deposited in high-energy sedimentary environment. This type of reservoir typically holds a high volume of natural gas (and/or crude oil) and generally has a long plateau period of production. These reservoirs often have edge water aquifer support.
Blowdown	The difference between set pressure and reseating pressure of a safety valve expressed in percent of the set pressure or in psi, bar, or kPa.
Blowout	An uncontrolled flow of gas, oil, or other well fluids into the atmosphere. A blowout, or gusher, occurs when formation pressure exceeds the pressure applied to it by the column of drilling fluid. A kick warns of an impending blowout. See: Kick.
Blowout preventer (BOP)	One of several valves installed at the wellhead to prevent the escape of pressure either in the annular space between the casing and the drill pipe or in open hole (i.e., hole with no drill pipe) during drilling or completion operations. Blowout preventers on land rigs are located beneath the rig at the land's surface; on jackup or platform rigs, at the water's surface; and on floating offshore rigs, on the seafloor.
Black shale:	A thinly bedded shale that is rich in carbon, sulfide, and organic material, formed by anaerobic (lacking oxygen) decay of organic matter; black shales occur in thin beds in many areas at various depths and are of interest both historically and economically.
Boiling point (boiling temperature)	The temperature at which the vapor pressure of the substance is equal to atmospheric pressure.
Bottomhole	The lowest or deepest part of a well; drillers use mud pumps to circulate drilling fluid through the central well annulus to the bottomhole of the peripheral well; these data sets could be acquired at any location in the wellbore from the bottomhole or bottom of the well to the wellhead.
Bottomhole pressure	The bottomhole pressure (BHP) is equal to the hydrostatic pressure (HP) on the annular side. Thus: Bottomhole pressure (BHP) = surface pressure (SP) + hydrostatic pressure (HP) If shut in on a kick, the bottomhole pressure is equal to the hydrostatic pressure in the annulus plus the casing (wellhead or surface pressure) pressure. See also: Kick, Statis bottomhole pressure.
Bottom-simulating reflector (BSR)	A seismic reflection at the sediment to clathrate stability zone interface caused by the different density between normal sediments and sediments laced with clathrates.
Bottom-supported offshore drilling rig	A type of mobile offshore drilling unit that has a part of its structure in contact with the seafloor when it is on site.
Brine	An aqueous solution of salts that occurs with gas and crude oil; seawater and saltwater are also known as brine.

Bright spot	A seismic phenomenon that shows up on a seismic, or record, section as a sound reflection that is much stronger than usual. A bright spot sometimes directly indicates natural gas in a trap.
Btu (British Thermal Unit)	A unit of measurement for energy; the amount of heat that is necessary to raise the temperature of one pound of water by 1 degree, Fahrenheit. See also: Btu, Bcf, Tcf, Quad.
Bubble-point curve	The curve that separates the pure liquid (oil) phase from the two-phase (natural gas and oil) region. At a given temperature, when pressure decreases and below the bubble-point curve, gas will be emitted from the liquid phase to the two-phase region.
Bubble-point pressure	At a given temperature, the pressure when crude oil releases a bubble of gas from solution.
Built-up back pressure	The pressure in the discharge header which develops as a result of flow after that the safety relief valve opens.
Bundled Service	Gas sales service and transportation service packaged together in a single transaction in which the pipeline, on behalf of the utility, buys gas from producers and then delivers it to the utility.
Calibration	The process of adjusting the instrument read-out so that it corresponds to the actual concentration value or a reference standard; involves checking the instrument with a known concentration of a gas or vapor to see that the instrument gives the proper response; calibration results in calibration factors or functions establishing the relationship between the analyzer response and the actual gas concentration introduced to the analyzer; an important element of quality assurance in emission control.
Cap gas	Natural gas trapped in the upper part of a reservoir and remaining separate from any crude oil, salt water, or other liquids in the well.
Caprock (cap rock)	A formation that prevents upward migration or movement of hydrocarbons in the subsurface of the Earth. Typically, the caprock is a shale formation or a carbonate formation with moderate or low porosities and very low permeabilities and are often water-saturated. Some caprocks are believed to be homogeneous solids without pores, including anhydrites and an occasional limestone
Carbonaceous material	A material which contains carbon as well as some hydrogen and other non-carbon and non-hydrogen elements such as nitrogen, oxygen, sulfur.
Carbonate rock	A rock consisting primarily of a carbonate mineral such as calcite or dolomite, the chief minerals in limestone and dolostone, respectively.
Carbonate washing	A chemical conversion process in which acid contaminants in natural gas are converted to compounds that are not objectionable or that can be removed from the stream with greater ease than the original constituents.
Carbon capture and storage	A combination of a number of existing technologies, with the potential to play a major role in the management and reduction of global carbon dioxide (CO_2) levels; the process allows for carbon dioxide emissions released during energy production to be captured and stored underground.
Carbon dioxide (CO_2)	An acid gas that does not support combustion and thus does not contribute to the gas.
Carbon dioxide fracturing	The use of gaseous carbon dioxide to fracture a formation.

Carbureting	Carbureting gas generally comprises passing it in contact with a liquid fuel and thereby mixing the air/gas and fuel.
Carbonyl sulfide (COS)	A compound that often appears in unrefined natural gas that has a high concentration of hydrogen sulfide; has the undesirable property of forming non-regenerable compounds with one of the most commonly used sweetening agents, monoethanolamine ($HOCH_2CH_2NH_2$) which causes increased chemical consumption of this agent; other sweetening agents such as diethanolamine can be used to absorb the carbonyl sulfide which is generally broken down (decomposed) in the regeneration step.
Casing	Steel pipe placed in an oil or gas well to prevent the wall of the hole from caving in, to prevent movement of fluids from one formation to another, and to aid in well control; also used to protect the surrounding earth and rock layers from being contaminated by petroleum, or the drilling fluids.
Casinghead gas (casing head gas)	Natural gas produced with oil in oil wells and is usually the flash gas from the oil reservoir. Casinghead gas collects in the annular space between the well tubing and casing of an oil well. The weight of the casinghead gas contributes to reducing the bottomhole pressure and lowering well production – specifically by holding a back-pressure against the formation; reducing the gas pressure on the well casing (annulus) can increase oil production. The goal is to maintain casinghead pressure as close to zero as possible. In areas or the country where it is allowed, these systems are often configured to pull a vacuum.
Catalytic oxidation	A chemical conversion process that is used predominantly for destruction of volatile organic compounds and carbon monoxide.
Cathodic Protection	The method of preventing corrosion in metal structures that involves using electric voltage to slow or prevent corrosion; used in natural gas pipelines to resist corrosion over an extended period of time.
Cenozoic era	The time period from 65 million years ago until the present. It is marked by rapid evolution of mammals and birds, flowering plants, grasses, and shrubs, and little change in invertebrates.
CFCs (Chlorofluorocarbons)	Gaseous compounds used for cooling; release into the atmosphere has produces ozone depletion.
Chadacryst	A crystal enclosed in another crystal.
Chemical waste	Natural gas production and processing produce chemical waste which, if not processed in a timely manner, can become a pollutant. Under some circumstances, chemical waste is reclassified as hazardous waste. See: Hazardous waste.
Chemisorption	See: Absorption.
Christmas tree	The series of pipes and valves that sits on top of a producing gas well; used in place of a pump to extract the gas from the well.
City gate	A location at which custody of gas passes from a gas pipeline company to a local distributor.
Clastic rock	A sedimentary rock composed of fragments of preexisting rocks. The principal distinction among clastic rocks is grain size; conglomerates, sandstones, and shale are clastic rocks.

Claus process	A sulfur recovery process recovering elemental sulfur from sour gas; a major producer of sulfur.
Clean Air Act Amendments of 1990	legislation to improve the quality of the atmosphere and curb acid rain promotes the use of cleaner fuels in vehicles and stationary sources.
Clinopyroxene	A subgroup name for monoclinic pyroxene group of minerals.
Coal	A carbonaceous, rocklike material that forms from the remains of plants that were subjected to biochemical processes, intense pressure, and high temperatures. It is used as fuel.
Coalbed methane (coal bed methane)	Methane from coal seams; released or produced from the seams when the water pressure within the seam is reduced by pumping from either vertical or inclined to horizontal surface holes. See also: Biogenic coal bed methane, Thermogenic coal bed methane.
Coalescer:	A mechanical process vessel with wettable, high-surface area packing on which liquid droplets consolidate for gravity separation from a second phase (for example gas or immiscible liquid).
Coal gas	A generic term for gaseous mixture (mainly hydrogen, methane, and carbon monoxide) made from coal by the destructive distillation (i.e., heating in the absence of air) of bituminous coal; also synonymous with *blue gas, producer gas, water gas, town gas, fuel gas manufactured gas*, and *syngas (synthetic natural gas*, SNG).
Coke oven gas	The mixture of permanent gases produced by the carbonization of coal in a coke oven at temperatures in excess of 1,000 °C (1,830 °F).
Composition	The make-up of a gaseous stream.
Compressed natural gas (CNG)	Natural gas compressed to a pressure at or above 2,900–3,600 psi and stored in high-pressure containers; used as a fuel for natural gas-powered vehicles.
Compressibility factor	The ratio of the molar volume V_m of the gas to the molar volume $V_m°$ of an ideal gas at the same pressure and temperature. Thus: $Z = V_m/V_m°$ The value of Z provides information on the dominant types of intermolecular forces acting in a gas. Thus, when $Z = 1$, there are no intermolecular forces, ideal gas behavior; when $Z < 1$, the attractive forces dominate, gas occupies a smaller volume than an ideal gas; when $Z > 1$, the repulsive forces dominate, gas occupies a larger volume than an ideal gas. All gases approach $Z=1$ at low pressures, when the spacing between particles is typically large.
Compression	Reduction in volume of natural gas is compressed during transportation and storage.
Concentration	The amount of a substance, expressed as mass, volume, or number of particles in a unit volume of a solid, liquid, or gaseous substance.
Concrete gravity rigid platform rig	A rigid offshore drilling platform built of steel-reinforced concrete and used to drill development wells. The platform is floated to the drilling site in a vertical position. At the site, one or more tall caissons that serve as the foundation of the platform are flooded so that the platform comes to rest on bottom. Because of the enormous weight of the platform, the force of gravity alone keeps it in place.
Condensate	A hydrocarbon liquid stream that consists of varying proportions of butane, propane, pentane, and heavier fractions, with little or no methane or ethane; separated from natural gas; higher molecular

weight hydrocarbons that exist in the reservoir as constituents of natural gas, but which are recovered as liquids in separators, field facilities or gas-processing plants. See: Gas condensate.

Condensate (lease condensate) Low-boiling liquid hydrocarbons recovered from lease separators or field facilities at associated and non-associated natural gas wells; mostly pentane derivative and higher molecular weight hydrocarbon derivatives that enter the crude oil stream after production.

Condensate reservoir A reservoir in which both condensate and gas exist in one homogeneous phase. When fluid is drawn from such a reservoir and the pressure decreases below the critical level, a liquid phase (condensate) appears.

Condensate Separator A unit for the removal of condensate from the gas stream at the wellhead through the use of mechanical separators. In most instances, the gas flow into the separator comes directly from the wellhead since the gas-oil separation process is not needed. Extracted condensate is routed to on-site storage tanks.

Coning A production problem in which cap gas or bottom water infiltrates the perforation zone in the near-wellbore area and reduces oil production; gas coning is distinctly different from, and should not be confused with, free-gas production caused by a naturally expanding gas cap. The term is used because, in a vertical well, the shape of the interface when a well is producing the second fluid resembles an upright or inverted cone. Important examples of coning include: (i) production of water in an oil well with bottom water drive, (ii) production of gas in an oil well overlain by a gas cap, and (iii) production of bottom water in a gas well. See: Gas coming, Water coning.

Connate water The natural water retained in a reservoir after gas or oil have entered the closure by water displacement. Since the original water may have been displaced over time, some prefer the use of the term interstitial water to eliminate questions of whether it was present at the time of rock deposition. See: Reservoir fluid, Reservoir water.

Condensate well A well that produces raw natural gas along with low-boiling hydrocarbon liquids; the gas is also *non-associated* gas and is often referred to as *wet gas*.

Consumption Natural gas consumed within the country, including imports but excluding amounts re-injected, flared and lost in shrinkage.

Contact In geology, any sharp or well-defined boundary between two different bodies of rock; a bedding plane or unconformity that separates formations. In a petroleum reservoir, a horizontal boundary where different types of fluids meet and mix slightly; for example, a gas-oil or oil-water contact; also called an interface.

Contaminant removal Removal of contaminants which includes the elimination of hydrogen sulfide, carbon dioxide, water vapor, helium, and oxygen. The most commonly used technique is to first direct the flow through a tower containing an olamine solution which absorbs sulfur compounds. After desulfurization, the gas flow is directed to the next section, which contains a series of filter tubes. As the velocity of the stream reduces in the unit, primary separation of remaining contaminants occurs due to gravity.

Continuous accumulations	Petroleum that occurs in extensive reservoirs and is not necessarily related to conventional structural or stratigraphic traps. these accumulations of oil and/or gas lack well-defined down-dip petroleum/water contacts and thus are not localized by the buoyancy of oil or natural gas in water.
Conventional gas	Natural gas that is extracted from underground reservoirs using traditional exploration and production methods. Conventional gas-liquid separator. A vertical or horizontal separator in which gas and liquid are separated by means of gravity settling with or without a mist eliminating device.
Conventional mud	A drilling fluid containing essentially clay and water; no special or expensive chemicals or conditioners are added.
Core	A cylindrical sample taken from a formation for geological analysis. Usually a conventional core barrel is substituted for the bit and procures a sample as it penetrates the formation. *V*: to obtain a solid, cylindrical formation sample for analysis.
Core analysis	Laboratory analysis of a core sample to determine porosity, permeability, lithology, fluid content, angle of dip, geological age, and probable productivity of the formation.
Coring	The process of cutting a vertical, cylindrical sample of the formations encountered as an oil well is drilled. The purpose of coring is to obtain rock samples, or cores, in such a manner that the rock retains the same properties that it had before it was removed from the formation.
Cretaceous	Of or relating to the geologic period from about 135 million to 65 million years ago at the end of the Mesozoic era, or to the rocks formed during this period, including the extensive chalk deposits for which it was named.
Cricondenbar pressure	The maximum pressure at which two phases can coexist.
Cricondentherm	The highest temperature at which liquid and vapor can coexist. That means the mixture will be gas irrespective of pressure when the temperature is larger than cricondentherm.
Cricondentherm temperature	The maximum temperature at which two phases can coexist.
Critical point	The end point of the pressure-temperature curve that designates conditions under which a liquid and its vapor can coexist. At higher temperatures, the gas cannot be liquefied by pressure alone. At the critical point, defined by the critical temperature, T_c, and the critical pressure, p_c, phase boundaries vanish. The critical point is also the pressure and temperature of a reservoir fluid where the bubblepoint pressure curve meets the retrograde dewpoint pressure curve representing a unique state where all properties of the bubblepoint oil are identical to the dewpoint gas.
Critical pressure	The pressure required to liquify a substance vapor at its critical temperature.
Critical temperature	The temperature which above, a substance cannot exist as a liquid, no matter how much pressure is applied.
Critical volume	The volume occupied by a certain mass, usually one gram molecule of a liquid or gaseous substance at its critical point; the numerical value

	of the critical volume depends upon the amount of gas under experiment.
Cross section	A geological or geophysical profile of a vertical section of the earth.
Crude oil	Unrefined liquid petroleum which ranges in gravity from 9° API to 55° API and in color from yellow to black, and may have a paraffin, asphalt, or mixed base. If a crude oil, or crude, contains a sizable amount of sulfur or sulfur compounds, it is called a sour crude; if it has little or no sulfur, it is called a sweet crude. In addition, crude oils may be referred to as heavy or light according to API gravity, the lighter oils having the higher gravities.
Crust	The outer layer of the earth, varying in thickness from 5 to 30 miles (10 to 50 kilometers). It is composed chiefly of oxygen, silicon, and aluminum.
Cryogenic process	A process involving low temperatures.
CSST (corrugated stainless-steel tubing)	Flexible piping used to install gas service in residential and commercial areas.
Cubic foot (ft^3)	The volume of a cube, all edges of which measure 1 foot. Natural gas in the United States is usually measured in cubic feet, with the most common standard cubic foot being measured at 60 °F and 14.65 pounds per square inch absolute, although base conditions vary from state to state.
Cubic meter (m^3)	A unit of volume measurement in the SI metric system, replacing the previous standard unit known as the barrel, which was equivalent to 35 imperial gallons or 42 U.S. gallons. The cubic meter equals approximately 6,2898 barrels.
Cumulative production	Volumes of oil and natural gas liquids that have been produced.
Cutting	A piece of rock or dirt that is brought to the surface of a drilling site as debris from the bottom of well; often used to obtain data for logging.
Cycles	The number of times a facility can be completely filled and emptied (or turned around).
Darcy's Law	Formulated as an equation that describes the flow of a fluid through a porous medium.
DEA	Diethanolamine; $HOCH_2CH_2NHCH_2CH_2OH$.
Dead crude oil	Crude oil in the reservoir with minimal or no dissolved associated gas; often difficult to produce as there is little energy to drive it.
Decline rate	The decline rate is the rate of change of production with respect to time and, for exponential decline, is constant for all time; the mathematical equation defining exponential decline has two constants, the initial production rate and the decline rate.
Deep water	In offshore operations, water depths greater than normal for the time and current technology.
Degradation products	Impurities in a treating solution which are formed both reversible and irreversible side reactions.
Delineation well	A well drilled in an existing field to determine, or delineate, the extent of the reservoir.
Dehydration	Removal of water from gas streams that is accomplished by several methods, such as the use of ethylene glycol (glycol injection) systems as an absorption mechanism to remove water and other solids from the gas stream. Alternatively, adsorption dehydration may be used,

	utilizing dry-bed dehydrators towers, which contain desiccants such as silica gel and activated alumina, to perform the extraction.
Dekatherm (dth)	A unit of energy used primarily to measure natural gas and was developed in about 1972 by the Texas Eastern Transmission Corporation – a natural gas pipeline company; the dekatherm is equal to 10 therms or 1,000,000 British thermal units (MMBtu) or, using the SI system, 1.055 gigajoules (GJ) and is also approximately equal to one thousand cubic feet (Mcf, Mft^3) of natural gas or exactly one thousand cubic feet of natural gas with a heating value of 1,000 Btu/ft^3.
Deliverability	The means by which storage (gas availability) is integrated with the transportation system.
Delivered gas	The physical transfer of natural, synthetic, and/or supplemental gas from facilities operated by the responding company to facilities operated by others or to consumers.
Delivery point (receipt point)	The point where natural gas is transferred from one party to another. The city gate is the delivery point for a pipeline or transportation company because this is where the gas is transferred to the LDC.
Density	The mass of a substance contained in a unit volume (mass divided by volume).
Deplete	To exhaust a supply; an oil and/or a gas reservoir is depleted when most or all economically recoverable hydrocarbons have been produced.
Depleted reservoirs	Reservoirs that have already been tapped of all their recoverable natural gas.
Depleted storage field	A sub-surface natural geological reservoir, usually a depleted gas or oil field, used for storing natural gas.
Depth	The distance to which a well is drilled, stipulated in a drilling contract as contract depth. Total depth is the depth after drilling is finished.
Derrick	A large load-bearing structure, usually of bolted construction. In drilling, the standard derrick has four legs standing at the corners of the substructure and reaching to the crown block; an assembly of heavy beams used to elevate the derrick and provide space to install equipment such as blowout preventers and casingheads; the standard derrick must be assembled piece by piece, it has largely been replaced by the mast, which can be lowered and raised without disassembly.
Desorption	See: Adsorption.
Destructive distillation	The process of pyrolysis conducted in a distillation apparatus to allow the volatile products to be collected.
Detection limit	The lowest amount of analyte in a sample which can be detected but not necessarily quantitated as an exact value; often called the *limit of detection* (*LOD*) which is the lowest concentration level that can be determined statistically different from a blank at a specified level of confidence; determined from the analysis of sample blanks. See: Method detection limit.
Development (of a gas or oil field)	All operations associated with the construction of facilities to enable the production of oil and gas.

Development drilling	Drilling that occurs after the initial discovery of hydrocarbons in a reservoir. Usually, several wells are required to adequately develop a reservoir.
Development well	A well drilled in proven territory in a field to complete a pattern of production; an exploitation well.
Deviation	Departure of the wellbore from the vertical, measured by the horizontal distance from the rotary table to the target. The amount of deviation is a function of the drift angle and hole depth. The term is sometimes used to indicate the angle from which a bit has deviated from the vertical during drilling.
Dew point curve	The curve that separates the pure gas phase from the two-phase region. It is the connected points of pressure and temperature at which the first liquid droplet is formed out of the gas phase.
Dew point reservoir	A reservoir in which condensation causes a liquid to leave the gas phase.
Dew point temperature	The temperature at which a vapor begins to condense and deposit as a liquid; the hydrocarbon dew point is the temperature (at a given pressure) at which the hydrocarbon constitunets of any hydrocarbon-rich gas mixture, such as natural gas, will start to condense out of the gaseous phase; the maximum temperature at which such condensation takes place is called the *cricondentherm*.
DGA	Diglycolamine.
Diagenesis	The physical and chemical changes occurring during the conversion of sediment to sedimentary rock. The changes that happen to the sediment after deposition can control the lateral continuity and vertical stacking of reservoir rock types.
Diamond bit	A drill bit that has small industrial diamonds embedded in its cutting surface. Cutting is performed by the rotation of the hard diamonds over the rock surface.
Dip	The angle at which it lies in relation to a flat line at the surface; often helps the geologist to locate possible traps (reservoirs).
DIPA	Diisopropanolamine.
Directional drilling	Intentional deviation of a wellbore from the vertical. Although wellbores are normally drilled vertically, it is sometimes necessary or advantageous to drill at an angle from the vertical; controlled directional drilling makes it possible to reach subsurface areas laterally remote from the point where the bit enters the earth. It often involves the use of deflection tools.
Discovery well	The first oil or gas well drilled in a new field that reveals the presence of a hydrocarbon-bearing reservoir. Subsequent wells are development wells.
Dissolved gas	Natural gas that is in solution with crude oil in the reservoir.
Dissolved natural gas	Gas that occurs in solution in the petroleum in a reservoir. See: Associated natural gas.
Dissolved water	Water in solution in oil at a defined temperature and pressure.
Distillate	Volatile products released from a solid charge during pre-distillation or during pyrolysis.
Drill	To bore a hole in the earth, usually to find and remove subsurface formation fluids such as oil and gas.

Drilling fluid	A circulating fluid, one function of which is to lift cuttings out of the wellbore and to the surface. It also serves to cool the bit and to counteract downhole formation pressure. Although a mixture of barite (a mineral composed of barium sulfate, $BaSO_4$), clay, water, and other chemical additives is the most common drilling fluid, wells can also be drilled by using air, gas, water, or oil-base mud as the drilling mud. Also called circulating fluid, drilling mud. See: Mud.
Drilling mud	Specially compounded liquid circulated through the wellbore during rotary drilling operations. See: Drilling fluid, Mud.
Drillship (drill ship)	A self-propelled floating offshore drilling unit that is a ship constructed to permit a well to be drilled from it.
Drip gas	A gas stream that can be drawn off the bottom of small chambers (called drip chambers or drips) sometimes installed in pipelines from gas wells; another name for natural gas condensate, a naturally occurring form of gasoline obtained as a byproduct of natural gas extraction; defined in the United States Code of Federal Regulations as consisting of butane, pentane, and hexane derivatives.
Dry gas	A gas stream which contains no or few or no hydrocarbon derivatives of higher molecular weight than methane that are not commercially recoverable – also called lean gas; may also refer to a gas stream in which the water content has been reduced by a dehydration process.
Dry-gas reservoir	A reservoir in which the hydrocarbon mixture exists as a gas both in the reservoir and in the surface facilities; the only liquid associated with the gas from a dry-gas reservoir is water; usually a system having a gas-oil ratio greater than 100,000 scf is considered to be a dry gas reservoir.
Dry gas well	A well that typically produces only raw natural gas that does not contain any hydrocarbon liquids; the gas is called *non-associated* gas.
Dry hole	Any well that does not produce oil or gas in commercial quantities. A dry hole may flow water, gas, or even oil, but not in amounts large enough to justify production.
Dry natural gas	Natural gas which remains after: (i) the liquefiable hydrocarbon portion has been removed from the gas stream and (ii) any volumes of non-hydrocarbon gases have been removed where they occur in sufficient quantity to render the gas unmarketable; also known as consumer-grade natural gas.
Ecology	Science of the relationships between organisms and their environment.
Eh/redox	A measure of the degree of oxygenation of a sediment.
Emissions	Gaseous, liquid, or solid by-products introduced into the environment as a result of exploration, recovery, processing, and transportation processes.
Endowment	The sum of cumulative production, remaining reserves, mean undiscovered recoverable volumes, and mean additions to reserves by field growth.
Environment	The sum of the physical, chemical, and biological factors that surround an organism; the water, air, and land and the interrelationship that exists among and between water, air, and land and all living things.

Environmental control	The use of various technologies to control and even prevent emissions from entering the environment.
Environmental effects The effects of refinery emissions on the flora and fauna in the various ecosystems. Environmental guidance	A document developed by a governmental agency that outlines a position on a topic or which gives instructions on how a procedure must be carried out. The document explains how to do something and provides governmental interpretations on a governmental act or policy.
Environmental policy	A requirement that specifies operating procedures that must be followed.
Environmental regulation	A legal mechanism that determines how the policy directives of an environmental law are to be carried out.
EPACT (Energy Policy Act of 1992)	Comprehensive energy legislation designed to expand natural gas use by allowing wholesale electric transmission access and providing incentives to developers of clean fuel vehicles.
EPIC (Engineering, Procurement, Installation, Commissioning)	An EPIC or turnkey contract integrates the responsibility going from the conception to the final acceptance of one or more elements of a production system. It can be awarded for all, or part, of a field development.
Equation of state (EOS)	A thermodynamic expression that relates pressure (P), temperature (T), and volume (V); used to describe the state of reservoir fluids at given conditions. The equation of state for a substance provides the additional information required to calculate the amount of work that the substance does in making a transition from one equilibrium state to another along some specified path. The equation of state is expressed as a functional relationship connecting the various parameters needed to specify the state of the system.
Erosion	The process by which material (such as rock or soil) is worn away or removed (as by wind or water).
Essexite	A dark gray or black holocrystalline plutonic igneous rock.
Ethanol	CH_3CH_2OH; produced through fermentation of agricultural raw materials (biomass), ethanol is used for various applications: drinks, pharmaceuticals, cosmetics, solvents, chemicals and more and more often in fuels, either in the form of an additive to gasoline (ETBE: Ethyl Tertiary-Butyl Ether) or blended directly with hydrocarbon-based gasoline.
Ethanolamine	$HOCH_2CH_2NH_2$; also called monoethanolamine; used in gas processing to remove acid gases (such as hydrgoen sulfide and carbon dioxide) from gas streams.
Estimated additional amount in place	The volume additional to the proved amount in place that is of foreseeable economic interest. Speculative amounts are not included.
Estimated additional reserves recoverable	The volume within the estimated additional amount in place which geological and engineering information indicates with reasonable certainty might be recovered in the future.
Estuary	A coastal indentation or bay into which a river empties and where fresh water mixes with seawater.
Evaporation ponds	Artificial ponds with large surface areas that are designed to allow the efficient evaporation of water through exposure to sunlight and ambient surface temperatures.

Exploration	The search for reservoirs of oil and gas, including aerial and geophysical surveys, geological studies, core testing, and drilling of wildcats.
Exploration well	A well drilled either in search of an as-yet-undiscovered pool of oil or gas (a wildcat well) or to extend greatly the limits of a known pool. It involves a relatively high degree of risk. Exploratory wells may be classified as (1) wildcat, drilled in an unproven area; (2) field extension or step-out, drilled in an unproven area to extend the proved limits of a field; or (3) deep test, drilled within a field area but to unproven deeper zones.
Exploratory well	A well drilled either in search of a new and as yet undiscovered accumulation of oil or gas, or in an attempt to significantly extend the limits of a known reservoir.
Extraction loss	The reduction in volume of natural gas due to the removal of natural gas liquid constituents such as ethane, propane, and butane at natural gas processing plant.
Extraction plant	A plant in which products, such as propane, butane, oil, ethane, or natural gasoline, which are initially components of the gas stream, are extracted or removed for sale.
Fabric filter	Collectors in which dust is removed from the gas stream by passing the dust-laden gas through a fabric of some type; commonly termed "bag filters" or "baghouses."
Fault	When part of the earth's crust fractures due to forces exerted on it by movement of plates on the earth's crust; of interest because they often form traps that are natural gas reservoirs; a break in the earth's crust along which rocks on one side have been displaced (upward, downward, or laterally) relative to those on the other side.
Fault plane	A surface along which faulting has occurred.
Fault trap	A subsurface hydrocarbon trap created by faulting, in which an impermeable rock layer has moved opposite the reservoir bed or where impermeable gouge has sealed the fault and stopped fluid migration.
Feldspathoid	Low silica igneous minerals that would have formed feldspars if only more silica (SiO_2) was present in the original magma.
Field	A geographical area in which a number of oil or gas wells produce from a continuous reservoir. A field may refer to surface area only or to underground productive formations as well. A single field may have several separate reservoirs at varying depths; a contiguous area consisting of a single reservoir or multiple reservoirs of petroleum, all grouped on, or related to, a single geologic structural and/or stratigraphic feature.
Fire point	The temperature to which gas must be heated under prescribed conditions of the method to burn continuously when the mixture of vapor and air is ignited by a specified flame.
Fischer-Tropsch process	The catalytic process by which synthesis gas (syngas; mixtures of carbon monoxide and hydrogen) is converted to hydrocarbon products.
Fixed platform	A structure made of steel or concrete, fixed to the bottom of the body of water in which it rests.

Flare	A tall stack equipped with burners used as a safety device at wellheads, refining facilities, gas processing plants, and chemical plants; used for the combustion and disposal of combustible gases.
Flare gas	Gas or vapor that is flared.
Flash point	The temperature to which gas must be heated under specified conditions to give of sufficient vapor to form a mixture with air that can be ignited momentarily by a specified flame; dependant on the composition of the gas and the presence of other hydrocarbon constituents.
Flash tank	A vessel used to separate the gas evolved from liquid flashed from a higher pressure to a lower pressure
Flexible flow line	Flexible pipe laid on the seabed for the transportation of production or injection fluids. It is generally an infield line, linking a subsea structure to another structure or to a production facility. Its length ranges from a few hundred meters to several kilometers.
Flexible riser	Riser constructed with flexible pipe. See: Riser.
Flexsorb process	A process that uses sterically hindered amines (olamines) in aqueous solutions or other physical solvents; the molecular structure hinders the carbon dioxide approach to the amine and preferentially removes hydrogen sulfide from the gas stream.
Floaters	Floating production units including floating platforms, and floating, production, storage and offloading units (FPSOs).
Floating offshore drilling rig	A type of mobile offshore drilling unit that floats and is not in contact with the seafloor (except with anchors) when it is in the drilling mode; floating units include barge rigs, drill ships, and semisubmersibles.
Floating production and system off-loader:	A floating offshore oil production vessel that has facilities for producing, treating, and storing oil from several producing wells and which puts (offloads) the treated oil into a tanker ship for transport to refineries on land; some floating, production, storage and offloading units are also capable of drilling, in case they are termed floating production, drilling, and system off-loaders (FPDSOs).
Flow rate	The rate that expresses the volume of fluid or gas passing through a given surface per unit of time (e.g., cubic feet per minute).
Flowing well	A well that produces oil or gas by its own reservoir pressure rather than by use of artificial means (such as pumps).
Fold	A flexure of rock strata (e.g., an arch or a trough) produced by horizontal compression of the earth's crust. See: Anticline, Syncline.
Formation	A bed or deposit composed throughout of substantially the same kind of rock; often a lithological unit; either a certain layer of the earth's crust, or a certain area of a layer; often refers to the area of rock where a reservoir is located; each formation is given a name, frequently as a result of the study of the formation outcrop at the surface and sometimes based on fossils found in the formation.
Formation volume factor (FVF)	The ratio of a phase volume (water, oil, gas, or gas plus oil) at reservoir conditions, relative to the volume of a surface phase (water, oil, or gas) at standard conditions resulting when the reservoir material is brought to the surface; denoted mathematically as B_w (bbl/STB), B_o (bbl/STB), B_g (ft^3/SCF), and B_t (bbl/STB).

Fossil	The remains or impressions of a plant or animal of past geological ages that have been preserved in or as rock.
FPSO unit (Floating, Production, Storage and Offloading unit)	A converted or custom-built ship-shaped floater, employed to process oil and gas and for temporary storage of the oil prior to trans-shipment.
FPU (Floating Production Unit)	A ship-shaped floater or a semi-submersible used to process and export oil and gas.
Fractionation	The process of separating the various natural gas liquids present in the remaining gas stream by using the varying boiling points of the individual hydrocarbons in the gas stream.
Fracturing	A method used by producers to extract more natural gas from a well by opening up rock formations using hydraulic or explosive force.
Free water	Water produced with oil. It usually settles out within five minutes when the well fluids become stationary in a settling space within a vessel.
FSHR (Free Standing Hybrid Riser)	A deepwater riser configuration (see: Riser) consisting of a vertical rigid pipe section between the seabed and a submerged buoy and a catenary flexible pipe jumper between the submerged buoy and the floater.
Fuel cell technology	The chemical interaction of natural gas and certain other metals, such as platinum, gold, and other electrolytes to produce electricity.
Future petroleum	The sum of the remaining reserves, mean reserve growth, and the mean of the undiscovered volume. cumulative production does not contribute to the future petroleum. the terms future oil, future liquid volume, or future endowment are sometimes used as variations of future petroleum to reflect those resources that are yet to be produced.
Gabbro	A large group of dark, often coarse-grained, mafic intrusive igneous rocks chemically equivalent to plutonic basalt; formed when molten magma is trapped beneath the surface of the Earth and slowly cools into a holo-crystalline mass.
Gas	A compressible fluid that completely fills any container in which it is confined. Technically, a gas will not condense when it is compressed and cooled, because a gas can exist only above the critical temperature for its particular composition. Below the critical temperature, this form of matter is known as a vapor, because liquid can exist, and condensation can occur; the terms "gas" and "vapor" are often used interchangeable. The latter, however, should be used for those streams in which condensation can occur and that originate from, or are in equilibrium with, a liquid phase. See: Vapor.
Gas cap	The gas trapped between the liquid petroleum and the impervious cap rock of the petroleum reservoir; a free-gas phase overlying an oil zone and occurring within the same producing formation as the oil. See also: Associated gas, Reservoir.
Gas-cap drive	Drive energy supplied naturally (as a reservoir is produced) by the expansion of the gas cap. In such a drive, the gas cap expands to force oil into the well and to the surface.
Gas condensate	A liquid condensed from natural gas that contains a significant amount of C_{5+} components which exhibits the phenomenon of retrograde condensation at reservoir conditions, in other words, as pressure

	decreases, increasing amounts of liquid condenses in the reservoir (down to about 2,000 psia), which results in a significant loss of in situ condensate reserves that may only be partially recovered by revalorization at lower pressures. See: Condensate, Gas condensate reservoir.
Gas condensate reservoir	A reservoir that exhibits producing gas-oil ratios from 2,500 to 50,000 SCF/STB (400 to 10 STB/MMSCF). See: Condensate, Gas condensate, Gas cycling project.
Gas conditioning	The removal of objectionable constituents and addition of desirable constituents.
Gas coming	Caused by a natural expanding gas cap; different from, and should not be confused with, free-gas production caused by a naturally expanding gas cap. See: Coming, Water Coning.
Gas correction factor	A correction factor (GCF) is used to indicate the ratio of flow rates of different gases for a given output voltage from a mass flow controller (MFC). The basis gas is nitrogen (N_2) which, by convention, has $GCFN_2 = 1$. To calculate the mass flow of a gas for a mass flow controller that is calibrated for a different gas, take the gas correction factor of the gas being used and divide that by the GCF of the gas that the mass flow controller was calibrated for.
Example	A mass flow controller is calibrated for Argon ($GCF_{Ar} = 1.39$) and the gas of interest is carbon dioxide ($GCF_{CO2} = 0.70$). The resulting effective gas conversion factor would be (GCF_{CO2}) (GCF_{Ar}) = 0.70 1.39 = 0.50.For a set point of 100 standard cubic centimeters per minute (cm^3/minute), the mass flow controller calibrated for argon will actually be flowing 100 cm^3/minute × 0.5 50 cm^3/minute of CO_2.If the gas is argon, the mass flow would be 100 cm^3/minute. If the gas is N_2, the gas flow would be: 100 cm^3/minute × ($GCFN_2/GCF_{Ar}$) = 100 cm^3/minute × (1.0 1.39) = 72 cm^3/minute
Gas cycling project	A project designed to avoid liquid loss from retrograde condensation can usually be justified for fluids with liquid content higher than about 50 to 100 STB/MMSCF; offshore, the minimum liquid content to justify cycling is approximately 100 STB/MMSCF. See: Condensate, Gas condensate, Gas condensate reservoir.
Gas deliverability	The deliverability of the gas from the storage facility (also referred to as the deliverability rate, withdrawal rate, or withdrawal capacity) is often expressed as a measure of the amount of gas that can be delivered (withdrawn) from a storage facility on a daily basis.
Gas detection analyzer	A device used to detect and measure any gas in the drilling mud as it is circulated to the surface.
Gas holder	A gas-tight receptacle or container in which gas is stored for future use. (i) at approximately constant pressure (low pressure containers) in which case the volume of the container changes; and (ii) in containers of constant volume (usually high-pressure containers) in which case the quantity of gas molecules stored varies with the pressure.
Gas hydrates	Solid, crystalline, wax-like substances composed of water, methane, and usually a small amount of other gases, with the gases being

	trapped in the interstices of a water-ice lattice; formed beneath permafrost and on the ocean floor under conditions of moderately high pressure and at temperatures near the freezing point of water.
Gasification	A partial oxidation process that converts materials such as coal, biomass or plastic waste into a gaseous mixture of carbon monoxide and hydrogen (also known as synthesis gas) by reacting the raw material at high temperatures with controlled amounts of oxygen and/or steam.
Gas in place	Natural gas that is estimated to exist in a reservoir but that has not been produced.
Gas, liquefied petroleum (LPG)	A gas containing certain specific hydrocarbons which are gaseous under normal atmospheric conditions but can be liquefied under moderate pressure at normal temperatures. Propane and butane are the principal examples.
Gas, manufactured	A gas obtained by destructive distillation of coal, or by the thermal decomposition of oil, or by the reaction of steam passing through a bed of heated coal or coke, or catalyst beds. Examples are coal gases, coke oven gases, producer gas, blast furnace gas, blue (water) gas, and carbureted water gas. Btu content varies widely.
Gas-liquid separator	A vertical or horizontal separators in which gas and liquid are separated by means of gravity settling with or without a mist eliminating device.
Gas-oil separator	Commonly a closed cylindrical shell, horizontally mounted with inlets at one end, an outlet at the top for removal of gas, and an outlet at the bottom for removal of oil.
Gasoline	A mixture of volatile, flammable liquid hydrocarbon derivatives produced from crude oil and used universally as a fuel for internal combustion, spark-ignition engines.
Gasoline plant	A plant in which hydrocarbon components common to the gasoline fractions are removed from "wet" natural gas, leaving a "drier" gas.
Gas pipeline	A transmission system for natural gas or other gaseous material. The total system comprises pipes and compressors needed to maintain the flowing pressure of the system.
Gas processing	The preparation of gas for consumer use by removal of the non-methane constituents; synonymous with gas refining; the separation of constituents from natural gas for the purpose of making salable products and also for treating the residue gas to meet required specifications.
Gas refining	The processes by which natural gas petroleum is processed and/or converted by application of physical and chemical processes to form a variety of products. See: Gas processing.
Gas reservoir	A geological formation containing a single gaseous phase. When produced, the surface equipment may or may not contain condensed liquid, depending on the temperature, pressure, and composition of the single reservoir phase; typically, if the reservoir temperature is above the critical temperature of the hydrocarbon system, the reservoir is classified as a natural gas reservoir and on the basis of the phase diagrams and the prevailing reservoir conditions, natural gases can be classified into four categories which are (i) retrograde gas-condensate reservoir, (ii) near-critical gas-condensate reservoir, (iii) wet gas reservoir, and (iv) dry gas reservoir.

Gas sand	A stratum of sand or porous sandstone from which natural gas is obtained.
Gas slippage	An effect that can help increase gas well productivity. Generally, a bigger slippage factor corresponds to bigger well productivity under the same production pressure difference. The larger the slippage factor is, the higher the well productivity. See also: Klinkenberg effect.
Gas-to-liquids (GTL)	Transformation of natural gas into liquid fuel (Fischer Tropsch technology).
Gas well	A well completed for production of natural gas from one or more gas zones or reservoirs.
Geochemistry	Study of the relative and absolute abundances of the elements of the earth and the physical and chemical processes that have produced their observed distributions.
Geological survey	The exploration for natural gas that involves a geological examination of the surface structure of the earth to determine the areas where there is a high probability that a reservoir exists.
Geologist	A scientist who gathers and interprets data pertaining to the rocks of the earth's crust.
Geology	The science of the physical history of the earth and its life, especially as recorded in the rocks of the crust.
Geophone	Equipment used to detect the reflection of seismic waves during a seismic survey.
Geophysical exploration	Measurement of the physical properties of the earth to locate subsurface formations that may contain commercial accumulations of oil, gas, or other minerals; to obtain information for the design of surface structures, or to make other practical applications. The properties most often studied in the oil industry are seismic characteristics, magnetism, and gravity.
Geophysics	The physics of the earth, including meteorology, hydrology, oceanography, seismology, volcanology, magnetism, and radioactivity.
Geothermal	Pertaining to heat within the earth.
Giammarco–Vetrocoke process	A process for hydrogen sulfide and/or carbon dioxide removal from natural gas.
Giant field	Recoverable reserves >500 million barrels of oil equivalents.
Girdler process	Amine (olamine) washing of natural gas to remove acid gases.
GK6 technology	A technology that improves the ethylene production of the furnaces by 35% compared to the original capacity. This increase in capacity is achieved by replacing the existing coil. Moreover, this technology allows the furnace to operate on a range of feedstocks from naphtha to heavy oils, with high selectivity and long on-stream time.
Global warming	An environmental issue that deals with the potential for global climate change due to increased levels of atmospheric greenhouse gases. See: Interglacial period.
Graben	A block of the earth's crust that has slid downward between two faults. Compare *horst*.
Gravel island	A man-made construction of gravel used as a platform to support drilling rigs and oil and gas production equipment.

Gravity	The attraction exerted by the earth's mass on objects at its surface; the weight of a body.
Gravity survey	An exploration method in which an instrument that measures the intensity of the earth's gravity is passed over the surface or through the water. In places where the instrument detects stronger- or weaker-than-normal gravity forces, a geologic structure containing hydrocarbons may exist.
Greenhouse effect	See Global warming.
Greenhouse gases	See Global warming.
Groundwater	Water that seeps through soil and fills pores of underground rock formations; the source of water in springs and wells.
Guard bed	A bed (usually alumina) which serves as a protector of a more expensive bed (e.g., molecular sieve); serves by the act of attrition and may be referred to as an attrition catalyst.
Gusher	An oil well that has come in with such great pressure that the oil jets out of the well like a geyser. In reality, a gusher is a blowout and is extremely wasteful of reservoir fluids and drive energy. In the early days of the oil industry, gushers were common, and many times were the only indication that a large reservoir of oil and gas had been struck. See: Blowout.
Hammerschmidt equation	See: Hydrate prevention.
Hazardous waste	Any gaseous, liquid, or solid waste material that, if improperly managed or disposed of, may pose hazards to human health and the environment. In some cases, the term chemical waste is used interchangeably (often incorrectly) with the term hazardous waste, but chemical waste is always hazardous and the correct use of the terms must be used.
HCR (Hybrid Catenary Riser)	Riser configuration comprised of two flexible sections of flexible pipe (at the top and the bottom) and a rigid section in the middle.
Heating value	The number of British thermal units per cubic foot (Btu/ft^3) of natural gas as determined from tests of fuel samples.
Heat of combustion (energy content)	The amount of energy that is obtained from burning natural gas; measured in British thermal units (Btu).
HCFC's (Hydrochlorofluorocarbons)	Gaseous compounds that meet current environmental standards for minimizing stratospheric ozone depletion.
High flash stocks	Liquids having a closed cup flash point of 55 °C (131 °F) or over (such as heavy fuel oil, lubricating oils, transformer oils etc.); does not include any stock that may be stored at temperatures above or within 8 °C (14.4 °F) of the flash point.
Hole	In drilling operations, the wellbore or borehole.
Holocrystalline	A rock that is completely crystalline.
Horizontal drilling	The method which allows producers to extend horizontal shafts into areas that could not otherwise be reached; especially useful in off shore drilling; categorized as short (extending only 20 to 40 feet from the vertical), medium (300 to 700 feet from the vertical) or long (1,000 to 4,500 feet from vertical) radius.
Horsehead (balanced conventional beam, sucker rod) pump	A common type of cable rod lifting equipment for recovery of oil and gas; so called because of the shape of the counter weight at the end of the beam.

Horst	A block of the earth's crust that has been raised (relatively) between two faults; compare Graben.
Hybrid riser	A riser configuration combining both flexible and rigid pipe technologies. See: Riser.
Hydrate prevention	Hydrate prevention can be achieved using the following methods (i) maintain the gas above the hydrate temperature, (ii) remove the water from the gas so that free water will not condense out, and (iii) add chemicals to the gas to combine with the water. the chemicals mostly used are methanol, glycols and sometimes ammonia. glycols are generally used for continuous processes while methanol is used for emergencies. The Hammerschmidt equation provides the methanol (CH3OH) or ethylene glycol (MEG) concentration in the aqueous phase. With that inhibitor concentration as a basis, the amount of inhibitor in the vapor or liquid hydrocarbon phases is estimated by: $$K = \exp\left(a - \frac{b}{T}\right)$$ In this equation, K is a function of inhibitor used and the phase into which it partitions [e.g., $(K_V)_{MeOH}$ or $(K_L)_{MEG}$], a and b are constants for the two most common inhibitors (MeOH and MEG), and T is the temperature in degrees Rankine (°R). The equation has been reworked by several investigators – for example, another version of the equation is: $d = (K_H(l)/100\ MW_l – MW – MW_l(l)]$ In this equation, K_H is equal to 4,000 for glycols, K_H is equal to 2,335 for methanol, d is the hydrate temperature, l is the weight percent inhibitor in the liquid phase, and MW_l is the molecular weight of the inhibitor.
Hydrates	Gas hydrates represent an operational hazard causing issues related to lost production and safety. The Hammerschmidt equation has been used to help predict the effect of thermodynamic hydrate inhibitors and the equation remains popular for its balance of accuracy with simplicity. See: Gas hydrates.
Hydraulic fracturing	A process through which small fractures are made in impermeable rock by a pressurized combination of water, sand and chemical additives; the small fractures are held open by a proppant (such as grains of sand) thereby allowing the natural gas to flow out of the rock and into the wellbore.
Hydrocarbon	An organic compound composed of hydrogen and carbon only; the density, boiling point, and freezing point increase as the molecular weight increases; the smallest molecules of hydrocarbons are gaseous; the largest are solids. Petroleum is a mixture of many different hydrocarbons.
Hydrocarbon, saturated	A chemical compound of carbon and hydrogen in which all the valence bonds of the carbon atoms are taken up with hydrogen atoms.
Hydrocarbon, unsaturated	A chemical compound of carbon and hydrogen in which not all the valence bonds of the carbon atoms are taken up with hydrogen atoms.
Hydrogen sulfide (H_2S)	A poisonous, corrosive compound consisting of two atoms of hydrogen and one of sulfur; gaseous in the natural state and a constituents of natural gas as well as a constituents of manufactured gas or process gas (refinery gas) and coal gas; the cause of the sourness in natural gas; one of the most dangerous of industrial gases.

Ice scour	The abrasion of material in contact with moving ice in a sea, ocean, or other body of water.
Ideal gas	A gas in which all collisions between atoms or molecules are perfectly elastic and in which there are no intermolecular attractive forces.
IFPEXOL process	A methanol-based process for water removal and hydrocarbon dew point control; also used for acid gas removal.
Igneous rock	A rock mass formed by the solidification of magma within the earth's crust or on its surface. It is classified by chemical composition and grain size. Granite is an igneous rock.
Ignition, automatic	A means which provides for automatic lighting of gas at the burner when the gas valve controlling flow is turned on and will affect relighting if the flame on the burner has been extinguished by means other than closing the gas burner valve.
Ignition, continuous	Ignition by an energy source which is continuously maintained through the time the burner is in service, whether the main burner is firing or not.
Ignition, intermittent	Ignition by an energy source which is continuously maintained through the time the burner is firing.
Ignition, interrupted	Ignition by an energy source which is automatically energized each time the main burner is fired and subsequently is automatically shut off during the firing cycle.
Ignition, manual	Ignition by an energy source which is manually energized and where the fuel to the pilot is lighted automatically when the ignition system is energized.
Ignition temperature	The temperature at which a substance, such as gas, will ignite and continue burning with adequate air supply.
Ilmenite	A weakly magnetic titanium-iron oxide mineral which is iron-black or steel-gray.
Impermeable	Preventing the passage of fluid. A formation may be porous yet impermeable if there is an absence of connecting passages between the voids within it. See: Permeability.
Impure natural gas	Natural gas as delivered from the well and before processing (refining).
Independent producer	A non-integrated company which receives nearly all of its revenues from production at the wellhead; by the IRS definition, a firm is an Independent if the refining capacity is less than 50,000 barrels per day in any given day or their retail sales are less than $5 million for the year.
Indirect vaporizer	A vaporizer in which heat furnished by steam, hot water, the ground, surrounding air or other heating medium is applied to a vaporizing chamber or to tubing, pipe coils, or other heat exchange surface containing the liquid LP gas to be vaporized; the heating of the medium used being at a point remote from the vaporizer.
Interglacial period	A geological interval of warmer global average temperature lasting thousands of years that separates consecutive glacial periods within an ice age; the current Holocene interglacial began at the end of the Pleistocene, approximately 11,700 years ago; alternatively known as interglacial, interglaciation.

Intermediate precision	The within-laboratory variations, such as different days, different analysts, and different equipment. See: Precision.
Interruptible Service contracts	Contracts that allow a distributing party to temporarily suspend delivery of gas to a buyer in order to meet the demands of customers who purchased firm service.
Interstice	A pore space in a reservoir rock.
Interstitial water	See: Reservoir fluid, Reservoir water.
IPB (Integrated Production Bundle)	A patented flexible riser assembly combining multiple functions of production and gas lift, which incorporates both active heating and passive insulation. Used for severe flow assurance requirements.
IRM (inspection, repair, and maintenance)	The term is typically applied to the routine inspection and servicing of offshore installations and subsea infrastructures; the term has also been applied to the routine inspection and servicing of onshore installations and subsea infrastructures.
Iron oxide process (iron sponge process, dry box method)	A process in which the gas is passed through a bed of wood chips impregnated with iron oxide to scavenge hydrogen sulfide and organic sulfur compounds (mercaptans) from natural gas streams.
Isopach map	A map that illustrates thickness variations within a tabular unit, layer or stratum. Isopachs are contour lines of equal thickness over an area and the maps are maps are utilized in hydrographic survey, stratigraphy, sedimentology, structural geology, and petroleum geology.
Isovol map	This type of map offers a simple method of showing the distribution of a geological unit in three-dimensions (3D) thickness of individual formations of reservoir rocks of groups of formations of intervals between unconformities and a normal stratigraphic contact or formation boundary, may be mapped in this manner.
Jackup drilling rig	A mobile bottom-supported offshore drilling structure with columnar or open-truss legs that support the deck and hull. When positioned over the drilling site, the bottoms of the legs penetrate the seafloor. A jackup rig is towed or propelled to a location with its legs up. Once the legs are firmly positioned on the bottom, the deck and hull height are adjusted and leveled. Also called self-elevating drilling unit.
J-Lay:	A vertical lay system for rigid pipes.
Joule–Thomson effect	The cooling which occurs when a compressed gas is allowed to expand in such a way that no external work is done. The effect is approximately 7 degrees Fahrenheit per 100 psi for natural gas.
Joule–Thomson expansion	The throttling effect produced when expanding a gas or vapor from a high pressure to a lower pressure with a corresponding drop in temperature.
Jumper	Short pipe (flexible or rigid) sometimes used to connect a flow line to a subsea structure or two subsea structures located close to one another.
Kelly	The heavy square or hexagonal steel pipe which goes through the rotary table and turns the drill string (also called grief stem).
Kick	In well drilling, a kick is an entry of water, gas, oil, or other formation fluid into the wellbore during drilling; it occurs because the pressure exerted by the column of drilling fluid is not great enough to overcome the pressure exerted by the fluids in the formation drilled.

Kitchen	The underground deposit of organic debris that is eventually converted to petroleum and natural gas.
Klinkenberg effect	This is due to slip flow of gas at pore walls which enhances gas flow when pore sizes are small; occurs because gas does not adhere to the pore walls as liquid does, and the slippage of gases along the pore walls gives rise to an apparent dependence of permeability on pressure; especially important in low-permeable rocks. See also: Gas slippage.
Klinkenberg permeability	The permeability of a sample to a gas varies with the molecular weight of the gas and the applied pressure, as a consequence of gas slippage at the pore wall.
Knock-out pot	A separator used for a bulk separation of gas and liquid, particularly when the liquid volume fraction is high.
Landfill gas	Gas produced during the in situ maturation of landfill materials which can include municipal waste and industrial waste.
Landman	A person in the petroleum industry or natural gas industry who negotiates with landowners for oil and gas leases, options, minerals, and royalties and with producers for joint operations relative to production in a field; also called a leaseman.
Lava	Magma that reaches the surface of the earth.
Lean gas	Natural gas in which methane is the major constituents; raw natural gas with a low concentration of higher molecular weight hydrocarbon derivatives; sometime referred to as dry gas.
Lease condensate	Low-boiling (low molecular weight) liquid hydrocarbons recovered from lease separators or field facilities at associated and non-associated natural gas wells. Mostly pentane derivatives and higher molecular weight hydrocarbon derivative; typically enters the stream after production.
Lease fuel	Natural gas used in well, field, and lease operations, such as gas used in drilling operations, heaters, dehydrators, and field compressors.
Lease operations	Any well, lease, or field operations related to the exploration for or production of natural gas prior to delivery for processing or transportation out of the field gas used in lease operations includes usage such as for drilling operations, heaters, dehydrator units, and field compressors used for gas lift.
Lease separation facility	A facility installed at the surface for the purpose of (i) separating gases from produced crude oil and water at the temperature and pressure conditions set by the separator and/or (ii) separating gases from that portion of the produced natural gas stream that liquefies at the temperature and pressure conditions set by the separator.
Lease separator	A facility installed at the surface for the purpose of separating the full well stream volume into two or three parts at the temperature and pressure conditions set by the separator; for gas wells, these parts include produced natural gas, lease condensate, and water.
Lenticular sand reservoir	A reservoir that has tapering edges and limited extension which has been deposited in low energy sedimentary environment. Because of their limited extension this type pf reservoir holds less reserves than a blanket sand reservoir and often has high pressure. At the beginning

	these reservoir produce natural gas (and/or crude oil) at a high rate but quickly die down and reserves become exhausted.
Limestone	A sedimentary rock rich in calcium carbonate that sometimes serves as a reservoir rock for petroleum.
Limit of quantitation (LOQ)	The level above which quantitative results may be determined with acceptable accuracy and precision.
Linearity	The ability of the method to elicit results that are directly proportional to analyte concentration within a given range.
Liquefied natural gas	The liquid form of natural gas; Natural gas (primarily methane) that has been liquefied by reducing its temperature to −260 °F at atmospheric pressure.
Liquefied petroleum gas (LPG)	The term applied to certain specific hydrocarbons and their mixtures, which exist in the gaseous state under atmospheric ambient conditions but can be converted to the liquid state under conditions of moderate pressure at ambient temperature.
Liquids, natural gas	Those liquid hydrocarbon mixtures which are gaseous at reservoir temperatures and pressures but are recoverable by condensation or absorption; gas condensate, natural gasoline, and liquefied petroleum gases are often included in this category.
Lithology	The study of rocks; important for exploration and drilling crews to understand lithology as it relates to the production of gas and oil.
LNG (liquefied natural gas)	Natural gas, liquefied through the reduction of its temperature to −162 °C (−260 °F), thus reducing its volume by 600 times, allowing its transport by LNG tanker.
Local Distribution Company	A retail gas distribution company that delivers natural gas to end users.
LO-CAT process	A wet oxidation process.
Log	A systematic recording of data, such as a driller's log, mud log, electrical well log, or radioactivity log. Many different logs are run in wells to discern various characteristics of downhole formation.
Logging	Lowering of different types of measuring instruments into the wellbore and gathering and recording data on *porosity*, *permeability*, and types of fluids present near the current well after which the data are used to construct subsurface maps of a region to aid in further exploration.
Lower explosive limit	The lower percent by volume of the gas vapor in air at which the gas will explode or inflame. See also: Upper explosive limit.
Lower flammability limit	The minimum concentration by volume of a combustible substance that is capable of propagating a flame under specified conditions.
Low-flash stocks	Liquids having a closed cup flash point under 55 °C (131 °F) such as gasoline, kerosene, jet fuels, some heating oils, diesel fuels and any other stock that may be stored at temperatures above or within 8 °C (14.4 °F) of the flash point.
Macrofauna	Benthic invertebrates that live on or in the sediment and that are retained on a mesh with an aperture of 0.5 mm.
Magma	The hot fluid matter within the earth's crust that is capable of intrusion or extrusion and that produces igneous rock when cooled.
Magnetic survey	An exploration method in which an instrument that measures the intensity of the natural magnetic forces existing in the earth's subsurface is passed over the surface or through the water. The instrumentation detects deviations in magnetic forces, and such

	deviations may indicate the existence of underground formations that favor the entrapment of hydrocarbons.
Magnetometer	A device to measure small changes in the earth's magnetic field at the surface, which indicates what kind of rock formations might be present underground.
Major field	Recoverable reserves >100 million barrels of oil equivalents.
Manufactured gas	Gas obtained by destructive distillation of coal or by the thermal decomposition of oil, or by the reaction of steam passing through a bed of heated coal or coke; examples are coal gases, coke oven gases, producer gas, blast furnace gas, blue (water) gas, carbureted water gas; the Btu content varies widely.
Marsh gas	See Natural gas.
Mass concentration	The concentration expressed in terms of mass of substance per unit volume (g substance/m^3 gas volume).
Material balance equations	Equations that are used for calculating volumes and densities when two or more insoluble materials are mixed together.
Mcf (Thousand Cubic Feet)	One thousand cubic feet; a unit of measure that is more commonly used in the low volume sectors of the gas industry, such as stripper well production. See also: Btu, Bcf, Tcf, Quad.
MDEA	Methyldiethanolamine.
MEA	Ethanolamine (often referred to as monoethanolamine.
Median	A statistical measure of the midmost value, such that half the values in a set are greater and half are less than the median.
Meiofauna	Benthic invertebrates that live in the sediment and that are retained on a mesh with an aperture of 0.062 mm.
MER	See Most Efficient Recovery rate.
Mercaptan	Typically, a hydrocarbon group (usually a methane, ethane, or propane) with a hydro-sulfur group (-SH) substituted on a terminal carbon atom.
Methane	A light, gaseous, colorless and naturally odorless flammable paraffinic hydrocarbon, CH_4, that has a boiling point of −32 °C (−25 °F) (and is the chief component of natural gas and an important basic hydrocarbon for petrochemical manufacture); commonly (often incorrectly) known as natural gas; burns efficiently without many by products.
Methane Separation	Cryogenic processing and absorption methods are some of the ways to separate methane from natural gas liquids. The cryogenic method is better at extraction of the lighter liquids, such as ethane, than is the alternative absorption method and consists of lowering the temperature of the gas stream to around −120 °F. The absorption method, on the other hand, uses a "lean" absorbing oil to separate the methane from the natural gas liquids. While the gas stream is passed through an absorption tower, the absorption oil soaks up a large amount of the natural gas liquids. The enriched absorption oil, now containing NGLs, exits the tower at the bottom. The enriched oil is fed into distillers where the blend is heated to above the boiling point of the natural gas liquids, while the oil remains fluid. The oil is recycled while the natural gas liquids are cooled and directed to a fractionator tower. Another absorption method that is often used is the

refrigerated oil absorption method where the lean oil is chilled rather than heated, a feature that enhances recovery rates somewhat.

Methanogens	Methane-producing microorganisms.
Method detection limit (MDL)	The minimum concentration of a substance than can be measured and reported with 99% confidence that the analyte concentration is greater than zero; determined from analysis of a sample in a given matrix containing the analyte. See: Detection limit.
Methyl Mercaptan and ethyl mercaptan	Methyl mercaptan (CH_3SH, also called methyl thiol) and ethyl mercaptan (CH_3CH_2SH, also called ethyl thiol) are the two mercaptan compounds most commonly found in natural gas.
MFO (mixed function oxygenases)	An enzyme system, located within the cells, which can assist in the metabolism and excretion of contaminants.
Microgabbro	Gabbro with finer grain crystals (<1 mm).
Micrometer	Also commonly known as the micron is 1×10^{-6} meter which is, one millionth of a meter (or one thousandth of a millimeter, 0.001 mm, or approximately 1 inch × 0.000039).
Migration	The movement of oil, gas or water through porous and permeable rock.
Mineral rights	The rights of ownership, conveyed by deed, of gas, oil, and other minerals beneath the surface of the earth.
Mist Extractor	A device installed in the top of scrubbers, separators, tray or packed vessels, etc. to remove liquid droplets entrained in a flowing gas stream.
Mixed gas	A fuel gas in which natural or liquefied petroleum gas is mixed with manufactured gas to give a product of better utility and higher heat content or Btu value.
MMBtu	A thermal unit of energy equal to 1,000,000 Btu, that is, the equivalent of 1,000 cubic feet of gas having a heating content of 1,000 Btu per cubic foot, as provided by contract measurement terms.
MMcf	A million cubic feet.
MMSCFD	Million standard cubic feet per day.
Mobile offshore drilling unit	A drilling rig that is used exclusively to drill offshore exploration and development wells and that floats upon the surface of the water when being used.
MODU (*plural*	MODUs): Mobile offshore drilling unit often used in conjunction with semi-submersibles and floating, production, storage, and offloading units, which do not have drilling rigs.
Moonpool	Opening in a vessel or a platform deck through which drilling, subsea pipe-laying, or construction is conducted. Several vessels are equipped with a moonpool allowing the use of the VLS for flexible pipe and Reel-Lay or J-Lay systems.
Most efficient recovery rate (MER)	The rate at which the greatest amount of natural gas may be extracted without harming the formation itself.
MSCC (Milli Second Catalytic Cracker)	A catalytic cracking unit of FCC type (FCC stands for *fluid catalytic cracking*).
Mud	The liquid circulated through the wellbore during rotary drilling and workover operations. In addition to its function of bringing cuttings to the surface, drilling mud cools and lubricates the bit and the drill stem, protects against blowouts by holding back subsurface pressures, and deposits a mud cake on the wall of the borehole to

	prevent loss of fluids to the formation; originally a suspension of earth solids (especially clay) in water, the mud used in modern drilling operations is a more complex, three-phase mixture of liquids, reactive solids, and inert solids – the liquid phase may be fresh water, diesel oil, or crude oil and may contain one or more conditioners. See: Drilling fluid.
Mud line	The sea bed, unless otherwise specified by the driller.
Mudstone	A fine-grained clastic sedimentary rock that is not laminated or fissile that is distinguished from shale by lack of fissility; the term mudstone is also used to describe carbonate rocks that are composed predominantly of carbonate mud; most definitions also include a requirement that the rock contain significant amounts of both silt- and clay-sized grains.
Multiple Completions	The result of drilling several different depths from a single well to increase the rate of production or the amount of recoverable gas.
Naphtha	An arbitrarily defined crude oil fraction containing primarily aliphatic (linear) hydrocarbons with boiling points ranging from 30 to 250 °C (86 to 482 °F); principal uses are for (i) gasoline blend stock, (ii) solvents, (iii) paint thinners, (iv) and as a feedstock for the production of organic chemicals, and (v) as a feedstock for the production of synthetic natural gas.
Native gas	Gas in place at the time that a reservoir was converted to use as an underground storage reservoir in contrast to injected gas volumes.
Natural gas	Also called *marsh gas*, *swamp gas*;, and *landfill gas*; a gaseous fossil fuel that is found in oil fields and natural gas fields, and in coal bedsa mixture of hydrocarbon and small quantities of non-hydrocarbons that exists either in the gaseous phase or is in solution in crude oil in natural underground reservoirs, and which is gaseous at atmospheric conditions of pressure and temperature.
Natural Gas Act	Passed in 1938 and give the Federal Power Commission (now the Federal Energy Regulatory Commission or FERC) jurisdiction over companies engaged in interstate sale or transportation of natural gas.
Natural gas co-firing:	The injection of natural gas with pulverized coal or oil into the primary combustion zone of a boiler.
Natural gas hydrates	Solid, crystalline, wax-like substances composed of water, methane, and usually a small amount of other gases, with the gases being trapped in the interstices of a water-ice lattice; formed beneath permafrost and on the ocean floor under conditions of moderately high pressure and at temperatures near the freezing point of water.
Natural gas in place	Natural gas that is estimated to exist in a reservoir but that has not been produced.
Natural gas liquids (NGLs)	A hydrocarbon liquid stream *gas condensate*); higher molecular weight hydrocarbons that exist in the reservoir as constituents of natural gas, but which are recovered as liquids in separators, field facilities or gas-processing plants.
Natural gasoline	A term used in the gas processing industry to refer to a mixture of liquid hydrocarbons (mostly pentane, including iso-pentane, and higher molecular weight hydrocarbons) extracted from natural gas.

Natural gas plant liquids	Hydrocarbons in natural gas that are separated as liquids at natural gas processing plants, fractionating and cycling plants, and in some instances, field facilities; the products obtained include liquefied petroleum gases (ethane, propane, and butanes), pentanes plus, and iso-pentane.
Natural Gas Policy Act of 1978	one of the first efforts to deregulate the gas industry and to determine the price of natural gas as dictated by market forces, rather than regulation.
Natural gas processing plant	A facility designed to recover natural gas liquids from a stream of natural gas that may or may not have passed through lease separators and/or field separation facilities; the facility controls the quality of the natural gas to be marketed.
Natural gas resource base	An estimate of the amount of natural gas available, based on the combination of proved reserves, and those additional volumes that have not yet been discovered, but are estimated to be "discoverable" given current technology and economics.
Natural gas shrinkage, natural gas	The reduction in volume of wet natural gas due to the extraction of some of its constituents, such as hydrocarbon products, hydrogen sulfide, carbon dioxide, nitrogen, helium, and water vapor.
Natural gas vehicle (NGV)	A car, bus, or truck that is powered by a natural gas, either in compressed or liquefied form, rather than the traditional gasoline or diesel fuel.
Near-critical gas-condensate reservoir	A gas reservoir in which the reservoir temperature is near the critical temperature and the hydrocarbon mixture is classified as a near-critical gas-condensate.
Nepheline magnetite	A sodium potassium aluminosilicate and magnetite – one of the two common naturally occurring iron oxides (chemical formula Fe_3O_4) and a member of the spinel group of minerals.
NES (National Energy Strategy)	A 1991 federal proposal that focused on national security, conservation, and regulatory reform, with options that encourage natural gas use.
Nitrogen (N_2)	An inert constituents of natural gas that does not contribute to the gas in any way; gases with a high nitrogen content often have high helium content.
Nitrogen extraction	Once the hydrogen sulfide and carbon dioxide are processed to acceptable levels, the stream is routed to a nitrogen rejection unit it is routed through a series of passes through a column and a brazed aluminum plate fin heat exchanger. Helium, if any, can be extracted from the gas stream through membrane diffusion in a pressure swing adsorption (PSA) unit.
Non-associated natural gas	Sometimes called *gas well gas*; gas produced from geological formations that typically do not contain much, if any, crude oil, or higher boiling hydrocarbon derivatives (*gas liquids*) than methane; can contain non-hydrocarbon gases such as carbon dioxide and hydrogen sulfide.
Non-combustible:	Material incapable of igniting or supporting combustion.
Non-hydrocarbon gases	Typical non-hydrocarbon gases that may be present in reservoir natural gas, such as carbon dioxide, helium, hydrogen sulfide, and nitrogen.

Nonporous	Containing no interstices; having no pores and therefore unable to hold fluids.
Non-standard method	A method that is not taken from authoritative and validated sources; includes methods from scientific journals and unpublished laboratory-developed methods.
Norite	A coarse-grained basic igneous rock dominated by essential calcic plagioclase and orthopyroxene. Norites also can contain up to 50% clinopyroxene and can be considered orthopyroxene dominated gabbro.
NO_x (nitrogen oxides)	Produced during combustion; precursors to acid deposition (acid rain).
NPC (National Petroleum Council)	An advisory body of appointed members whose purpose is to advise the Secretary of Energy.
Off Peak Period	The time during a day, week, month or year when gas use on a particular system is not at its maximum.
Offshore	That geographic area that lies seaward of the coastline. In general, the term "coastline" means the line of ordinary low water along that portion of the coast that is in direct contact with the open sea or the line marking the seaward limit of inland waters.
Offshore drilling	Drilling for oil or gas in an ocean, gulf, or sea. A drilling unit for offshore operations may be a mobile floating vessel with a ship or barge hull, a semisubmersible or submersible base, a self-propelled or towed structure with jacking legs (jackup drilling rig), or a permanent structure used as a production platform when drilling is completed. In general, wildcat wells are drilled from mobile floating vessels or from jackups, while development wells are drilled from platforms or jackups.
Offshore oil and gas installation	Subsea or surface (platform) oil and gas drilling facilities.
Offshore production platform	An immobile offshore structure from which wells are produced.
Offshore rig	Any of various types of drilling structures designed for use in drilling wells in oceans, seas, bays, gulfs, and so forth. Offshore rigs include platforms, jackup drilling rigs, semisubmersible drilling rigs, submersible drilling rigs, and drill ships.
Oikocrysts	Small, randomly orientated, crystals are enclosed within larger crystals of another mineral; the term is most commonly applied to igneous rock textures. The smaller enclosed crystals are known as chadacrysts, whilst the larger crystals are known as oikocrysts.
Oil	S simple or complex liquid mixture of hydrocarbons that can be refined to yield gasoline, kerosene, diesel fuel, and various other products.
Oil field (oilfield)	A field producing oil and gas is termed an oil field when the petroleum contained within has a gas-oil ratio (GOR) of less than 20,000 cubic feet per barrel. if the gas-oil ratio >20,000 cubic feet per barrel, the field is called a gas field.
Oil in place	Crude oil that is estimated to exist in a reservoir but that has not been produced.
Oil patch	(slang) The oilfield.
Oil pool	A loose term for an underground reservoir where oil occurs. Oil is actually found in the pores of rocks, not in a pool.

Oil seep	A surface location where oil appears, the oil having permeated its subsurface boundaries and accumulated in small pools or rivulets. Also called oil spring.
Oil shale	A shale containing hydrocarbons that cannot be recovered by an ordinary oil well but that can be extracted by mining and processing.
Oil slick	A film of oil floating on water; considered a pollutant.
Oil spill *n*	A quantity of oil that has leaked or fallen onto the ground or onto the surface of a body of water.
Oil well	A well from which oil is obtained.
Oil wet	Reservoir rock that which is preferentially in contact with crude oil, which occupies the small pores and contacts the majority of the rock surface.
Oil zone	A formation or horizon of a well from which oil may be produced. The oil zone is usually immediately under the gas zone and on top of the water zone if all three fluids are present and segregated.
Olamine process	A process that used an amine derivative (an olamine) to remove acid gas form natural gas streams.
Olamines	Compounds such as ethanolamine (monoethanolamine, MEA), diethanolamine (DEA), triethanolamine (TEA), methyldiethanolamine (MDEA), diisopropanolamine (DIPA), and diglycolamine (DGA) that are widely used in gas processing.
Olefins	Family of molecules including in particular ethylene and propylene, which constitutes the raw material allowing for the manufacture of many plastics.
Oleophilic	Preferentially oil-wet.
Olivine	A name for a series between two end members, fayalite and forsterite; fayalite is the iron-rich member (Fe_2SiO_4.) whereas forsterite is the magnesium-rich member (Mg_2SiO_4).
Onshore oil and gas installation	Onshore oil and gas exploration/production.
Order 636	The Federal Energy Regulatory Commission's 1992 order that required pipelines to unbundle their transportation, sales and storage services.
Organic compounds	Chemical compounds that contain carbon atoms, either in straight chains or in rings, and hydrogen atoms. They may also contain oxygen, nitrogen, or other atoms.
Organic rock	Rock materials produced by plant or animal life (coal, petroleum, limestone, and so on).
Orthopyroxene	An essential constituent of various types of igneous rocks and metamorphic rocks.
Outcrop	Part of a formation exposed at the earth's surface. *V*: to appear on the earth's surface (as a rock).
Overburden pressure	The pressure exerted by the total weight of overlying formations above the point of interest.
PAH	polycyclic aromatic hydrocarbons, which contain more than one fused benzene ring. See: PNA.
Paraffin hydrocarbons	Saturated hydrocarbon compounds with the general formula C_nH_{2n+2} containing only single bonds and carbon and hydrogen only; sometimes referred to as alkanes or natural gas liquids.
Paragenesis	A petrologic concept meaning an equilibrium sequence of mineral phases that is used in studies of igneous and metamorphic rock

	genesis and in studies of the hydrothermal deposition of ore minerals and the rock alteration associated with ore mineral deposits.
Part concentration	The concentration expressed as number of particles of substance per a certain number of particles.
Peak load requirements (peak load storage)	The design to have high deliverability for short periods of time during which the natural gas can be withdrawn from storage quickly as the need arises.
Peak Shaving	Use of natural gas from storage to supplement the normal amounts delivered to customers during peak-use periods.
Peak Use Period	The period of time when gas use on a particular system is at its maximum.
Peat	An organic material that forms by the partial decomposition and disintegration of vegetation in tropical swamps and other wet, humid areas. It is believed to be the precursor of coal.
Pegmatite	A holocrystalline, intrusive igneous rock composed of interlocking phaneritic crystals usually larger than 2.5 cm in size; such rocks are referred to as *pegmatitic*.
Permeability (μ)	A measure of the ease that a fluid can pass through a section of rock through the connecting pore spaces of rock or cement – the unit of measurement is the millidarcy; fluid conductivity of a porous medium; the ability of a fluid to flow within the interconnected pore network of a porous medium; a factor that quantifies how hard or how easy it is for the fluid to flow through the reservoir to the oil producing well.
Permeability (μ_{abs}) (absolute. μ	The permeability of that rock to water.
Permeability (μ_{eff}) (effective, μ	The permeability of a reservoir rock to any one fluid in the presence of others; dependent upon the values of fluid saturations.
Permeability (relative, μ_{rel})	The ratio of the material permeability to the permeability of free space (or vacuum): $\mu_{rel} = \mu / \mu_o$
Permeable rock	A porous rock formation in which the individual pore spaces are connected, allowing fluids to flow through the formation.
Petroleum	Crude oil; a flammable naturally occurring liquid that may vary from almost colorless to black, occurs in many places in the upper strata of the earth; a complex mixture of hydrocarbons with small amounts of other substances, and is prepared for use as naphtha, gasoline, and other products by various refining processes.
Petroleum geology	The study of oil and gas-bearing rock formations. It deals with the origin, occurrence, movement, and accumulation of hydrocarbon fuels.
Petroleum reservoir	A rock formation that holds oil and gas.
Petroleum rock	Sandstone, limestone, dolomite, fractured shale, and other porous rock formations where accumulations of oil and gas may be found.
Petroleum window	The conditions of temperature and pressure under which petroleum will form; also called oil window.
pH	A measure of the acidity or alkalinity of a solution.
Phase envelope (two-phase region)	The region enclosed by the bubble-point curve and the dew-point curve wherein gas and liquid coexist in equilibrium, is identified as the phase envelope of the hydrocarbon system.
Phase rule	The Gibbs phase rule is a very important tool in studying phase behavior and the rule applies to systems at equilibrium. The

equilibrium between phases can be reached by isolating a system at constant temperature and pressure for long periods of time or by intimate mixing of the parts of a system. At equilibrium, the composition of each phase is uniform throughout, and no net transfer of components occurs between phases when they are permitted to remain in contact over long periods of time. Many natural gas/liquid systems in industry approach equilibrium, but, even for systems that are not in a state of equilibrium, the laws governing equilibrium impose limiting conditions. When phases are in equilibrium with each other, they are said to be saturated.

Pig
A robotic agent that is used to inspect pipeline interior walls for corrosion and defects, measure pipeline interior diameters, remove accumulated debris and for other specialty tasks.

Pip (pipe-in-pipe):
Steel pipes assembly consisting of a standard production pipe surrounded by a so-called carrier pipe. The gap between the carrier and production pipes is filled with an insulation material. As the insulation is protected from the external pressure by the carrier pipe, a high thermal performance material can be used.

Pipeline (natural gas)
A continuous pipe conduit, complete with such equipment as valves, compressor stations, communications systems, and meters for transporting natural and/or supplemental gas from one point to another, usually from a point in or beyond the producing field or processing plant to another pipeline or to points of utilization. Also refers to a company operating such facilities.

Plagioclase
A large group of dark, often phaneritic (coarse-grained), mafic intrusive igneous rocks, chemically equivalent to plutonic basalt; formed when molten magma is trapped beneath the surface of the Earth and slowly cools into a holocrystalline mass.

Plate tectonics
Movement of great crustal plates of the earth on slow currents in the *plastic* mantle.

Platform
The structure that supports production and drilling operations. The types of offshore platforms can be either floating or fixed, depending on the location, water depth, climate, and the facility's size.

Play
The extent of a petroleum-bearing formation; the activities associated with petroleum development in an area.

Plutonic rock
An intrusive igneous rock that is crystallized from magma slowly cooling below the surface of the Earth.

PNA
Polynuclear aromatic hydrocarbons, which contain more than one fused benzene ring. See: PAH.

Pollutant
A substance present in a particular location (ecosystem) – usually it is not indigenous to the location or is present in a concentration greater than the concentration that occurs naturally. The substance is often the product of human activity and has a detrimental effect on the environment, in part or in toto. Pollutants can also be subdivided into two classes: (i) primary pollutants and (ii) and secondary pollutants. See also: Primary pollutant, Secondary pollutant.

Polypropylene
Due to its exceptional shock resistance properties, polypropylene is a plastic material used in a wide range of industries including automobile parts, household goods, fibers and films.

Pool	A reservoir or group of reservoirs. The term is a misnomer in that hydrocarbons seldom exist in pools, but, rather, in the pores of rock. *V*: to combine small or irregular tracts into a unit large enough to meet state spacing regulations for drilling.
Porosity	The condition of being porous (such as a rock formation); also, the ratio of the volume of empty space to the volume of solid rock in a formation, indicating how much fluid rock can hold; the spaces between grains of sediment in sedimentary rock; a measure, expressed in percent, of the void space in the rock that is filled with the reservoir fluid.
Potential	The maximum volume of oil or gas that a well is capable of producing, calculated from well test data.
Precision	The agreement between a set of replicate measurements without assumption of knowledge of the true value. See: Intermediate precision, Repeatability, Reproducibility.
Pressure, absolute (psia)	Pressure in excess of a perfect vacuum; absolute pressure is obtained by algebraically adding gauge pressure to atmosphere pressure. Pressures reported in "atmospheres" are understood to be absolute. Absolute pressure must be used in equations of state and in all gas-law calculations. Gauge pressures below atmospheric pressure are referred to under the term called vacuum.
Primary pollutant	A pollutant that is emitted directly from the source. In terms of atmospheric pollutants from natural gas, examples are carbon oxides, sulfur dioxide, and nitrogen oxides from combustion operations. See also: Pollutant, Secondary pollutant.
Producer	The company generally involved in exploration, drilling, and refining of natural gas.
Producer gas	A gas mixture containing carbon monoxide (CO), hydrogen (H_2), carbon dioxide (CO_2) and nitrogen (N_2); also a generic term referring to wood gas, town gas or syngas. In the process, air is passed over the red-hot fuel and carbon monoxide is produced in an exothermic reaction $2C + O_2 \rightarrow 2CO$. The nitrogen in the air remains unchanged and dilutes the gas, so it has a low calorific value. The gas may be used to power gas turbines which are suited to fuels of low calorific value.
Producing sand	A rock stratum that contains recoverable oil or gas.
Producing zone	The interval of rock actually producing oil or gas.
Production rate	The rate of production of oil and/or gas from a well; usually given in barrels per day (bbls/day) for oil or standard cubic feet ($scft^3$/day) for gas.
Propping agents	Sand, glass beads, epoxy, or silica sand that serve to prop open the newly widened fissures in the formation.
Proved reserves of crude oil	According to API standard definitions, proved reserves of crude oil as of December 31 of any given year are the estimated quantities of all liquids statistically defined as crude oil that geological and engineering data demonstrate with reasonable certainty to be recoverable in future years from known reservoirs under existing economic and operating conditions.

Proved resources	Part of the resource base that includes the working inventory of natural gas; volumes that have already been discovered and are readily available for production and delivery.
Proved amount in place	The volume originally occurring in known natural reservoirs which has been carefully measured and assessed as exploitable under present and expected local economic conditions with existing available technology.
Proved recoverable reserves	The volume within the proved amount in place that can be recovered in the future under present and expected local economic conditions with existing available technology.
PSI (Pounds Per Square Inch)	Pressure measured with respect to that of the atmosphere.
PUHCA (Public Utility Holding Company Act of 1935)	Amended by EPACT, to allow power generation by independent power producers (IPPs) without restrictions on corporate structure.
Pyrolysis	The chemical decomposition of organic materials by heating in the absence of oxygen or any other reagents, except possibly steam. Pyrolysis is somewhat endothermic and the products can be gases, liquids, and/or solids (e.g., coke and volatiles produced by coking coal).
Pyroxene	A group of rock-forming inosilicate minerals found in many igneous and metamorphic rocks; variable composition, among which calcium-, magnesium-, and iron-rich varieties predominate.
Quad	An abbreviation for a quadrillion (1,000,000,000,000,000) Btu; roughly equivalent to one trillion (1,000,000,000,000) cubic feet, or 1 Tcf. See also: Bcf, Mcf, Tcf.
Quartzgabbro	A coarse-grained igneous rock dominated by plagioclase (50% v/v), orthopyroxene (15%), clinopyroxene (25%) and quartz (<5%) with minor biotite and accessory apatite.
Quenching	Cooling of a hot vapor by mixing it with another fluid or by partially vaporizing another liquid.
Range	The interval between the upper and lower concentration of analyte in sample for which it has been demonstrated that the analytical procedure has an acceptable level of accuracy, precision, and linearity.
Raw natural gas	Impure natural gas as delivered from the well and before processing (refining).
Recovery rate	The rate at which natural gas can be removed from a reservoir.
Rectisol process	A process that used a physical (non-reactive) solvent for gas cleaning.
Redox process	A sulfur recovery process that involves liquid-phase oxidation; uses a dilute aqueous solution of iron or vanadium to remove hydrogen sulfide selectively by chemical absorption from sour-gas streams; can be used on relatively small or dilute hydrogen sulfide stream to recover sulfur from the acid gas stream or, in some cases, they can be used in place of an acid gas removal process.
Reeled pipe	An installation method based on the onshore assembly of long sections of rigid steel pipeline, approximately 0.5 mile long, which are welded together as they are spooled onto a vessel-mounted reel for transit and subsequent cost-effective unreeling onto the seabed. Minimum welding is done at sea.

Refinery gas	Non-condensable gas collected in petroleum refineries.
Regulations	The laws by which environmental emissions are controlled. See also: Environmental control, Environmental effects, Environmental guidance.
Reid vapor pressure (RVP)	The pressure of the vapor in equilibrium with liquid at 37.8 °C (100 °F).
Remaining reserves	Recoverable volumes of crude oil and natural gas liquids that were originally present and have not yet been produced.
Repeatability	The precision under the same operating conditions over a short period of time. See Precision.
Reproducibility	The precision between laboratories. See: Precision.
Reserve growth (field growth)	The increases of estimated petroleum volume that commonly occur as oil and gas fields are developed and produced.
Reserves	The amount of a resource available for recovery and/or production; the recoverable amount is usually tied to economic aspects of production.
Reservoir	A subsurface, porous, permeable, or naturally fractured rock body in which oil or gas are stored. Most reservoir rocks are limestone, dolomites, sandstones, or a combination of these. The four basic types of hydrocarbon reservoirs are oil, volatile oil, dry gas, and gas condensate. An oil reservoir generally contains three fluids – gas, oil, and water-with oil the dominant product. In the typical oil reservoir, these fluids become vertically segregated because of their different densities. Gas, the lightest, occupies the upper part of the reservoir rocks; water, the lower part; and oil, the intermediate section. In addition to its occurrence as a cap or in solution, gas may accumulate independently of the oil; if so, the reservoir is called as gas reservoir. Associated with the gas, in most instances, are salt water and some oil. Volatile oil reservoirs are exceptional in that during early production they are mostly productive of light oil plus gas, but, as depletion occurs, production can become almost totally completely gas. Volatile oils are good candidates for pressure maintenance, which can result in increased reserves. In the typical dry gas reservoir natural gas exists only as a gas and production is only gas plus fresh water that condenses from the flow stream reservoir. In a gas condensate reservoir, the hydrocarbons may exist as a gas, but, when brought to the surface, some of the heavier hydrocarbons condense and become a liquid.
Reservoir rock	A permeable rock that may contain oil or gas in appreciable quantity and through which petroleum may migrate.
Residue gas	Natural gas from which the higher molecular weight hydrocarbons have been extracted; mostly methane.
Resource	A concentration of naturally occurring solid, liquid, or gaseous hydrocarbons in or on the crust of the Earth, some of which is currently or potentially economically extractable.
Retrograde condensation	Occurs when gas is compressed beyond the point of condensation with the effect that the liquid evaporates again.
Retrograde dewpoint pressure	At a given temperature, the pressure when a gas condenses, and the pressure drops below the dewpoint.
Rich gas	A gaseous stream is traditionally rich in natural gas liquids (NGLs). See: Natural gas liquids.

Reburning	An effective and economic means of reducing NO_x emissions from all types of industrial and electric utility boilers. Reproducibility (of an analytical method): A measure of the repeatability of the data using an analytical method.
Reserve Additions	Volumes of the resource base that are continuously moved from the resource category to the proved resources category.
Reservoir	A geological formation that retains or *traps* the gas; a section of porous rock (beneath an impervious layer of rock) that has collected high concentrations of natural gas in the minute void spaces that weave through the rock. The gas, along with some water is trapped beneath the impervious rock.
Reservoir characterization	Determination of the physical properties of a reservoir (such as porosity, permeability, fluid saturation, etc.) and changes in the distribution of these properties throughout the reservoir.
Reservoir deliverability	The oil or gas production rate achievable from a reservoir at a given bottomhole pressure; a major factor affecting well deliverability which deliverability determines types of completion and the selection of an artificial lift method.
Reservoir energy	The underground pressure in a reservoir that will push the petroleum and natural gas up the wellbore to the surface.
Reservoir fluid	An all-encompassing term that related to the gas, liquid, and solid constituents of the reservoir. Reservoir fluids are segregated into phases according to the density of the fluid. The specific gravity of the gas (γg) is the ratio of the density of natural gas to the density of air while the specific gravity of the oil (γo) is the ratio of the density of oil to the density of water. Since the density of gas is less than that of the oil and both are less than water, the gas rests at the top of the reservoir, followed by oil and finally water. Usually the interface between two reservoir fluid phases is horizontal and is called a *contact*. Between gas and oil is a gas-oil contact, between oil and water is an oil-water contact, and between gas and water is a gas-water contact if no oil phase is present. A small volume of water (*connate* or *interstitial* water) remains in the oil and gas zones of the reservoir.
Reservoir water	The water found in petroleum reservoirs is usually a *brine* consisting mostly of sodium chloride (NaCl) in quantities from 10 to 350 ppt (‰); seawater has about 35 ppt. Other compounds (electrolytes) found in reservoir brines include calcium (Ca), magnesium (Mg), sulfate (SO_4), bicarbonate (HCO_3), iodide (I), and bromide (Br); sometimes referred as *brine* or *connate water or interstitial water*. See: Reservoir fluid.
Resources	Concentrations of naturally occurring liquid or gaseous hydrocarbons in the earth's crust, some part of which are currently or potentially economically extractable.
Retrograde gas condensate reservoir	A reservoir in which the temperature T lies between the critical temperature and cricondentherm of the reservoir fluid, the reservoir is classified as a retrograde gas condensate reservoir. This category of gas reservoir is a unique type of hydrocarbon accumulation in that the special thermodynamic behavior of the reservoir fluid is the controlling factor in the development and the depletion process of the reservoir. When the pressure is decreased on these mixtures, instead

	of expanding (if a gas) or vaporizing (if a liquid) as might be expected, they vaporize instead of condensing.
Rich gas	Sometime referred to as wet gas; raw natural gas with relatively high concentration of higher molecular weight hydrocarbon derivatives.
Rig	The drilling equipment used to drill the well that can either be installed on a platform or a MODU.
Riser	A pipe or assembly of pipes used to transfer produced fluids from the seabed to the surface facilities or to transfer injection fluids, control fluids or lift gas from the surface facilities and the seabed.
Risk analysis	The activity of assigning probabilities to the expected outcomes of drilling venture.
Rod pumping	The use of a lifting (pumping) method to recover oil from a reservoir.
Rotary bit	The cutting tool attached to the lower end of the drill pipe of a rotary drilling rig; the bit does the actual drilling of the hole through the formation.
Rotary drilling	A method for drilling wells using a cutting bit attached to a revolving drill pipe.
ROV (remotely operated vehicle)	An unmanned subsea vehicle remotely controlled from a vessel or an offshore platform. It is equipped with manipulator arms that enable it to perform simple operations.
R/P (reserves/production) ratio	Calculated by dividing proved recoverable reserves by production (gross less re-injected) in a given year.
RSCR (Reeled Steel Catenary Riser)	Installation of the SCR by the reel-lay method which, compared to conventional installation solutions, allows most of the welding to be performed onshore in a controlled environment, thereby reducing offshore welding which brings many benefits, particularly for fatigue sensitive components of the pipeline.
Ruggedness or robustness	A measure of an analytical procedure's capacity to remain unaffected by small, but deliberate variations in method parameters and provides an indication of its reliability during normal usage.
Salt caverns	Caverns formed out of existing salt deposits.
Sandstone	A sedimentary rock composed of individual mineral grains of rock fragments between 0.06 and 2 millimeters (0.002 and 0.079 inches) in diameter and cemented together by silica, calcite, iron oxide, and so forth. Sandstone is commonly porous and permeable and therefore a likely type of rock in which to find a petroleum reservoir.
Satellite well	Usually a single well drilled offshore by a mobile offshore drilling unit to produce hydrocarbons from the outer fringes of a reservoir.
Saturated condition	A condition where an oil and gas are in thermodynamic equilibrium, that is, the chemical force exerted by each component in the oil phase is equal to the chemical force exerted by the same component in the gas phase, thereby eliminating mass transfer of components from one phase to the other. See: Undersaturated condition.
Saturation limit	The point at which a substance will receive no more of another substance in solution or in chemical combination.
Saturation pressure	An oil at its bubblepoint pressure or a gas at its dewpoint pressure.
SCADA (Supervisory Control and Data Acquisition)	Remote controlled equipment used by pipelines and LDCs to operate their gas systems.

SCOT (Shell Claus Off-gas Treating) process	A tail gas treating process.
SCR (Steel Catenary Riser)	A deepwater steel riser (see: Riser) suspended in a single catenary from a platform (typically a floater) and connected horizontally on the seabed.
Scrubber	A unit that has been designed for the removal of contaminates from a gas stream.
Seafloor	The bottom of the ocean; the seabed.
Secondary pollutant	A pollutant that is produced by the interaction of a primary pollutant with another chemical. A secondary pollutant may also be produced by dissociation of a primary pollutant, or other effects within a particular ecosystem; using the atmosphere as an example, the formation of the constituents of acid rain is an example of the formation of secondary pollutants. In many cases, these secondary pollutants can have significant environmental effects, such as participation in the formation of acid rain and smog. See also: Pollutant, Primary pollutant.
Sedimentary rock	A rock composed of materials that were transported to their present position by wind or water. Sandstone, shale, and limestone are sedimentary rocks.
Seep	The surface appearance of oil or gas that results naturally when a reservoir rock becomes exposed to the surface, thus allowing oil or gas to flow out of fissures in the rock.
Seismic	Of or relating to an earthquake or earth vibration, including those artificially induced.
Seismic data	Detailed information obtained from earth vibration produced naturally or artificially (as in geophysical prospecting).
Seismic method	A method of geophysical prospecting using the generation, reflection, refraction detection, and analysis of sound waves in the earth.
Seismic survey	An exploration method in which strong low-frequency sound waves are generated on the surface or in the water to find subsurface rock structures that may contain hydrocarbons. The sound waves travel through the layers of the earth's crust; however, at formation boundaries some of the waves are reflected back to the surface where sensitive detectors pick them up. Reflections from shallow formations arrive at the surface sooner than reflections from deep formations, and since the reflections are recorded, a record of the depth and configuration of the various formations can be generated. Interpretation of the record can reveal possible hydrocarbon-bearing formations.
Seismic wave	The record of an earth tremor by a seismograph.
Seismograph	An instrument used to detect and record earthquakes, is able to pick up and record the vibrations of the earth that occur during an earthquake; when seismology is applied to the search for natural gas, seismic waves, emitted from a source, are sent into the earth and the seismic waves interact differently with the underground formation (underground layers), each with its own properties.
Seismology	The study of the movement of energy, in the form of seismic waves, through the Earth's crust.

Selexol process	A process that used a physical (non-reactive) solvent for gas cleaning.
Semisubmersible drilling rig	A floating offshore drilling unit that has pontoons and columns that, when flooded, cause the unit to submerge to a predetermined depth. Living quarters, storage space, and so forth are assembled on the deck. Semisubmersible rigs are self-propelled or towed to a drilling site and anchored or dynamically positioned over the site, or both. In shallow water, some semisubmersibles can be ballasted to rest on the seabed. Semisubmersibles are more stable than drill ships and ship-shaped barges and are used extensively to drill wildcat wells in rough waters such as the North Sea. Two types of semisubmersible rigs are the bottle-type and the column-stabilized.
Separator Tank	Tanks are usually located at the well site to separate oil, gas, and water before sending each off to be processed at different locations.
Sewage gas	A gas produced by the fermentation of sewage sludge low in heating value due to dilution with carbon dioxide, and nitrogen.
Shale	A fine-grained, sedimentary rock composed of clay minerals and tiny fragments (silt-sized particles) of other materials.
Shale basin	An underground rock formation that serves both as a natural gas generator and a natural gas reservoir.
Shale gas	Gas that occurs in low permeability shale. See: Unconventional gas.
Shale oil	A liquid produced by the thermal decompositon of the kerogen component of oil shale.
Shallow gas	Natural gas deposit located near enough to the surface that a conductor or surface hole will penetrate the gas-bearing formations. Shallow gas is potentially dangerous because, if encountered while drilling, the well usually cannot be shut in to control it. Instead, the flow of gas must be diverted.
Shift converter	A reactor in which carbon monoxide and water are catalytically converted to hydrogen and carbon dioxide.
Show	The appearance of oil or gas in cuttings, samples, or cores from a drilling well.
Shrinkage, natural gas	The reduction in volume of wet natural gas due to the extraction of some of its constituents, such as hydrocarbon products, hydrogen sulfide, carbon dioxide, nitrogen, helium, and water vapor.
Skin	A measure of the amount of damage or improvement to the formation near the wellbore.
Skin effect	A dimensionless quantity and is defined as the difference between the actual and the ideal dimensionless pressure drop in a reservoir or pressure drop due to skin.
Skin factor	A constant that is used to adjust the flow equation derived from the ideal condition (homogeneous and isotropic porous media) to suit the applications in non-ideal conditions. The skin factor does affect the shape of the pressure buildup data. In fact, an early-time deviation from the straight line can be caused by skin factor as well as by wellbore storage. A positive skin factor indicates a flow restriction, i.e., wellbore damage, whereas a negative skin factor indicates stimulation.
Skin zone	A region of altered permeability due to wellbore damage; this zone can extend from a few inches to several feet from the wellbore. Many other

wells are stimulated by acidizing or fracturing, which in effect increases the permeability near the wellbore. Thus, the permeability near the wellbore is always different from the permeability away from the well where the formation has not been affected by drilling or stimulation. The effect of the skin zone is to alter the pressure distribution around the wellbore – in the case of wellbore damage, the skin zone causes an additional pressure loss in the formation while in the case of wellbore improvement, the opposite effect to that of wellbore damage occurs.

Slickwater	Slickwater fracturing is a method or system of hydraulic fracturing that involves adding chemicals to water to reduce friction and increase the fluid flow; slickwater increases the speed at which the pressurized fluid can be pumped into the wellbore.
Slurry process	A process that uses a slurry of iron oxide to selectively absorb hydrogen sulfide.
SO_2 (sulfur dioxide)	A precursor to acid deposition (acid rain); produced when sulfur is combusted to sulfur dioxide.
Solution gas	Raw natural gas dissolved in crude under reservoir conditions – a type of associated gas.
Solution gas-oil ratio (GOR)	The amount of surface gas that can be dissolved in a stock tank oil when brought to a specific pressure and temperature; denoted mathematically as R_s (SCF/STB).
Solution oil-gas ratio (OGR)	The amount of surface condensate that can be vaporized in a surface gas at a specific pressure and temperature; sometimes referred to as liquid content. Denoted mathematically as r_s (STB/MMSCF).
Sour gas	Natural gas that contains hydrogen sulfide and/or other acid gases such as mercaptans (also called thiols).
SPAR	Deep draft surface piercing cylinder type of floater, particularly adapted to deepwater, which accommodates drilling, top tensioned risers and dry completions.
Specifications	A feedstock specification or product specification is the data that give adequate control of feedstock behavior in a refinery or product quality; the specifications are derived from the set of tests and data limits applicable (in the context of this book) to the natural gas or to a finished product in order to ensure that every batch is of satisfactory and consistent quality at release for sales; the specifications should include all critical parameters in which variations would be likely to affect the safety and in-service use of the product.
Specific gravity (API)	A common measure of oil specific gravity, defined by $\gamma_{API} = (141.5/\gamma_o) - 131.5$, with units in °API (degrees API).
Specific gravity (gas)	The ratio of density of any gas at standard conditions (14.7 psia and 60 °F) to the density of air at standard conditions; based on the ideal gas law (pV = nRT), gas gravity is also equal to the gas molecular weight divided by air molecular weight (M_{air} = 28.97); denoted mathematically as γ_g (where air = 1)
Specific gravity (liquid)	The ratio of density of any liquid measured at standard conditions (usually 14.7 psia and 60 °F) to the density of pure water at the same standard conditions; denoted mathematically as γ_o (where water = 1).

Specificity	The ability to assess unequivocally the analyte in the presence of components which may be expected to be present.
Specific volume	The volume of a unit weight of a substance at specific temperature and pressure conditions.
Specific weight	Weight per unit volume of a substance.
Spinel group	A class of minerals of general formulation $A^{2+}B^{3+}{}_2O^{2-}{}_4$ which crystallize in the cubic (isometric) crystal system, with the oxide anions arranged in a cubic close-packed lattice and the cations A and B occupying some or all of the octahedral and tetrahedral sites in the lattice; a and b can be divalent, trivalent, or quadrivalent cations, including magnesium, zinc, iron, manganese, aluminum, chromium, titanium, and silicon.
Spool	Short-length pipe connecting a subsea pipeline and a riser, or a pipe and a subsea structure.
Spot Market	A method of contract purchasing whereby commitments by the buyer and seller are of a short duration at a single volume price.
Spot Purchase	Gas purchased on the spot market, which involves short-term contracts for specified amount of gas, at a one-time purchase price.
Spud	To begin operations on a well.
Spud cans	Cylindrically shaped steel shoes with pointed ends.
SRB	Sulfate reducing bacteria.
Standard condition	A temperature of 15 °C (59 °F) and a pressure of one atmosphere (14.7 psi, 101.325 kPa) which also is known as Standard Temperature and Pressure (STP).
Standard method	A method that is traceable to a recognized, validated method.
Static bottomhole pressure (SBHP)	If no fluid is moving, the well is static; the bottomhole pressure (BHP) is equal to the hydrostatic pressure (HP) on the annular side; if shut in on a kick, the bottomhole pressure is equal to the hydrostatic pressure in the annulus plus the casing (wellhead or surface pressure) pressure. See also: Kick, Bottomhole pressure.
Steady-state flow	A flow condition under which the pressure at any point in the reservoir remains constant over time.
Stock tank liquid (stock tank oil)	A term that is used both as a measure for oil well performance and in commercial pricing of oil; crude oil from which all of the associated natural gas has been stripped from the liquid at one atmosphere pressure.
Storage capacity	The ability of a system to store natural gas; usually thought of in terms of the base capacity, the working capacity, and the total capacity; measured in billion cubic feet (Bcf or Bft^3).
Storage measures	Several volumetric measures are used to quantify the fundamental characteristics of an underground storage facility and the gas contained within it. For some of these measures, it is important to distinguish between the characteristic of a facility, such as it's the storage capacity, and the characteristic of the natural gas within the facility such as the actual inventory level.
Stratigraphic test	A borehole drilled primarily to gather information on rock types and sequence.

Stratigraphic trap	A petroleum trap that occurs when the top of the reservoir bed is terminated by other beds or by a change of porosity or permeability within the reservoir itself. See: Structural trap.
Stretford process	A wet oxidation process.
Stripper Wells	Natural gas wells that produce less than 60,000 cubic feet of gas per day.
Structural trap	A petroleum trap that is formed because of deformation (such as folding or faulting) of the reservoir formation. Compare *stratigraphic trap*.
Structure	A geological formation of interest to drillers. For example, if a particular well is on the edge of a structure, the wellbore has penetrated the reservoir (structure) near its periphery.
Subduction zone	A deep trench formed in the ocean floor along the line of convergence of oceanic crust with other oceanic or continental crust when one plate (always oceanic) dives beneath the other. The plate that descends into the hot mantle is partially melted. Magma rises through fissures in the heavier, non-liquid (unmelted) crust above, creating a line of plutons and volcanoes that eventually form an island arc parallel to the trench.
Subsea Technology	All products and services required to install and operate production installations on the seabed.
SulfaTreat process	A batch-type process for the selective removal of hydrogen sulfide and mercaptans from natural gas; the process is dry, using no free liquids, and can be used for natural gas applications where a batch process is suitable.
Sulfinol process	A combination process that uses a mixture of amines and a physical solvent (an aqueous amine and sulfolane).
SulFerox process	A wet oxidation process.
Super compressibility factor	A factor used to account for the following effect: Boyle's law for gases states that the specific weight of a gas is directly proportional to the absolute pressure, the temperature remaining constant. All gases deviate from this law by varying amounts, and within the range of conditions ordinarily encountered in the natural gas industry, the actual specific weight under the higher pressure is usually greater than the theoretical. The factor used to reflect this deviation from the ideal gas law in gas measurement with an orifice meter is called the *super compressibility factor, Fpv*. The factor is used to calculate actual volumes from volumes at standard temperatures and pressures from actual volumes. The factor is of increasing importance at high pressures and low temperatures.
Supergiant field	Recoverable reserves >5 billion barrels of oil equivalents.
Supplemental gaseous fuels supplies	Synthetic natural gas, propane-air, coke oven gas, refinery gas, biomass gas, air injected for Btu stabilization, and manufactured gas commingled and distributed with natural gas.
SURF	Subsea Umbilicals Risers Flowlines.
Swamp gas	See Natural gas.
Sweet crude oil	Oil containing little or no sulfur, especially little or no hydrogen sulfide.
Sweetening process	A process for the removal of hydrogen sulfide and other sulfur compounds from natural gas.

Sweet gas	Natural gas that contains little, if any, hydrogen sulfide.
Sweet spot	A general expression for a target location or area within a play or a reservoir that represents a relatively high the best production potential. Reservoir engineers attempt to identify sweet spots to enable wellbores to be placed in the most productive areas of the reservoir. A sweet spots in tight (shale) reservoirs may be defined by (i) source rock richness or thickness, (ii) natural fractures, or (iii) other factors, using geological data such as core analysis, well log data, or seismic data.
Syncline	A trough-shaped configuration of folded rock layers. See: Anticline.
Synthesis gas (syngas)	A gas mixture that contains varying amounts of carbon monoxide (CO) and hydrogen (H_2) generated by the gasification of a carbon-containing material to a gaseous product with a heating value (but less than half the energy density of natural gas). When used as a fuel, it is produced by gasification of coal or municipal waste by the following reactions: $$C + O_2 \rightarrow CO_2$$ $$CO_2 + C \rightarrow 2CO$$ $$C + H_2O \rightarrow CO + H_2$$ The name comes from the use of the gas as an intermediate in creating synthetic natural gas (SNG) and in producing ammonia or methanol.
Synthetic natural gas (SNG)	A manufactured product, chemically similar in most respects to natural gas, resulting from the conversion or reforming of hydrocarbons that may easily be substituted for or interchanged with pipeline-quality natural gas; also referred to as substitute natural gas.
Tail gas	The residue gas left after the completion of a treating process designed to remove certain liquids or liquefiable hydrocarbons.
Tail gas treating	The removal of the remaining sulfur compounds from gases remaining after sulfur recovery.
Tar sand	A sandstone that contains chiefly heavy, tarlike hydrocarbons. Tar sands are difficult to produce by ordinary methods; thus, it is costly to obtain usable hydrocarbons from them.
Tcf (trillion cubic feet)	Gas measurement approximately equal to one quadrillion (1,000, 000,000,000,000) Btu's. See also: Bcf, Mcf, Quad.
TEA	Triethanolamine; $(HOCH_2CH_2)_3N$.
Tectonic	Of or relating to the deformation of the earth's crust, the forces involved in or producing such deformation, and the resulting rock forms.
Temperature, dew-point	The temperature at which a vapor begins to condense and deposit as a liquid.
Terminal velocity (drop-out velocity)	The velocity at which a particle or droplet will fall under the action of gravity, when drag force just balance gravitational force and the particle (or droplet) continues to fall at constant velocity.
Teta wire	Wire with a specific, patented, T-shape used in flexible pipe to resist the radial effect of the internal pressure. Used for high pressure and harsh environments.
Therm	A unit of heating value equivalent to 100,000 British thermal units (Btu).

Thermal decomposition	The breakdown of a compound or substance by temperature into simple substances or into constituent elements.
Thermal maturity	The amount of heat, in relative terms, to which a rock has been subjected; a thermally immature rock has not been subjected to enough heat to begin the process of converting organic material into oil and/or natural gas; a thermally over-mature rock has been subjected to enough heat to convert organic material to graphite. However, these are the two extremes, and there are many intermediate stages of thermal maturity.
Thermogenic	Generated or formed by heat, especially via physiological processes.
Thermogenic coal bed methane	Methane formed in coal seams by the action of increasing temperature. See: Coal bed methane.
Thermogenic gas	Gas formed by pressure effects and temperature effects on organic debris.
Three dimensional (3D) Seismic Survey	Allows producers to see into the crust of the Earth to find promising formations for retrieval of gas.
	Threshold limit value (TLV); The amount of a contaminant to which a person can have repeated exposure for an eight-hour day without adverse effects.
Tie-back:	Connection of a satellite subsea development to an existing infrastructure.
Tight formation	A petroleum- or water-bearing formation of relatively low porosity and permeability.
Tight gas	Natural gas found trapped in impermeable rock and non-porous sandstone or limestone formations, typically at depths greater than 10,000 feet below the surface.
Tight oil	Oil produced from petroleum-bearing formations with low permeability such as the Eagle Ford, the Bakken, and other formations that must be hydraulically fractured to produce oil at commercial rates.
TLP (tension leg platform)	A floating production unit anchored to the seabed by taut vertical cables, which considerably restrict its heave motion, making it possible to have the wellheads on the platform.
Total organic carbon (TOC)	The concentration of material derived from decaying vegetation, bacterial growth and metabolic activities of living organisms or chemicals in the source rocks.
Town gas	Also known as coal gas, and contains hydrogen (H_2), carbon monoxide (CO), carbon dioxide (CO_2), methane (CH_4), nitrogen (N_2) and volatile hydrocarbons. It is made by blowing air and steam over an incandescent fuel bed, usually of coke or coal. The words "coal gas" could also be used to mean gas made by the destructive distillation of coal. The gas was used inter alia for lighting before the advent of electric lighting, and for heating and cooking before natural gas became widely available.
Trap	A body of permeable oil-bearing rock and/or gas-bearing rock surrounded or overlain by an impermeable barrier that prevents oil from escaping; the types of traps are structural, stratigraphic, or a combination of these.
Trillion cubic feet	A volume measurement of natural gas. Approximately equivalent to one Quad. See also: Btu's, Bcf, Mcf.

Triple point	The temperature and pressure at which the three phases (gas, liquid, and solid) of a substance coexist in thermodynamic equilibrium.
Ultimate analysis	The determination of the elements contained in a compound, i.e., carbon, hydrogen, oxygen, nitrogen, sulfur, and other components.
Ulvospinel	A mineral from the spinel group of minerals being iron and titanium oxide of the formula: Fe_2TiO_4 (ortho-titanate iron).
Umbilicals	An assembly of hydraulic hoses which can also include electrical cables or optic fibers, used to control subsea structures from a platform or a vessel.
Unassociated gas	Natural gas unaccompanied by crude oil when produced; also called non-associated gas or gas well gas.
Unbundled Services	Unbundling, or separating, pipeline transmission, sales and storage services, along with guaranteeing space on the pipelines for all gas shippers.
Unconventional gas	Gas that occurs in tight sandstones, siltstones, sandy carbonates, limestone, dolomite, and chalk; natural gas that cannot be economically produced using current technology. See also: Shale gas.
Underground gas storage	The use of sub-surface facilities for storing gas that has been transferred from its original location for the primary purpose of load balancing; usually natural geological reservoirs, such as depleted oil or gas fields or water-bearing sands on the top by and impermeable cap rock.
Undersaturated condition	A condition when an oil or a gas is in a single phase but not at its saturation point (bubblepoint or dewpoint), that is, the mixture is at a pressure greater than its saturation pressure. See: Saturated condition.
Undiscovered resources	Resources inferred from geologic information and theory to exist outside of known oil and gas fields.
Unsaturated compound	Any compound having a double bond or triple bond between two adjacent carbon atoms.
Upper explosive limit	The higher percent by volume of the gas vapor in air at which the gas will explode or inflame. See: Lower explosive limit,
Upper flammability limit	The maximum concentration by volume of a combustible substance that is capable of continued propagation of a flame under the specified conditions.
Vacuum	A pressure less than atmospheric pressure, measured either from the base of zero pressure or from the base of atmospheric pressure.
Validation	The process of establishing the performance characteristics and limitations of a method and the identification of the influences which may change these characteristics and to what extent.
Vapor	When the gas phase of a substance is present under conditions when the substance would normally be a solid or liquid (e.g., below the boiling point of the substance), this is referred to as the vapor phase. See: Gas.
Vapor density	The density of any gas compared to the density of air with the density of air equal to unity.
Vaporizer	A device other than a container which receives gas in liquid form and adds sufficient heat to convert the liquid to a gaseous state.

Vent Stack	The elevated vertical termination of a disposal system that discharges vapors into the atmosphere without combustion or conversion of the relieved fluid.
Verification	The confirmation by examination and provision of objective evidence that specified requirements have been fulfilled.
Viscosity	The measure of a fluid's thickness, or how well it flows.
VLS (Vertical Lay System)	A Technip proprietary technology for installation of flexible pipes.
Volatile matter	Matter which is readily vaporizable at a relatively low temperature.
Volume, specific	The volume of a unit weight of a substance at specific temperature and pressure conditions.
Volume concentration	The concentration expressed in terms of gaseous volume of substance per unit volume (cm^3 substance/m^3 gas volume).
Vugs	Small unfilled cavities in a rock formation.
Wash	The removal of impurities from a gas or vapor by passing the gas through water or other liquid which retains or dissolves the impurity.
Washer	A shell with internal baffler or packing, so arranged that gas to be cleaned passes up through the baffles counter-current to the flow of scrubbing liquid down through the washer. The baffler or packing causes intimate contact and mixing of the gas with the liquid stream.
Washer-cooler:	A washer in the form of a tall tower in which the washing liquid is sprayed in at top is collected in the bottom of the tower and then is cooled and recycled through the tower. Serves a dual purpose of washing the gas free of impurities and also cooling the gas.
Water coning	Should not be confused with water production caused by a rising water/oil contact from water influx. See: Coning, Gas coning.
Water gas	A mixture of carbon monoxide (CO) and hydrogen (H_2) produced by passing steam over red-hot coke using the endothermic reaction: $C + H_2O \rightarrow CO + H_2$ This product had a lower calorific value than coal gas so the gas was often passed through a heated retort into which oil was sprayed; the resulting mixed gas was called carbureted water gas.
Water-producing interval	The portion of an oil or gas reservoir from which water or mainly water is produced.
Weathering	The breakdown of large rock masses into smaller pieces by physical and chemical climatological processes; the evaporation of liquid by exposing it to the conditions of atmospheric temperatures and pressure.
Weight, specific	Weight per unit volume of a substance.
Well	The hole made by the drilling bit, which can be open, cased, or both. Also called borehole, hole or wellbore.
Wellbore (well bore)	The channel created by the drill bit.
Well casing	A series of metal tubes installed in the freshly drilled hole; serves to strengthen the sides of the well hole, ensure that no oil or natural gas seeps out of the well hole as it is brought to the surface, and to keep other fluids or gases from seeping into the formation through the well.
Well completion	The process for completion of a well to allow for the flow of petroleum or natural gas out of the formation and up to the surface; includes strengthening the well hole with casing, evaluating the pressure and

	temperature of the formation, and then installing the proper equipment to ensure an efficient flow of natural gas out of the well.
Well control	The methods used to control a kick and prevent a well from blowing out. Such techniques include, but are not limited to, keeping the borehole completely filled with drilling mud of the proper weight or density during all operations, exercising reasonable care when tripping pipe out of the hole to prevent swabbing, and keeping careful track of the amount of mud put into the hole to replace the volume of pipe removed from the hole during a grip.
Wellhead (well head)	The pieces of equipment mounted at the opening of the well to regulate and monitor the extraction of hydrocarbons from the underground formation; prevents leaking of oil or natural gas out of the well and prevents blowouts due to high pressure formations.
Well intervention (well work)	Any operation carried out on a crude oil or natural gas well during or at the end of its productive life, which alters the state of the well and/or well geometry, provides well diagnostics, or manages the production of the well.
Well logging	A method used for recording rock and fluid properties to find gas and oil containing zones in subterranean formations; the recording of information about subsurface geologic formations, including records kept by the driller and records of mud and cutting analyses, core analysis, drill stem tests, and electric, acoustic, and radioactivity procedures.
Well servicing	Intervention in subsea production wells carried out from a floating rig or a dynamically positioned vessel.
Wet gas	Typically, a natural gas stream that contains considerable amounts of higher molecular weight hydrocarbons other than methane; may also refer to the water content of the gas stream; sometime referred to as dry gas.
Water wet	Refers to reservoir rock that is preferentially in contact with water. Wet-gas reservoir. A reservoir in which the reservoir temperature is above the cricondentherm of the hydrocarbon mixture and because the reservoir temperature exceeds the cricondentherm of the hydrocarbon system, the reservoir fluid will always remain in the vapor phase region as the reservoir is depleted isothermally,
Wet oxidation process	Based on reduction-oxidation (Redox) chemistry to oxidize the hydrogen sulfide to elemental sulfur in an alkaline solution containing an oxygen carrier; vanadium and iron are the two oxygen carriers that are used.
Wettability	The wettability of reservoirs rocks refers to the tendency of the fluid (e.g., crude oil) to spread on or adhere to a solid surface in the presence of other immiscible fluids and is determined by complex interface boundary conditions acting within pore space of sedimentary rocks.
Wildcat	A well drilled in an area where no oil or gas production exists.
Wireline	A cable technology used by operators of crude oil and natural gas wells to lower equipment or measuring devices into the well for the purposes of well intervention, reservoir evaluation, and pipe recovery.

Wobbe Index (Wobbe Number)	A number which indicates interchangeability of fuel gases and is obtained by dividing the heating value of a gas by the square root of its specific gravity.
Wood gas	The product of thermal gasification of biomass (such as wood, wood chips, and sawdust) in a gasifier or in a wood gas generator. The gas is the result of a high temperature reaction (>700 °C, >1,290 °F) where carbon reacts with steam or a limited amount of air producing carbon monoxide (CO), carbon dioxide (CO_2), hydrogen (H_2) and methane (CH_4). The gas can be filtered, purified or scrubbed and used to power internal combustion engines, gas turbines, Stirling engines, or fuel cells.
Working gas	The quantity of natural gas in the reservoir (reported in thousand cubic feet at standard temperature and pressure) that is in addition to the cushion or base gas; may or may not be completely withdrawn during any particular withdrawal season; conditions permitting, the total working capacity could be used more than once during any season; the volume of gas in the reservoir above the level of base gas and is, simply, the natural gas is that is available for withdrawal and sales. I
Working gas capacity	The total gas storage capacity minus base gas.
Yield point	The stress at which a material exceeds its elastic limit; below this stress, the material will recover its original size on removal of the stress; above this stress, it will not recover its original size on removal of the stress.
Zero gas	Gas at atmospheric pressure.
Zone	A geographical area. A geological zone, however, means an interval of strata of the geologic column that has distinguishing characteristics from surrounding strata; also, a space or group of spaces within a building with heating and/or cooling requirements sufficiently similar so that comfort conditions can be maintained by a single controlling device.

Appendix A
Unconventional gas resources

Unconventional gas resources are included in this test because of the potential (in some cases, already a reality) of blending unconventional gas streams with conventional natural gas as the feedstock for gas cleaning processes and for the production of fuel gas streams. Because of this, new technologies are constantly being introduced that allows for the more economic extraction of non-traditional gas that may have been previously impossible to obtain. Development of these unconventional gas resources has significant economic potential for the production of gaseous fuels (Table A.1).

Table A.1: The composition of gas from various carbonaceous fuels.

Composition	Coal gas	Coke oven gas	Biogas	Digester gas	Landfill gas	Natural gas
Hydrogen (H_2)	14.0%	51.9%	18.0%		0.1%	–
Carbon monoxide (CO)	27.0%	2.0%	24.0%		0.1%	–
Carbon dioxide (CO_2)	4.5%	5.5%	6.0%	30.0%	47.0%	–
Oxygen (O_2)	0.6%	0.3%	0.4%	0.7%	0.8%	–
Methane (CH_4)	3.0%	32.0%	3.0%	64.0%	47.0%	90.0%
Nitrogen (N_2)	50.9%	4.8%	48.6%	2.0%	3.7%	5.0%
Ethane (C_2H_6)	–		–			5.0%

The boundary between conventional gas and unconventional gas resources is not well defined, because they result from a variety of sources under different conditions and have varying compositions (Table A.2). In addition and with the onset of renewable energy programs, it would be remiss not to mention these gases because of the potential for increasing use as well as the obvious potential of being sent to gas processing units as blends with natural gas.

A.1 Biogas

Biogas (often called biogenic gas and sometimes incorrectly known as swamp gas) is a combustible gas (predominantly methane) produced by (i) anaerobic digestion with anaerobic organisms, which digest material inside a closed system or (ii) fermentation of biodegradable organic matter including manure, sewage sludge, domestic

https://doi.org/10.1515/9783110691023-010

Table A.2: Composition of unconventional gases compared to natural gas.

Gas	Carbon dioxide	Carbon monoxide	Methane	Butane	Ethane	Propane	Hydrogen	Hydrogen sulfide	Oxygen	Nitrogen
								Composition, % v/v		
Coal gas	3.8	28.4	0.2				17.0			50.6
Coke oven gas	2.0	5.5	32				51.9		0.3	4.8
Digester gas	30		64				0.7	0.8		2.0
Landfill gas	47	0.1	47				0.1	0.01	0.8	3.7
Natural gas	0–0.8	0–0.45	82–93		0–15.8		0–1.8	0–0.18	0–0.35	0.5–8.4

solid waste, municipal solid waste, biodegradable waste, or any other biodegradable feedstock, under anaerobic conditions.

The composition of biogas is highly variable (Table A.3). For example, biogas from sewage digesters usually contains 55–65% v/v methane, 35–45% v/v carbon dioxide, and <1% nitrogen whereas biogas from organic waste digesters usually contains 60–70% methane, 30–40% carbon dioxide, and <1% v/v nitrogen, while biogas from landfills typically contains 45–55% v/v methane, 30–40% v/v carbon dioxide, and 5–15% v/v nitrogen.

Table A.3: Examples of biogas composition.

Constituents	Source 1	Source 2	Source 3
Methane (CH_4), % v/v	50–60	60–75	60–75
Carbon dioxide (CO_2), % v/v	38–34	33–19	33–19
Nitrogen (N_2), % v/v	5–0	1–0	1–0
Oxygen (O_2) % v/v	1–0	< 0,5	< 0,5
Water (H_2O) % v/v	6 (at 40 °C)	6 (at 40 °C)	6 (at 40 °C)
Hydrogen sulfide (H_2S) mg/m^3	100–900	1,000–4,000	3,000–10,000
Ammonia (NH_3) mg/m^3	–	–	50–100

Source 1: Household waste
Source 2: Wastewater treatment plant sludge
Source 3: Agricultural waste

In addition to the components enumerated above, biogas also contains hydrogen sulfide and other sulfide compounds (such as mercaptan derivatives, RSH, also called thiols), aromatic compounds, halogenated compounds, and siloxane derivatives. The latter two classes of compounds (i.e., halogenated compounds and siloxane derivatives) are more common in biogas from a landfill than in biogas from the anaerobic digestion of manure (unless the livestock has been on a very mysterious diet).

Biogas production (typically an anaerobic process) is a multi-step process in which originally complex organic (liquid or solid) wastes are progressively transformed into low-molecular-weight products by different bacteria strains. Biogas can also be produced by pyrolysis of biomass (freshly harvested or as a biomass waste). Thus, the name "biogas" gathers a large variety of gases under the name resulting from specific treatment processes, starting from various organic wastes such as livestock manure, food waste, and sewage are all potential sources of biogenic gas, or biogas, which is usually considered as a form of renewable energy and is often categorized according to the source. In spite of the potential differences in composition, biogas can be processed (upgraded) to the standards required for natural gas through choice of the relevant processing sequence depending upon the composition of the gas.

During the combustion biomass, various kinds of impurities are generated and some of them occur in the flue gas and most of the contaminants in the flue gas are related to the composition of the biomass. If the combustion is incomplete (i.e., carried out in a deficiency of oxygen), soot, unburned matter, and toxic dioxin derivatives may also occur in the flue gas.

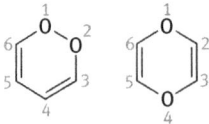

1,2-Dioxin 1,4-Dioxin

In addition, metals, such as lead (Pb), also occur in the ash and may even evaporate during combustion and react, condense, and/or sublime during cooling in the boiler. Upstream of the gas cleaning installation, normally at a temperature <200 °C (<390 °F), all metals will occur as solid particles, except mercury which evaporates during combustion and reacts in the boiler but remains mainly in its gaseous form. The impurities in the biogas are harmful if they are emitted to the atmosphere and gas cleaning units must be installed to eliminate or at least reduce this problem. The degree of cleaning depends on federal, regional, and local regulations but regional and local authorities, organizations, and individuals have often an opinion on an actual plant due to its size and location.

More generally, contaminants aside, in terms of composition, biogas is primarily a mixture of methane (CH_4) and inert carbonic gas (CO_2) but variations in the composition of the source material lead to variations in the composition of the gas. Water (H_2O), hydrogen sulfide (H_2S), and particulates are removed if present at high levels or if the gas is to be completely cleaned. Carbon dioxide is less frequently removed, but it must also be separated to achieve pipeline quality gas. If the gas is to be used without extensive cleaning, it is sometimes cofired with natural gas to improve combustion. Biogas cleaned up to pipeline quality is called renewable natural gas.

Finally, whereas natural gas is classified as fossil fuel, whereas biomethane is defined as a non-fossil fuel and is further characterized or described as a green source of energy. Noteworthy at this point is that methane, whatever the source (fossil fuel or non-fossil fuel) and when released into the atmosphere, is approximately 20 times more potent as a greenhouse gas than carbon dioxide. Organic matter from which biomethane is produced would release carbon dioxide into the atmosphere if simply left to decompose naturally, while other gases that are produced during the decomposition process such as, for example, nitrogen oxide(s) would make an additional contribution to the greenhouse effect.

A.2 Coalbed methane

Just as natural gas is often located in the same reservoir as with crude oil, a gas (predominantly methane) can also be found trapped within coal seams where it is often referred to as coalbed methane (or coal bed methane, CBM, sometimes referred to as coalmine methane, CMM). The gas occurs in the pores and cracks in the coal seam and is held there by underground water pressure. To extract the gas, a well is drilled into the coal seam and the water is pumped out (dewatering) which allows the gas to be released from the coal and brought to the surface.

However, the occurrence of methane in coal seams is not a new discovery and methane (also called firedamp) was known to coal miners for at least 150 years (or more) before it was rediscovered and developed as CBM. To the purist, CMM is the fraction of CBM that is released during the mining operation (referred to in the older literature as firedamp by miners because of its explosive nature). In practice, the terms CBM and CMM may usually refer to different sources of gas – both forms of gas, whatever the name, are equally dangerous to the miners.

CBM is relatively pure compared to conventional natural gas, containing only small proportions of higher molecular weight hydrocarbon derivatives such as ethane and butane and other gases (such as hydrogen sulfide and carbon dioxide). Because coal is a solid, high-carbon content mineral, there are usually no liquid hydrocarbon derivatives contained in the produced gas. The coal bed (coal seam) must first be de-watered to allow the trapped gas to flow through the formation to produce the gas. Consequently, CBM usually has a lower heating value, and elevated levels of carbon dioxide, oxygen, and water that must be treated to an acceptable level, given the potential to be corrosive.

The gas from coal seams can be extracted by using technologies that are similar to those used to produce conventional gas, such as using well-bores. However, complexity arises from the fact that the coal seams are generally of low permeability and tend to have a lower flow rate (or permeability) than conventional gas systems; gas is only sourced from close to the well and as such a higher density of wells is required to develop a CBM resource as an unconventional resource (such as tight gas) than a conventional gas resource. Technologies such as horizontal and multilateral drilling with hydraulic fracturing are sometimes used to create longer more open channels that enhance well productivity but not all coal seam gas wells require application of this technique.

Unlike natural gas from conventional reservoirs, CBM contains little higher molecular weight hydrocarbon derivatives such as propane or butane or condensate. It often contains up to a few percent carbon dioxide. Some coal seams contain little methane, with the predominant coal seam gas being carbon dioxide. In fact, CBM has been suggested with sufficient justification that the materials comprising a coalbed fall broadly into the following two categories: (i) volatile low-molecular-weight materials that can be liberated from the coal by pressure reduction, mild heating, or solvent extraction; and (ii) materials that will remain in the solid state after the separation of volatile components.

Typically, with some exceptions, coalbed gas is typically in excess of 90% v/v methane and, subject to gas composition data, may be suitable for introduction into a commercial pipeline with little or no treatment. Methane within coalbeds is not structurally trapped by overlying geologic strata, as in the geologic environments typical of conventional gas deposits. Only a small amount (on the order of 5–10% v/v) of the CBM is present as free gas within the joints and cleats of coalbeds. Most of the CBM is contained within the coal itself (adsorbed to the sides of the small pores in the coal).

A.3 Coal gas

Coal gas is any gaseous product that is produced by gasification of coal or by carbonization of coal.

Coal carbonization is used for processing of coal to produce coke using metallurgical-grade coal. The process involves heating coal in the absence of air to produce coke and is a multistep complex process, and variety of solid liquids and gaseous products are produced which contain many valuable products. The various products from coal carbonization in addition to coke are (i) coke oven gas, (ii) coal tar, (iii) low boiling oil, also called light oil, and (iv) aqueous solution of ammonia and ammonium salts. With the development of the steel industry, there was a continuous development in coke oven plants during the latter half of nineteenth century to improve the process conditions, recovery of chemicals, and this continued during the twentieth century to adapt to environmental pollution control strategies and energy consumption measures.

The carbonization can be carried out at various temperatures, although low temperature or high temperature is preferred. Low-temperature carbonization is used to produce liquid fuels while high-temperature carbonization is used to produce gaseous products. Low-temperature carbonization (approximately 450–750 °C, 840–1,380 °F) is used to produce liquid fuels (with smaller amounts of gaseous products) while the high-temperature carbonization process (approximately 900 °C, 1650 °F) is used to produce gaseous products. In the high-temperature carbonization process, gaseous product is less while liquid products are large and the production of tar is relatively low because of the cracking of the secondary (liquid products and tar) products.

Most of these products have been replaced by natural gas (the city and town gas works is an industry of the past) but still find use in some parts of the world. The products of coal gasification are varied insofar as the gas composition varies with the system employed. It is emphasized that the gas product must be first freed from any pollutants such as particulate matter and sulfur compounds before further use, particularly when the intended use is a water gas shift or methanation.

Gases of high calorific value are obtained by low-temperature or medium-temperature carbonization of coal. The gases obtained by the carbonization of any given coal change in a progressive manner with increasing temperature. The composition of coal gas also changes during the course of carbonization at a given temperature, and secondary reactions of the volatile products are important in determining gas composition.

Low heat-content gas (low-Btu gas) is produced during the gasification of when the oxygen is not separated from the air and, thus, the gas product invariably has a low heat content (on the order of 150–300 Btu/ft^3). Low heat-content gas is also the usual product of in situ gasification of coal which is used essentially as a technique for obtaining energy from coal without the necessity of mining the coal. The process is a technique for utilization of coal which cannot be mined by other techniques.

The nitrogen content of low heat-content gas ranges from somewhat less than 33% v/v to slightly more than 50% v/v and cannot be removed by any reasonable means; the presence of nitrogen at these levels renders the product gas to be low heat content. The nitrogen also strongly limits the applicability of the gas to chemical synthesis. Two other noncombustible components (water, H_2O, and carbon dioxide (CO_2)) lower the heating value of the gas further. Water can be removed by condensation and carbon dioxide by relatively straightforward chemical means.

The two major combustible components are hydrogen and carbon monoxide; the H_2/CO ratio varies from approximately 2:3 to approximately 3:2 but methane may also make an appreciable contribution to the heat content of the gas. Of the minor components, hydrogen sulfide is the most significant and the amount produced is, in fact, proportional to the sulfur content of the feed coal. Any hydrogen sulfide present must be removed by one, or more, of several procedures. Low heat-content gas is of interest to industry as a fuel gas or even, on occasion, as a raw material from which ammonia, methanol, and other compounds may be synthesized.

Medium heat-content gas (medium-Btu gas) has a heating value in the range 300–550 Btu/ft^3 and the composition is much like that of low heat-content gas, except that there is virtually no nitrogen. The primary combustible gases in medium heat-content gas are hydrogen and carbon monoxide. Medium heat-content gas is considerably more versatile than low heat-content gas; like low heat-content gas, medium heat-content gas may be used directly as a fuel to raise steam or used through a combined power cycle to drive a gas turbine, with the hot exhaust gases employed to raise steam.

Medium heat-content gas is especially amenable to the production of (i) methane, (ii) higher molecular weight hydrocarbon derivatives by the Fischer–Tropsch synthesis, (iii) methanol, and (iv) a variety of synthetic chemicals. The reactions used to produce medium heat-content gas are the same as those employed for low heat-content gas synthesis, the major difference being the application of a nitrogen barrier (such as the use of pure oxygen) to keep diluent nitrogen out of the system.

In medium heat-content gas, the hydrogen–carbon monoxide ratio varies from 2:3 to ca. 3:1, and the increased heating value correlates with higher methane and hydrogen contents as well as with lower carbon dioxide contents. In fact, the nature of the gasification process used to produce the medium heat-content gas has an effect on the ease of subsequent processing. For example, the carbon dioxide-acceptor product is available for use in methane production because it has (i) the desired hydrogen–carbon dioxide ratio just exceeding 3:1, (ii) an initially high methane content, and (iii) relatively low carbon dioxide content and low water content.

High heat-content gas (high-Btu gas) is almost pure methane and often referred to as synthetic natural gas or substitute natural gas (SNG). However, to qualify as SNG, a product must contain at least 95% methane; the energy content on the order of 980–1,080 Btu/ft^3.

The commonly accepted approach to the synthesis of high heat-content gas is the catalytic reaction of hydrogen and carbon monoxide:

$$3H_2 + CO \rightarrow CH_4 + H_2O$$

To avoid catalyst poisoning, the feed gases for this reaction must be quite pure and, therefore, impurities in the product are rare. The water produced by the reaction is removed by condensation and recirculated as very pure water through the gasification system. The hydrogen is usually present in slight excess to ensure that the toxic carbon monoxide is reacted.

The carbon monoxide–hydrogen reaction is not the most efficient way to produce methane because of the exothermicity of the reaction. Also, the methanation catalyst is subject to poisoning by sulfur compounds, and the decomposition of metals can destroy the catalyst. Hydrogasification may be employed to minimize the need for methanation:

$$C_{coal} + 2H_2 \rightarrow CH_4$$

The product of this reaction is not pure methane and additional methanation is required after hydrogen sulfide and other impurities are removed.

A.3.1 Blue gas

Blue gas is a mixture of gases and consists predominantly of carbon monoxide and hydrogen formed by action of steam on hot coal or coke. The mixture has a tendency to burn with a blue flame, hence the name of the gas.

Carbureted blue gas is a mixture of carbon monoxide and hydrogen formed by the action of air and then steam on hot coal or coke and enriched with hydrocarbon gases during its manufacture by a simultaneous process when low-boiling distillate, gas oil, or fuel oil is gasified. The gas has a gross heat content of approximately 500–550 Btu/ft^3.

A.3.2 Carbureted water gas

By way of introduction, water gas (carbureted blue gas) is a mixture of carbon monoxide and hydrogen formed by the action of air and then steam on hot coal or coke and enriched with hydrocarbon gases from the pyrolysis of oils.

On the other hand, semi-water gas is a mixture of water gas and producer gas that is made by passing a mixture of air and steam through heated coke. The heat generated when producer gas is formed keeps the temperature of the coke high enough to allow water gas to be formed.

The Lowe's water gas process is a process by which large amounts of hydrogen gas could be generated for residential and commercial use in heating and lighting. This gas provides a more efficient heating fuel than the common coal gas, or coke gas, which was used in municipal service. The process used the water gas shift reaction:

$$CO + H_2O \rightarrow CO_2 + H_2$$

The process was discovered by passing high-pressure steam over hot coal, the major source of coke gas. Lowe's process improved upon the chimney systems by which the coal could remain superheated, thereby maintaining a consistently high supply of the gas. The reaction produced carbon dioxide and hydrogen, which, after a process of cooling and scrubbing, produced hydrogen gas.

Carbureted water gas is the result of combining the water gas and oil gas methods. Briefly, oil gas is the gas formed by the thermal cracking of crude oil. If oil is sprayed onto heated checker work (refractory), it cracks to form lower gaseous hydrocarbon derivatives. The types of hydrocarbon derivatives depend entirely on the type of feedstock but do cause an increase in the heat content of the product gas.

In the oil gas method, an oil product (typically heavy fuel oil, as described above) is sprayed into the hot water gas chamber to produce a better quality (higher heat content) gas. The amount of the heavy fuel ratio is a determinant of the quality of the gas. In the nineteenth century and early twentieth century, this was the method used to produce the town gas but has largely been superseded by natural gas in countries with an abundant supply. As supplies of natural gas diminish and various carbonaceous feedstocks are used for gas production, the process may once again rise in importance.

A.3.3 Producer gas

Producer gas is a low Btu gas obtained from a coal gasifier (fixed-bed) upon introduction of air instead of oxygen into the fuel bed. Producer gas is mainly carbon monoxide with smaller amounts of hydrogen, methane, and variable amounts of nitrogen (Table A.4), obtained from partial combustion of coal or coke in air or oxygen, having a heat content of 110–160 Btu/ft^3 (air combustion) or 400–500 Btu/ft^3 (oxygen combustion).

Table A.4: Typical composition of producer gas.

Source	Coal/coke	Coal/coke
	Gas constituent	
	%, v/v	%, v/v
Hydrogen	20–22	12–15
Carbon monoxide	19–20	20–22
Carbon dioxide	12–14	9–11
Methane	2–5	2–3
Nitrogen	50–60	50–54

The gas is produced by blowing air and sometimes steam through an incandescent fuel bed (the process is self-heating). The reaction with air is exothermic but insufficient air is added, hence carbon monoxide is produced. Steam addition results in the formation of hydrogen by the water gas reaction. This is endothermic and hence balances out the exothermic air reaction. Producer gas is low calorific value and is typically used as an on-site fuel gas.

In the process – using coal as the example, of the carbonaceous feedstock – air is blown through an incandescent bed of coal. The amount of air is sufficient to maintain the process temperature but insufficient to complete the combustion reaction. The gas products is a mixture containing approximately 50% v/v nitrogen, form the air, 29% v/v carbon monoxide, due to incomplete combustion of the feedstock, and 4% v/v carbon dioxide, as well as sundry other products, including hydrocarbon derivatives. The heat content of the gas is low and, although not suitable for distribution to consumers, can be used on site as a fuel gas. In order to modify the product distribution, steam can be used in the process, and endothermic reactions (such as the water gas reaction) result in the formation of hydrogen.

Initially, the oxidation of the feedstock (coal) in the presence of the available oxygen occurs after which there is competition for oxygen between the carbon and the water gas reactions occur:

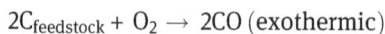

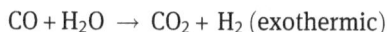

$$2C_{feedstock} + O_2 \rightarrow 2CO \text{ (exothermic)}$$

$$C_{feedstock} + H_2O \rightarrow CO + H_2 \text{ (endothermic)}$$

$$C_{feedstock} + 2H_2O \rightarrow CO_2 + 2H_2 \text{ (endothermic)}$$

$$CO + H_2O \rightarrow CO_2 + H_2 \text{ (exothermic)}$$

The ratio of steam and air may be used to modify the composition of the gas that is produced.

A.4 Flue gas

Flue gas (sometimes called exhaust gas or stack gas) is the gas that emanates from combustion plants and which contains the reaction products of fuel and combustion air and residual substances such as particulate matter (dust), sulfur oxides, nitrogen oxides, and carbon monoxide (Table A.5). When burning coal and/or waste materials, hydrogen chloride and hydrogen fluoride may be present in the flue gas as well as hydrocarbon derivatives and heavy metal derivatives. In many countries, as part of a national environmental protection program, exhaust gases must comply with strict governmental regulations regarding the limit values of pollutants such as dust, sulfur and nitrogen oxides, and carbon monoxide. To meet these limit values, combustion plants are equipped with flue gas cleaning systems such as gas scrubbers and dust filters.

Table A.5: Constituents of flue gas.

Constituents	Comment
Nitrogen	The main constituent (79 Vol.%) of air; fed to the combustion as part of the combustion air but is not involved directly in the combustion process. Minor quantities of this combustion air-related nitrogen, together with the nitrogen released from the fuel, are responsible for the formation of nitrogen oxides.
Carbon dioxide	A colorless and odorless gas with a slightly sour taste; produced during all combustion processes including respiration.
Water vapor	Hydrogen contained in the fuel will react with oxygen and form water (H_2O); this, together with the water content of the fuel and the combustion air, exists either as flue gas humidity (at higher temperatures) or as condensate (at lower temperatures).
Oxygen	The oxygen that has not been consumed by the combustion process remains as part of the flue gas and is a measure for the efficiency of the combustion.
Carbon monoxide	A colorless, odorless, toxic gas; formed predominantly during incomplete combustion of carbonaceous fuels.
Oxides of nitrogen	The nitrogen from the fuel and al;so the nitrogen from the combustion air reacts to a certain amount with oxygen of the combustion air and forms first nitric oxide (fuel-NO_x and thermal-NO_x).
Sulfur dioxide	A colorless, toxic gas with a pungent smell. It is formed through oxidation of sulfur that is present in the fuel; forms sulfurous acid and sulfuric acid with water or condensate.
Hydrogen sulfide	A toxic odorous gas which is a component of crude oil and natural gas and is therefore present in refineries and natural gas plants; also generated during some other industrial processes.

Table A.5 (continued)

Constituents	Comment
Hydrocarbons	An extensive group of chemical compounds that are composed of hydrogen and carbon; occur in crude oil, natural gas, and coal; formed through incomplete combustion processes.
Hydrocyanic acid	A very toxic liquid with a boiling point of only 25.6 °C (78.1 °F); it may exist in flue gases of incineration plants.
Ammonia	Relevant in flue gases in connection with denitrification plants.
Hydrogen halides	Occur in flue gas from the combustion of coal and/or waste material; the formation of hydrogen halides such as hydrogen chloride and hydrogen fluoride may occur, which form aggressive acids in humid atmospheres.
Solids (dust, soot)	Solid pollutants in flue gases originate from the incombustible components of solid or liquid fuels; include oxides of silica, aluminum, and calcium in case of coal.

As with other gases, the composition of flue gas depends on the type of fuel and the combustion conditions, for example, the air ratio value. Many flue gas components are air pollutants and must therefore, due to governmental regulations, be eliminated or minimized by special cleaning procedures before the gas is released to the atmosphere. For example, flue gas produced by the combustion of fossil fuels such as coal and crude oil can contain a significant amount of sulfur derivatives. In fact, when carbonaceous fuels, such as fossil fuels, are burned, approximately 90%+ w/w of sulfur in the feedstock is converted to sulfur dioxide (SO_2), which occurs under normal conditions of temperature in the furnace and the oxygen fed to the combustor. However, when there is an excess of oxygen present, the sulfur dioxide is oxidized to sulfur trioxide (SO_3) and at higher temperatures (approximately 800 °C, 1470 °F), the formation of sulfur trioxide is favored.

As with other gases, the composition of flue gas depends upon the properties of the fuel being combusted but, nevertheless, the composition will usually consist predominantly of nitrogen (usually on the order of 60% v/v and higher) derived from the combustion process using air as the oxidant, carbon dioxide, and water vapor, as well as any excess oxygen (also derived from the air). Flue gas can also contain a small percentage of a number of pollutants, such as (i) particulate matter, for example soot, (ii) carbon monoxide, (iii) nitrogen oxides, NO_x, and (iv) sulfur oxides, SO_x. The potential for the presence of hydrocarbon derivatives is low unless the combustion process uses a minimal of air and then also uses a hydrocarbonaceous feedstock. The particulate matter is composed of small particles of solid materials and small liquid droplets which give flue gases their smoky appearance. The nitrogen oxides are derived from nitrogen in the ambient air as well as from any nitrogen-containing compounds in the fuel. The sulfur dioxide is derived from any sulfur-containing compounds in the fuel.

At power plants, flue gas is often treated with a series of chemical processes and scrubbers, which remove pollutants such as sulfur dioxide and sulfur trioxide. Flue gas desulfurization units capture the sulfur dioxide (and the sulfur trioxide, if present) and electrostatic precipitators or fabric filters remove particulate matter produced by burning fossil fuels, particularly coal. Nitrogen oxides are treated either by modifications to the combustion process to prevent their formation, or by high temperature or catalytic reaction with ammonia (NH_3) or urea (H_2NCONH_2). In either case, the aim is to produce nitrogen gas, rather than nitrogen oxides. In the United States, there is a rapid deployment of technologies to remove mercury from flue gas. This is typically accomplished by absorption of sorbents or by capture in inert solids as part of the glue gas desulfurization process. Such scrubbing can lead to meaningful recovery of sulfur for further industrial use.

Examples of common flue gas cleaning processes are: (i) a wet scrubbing process, which uses a slurry of alkaline sorbent, usually limestone or lime, or seawater to scrub the gases; (ii) a spray-dry scrubbing process, which uses similar sorbent slurries as described in the first category; (iii) a wet sulfuric acid process, which allows the recovery of sulfur in the form of commercial quality sulfuric acid; (iv) a process commonly referred to as a SNOX flue gas desulfurization process, which removes sulfur dioxide, nitrogen oxides, and particulate matter from flue gases; and (v) a dry sorbent injection process, which introduces powdered hydrated lime or other sorbent material into the exhaust ducts to eliminate sulfur dioxide and sulfur trioxide from the process emissions.

In terms of flue gas (and biogas) cleaning, the separation of acid gaseous pollutants, the separation of hydrogen chloride, sulfur dioxide, and hydrogen fluorides (and any vestiges thereof) is essential. These gases form the greater part of the polluting constituents affected by absorption, preferentially by means of lime products [CaO, $Ca(OH)_2$] and also by sodium-based products ($NaOH$, $NaHCO_3$, and Na_2CO_3). The available gas cleaning processes for the absorption of the acid gaseous pollutants can be classified into the following three groups: (i) dry sorption, (ii) spray absorption/drying, and (iii) wet scrubbing.

The separation of the fly ash and the metals occurring in the form of particulate matter at the boiler outlet takes place likewise via filtration, so that this process step can be integrated into the absorption process for the acid gaseous pollutants. The separation of dioxins, furans, and those metals, in particular mercury, present in gaseous form at the boiler outlet generally takes place by adsorption on activated carbon, zeolites, open hearth furnace coke, or bentonite. For adsorption, either static or a moving bed adsorber unit or a filter layer adsorber unit may be employed, which offers the possibility of integration of the adsorption process with the filtering out of fly ash and reaction products from various gas cleaning concepts (Table A.6). Removal of nitrogen oxides (NO_x) from the flue gases can be achieved in conjunction with the above pollution control systems by the use of selective non-catalytic reduction.

Table A.6: Summary of processes used for gas cleaning.

Separation concept	Process	Principle
Adsorption	Pressure swing adsorption	Adsorption of carbon dioxide using a molecular sieve
Physical absorption	Pressurized water wash	Dissolution of carbon dioxide
	Selexol, Rectisol, Purisol processes	Dissolution of carbon dioxide in a specialized solvent
Chemical absorption	Alkanolamine wash	Chemical reaction of carbon dioxide with the alkanolamine
Membrane separation	Polymer membrane (dry separation)	Membrane permeability of hydrogen sulfide and carbon dioxide is higher than methane
Cryogenic process	Low-temperature process	Phase transformation of carbon dioxide to liquid methane remains gaseous

A.5 Landfill gas

Landfill gas, which is often included under the umbrella definition of biogas, is also produced from the decay of organic wastes (such as municipal solid waste that contains organic materials) but these wastes may not be biomass-type materials. Landfill sites offer another under-utilized source of biogas. When municipal waste is buried in a landfill, bacteria break down the organic material contained in garbage such as newspapers, cardboard, and food waste, producing gases such as carbon dioxide and methane. Rather than allowing these gases to go into the atmosphere, where they contribute to global warming, landfill gas facilities can capture them, separate the methane, and combust it to generate electricity, heat, or both.

Landfill gas is produced by wet organic waste decomposing under anaerobic conditions in a biogas. In fact, landfill gas is a product of three processes: (i) evaporation of volatile organic compounds such as low-boiling solvents, (ii) chemical reactions between waste components, and (iii) microbial action, especially methanogenesis. The first two processes depend strongly on the nature of the waste – the most dominant process in most landfills is the third process whereby anaerobic bacteria decompose organic waste to produce biogas, which consists of methane and carbon dioxide together with traces of other compounds. Despite the heterogeneity of waste, the evolution of gases follows well-defined kinetic pattern in which the formation of methane and carbon dioxide commences approximately 6 months after depositing the landfill material. The evolution of landfill gases reaches a maximum at approximately 20 years, then declines over the course of several decades.

Thus, landfill gas is another source of gas generated by microorganisms from the decomposition of the biodegradable organic fraction of municipal solid waste. This generally occurs under semi-controlled conditions in a landfill; its constituents depend on the composition and age of the waste. Large variations may exist in the composition of landfill gas due to differences in sources of municipal solid waste and operating conditions at the landfill. The three main gas constituents (methane (CH_4), carbon dioxide (CO_2), and hydrogen sulfide (H_2S)) are used to characterize landfill gas.

In the process, the waste is covered and mechanically compressed by the weight of the material that is deposited above. This material prevents oxygen exposure, thus allowing anaerobic microbes to thrive. Biogas builds up and is slowly released into the atmosphere if the site has not been engineered to capture the gas. Landfill gas released in an uncontrolled way can be hazardous since it can become explosive when it escapes from the landfill and mixes with oxygen.

As should be expected, the amount of methane that is produced varies significantly based on composition of the waste (Table A.7). The gas is a complex mix of different gases created by the action of microorganisms within a landfill. Typically, landfill gas is composed of 45–60% v/v methane, 40–60% v/v carbon dioxide, 0–1.0% v/v hydrogen sulfide, 0–0.2% v/v hydrogen (H_2), trace amounts of nitrogen (N_2), low-molecular-weight hydrocarbon derivatives (dry volume basis) and water vapor (saturated). The specific gravity of landfill gas is approximately 1.02 to 1.06. Trace amounts of other volatile organic compounds comprise the remainder (typically, 1–2% v/v or less) and these trace gases include a large array of species, such as low-molecular-weight hydrocarbon derivatives. Other minor components include hydrogen sulfide, nitrogen oxides, sulfur dioxide, non-methane volatile organic compounds, polycyclic aromatic hydrocarbon derivatives, polychlorinated dibenzodioxin derivatives, and polychlorinated dibenzofuran derivatives. All of the aforementioned agents are harmful to human health at high doses.

Table A.7: General composition of landfill gas.

Component	Composition, %v/v
Methane (CH_4)	35–55
Carbon dioxide (CO_2)	30–44
Nitrogen (N_2)	5–25
Oxygen (O_2)	0–6
Hydrogen sulfide (H_2S)	0–5
Hydrogen	0–0.2
Carbon monoxide	0–0.2

Table A.7 (continued)

Component	Composition, %v/v
Non-methane organic compounds	0.01–0.6
Siloxane derivatives	Trace-0.2
Heavy metals	Trace-0.2
Halogenated hydrocarbon compounds	Trace-0.2
Water vapor	Saturated

Landfill gas collection is typically accomplished through the installation of wells installed vertically and/or horizontally in the waste mass. Design heuristics for vertical wells call for about one well per acre of landfill surface, whereas horizontal wells are normally spaced about 50–200 feet apart on center. Efficient gas collection can be accomplished at both open and closed landfills, but closed landfills have systems that are more efficient, owing to greater deployment of collection infrastructure since active filling is not occurring. On average, closed landfills have gas collection systems that capture approximately 84% v/v of produced gas, compared to approximately 67% v/v for open landfills.

Landfill gas can also be extracted through horizontal trenches instead of vertical wells. Both systems are effective at collecting. Landfill gas is extracted and piped to a main collection header, where it is sent to be treated or flared. The main collection header can be connected to the leachate collection system to collect condensate forming in the pipes. A blower is needed to pull the gas from the collection wells to the collection header and further downstream.

The gas produced within a landfill site can be collected for various uses, such as direct utilization on site in a boiler or any type of combustion system, providing heat. Electricity can also be generated on site through the use of microturbines, steam turbines, or fuel cells. The landfill gas can also be sold off site and sent into natural gas pipelines. This approach requires the gas to be processed into pipeline quality, for example, by removing various contaminants and components. As should be expected, the amount of methane that is produced varies significantly based on composition of the waste. The efficiency of gas collection at landfills directly impacts the amount of energy that can be recovered – closed landfills (those no longer accepting waste) collect gas more efficiently than open landfills (those that are still accepting waste).

The gases produced within a landfill site can be collected for various uses, such as direct utilization on site in a boiler or any type of combustion system, providing heat. Electricity can also be generated on site through the use of microturbines, steam turbines, or fuel cells. The landfill gas can also be sold off-site and sent into natural gas pipelines. This approach requires the gas to be processed into pipeline quality, for example, by removing various contaminants and components. As should be expected, the amount of methane that is produced varies significantly based on composition of the waste. The efficiency of gas collection at landfills directly impacts the amount of energy that can be recovered – closed landfills (those no longer accepting waste) collect gas more efficiently than open landfills (those that are still accepting waste).

However, landfill gas cannot be distributed through utility natural gas pipelines unless it is cleaned up to less than 3% carbon dioxide and a few parts per million of hydrogen sulfide, because carbon dioxide and hydrogen sulfide corrode the pipelines. Thus, landfill gas must be treated to remove impurities, condensate, and particulates. Hence the need for analysis to determine the composition of the gas; however, the treatment system depends on the end use: (i) minimal treatment is needed for the direct use of gas in boiler, furnaces, or kilns; and (ii) using the gas in electricity generation typically requires more in-depth treatment.

Treatment systems are divided into primary and secondary treatment processing. Primary processing systems remove moisture and particulates. Gas cooling and compression are common in primary processing. Secondary treatment systems employ multiple cleanup processes, physical and chemical, depending on the specifications of the end use. Two constituents that may need to be removed are siloxane derivatives and sulfur-containing compounds, which are damaging to equipment and significantly increase maintenance cost. Adsorption and absorption are the most common technologies used in secondary treatment processing. Also, landfill gas can be converted to high-Btu gas by reducing the amount of carbon dioxide, nitrogen, and oxygen in the gas.

The high-Btu gas can be piped into existing natural gas pipelines or in the form of compressed natural gas or liquid natural gas. Compressed natural gas and liquid natural gas can be used on site to power hauling trucks or equipment or sold commercially. Three commonly used methods to extract carbon dioxide from gas are membrane separation, molecular sieve, and amine scrubbing. Oxygen and nitrogen are controlled by the design and operation of the landfill since the primary cause for oxygen or nitrogen in the gas is intrusion from outside into the landfill because of a difference in pressure.

Landfill gas condensate is a liquid that is produced in landfill gas collection systems and is removed as the gas is withdrawn from landfills. Production of condensate may be through natural or artificial cooling of the gas or through physical processes such as volume expansion. The condensate is composed principally of water and organic compounds. Often the organic compounds are not soluble in water and the condensate separates into a watery (aqueous) phase and a floating organic (hydrocarbon) phase which may constitute up to 5% v/v of the liquid.

A large number of acid and base/neutral compounds are typically present in the aqueous phase but, as expected, are dependent upon the compound types in the landfill. The organic phase can consist of hydrocarbon derivatives, xylene isomers, chloroethane derivatives, chloroethylene derivatives, benzene, toluene, other priority pollutants, and trace moisture.

A.6 Manufactured gas

Manufactured gas is a fuel-gas mixture made from other solid, liquid, or gaseous materials, such as coal, coke, or crude oil and should not be confused with natural gas. The principal types of manufactured gas are retort coal gas, coke oven gas, water gas, carbureted water gas, producer gas, oil gas, reformed natural gas, and reformed propane or liquefied petroleum gas (LPG). Several processes for making SNG from coal have been developed. Most of the manufactured gas, as it fits into the context of this book, is produced by any one of three processes which, at the present time, find limited use but are still active processes on some areas and these are: (i) coal carbonization process, (ii) the carbureted water gas process, and (iii) oil gas process.

The coal carbonization process was the primary commercial mode of manufacturing gas from ca. 1816 to 1875. After 1875, newer processes and technologies gradually replaced coal carbonization. Coal gas was produced through the distillation of bituminous coal in heated, anaerobic vessels called retorts. In this process, coal was broken down into its volatile components through the action of heat in an anaerobic environment. During the retorting phase, approximately 40% w/w of the coal is converted into volatile gases and liquids, while the remainder of the coal is converted into solids, primarily coke.

From the retort, some of the gaseous products are condensed (to liquids) and others remain in the gaseous state. The liquids (also called *liquors*) consist of water and coal tar. The products remaining in the gaseous phase are cooled to produce additional coal tar after which the gas is cooled further to remove any other non-gaseous impurities, such as ammonia and sulfur compounds. These are removed by washing the gas in water and by running the gas through beds of moist lime or moist iron oxides. After this final purification process, the coal gas is sent to storage.

The carbureted water gas process consists of enriching a form of coal gas, known as water gas (blue gas), to increase the heat content. Thus, by injecting oil into a vessel containing heated water gas, the oil and vapor combined, forms a gaseous fuel with a thermal content of approximately 300–350 Btu/ft^2. Typically, a carbureted water gas plant consists of a brick-lined, cylindrical, steel vessel, the generator, the carburetor (carbureted), and a super-heater. As the gas exits the generator, it is passed into the carburetor where the oil is introduced into the vapor.

The oil gas process is similar to the carbureted water gas process but consists of steam-cracking the oil in a steam environment to produce the raw gas rather than by distilling coal. From the generator, the gas is passed to a vaporizer, where it is enriched with additional injections of oil and then routed through a super-heater. After exiting the super-heater, the gas is scrubbed and processed for distribution in much the same way as was carbureted water gas. Many of the same waste products associated with the production of coal gas, notably tars containing polynuclear aromatic hydrocarbons (or polyaromatic hydrocarbons), were also generated during gas manufacture.

A.7 Refinery gas

The term refinery gas (also known as crude oil gas or petroleum gas) is often used to identify the gas that emanates as light ends (gases and volatile liquids) from the atmospheric distillation unit or from any one of several other refinery processes. Refinery gas is included here because of the frequent occurrences of co-processing the gas with natural gas. Basically, refinery gas is for industrial use as fuel and petrochemical raw materials; street gas is used as a residential and commercial fuel (LPG); in addition to its residential and commercial use, it is used as an industrial fuel.

For the purpose of this text, refinery gas not only describes LPG but also natural gas and refinery gas. In this section, each gas is, in turn, referenced by its name rather than the generic term *petroleum gas*. However, the composition of each gas varies and recognition of this is essential before the relevant testing protocols are selected and applied. Thus, refinery gas (fuel gas) is the non-condensable gas that is obtained during distillation of crude oil or treatment (cracking, thermal decomposition) of petroleum.

Refinery gas is produced in considerable quantities during the different refining processes and is used as fuel for the refinery itself and as an important feedstock for the production of petrochemicals. It consists mainly of hydrogen (H_2), methane (CH_4), ethane (C_2H_6), propane (C_3H_8), butane (C_4H_{10}), and olefins ($RCH=CHR^1$, where R and R^1 can be hydrogen or a methyl group) and may also include off-gases from petrochemical processes (Table A.8). Olefins such as ethylene ($CH_2=CH_2$, boiling point: −104 °C, −155 °F), propene (propylene, $CH_3CH=CH_2$, boiling point: −47 °C, −53 °F), butene (butene-1, $CH_3CH_2CH=CH_2$, boiling point: −5 °C, 23 °F), isobutylene (($CH_3)_2C=CH_2$, −6 °C, 21 °F), *cis-* and *trans*-butene-2 ($CH_3CH=CHCH_3$, boiling point: ca. 1 °C, 30 °F), and butadiene ($CH_2=CHCH=CH_2$, boiling point: −4 °C, 24 °F) as well as higher boiling olefins are produced by various refining processes.

Still gas is a broad terminology for low-boiling hydrocarbon mixtures and is the lowest boiling fraction isolated from a distillation (*still*) unit in the refinery. If the distillation unit is separating light hydrocarbon fractions, the still gas will be almost entirely methane with only traces of ethane (CH_3CH_3) and ethylene ($CH_2=CH_2$). If the distillation unit is handling higher boiling fractions, the still gas might also contain propane ($CH_3CH_2CH_3$), butane ($CH_3CH_2CH_2CH_3$), and their respective isomers. *Fuel gas* and still gas are terms that are often used interchangeably but the term *fuel gas*

Table A.8: Origin of crude oil-related gases.

Gas	Origin
Natural gas	Occurs naturally with or without crude oil
	A gaseous combination of hydrocarbons
	Predominantly C1 through C4 hydrocarbons
	May also contain gas condensate or natural gasoline
Refinery gas (process gas)	A combination of gases produced by distillation of crude oil
	Products from cracking of crude oil
	Consists of C2 to C4 hydrocarbons including olefin derivatives
	Boiling range of approximately −51 to −1 °C (−60 to 30 °F)
Tail gas	A combination of hydrocarbons from distillation of products from catalytically cracked feedstocks
	Predominantly of C1 to C4 hydrocarbons

is intended to denote the product's destination to be used as a fuel for boilers, furnaces, or heaters.

A group of refining operations that contributes to gas production are the thermal cracking and catalytic cracking processes. The thermal cracking processes (such as the coking processes) produce a variety of gases, some of which may contain olefin derivatives ($>C=C<$). In the visbreaking process, fuel oil is passed through externally fired tubes and undergoes liquid phase cracking reactions, which result in the formation of lower boiling fuel oil components. Substantial quantities of both gas and carbon are also formed in coking (both fluid coking and delayed coking) in addition to the middle distillate and naphtha. When coking a residual fuel oil or heavy gas oil, the feedstock is preheated and contacted with hot carbon (coke), which causes extensive cracking of the feedstock constituents of higher molecular weight to produce lower molecular weight products ranging from methane, via LPG and naphtha, to gas oil and heating oil. Products from coking processes tend to be unsaturated, and olefin-type components predominate in the tail gases from coking processes.

The various catalytic cracking processes convert higher boiling gas oil fractions to gaseous products, various naphtha fractions, fuel oil, and coke by contacting the feedstock with the hot catalyst. Thus, both catalytic and thermal cracking processes, the latter being now largely used to produce chemical raw materials, result in the formation of unsaturated hydrocarbon derivatives, particularly ethylene ($CH_2=CH_2$), but also propylene (propene, $CH_3CH=CH_2$), isobutylene [isobutene, $(CH_3)_2C=CH_2$], and n-butenes ($CH_3CH_2CH=CH_2$ and $CH_3CH=CHCH_3$) in addition to hydrogen (H_2), methane (CH_4) and smaller quantities of ethane (CH_3CH_3), propane ($CH_3CH_2CH_3$), and butane isomers [$CH_3CH_2CH_2CH_3$, $(CH_3)_3CH$]. Diolefins such as butadiene ($CH_2=CHCH=CH_2$) and are also present.

In a series of reforming processes, distillation fractions that include paraffin derivatives and naphthene derivatives (cyclic non-aromatic) are treated in the presence of hydrogen and a catalyst to produce lower molecular weight products or are isomerized to more highly branched hydrocarbon derivatives. Also, the catalytic reforming process not only results in the formation of a liquid product of higher octane number but also produce substantial quantities of gaseous products. The composition of these gases varies in accordance with process severity and the properties of the feedstock. The gaseous products are not only rich in hydrogen but also contain hydrocarbon derivatives from methane to butane derivatives, with a preponderance of propane ($CH_3CH_2CH_3$), n-butane ($CH_3CH_2CH_2CH_3$), and isobutane [$(CH_3)_3CH$]. Since all catalytic reforming processes require substantial recycling of a hydrogen stream, it is normal to separate reformer gas into a propane ($CH_3CH_2CH_3$) and/or a butane [$CH_3CH_2CH_2CH_3$/ or $(CH_3)_3CH$] stream, which becomes part of the refinery LPG production (Table A.9), and a lower boiling gaseous fraction, a part, is recycled.

A further source of refinery gas is produced by the hydrocracking process which is a high-pressure pyrolysis process carried out in the presence of fresh and recycled hydrogen. The feedstock is again heavy gas oil or residual fuel oil, and the process is mainly directed at the production of additional middle distillates and gasoline. Since hydrogen is to be recycled, the gases produced in this process again must be separated into lighter and heavier streams; any surplus recycle gas and the LPG from the hydrocracking process are both saturated.

Table A.9: Properties of propane and butane – the constituents of liquefied petroleum gas (LPG).

	Propane	n-Butane
Formula	C_3H_8	C_4H_{10}
Boiling point	−44 °F	32 °F
Specific gravity – gas (air = 1.00)	1.53	2.00
Specific gravity – liquid (water = 1.00)	0.51	0.58
Lbs./gallon – liquid @ 60 °F	4.24	4.81
BTU/gallon – gas @ 60 °F	91,690	102,032
BTU/lb – gas	21,591	21,221
BTU/ft³ – gas @ 60 °F	2,516	3,280
Cubic feet of vapor @ 60 °F/gal of liquid @ 60 °F	36.39	31.26
Cubic feet of vapor @ 60 °F/lb of liquid @ 60 °F	8.547	6.506

Table A.9 (continued)

	Propane	n-Butane
Flash point, F	−156	−76
Limits of inflammability % of gas in air mixture:		
At lower limit – %	2.4	1.9
At upper limit – %	9.6	8.6
Ignition temperature in air, F	920–1,020	900–100

Both hydrocracker and catalytic reformer tail gases are commonly used in catalytic desulfurization processes. In the latter, feedstocks ranging from light to vacuum gas oils are passed at pressures on the order of 500–1,000 psi with hydrogen over a hydrofining catalyst. This results mainly in the conversion of organic sulfur compounds to hydrogen sulfide:

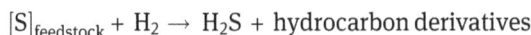

$$[S]_{feedstock} + H_2 \rightarrow H_2S + \text{hydrocarbon derivatives}$$

The process also has the potential to produce lower-boiling hydrocarbon derivatives by hydrocracking.

Olefin derivatives are not typical constituents of natural gas but do occur in refinery gases, which can be complex mixtures of hydrocarbon gases and non-hydrocarbon gas (Table 1.2). Some gases may also contain inorganic compounds, such as hydrogen, nitrogen, hydrogen sulfide, carbon monoxide and carbon dioxide. Many low–molecular-weight olefins (such as ethylene and propylene) and diolefins (such as butadiene) which are produced in the refinery are isolated for petrochemical use. The individual products are (i) ethylene, (ii) propylene, and (iii) butadiene.

Ethylene (C_2H_4) is a normally gaseous olefinic compound having a boiling point of approximately −104 °C (−155 °F) which may be handled as a liquid at high pressures and low temperatures. Ethylene is made normally by cracking an ethane or naphtha feedstock in a high-temperature furnace and subsequent isolation from other components by distillation. The major uses of ethylene are in the production of ethylene oxide, ethylene dichloride, and the polyethylene polymers. Other uses include the coloring of fruit, rubber products, ethyl alcohol, and medicine (anesthetic).

Propylene concentrates are mixtures of propylene and other hydrocarbons, principally propane and trace quantities of ethylene, butylenes, and butanes. Propylene concentrates may vary in propylene content from 70 mol% to over 95 mol% and may be handled as a liquid at normal temperatures and moderate pressures. Propylene concentrates are isolated from the furnace products mentioned in the preceding paragraph on ethylene. Higher purity propylene streams are further purified by distillation and extractive techniques. Propylene concentrates are used in the production of propylene oxide, isopropyl alcohol. polypropylene, and the synthesis of isoprene. As is the case for ethylene, moisture in propylene is critical.

Butylene concentrates are mixtures of butene-1, *cis-* and *trans-*butene-2, and, sometimes, isobutene (2-methyl propylene) (C$_4$H$_8$).

Butene-1

cis-Butene-2

trans-Butene-2

Isobutene (2-methylpropene, 2-methyl propylene)

These products are stored as liquids at ambient temperatures and moderate pressures. Various impurities such as butane, butadiene, and the C$_5$ hydrocarbons are generally found in butylene concentrates. The majority of the butylene concentrates are used as a feedstock for either: (i) an alkylation plant, where isobutane and butylenes are reacted in the presence of either sulfuric acid or hydrofluoric acid to form a mixture of C$_7$–C$_9$ paraffins used in gasoline, or (ii) butylene dehydrogenation reactors for butadiene production.

Butadiene (C$_4$H$_6$, CH$_2$=CHCH=CH$_2$) is a normally gaseous hydrocarbon having a boiling point of −4.38 °C (24.1 °F) which may be handled as a liquid at moderate pressure. Ambient temperatures are generally used for long-term storage due to the easy formation of butadiene dimer (4-vinyl cyclohexene-l). Butadiene is produced by two major methods: the catalytic dehydrogenation of butane or butylenes or both, and as a by-product from the production of ethylene. In either case, the butadiene must be isolated from other components by extractive distillation techniques and subsequent purification to polymerization-grade specifications by fractional distillation. The largest end use of butadiene is as a monomer for production of GR-S synthetic rubber. Butadiene is also chlorinated to form 2-chlorobutadiene (chloroprene) (CH_2=CHCCl=CH_2) that is a feedstock used to produce neoprene (a polychloroprene rubber).

The major quality criteria for butadiene are the various impurities that may affect the polymerization reactions for which butadiene is used. The gas chromatographic examination of butadiene can be employed to determine the gross purity as well as C_3, C_4, and C, impurities. Most of these hydrocarbons are innocuous to polymerization reactions, but some, such as butadiene-1,2 and pentadiene-1,4, are capable of polymer cross-linking.

A.8 Synthesis gas

Synthesis gas (also commonly referred to as syngas) is the name given to a gas mixture that contains varying amounts of carbon monoxide and hydrogen which is generated by the gasification of a carbonaceous (carbon-containing) feedstock. The name comes from the use of the gas as intermediates in creating SNG and for producing ammonia or methanol as well as a host of other chemicals via the Fischer–Tropsch process (Table A.10).

Table A.10: Illustration of the role of synthesis gas in the production of various products.

	Stage 1	Stage 2	Stage 3	Stage 4	Products
Feedstock					
	Gasification				
	Synthesis gas				
			Gas cleaning		
				Fischer–Tropsch process	
					Naphtha*
					Gasoline
					Kerosene
					Diesel fuel
					Jet fuel
					Wax

*Gasoline blend stock.

Gasification to produce synthesis gas can proceed from most types of carbonaceous feedstocks, including biomass and plastic waste. The resulting synthesis gas burns cleanly into water vapor and carbon dioxide:

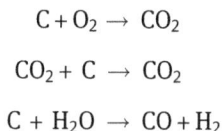

$$C + O_2 \rightarrow CO_2$$

$$CO_2 + C \rightarrow CO_2$$

$$C + H_2O \rightarrow CO + H_2$$

Examples of such processes include (i) steam reforming of natural gas or higher molecular weight hydrocarbon derivatives, (ii) gasification of crude oil residues, (iii) the gasification of extra heavy oil and tar sand bitumen that may or may not have been topped to remove constituents boiling below, say, 345 °C (650 °F), (iv) gasification of coal, and (v) gasification of organic waste in some types of waste-to-energy facilities or waste formerly destined for a landfill site.

Currently, in many refineries, hydrogen is produced from a variety of carbonaceous feedstocks by gasification and the subsequent processing of the resulting gaseous product to produce synthesis gas. In its simplest form, coal gasification works by first reacting coal with oxygen and steam under high pressures and temperatures to form a synthesis gas consisting primarily of carbon monoxide and hydrogen. This synthesis gas is cleaned of virtually all of its impurities and shifted to produce additional hydrogen. The clean gas is sent to a separation system to recover hydrogen.

Natural gas is the most common feedstock for hydrogen production since it meets all the requirements for reformer feedstock. Natural gas typically contains more than 90 percent methane and ethane with only a few percent of propane and higher boiling hydrocarbon derivatives. Natural gas may (or most likely will) contain traces of carbon dioxide with some nitrogen and other impurities. Purification of natural gas, before reforming, is usually relatively straightforward. Traces of sulfur must be removed to avoid poisoning the reformer catalyst; zinc oxide treatment in combination with hydrogenation is usually adequate.

In the steam reforming process, natural gas (methane) reacts with steam to form hydrogen:

$$CH_4 + H_2O \rightarrow 3H_2 + CO, \quad \Delta H_{298\,K} = +97,400 \text{ Btu/lb}$$

ΔH_{298K} is the heat of reaction. A more general form of the equation that shows the chemical balance for higher boiling hydrocarbon derivatives is

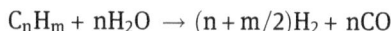

$$C_nH_m + nH_2O \rightarrow (n + m/2)H_2 + nCO$$

The reaction is typically carried out at approximately 815 °C (1500 °F) over a nickel catalyst packed into the tubes of a reforming furnace. The high temperature also causes the hydrocarbon feedstock to undergo a series of cracking reactions, plus the reaction of carbon with steam:

$$CH_4 \rightarrow 2H_2 + C$$

$$C + H_2O \rightarrow CO + H_2$$

Carbon is produced on the catalyst at the same time that hydrocarbon is reformed to hydrogen and carbon monoxide. With natural gas or similar feedstock, reforming predominates and the carbon can be removed by reaction with steam as fast as it is formed. When higher boiling feedstocks are used, the carbon is not removed fast enough and builds up thereby requiring catalyst regeneration or replacement. Carbon buildup on the catalyst (when high-boiling feedstocks are employed) can be avoided by addition of alkali compounds, such as potash, to the catalyst thereby encourage or promoting the carbon–steam reaction.

However, even with an alkali-promoted catalyst, feedstock cracking limits the process to hydrocarbon derivatives with a boiling point less than 180 °C (350 °F). Natural gas, propane, butane, and low-boiling naphtha are most suitable. Pre-reforming, a process that uses an adiabatic catalyst bed operating at a lower temperature, can be used as a pretreatment to allow heavier feedstocks to be used with lower potential for carbon deposition (coke formation) on the catalyst.

After reforming, the carbon monoxide in the gas reacts with steam to form additional hydrogen (the *water gas shift* reaction):

$$CO + H_2O \rightarrow CO_2 + H_2, \Delta H_{298\,K} = -16,500 \text{ Btu/lb}$$

This leaves a mixture consisting primarily of hydrogen and carbon monoxide that is removed by conversion to methane:

$$CO + 3H_2O \rightarrow CH_4 + H_2O$$

$$CO_2 + 4H_2 \rightarrow CH_4 + 2H_2O$$

The critical variables for steam reforming processes are (i) temperature, (ii) pressure, and (iii) the steam/hydrocarbon ratio. Steam reforming is an equilibrium reaction, and conversion of the hydrocarbon feedstock is favored by high temperature, which in turn requires higher fuel use. Because of the volume increase in the reaction, conversion is also favored by low pressure, which conflicts with the need to supply the hydrogen at high pressure. In practice, materials of construction limit temperature and pressure.

On the other hand, and in contrast to reforming, shift conversion is favored by low temperature. The gas from the reformer reacts over iron oxide catalyst at 315–370 °C (600–700 °F) with the lower limit being dictated activity of the catalyst at low temperature.

Hydrogen can also be produced by partial oxidation (POX) of hydrocarbon derivatives in which the hydrocarbon is oxidized in a limited or controlled supply of oxygen:

$$2CH_4 + O_2 \rightarrow CO + 4H_2, \quad \Delta H_{298\,K} = -10,195 \text{ Btu/lb}$$

The shift reaction also occurs and a mixture of carbon monoxide and carbon dioxide is produced in addition to hydrogen. The catalyst tube materials do not limit the reaction temperatures in POX processes and higher temperatures may be used that enhance the conversion of methane to hydrogen. Indeed, much of the design and operation of hydrogen plants involves protecting the reforming catalyst and the catalyst tubes because of the extreme temperatures and the sensitivity of the catalyst. In fact, minor variations in feedstock composition or operating conditions can have significant effects on the life of the catalyst or the reformer itself. This is particularly true of changes in molecular weight of the feed gas, or poor distribution of heat to the catalyst tubes.

Since the high temperature takes the place of a catalyst, POX is not limited to the lower boiling feedstocks that are required for steam reforming. POX processes were first considered for hydrogen production because of expected shortages of lower boiling feedstocks and the need to have available disposal method for higher boiling, high-sulfur streams such as asphalt or crude oil coke.

The fuels petrochemical industry, as the name implies, is based upon the production of chemicals from, initially, crude oil. However, there is more to the industry than just crude oil products. The petrochemical industry also deals with chemicals manufactured from the by-products of crude oil refining, such as natural gas, natural gas liquids, and (in the context of this book) other feedstocks such as coal, oil shale, and biomass. The structure of the industry is extremely complex, involving thousands of chemicals and processes, and there are many inter-relationships within the industry with products of one process being the feedstocks of many others. For most chemicals, the production route from feedstock to final products is not unique, but includes many possible alternatives. As complicated as it may seem, however, this structure is comprehensible, at least in general form.

At the beginning of the production chain are the raw feedstocks such as crude oil, natural gas, and alternate carbonaceous feedstocks tar. From these are produced a relatively small number of important building blocks which include primarily, but not exclusively, the lower boiling olefins and aromatic derivatives, such as ethylene, propylene, butylene isomers, butadiene, benzene, toluene, and the xylene isomers. These building blocks are then converted into a complex array of thousands of intermediate chemicals. Some of these intermediates have commercial value in and of themselves, and others are purely intermediate compounds in the production chains. The final products of the petrochemical industry are generally not consumed directly by the public, but are used by other industries to manufacture consumer goods.

Thus, on a scientific basis, as might be expected, the petrochemical industry is concerned with the production and trade of petrochemicals that have a wide influence on lifestyles, though the production of commodity chemicals and specialty chemicals has a marked influence on lifestyles:

Natural gas → bulk chemicals (commodity chemicals) → specialty chemicals

The basis of the petrochemical industry and, therefore, petrochemical production consists of two steps: (i) feedstock production from primary energy sources to feedstocks and (ii) and petrochemical production from feedstocks:

Natural gas → feedstock production → petrochemical products

This simplified equation encompasses the multitude of production routes available for most chemicals. In the actual industry, many chemicals are products of more than one method, depending upon local conditions, corporate polices, and desired by-products. There are also additional methods available, which have either become obsolete and are no longer used, or which have never been used commercially but could become important as technology, supplies, and other factors change. Such versatility, adaptability, and dynamic nature are three of the important features of the modern petrochemical industry.

Thus, the fuels and petrochemical industry began as suitable by-products became available through improvements in the refining processes. As the decades of the 1920s and 1930s closed, the industry had developed in parallel with the crude oil industry and has continued to expand rapidly since the 1940s as the crude oil refining industry was able to provide relatively cheap and plentiful raw materials (Speight, 2002; Hsu and Robinson, 2006; Gary et al., 2007; Lee et al., 2007; Speight, 2011). The supply–demand scenario as well as the introduction of many innovations have resulted in basic chemicals and plastics which become the key building blocks for manufacture of a wide variety of durable and nondurable consumer goods. Chemicals and plastic materials provide the fundamental building blocks that enable the manufacture of the vast majority of consumer goods. Moreover, the demand for chemicals and plastics is driven by global economic conditions, which are linked to demand for consumer goods.

Appendix B
Illustration of the terms used to describe the resources and reserves of natural gas

Contingent resources: The quantities of natural gas estimated, as of a given date, to be potentially recoverable from known accumulations, but the applied project(s) are not yet considered mature enough for commercial development due to one or more contingencies. Contingent resources may include, for example, projects for which there are currently no viable markets, or where commercial recovery is dependent on technology under development or where evaluation of the accumulation is insufficient to clearly assess commerciality. Contingent resources are further categorized in accordance with the level of certainty associated with the estimates and may be subclassified based on project maturity and/or characterized by their economic status.

Developed reserves: Reserves that are expected to be recovered from existing wells (including reserves behind pipe). Improved recovery reserves are considered developed only after the necessary equipment has been installed, or when the costs to do so are relatively minor. Developed reserves may be subcategorized as producing or nonproducing.

Discovered natural gas initially in place: The quantity of natural gas that is estimated, as of a given date, to be contained in known accumulations prior to production.

Estimated ultimate recovery: A term that may be applied to any accumulation or group of accumulations (discovered or undiscovered) to define those quantities of natural gas estimated, as of a given date, to be potentially recoverable under defined technical and commercial conditions plus those quantities already produced (total of recoverable resources). May also be termed basin potential.

Gas reserve: The gas reserve applied to the prospect gas in the reservoir, generally, means that the future production of gas can be expected from the reservoir; often the word reserve is used to denote ultimate recovery.

Nonproducing reserves: These include shut-in and behind-pipe reserves. Shut-in reserves are expected to be recovered from completion intervals open at the time of the estimate, but have not started producing, or were shut in for market conditions of pipeline connection, or were not capable of production for mechanical reasons, and the time when sales will start is uncertain. Behind-pipe reserves are expected to

---

Note: Listed alphabetically.

https://doi.org/10.1515/9783110691023-011

be recovered from zones behind casing in existing wells, which will require additional completion work or a future recompletion prior to the start of production.

Possible reserves: Reserves that are associated with known accumulations and are less certain to be recovered than probable reserves. In general, possible reserves may include (i) reserves indicated by structural and/or stratigraphic extrapolation from developed areas, (ii) reserves located where reasonably definitive geophysical interpretations indicate an accumulation larger than could be included within the proved and probable limits, (iii) reserves in formations that have favorable log characteristics but questionable productivity, (iv) reserves in untested fault segments adjacent to proven reservoirs where a reasonable doubt exist as to whether such fault segment contains recoverable hydrocarbons, (v) incremental reserves attributable to infill drilling that are subject to technical or regulatory uncertainty, and (vi) reserves from a planned.

Producing reserves: Reserves that are expected to be recovered from completion intervals open at the time of the estimate and producing to market. Improved recovery reserves are considered to be producing only after an improved recovery project is in operation. Unproved (probable or possible) producing reserves are in addition to proved producing reserves, such as (i) reserves that may be recovered from portions of the reservoir downdip from proved reserves or (ii) reserves that may be recovered if a higher recovery factor is realized than was used in the estimate of proved reserves.

Production: The cumulative quantity of natural gas that has been recovered at a given date. While all recoverable resources are estimated and production is measured in terms of the sales product specifications, raw production (sales plus non-sales) quantities are also measured and required to support engineering analyses based on reservoir voidage. Multiple development projects may be applied to each known accumulation, and each project will recover an estimated portion of the initially-in-place quantities. The projects are subdivided into *commercial* and *sub-commercial*, with the estimated recoverable quantities being classified as reserves and contingent resources, respectively, which are defined as follows.

Probable reserves: Reserves that are attributed to known accumulations and are less certain to be recovered than proved reserves. In general, probable reserves may include (i) reserves that appear to exist a reasonable distance beyond the proved limits of productive reservoirs, where fluid contacts have not been determined and proved limits are established by the lowest known structural occurrence of hydrocarbons, (ii) reserves in formations that appear to be productive from core and/or log characteristics only, but that lack definitive tests or analogous producing reservoirs in the area, (iii) reserves in a portion of a formation that has been proved productive in other areas in a field, but that is separated from the proved area by faults, (iv) reserves obtainable by improved recovery methods and located where an improved recovery method (that has yet to be established through repeated commercially successful operation) is planned but not yet in operation, and where a successful pilot test has not been performed but reservoir and formation characteristics appear favorable for its success, (v) reserves in the same reservoir as proved reserves that would be recoverable if a more efficient primary recovery mechanism were to develop than that assumed in estimating proved reserves, (vi) incremental reserves attributable to infill drilling where closer statutory spacing had not been approved at the time of the estimate, and (vii) reserves that are dependent for recovery on a successful workover, treatment, retreatment, change of equipment, or other mechanical procedures, when such procedures have not been proved successful in wells exhibiting similar behavior in analogous formations.

Prospective resources: The quantities of natural gas estimated, as of a given date, to be potentially recoverable from undiscovered accumulations by application of future development projects. Prospective resources have both an associated chance of discovery and a chance of development. Prospective resources are further subdivided in accordance with the level of certainty associated with recoverable estimates, assuming their discovery and development and may be subclassified based on project maturity.

Proved reserves (proven reserves): Reserves that are attributed to known reservoirs, proved reserves can be estimated with reasonable certainty. In general, reserves are considered proved if commercial producibility of the reservoir is supported by actual production or formation tests. The term proved refers to the estimated volume of reserves and not just to the productivity of the well or reservoir. In certain instances, proved reserves may be assigned on the basis of a combination of core analysis and/or electrical and other type logs that indicate the reservoirs are analogous to reservoirs in the same areas that are producing, or have demonstrated the ability to produce in a formation test. The area of a reservoir considered proved includes (i) the area delineated by drilling and defined by fluid contacts, if any, and (ii) the undrilled areas that can be reasonably judged as commercially productive on the basis of available geological and engineering data. In the absence of data on fluid contacts, the lowest known structural occurrence of hydrocarbons controls the proved limit unless otherwise indicated by definitive engineering or performance data.

Proved undeveloped reserves: Reserves that are assigned to undrilled locations that satisfy the following conditions: (i) the locations are direct offsets to wells that have indicated commercial production in the objective formation, (ii) it is reasonably certain that the locations are within the known proved productive limits of the objective formation, (iii) the locations conform to existing well spacing regulation, if any, and (iv) it is reasonably certain that the locations will be developed.

Reserves for other undrilled locations are classified as proved undeveloped only in those cases where interpretation of data from wells indicates that the objective formation is laterally continuous and contains commercially recoverable hydrocarbons at locations beyond direct offsets.

Recovery estimates during reservoir life: There are three periods which are described as follows:

Period I: The period before drilling the well, and the existence of gas is determined based on geological data.

Period II: The period after drilling the well, and depending on the well data, the gas volume is determined in the reservoir using the volumetric method.

Period III: The period that depends on the production data, and future recovery is estimated using the material balance method.

Reserves: The quantities of natural gas anticipated to be commercially recoverable by application of development projects to known accumulations from a given date forward under defined conditions. Reserves must further satisfy four criteria: they must be discovered, recoverable, commercial, and remaining (as of the evaluation date), based on the development project(s) applied. Reserves are further categorized in accordance with the level of certainty associated with the estimates and may be subclassified based on project maturity and/or characterized by development and production status.

Reserve status categories: These categories define the development and producing status of wells and/or reservoirs. They may be applied to proven reserves or unproven (probable or possible) reserves.

Total natural gas initially-in-place: The quantity of natural gas that is estimated to exist originally in naturally occurring accumulations. It includes the quantity of natural gas that is estimated, as of a given date, to be contained in known accumulations prior to production, plus those estimated quantities in accumulations yet to be discovered (equivalent to total resources).

Undeveloped reserves: Reserves that are expected to be recovered: (i) from new wells on undrilled acreage, (ii) from deepening existing wells to a different reservoir, or (iii) where a relatively large expenditure is required to recomplete an existing well, or to install production or transportation facilities for primary or improved recovery projects. Undeveloped reserves usually will be distinguished from developed reserves. The ownership status of reserves may change due to the expiration of a production license or contract; when relevant to reserve assignment, such changes should be identified for each reserve classification.

Undiscovered natural gas initially-in-place: This is the quantity of natural gas estimated, as of a given date, to be contained within accumulations yet to be discovered.

Unproved reserves: Reserves based on geologic and/or engineering data similar to
that used in estimates of proved reserves, but technical, contractual, or regulatory
uncertainties preclude such reserves being classified as proved. Estimates of unproved reserves may be made for internal planning of special evaluations, but are
not routinely complied. Unproved reserves are not to be added to proved reserves
because of different levels of uncertainty. Unproved reserves may be divided into
two subclassifications: probable and possible.

Unrecoverable reserves: Reserves that are a part of the discovered reserves or undiscovered natural gas initially-in-place quantities that are estimated, as of a given
date, not to be recoverable by future development projects. A portion of these quantities may become recoverable in the future as commercial circumstances change or
technological developments occur; the remaining portion may never be recovered
due to physical/chemical constraints represented by subsurface interaction of fluids
and reservoir rock.

Volumetrics: Gas volumetrics is a static measurement based on a geologic model
that uses geometry to describe the volume of gas in the reservoir. Two main techniques are commonly employed in gas volumetrics, namely (i) the volumetric methods and (ii) the material balance method.

In the volumetric method, the gas volume is calculated using the following equation:

$$G_i = 43.56 \times V_s \times \varnothing \times (1 - S_{wi}) \times (1/B_g)$$

In this equation, G_i is the initial gas in place, V_s is the sand volume, $\varnothing$ is the porosity, S_{wi} is the water saturation, and B_g is the gas formation volume factor.

In the material balance method, the gas volume is calculated using the following equation:

$$G_i = G_P + G_R$$

In this equation, G_i is the initial gas in place, G_P is the produced gas, and G_R is the
remaining gas.

In terms of recovery estimates during the life of the reservoir, there are three
periods: (i) Period I, which is the period before drilling the well, and the existence
of gas is determined based on geological data; (ii) Period II, which is the period
after drilling the well, and depending on the well data, the gas volume is determined in the reservoir using the volumetric method; and (iii) Period III, which is the
period that depends on the production data – production is done and future recovery is estimated using the material balance method.

Appendix C
Properties of natural gas

Natural gas engineers invariably deal with gas mixtures and rarely with single-component gases. Since natural gas is a mixture of hydrocarbon compounds and because this mixture is varied in types as well as the relative amounts of the individual components, the overall physical properties will vary which determines the behavior of the gas under various processing conditions. If the composition of the gas mixture is known, the overall physical properties can be established from the physical properties of each pure component in the mixture using Kay's mixing rule which involves the use of a pseudo-critical pressure and pseudo-critical temperature for the mixture, defined in terms of the critical pressures and temperatures of the mixtures components as:

$$T'c = y_A T_{cA} + y_B T_{cB} + \cdots$$

$$P'c = y_A P_{cA} + y_B P_{cB} + \cdots$$

$$V'_r[\text{ideal}] = V/[RT'c)/P'c]$$

Using the same procedure as with single-component systems, the pseudo-reduced quantities can be derived and obtain the (generalized) compressibility factor from a chart can be obtained.

Physical properties that are most useful in natural gas processing are molecular weight, boiling point, freezing point, density, critical temperature, critical pressure, critical volume, heat of vaporization, and specific heat.

The composition of a natural gas mixture may be expressed as either the mole fraction, volume fraction, or weight fraction of its components. These may also be expressed as mole percent, volume percent, or weight percent by multiplying the fractional values by 100. Volume fraction is based on gas component volumes measured at standard conditions, so that volume fraction is equivalent to mole fraction. The mole fraction, Y_i, is defined as:

$$Y_i = n_i/\Sigma n_i$$

where n_i is the mole fraction of component i, n_i is the number of moles of component I, Σn_i is the total number of moles of all components in the mixture. The volume fraction is defined as:

$$(\text{volume fraction})_i = Vi/\Sigma Vi = y_i$$

where Vi is the volume occupied by component i at standard conditions and ΣVi is the volume of total mixture measured at standard conditions

The weight fraction, ω_i, is defined as

https://doi.org/10.1515/9783110691023-012

$$\omega_i = Wi/\Sigma Wi$$

where ω_i is the weight fraction of component i, Wi is the weight of component I, and ΣWi is the total weight of the mixture.

There is no single composition of components which might be termed *typical* natural gas. Methane and ethane often constitute the bulk of the combustible components; carbon dioxide (CO_2) and nitrogen (N_2) are the major non-combustible (inert) components. Thus, sour gas is natural gas that occurs mixed with higher levels of sulfur compounds (such as hydrogen sulfide, H_2S, and mercaptans or thiols, RSH) and which constitutes a corrosive gas. The sour gas requires additional processing for purification. Olefins are also present in the gas streams from various refinery processes and are not included in liquefied petroleum gas but are removed for use in petrochemical operations.

The analysis of any gas streams, compared to liquid streams, is relatively simple because bulk characterization of a single phase is implicit. However, when gas condensate is present, the analysis is more complicated. In the case of gas condensate, besides bulk analysis, there may be interest on surface composition (often quite distinct to that of the bulk phase). Compositional analysis in which the components of the mixture are identified, may be achieved by (i) physical means, which is measurement of physical properties; (ii) pure chemical means, which is measurement of chemical properties; or more commonly (iii) by physico-chemical means. Gas analysis is even more may be dangerous and difficult if the composition is a complete unknown. However, when some main constituents are known to occur, the analysis gains accuracy (and may be easier) if the known component is removed; this is particularly important in the case of water vapor, which may condense on the instruments, or constituents when the molecular behavior may complicate spectral analyses.

Knowledge of the properties of natural gas is essential for solving the problems related to gas engineering. Since the properties of gas streams vary with different pressure and temperature during the whole process of gas production and use. The following sections present an outline of the properties that play important role in gas production, prediction of gas behavior, and evaluation. These include the (i) phase behavior, (ii) whether or not the gas is an ideal gas, (iii) the compressibility of the gas, (v) the specific gravity of the gas, often compared to air, (vi) the gas deviation factor, (vii) the formation volume factor, and (viii) the viscosity of the gas.

C.1 Phase behavior

Conventional gas reservoirs have been characterized in many different ways but most commonly on the basis of the surface-producing gas-oil ratio (GOR). Using this method, any well (or field) that produces at a GOR in excess of 100,000 cu ft per barrel of oil (standard cubic feet per stock tank barrel (scf/STB) is considered a gas well; one producing with a GOR of 5000 to 100,000 scf/STB, a gas-condensate well; and one producing with a GOR of zero to several thousands scf/STB, an oil well. In practice, similar surface GORs have been obtained for reservoirs containing a variety of hydrocarbon fluid compositions, existing over a wide range of reservoir pressures and temperatures, and producing with natural or artificial mechanisms. This has resulted in both technical and legal misunderstanding of the nature of conventional gas reservoirs. Therefore, the simplified classification described above is considered inadequate.

Conventional gas reservoirs should be defined on the basis of their initial reservoir pressure and temperature on the usual pressure–temperature (P–T) phase diagram. P–T phase diagrams show the effects of pressure and temperature on the physical state of a hydrocarbon system.

If the phase diagram is such that the separator conditions lies outside the two-phase envelope in the single-phase (gas) region, then only gas will exist on the surface. No liquid will be formed in the reservoir or at the surface and the gas is called dry natural gas. The word *dry* indicates that the liquid does not contain enough of the heavier hydrocarbons to form a liquid at surface conditions. Nevertheless, it may contain liquid fractions, which can be removed by low-temperature separation or by natural gasoline plants.

A multicomponent mixture exhibits an envelope for liquid/vapor phase change in the pressure/temperature diagram, which contains a bubble-point line and a dew-point line, compared with only a phase-change line for a pure component. For a pure substance, a decrease in pressure causes a change of phase from liquid to gas at the vapor–pressure line; likewise, in the case of a multicomponent system, a decrease in pressure causes a change of phase from liquid to gas at temperatures below the critical temperature.

The pressure–volume relationships obtained can be plotted on a pressure–volume diagram with the bubble point and dew point locus also included (Figure C.1). The bubble point and dew point curves join at a point known as the critical point. The region under the bubble point–dew point envelope is the region where the vapor phase and liquid phase can coexist, and hence have an interface (the surface of a liquid drop or of a vapor bubble). The region above this envelope represents the region where the vapor phase and liquid phase do not coexist. The bubble point, dew point, and single-phase regions (Figure C.1) are often used to classify reservoirs. At temperatures greater than the cricondentherm, which is the maximum temperature for the formation of two phases, only one phase occurs at any pressure. For instance, if the hydrocarbon mixture (Figure C.1) was to occur in a reservoir at temperature T_A and pressure p_A (point A), a decline in pressure at approximately constant temperature caused by removal of fluid from the reservoir would not cause the formation of a second phase.

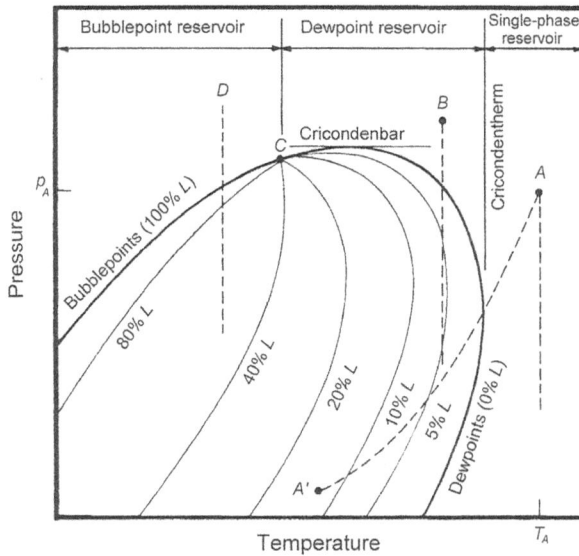

Figure C.1: A pressure–temperature phase diagram for reservoir systems.

Dry gas is predominantly methane and is in the gas phase under all conditions of pressure and temperature encountered during the production phases from reservoir conditions involving transport and process conditions. In particular, no hydrocarbon-based liquids are formed from the gas although liquid water can condense. Dry gas reservoirs have temperatures above the cricondentherm. During production, the fluids are reduced in temperature and pressure. The temperature–pressure path followed during production does not penetrate the phase envelope, resulting in the production of gas at the surface with no associated liquid phase.

A wet gas exists in a pure gas phase in the reservoir but becomes a liquid/gas two-phase mixture in a flowline from the well tube to the separator at the topside platform. During the pressure drop in the flow line, liquid condensate appears in the wet gas. In a wet gas reservoir, the reservoir temperature is just above the cricondentherm. During production, the fluids are reduced in temperature and pressure. The temperature–pressure path followed during production just penetrates the phase envelope, resulting in the production of gas at the surface with a small associated liquid phase.

Retrograde gas is the name of a fluid that is gas at reservoir pressure and temperature. However, as pressure and temperature decrease, large quantities of liquids are formed due to retrograde condensation. Retrograde gases are also called retrograde gas condensates, gas condensates, or condensates. In a condensate reservoir, the reservoir temperature is such that it falls between the temperature of the critical point and the cricondentherm. The production path then has a complex history. Initially, the fluids are in an indeterminate vapor phase, and the vapor expands as the pressure and temperature drop. This occurs until the dewpoint line is reached, whereupon increasing amounts of liquids are condensed from the vapor phase. If the pressures and temperatures reduce further, the condensed liquid may reevaporate, although sufficiently low pressures and temperatures may not be available for this to happen.

In summary, at any given constant low fluid pressure, reduction of fluid volume will involve the vapor condensing to a liquid via the two-phase region, where both liquid and vapor coexist. But at a given constant high fluid pressure (higher than the critical point), a reduction of fluid volume will involve the vapor phase turning into a liquid phase without any fluid interface being generated (i.e., the vapor becomes denser and denser until it can be considered as a light liquid). Thus, the critical point can also be viewed as the point at which the properties of the liquid and the gas become indistinguishable (i.e., the gas is so dense that it looks like a low-density liquid and vice versa).

While the fluid in the reservoir remains a single phase, the produced gas splits into two phases as it cools and expands to surface temperature and pressure at point A′. Thus, some condensate would be collected at the surface even though only one phase is present in the formation. The amount of condensate collected depends on the operating conditions of the separator. The lower the temperature at a given pressure, the larger the volume of condensate collected.

Depending on the type of reservoir, the composition of natural gas vary widely. Generally, it contains primarily methane (CH_4) with decreasing quantities of ethane (C_2H_6), propane (C_3H_8), butane (C_4H_{10}), and pentane (C_5H_{12}). Some natural gas mixtures can also contain nonhydrocarbon gases such as carbon dioxide (CO_2), oxygen (O_2), nitrogen (N_2), hydrogen sulfide (H_2S), and traces of rare gases (argon, Ar; helium, He; neon, Ne; and xenon, Xe). No matter what the natural composition of gas is, the product delivered and finally used by the consumers is almost pure methane.

The phase behavior of natural gas (or other gas streams, depending upon the composition) is a function of pressure, temperature, and volume (Figure C.1). In fact, gases can be converted to liquids by compressing the gas at a suitable temperature. However, they become more difficult to liquefy as the temperature increases because the kinetic energies of the particles that make up the gas also increase. At the critical temperature, they cannot longer be liquified.

By way of definition, the critical temperature is the temperature above which a substance cannot exist as a liquid, no matter how much pressure is applied, every substance has a critical temperature. The critical pressure is the pressure required to liquify a substance vapor at its critical temperature. The critical point is the end point of the pressure–temperature curve that designates conditions under which a liquid and the corresponding vapor can coexist. At higher temperatures, the gas cannot be liquefied by pressure alone. At the critical point, defined by the critical temperature t_c and the critical pressure, p_c, phase boundaries vanish. The triple point is the temperature and pressure at which the three phases (gas, liquid, and solid) of a substance coexist in thermodynamic equilibrium. Finally, the critical volume is the volume occupied by a certain mass, usually 1 g molecule of a liquid or gaseous substance at its critical point; the numerical value of the critical volume depends upon the amount of gas under the experiment.

Therefore, it is very often illustrated by the PVT diagram or phase behavior envelope. Understanding phase behavior is critical to the hydrocarbon recovery mechanism and production prediction. Certain concepts associated with phase envelopes are worth introducing at this point. As part of this diagram, the cricondentherm is the highest temperature at which liquid and vapor can coexist, that means the mixture will be gas irrespective of pressure when the temperature is larger than cricondentherm. On the other hand, the cricondenbar is the highest pressure at which a liquid and vapor can coexist.

The bubble-point curve is the curve that separates the pure liquid (oil) phase from the two-phase (natural gas and oil) region (Figure C.1). This means that at a given temperature, when pressure decreases and below the bubble-point curve, gas will be emitted from the liquid phase to the two-phase region. On the other hand, the dew point curve is the line that separates the pure gas phase from the two-phase region (Figure C.1). It is the connected points of pressure and temperature at which the first liquid droplet is formed out of the gas phase.

The critical point is the point on the phase envelope where the bubble-point curve meets the dew point curve. At that given pressure and temperature, gas properties are identical to liquid properties. The pressure and temperature at the critical point are called critical pressure and temperature, respectively (Table C.1).

Table C.1: Critical properties of the common constituents of natural gas.

Compound	Formula	Molecular weight	Critical pressure (psia)	Critical temperature (oF)	Critical volume (ft³/lb)	Liquid gravity (water = 1)	Gas specific (air = 1)
*Hydrocarbons**							
Methane	CH_4	16.042	667.0	−116.66	0.0985	(0.3)*	0.55400
Ethane	C_2H_6	30.069	706.6	89.92	0.0775	0.35643	1.03830
Propane	C_3H_8	44.096	615.5	205.92	0.0728	0.50738	1.52270
n-Butane	C_4H_{10}	58.122	550.9	305.55	0.0703	0.58408	2.00710
n-Pentane	C_5H_{12}	72.149	488.8	385.8	0.0676	0.63113	2.49140
n-Hexane	C_6H_{14}	86.175	436.9	453.8	0.0688	0.66404	2.97580
n-Heptane	C_7H_{16}	100.202	396.8	512.9	0.0682	0.68819	3.46020
n-Octane	C_8H_{18}	114.229	360.7	564.2	0.0673	0.70698	3.94450

Table C.1 (continued)

Compound	Formula	Molecular weight	Critical pressure (psia)	Critical temperature (oF)	Critical volume (ft³/lb)	Liquid gravity (water = 1)	Gas specific (air = 1)
n-Nonane	C_9H_{20}	128.255	330.7	610.8	0.0693	0.72186	4.42890
n-Decane	$C_{10}H_{22}$	142.282	304.6	652.2	0.0703	0.73406	4.91330
Non-hydrocarbons							
Carbon monoxide	CO	28.01	506.7	−220.63	0.0527	0.79265	0.96720
Carbon dioxide	CO_2	44.01	1070.0	87.76	0.0343	0.82203	1.51970
Hydrogen sulfide	H_2S	34.082	1306.5	212.81	0.0462	0.80269	1.17690
Oxygen	O_2	31.9988	731.4	−181.43	0.0367	1.14230	1.10500
Nitrogen	N_2	28.0135	492.5	−232.53	0.0511	0.80687	0.96740
Water	H_2O	18.0183	3200.1	705.1	0.04975	1.00000	0.62210

*Because of the relative dearth of non-linear hydrocarbon derivatives in natural gas, only the n-derivatives are presented here.
**Critical pressure (psi).
***Critical temperature (°Rankine; °R = °C + 273.15, °R = °F + 459.67, and °R = 1.8oK).

C.1.1 Dry gas and wet gas phase behavior

Dry gas is in the gaseous phase under reservoir conditions and contains primarily methane with small amounts of ethane, propane, and butane, with little or no heavier compounds. When it is produced at the surface, it is maintained in the gaseous phase with surface temperature falling outside the two-phase envelope. Therefore, it will not form any liquids, which are at times referred to as natural gas liquids (NGL).

Wet gas, on the other hand, will have liquid dropped out once it reaches the surface, which means that the surface conditions of pressure and temperature will fall inside the two-phase region.

C.1.2 Retrograde condensate gas-phase behavior

Retrograde condensate systems and reservoirs are a unique phenomenon that appears only among hydrocarbon mixtures. No other mixtures of gases exhibit such behavior. As pressure decreases, the amount of liquid in the reservoir increases. Many natural gas reservoirs behave in this manner. During production from such reservoirs, the pressure gradient formed between the reservoir pressure and the flowing bottomhole pressure may result in liquid condensation and form a conden-sate bank around the wellbore, reduce gas relative permeability, and remain unre-coverable. Sometimes the occurrence of this phenomenon could prevent production of the gas (McCain, 1973; Wang and Economides, 2009).

One way to prevent the formation of condensate is to maintain the flowing well bottomhole pressure above the dew point pressure. This is often not satisfactory be-cause the drawdown (reservoir pressure minus flowing bottomhole pressure) may not be sufficient enough for the economic production rate. An alternative technique is to allow the formation of condensate, but occasionally to inject methane gas into the production well. The gas dissolves and sweeps the liquid condensate into the reservoir. The well is then put back in production. This approach is repeated several times in the life of the well and is often referred to as gas cycling. Another way is to inject both nitrogen and methane, which develops a miscible displacement process and results in high condensate recoveries.

Removing the bank of condensate from the near-wellbore region is still a chal-lenge for the oil and gas industry. Understanding the near-wellbore gas-condensate flow is thus important to optimize production of gas-condensate reservoirs.

C.1.3 Associated gas-phase behavior

Under reservoir conditions, gas is often dissolved in the oil phase as associated gas. As it is produced in the surface under lower pressure and temperature, gas will come out from the oil phase. An oil reservoir whose pressure is above the bubble point is usually referred to as undersaturated. If the pressure is inside the two-phase envelope, it is called a saturated, or two-phase, reservoir and may form a gas-cap on top of the oil zone.

C.2 Gas laws

The behavior of a gas is dictated by the gas laws. There are two types of gases: an ideal gas and a non-ideal gas. An ideal gas has the following properties: (i) there are no intermolecular forces between the gas particles, (ii) the volume occupied by the particles is negligible compared to the volume of the container they occupy, and (iii) the only interactions between the particles and with the container walls are per-fectly elastic collisions.

An ideal gas is a fluid in which the volume occupied by the molecules is insig-nificant with respect to the volume occupied by the total fluid, there are no attrac-tive or repulsive forces between the molecules or between the molecules and the walls of the container, and all collisions of molecules are perfectly elastic, that is, there is no loss in internal energy upon collision.

At low pressures, most gases behave like the ideal gas. In addition, under normal distribution pressures, natural gas follows the ideal gas laws quite closely. Under these conditions, it is not normally necessary, therefore, that an accurate determination of any deviation from these laws be made. However, when gas pressures increase, a wide variation between the actual and ideal volumes of the gas may occur. To understand fully what happens when natural gas is subjected to changes in pressure and temperature, the fundamental gas laws must be reviewed.

The nomenclature is as follows:

V_i = volume of gas under original conditions, ft^3

V_2 = volume of gas under changed conditions, ft^3

T_1 = absolute temperature of the gas under original condition, °R (°F + 460)

T_2 = absolute temperature of the gas under changed conditions, °R

p_1 = absolute pressure of the gas under original conditions, psia

p_2 = absolute pressure of the gas under changed conditions, psia

An ideal gas is a gas in which all collisions between atoms or molecules are perfectly elastic and in which there are no intermolecular attractive forces. An ideal gas can be characterized by three variables which are (i) the absolute pressure, P, (ii) the volume, V, and (iii) the absolute temperature, T. Thus:

$$PV = nZRT$$

where P is the pressure, V is the volume, T is the absolute temperature (degree Kelvin), Z is the compressibility, n is the number of kilo-moles of the gas, and R is the gas constant. For example, if all other factors remained constant, when the volume of a certain mass of gas is reduced by 50%, the pressure would double and so on. As a gas, it would expand to fill any volume it is in. However, the compressibility, Z, is the factor that differentiates natural gas from an ideal gas. For methane, Z is 1 at 1 atmosphere pressure (14.7 psi). The relationship may also be expressed in the form

$$PV = nRT = NkT$$

In this equation, n is the number of moles, R is the universal gas constant (8.3145 J/mol K), N is the number of molecules, and k is the Boltzmann constant (1.38066×10^{-23}). Thus, J/K = 8.617385×10^{-5} eV/K and k = R/N_A, where N_A is the Avogadro number (6.0221×10^{23} /mol).

The ideal gas law can arise from the pressure of gas molecules colliding with the walls of a container. And 1 mol of an ideal gas at standard temperature and pressure (STP) occupies 22.4 L. However, natural gas is a non-ideal gas and does not obey the ideal gas law but obeys the modified gas law:

$$PV = nZRT$$

where P is the pressure, V is the volume, T is the absolute temperature (degree Kelvin), Z is the compressibility, n is the number of kilo-moles of the gas, and R is the gas constant.

For example, if all other factors remained constant, when the volume of a certain mass of gas is reduced by 50%, the pressure would double and so on. As a gas, it would expand to fill any volume it is in. However, the compressibility, Z, is the factor which differentiates natural gas from an ideal gas. For methane, Z is 1 at 1 bar but decreases to 0.85 at 100 atmospheres, both at 25 °C, that is, it compresses to a smaller volume than the proportional relationship.

Thus, an ideal gas is a gas in which all collisions between atoms or molecules are perfectly elastic and in which there are no intermolecular attractive forces. An ideal gas can be characterized by three variables: (i) absolute pressure, P, (ii) volume, V, and (iii) absolute temperature, T. The relationship between them is called the *ideal gas law*:

$$PV = nRT = NkT$$

In this equation, n is the number of moles, R is the universal gas constant (= 8.3145 J/mol K), N is the number of molecules, k is the Boltzmann constant (= 1.38066×10^{-23} J/K = 8.617385×10^{-5} eV/K), k is the R/N_A, and N_A is the Avogadro number = 6.0221×10^{23} /mol.

The gas deviation factor, z, is close to 1 at low pressures and high temperatures which means the gas behaves as an ideal gas at these conditions. At standard or atmospheric conditions, the gas z factor is always approximately 1. As the pressure increases, the z factor first decreases to a minimum, which is approximately 0.27 for the critical temperature and critical pressure. For temperatures of 1.5 times, the critical the minimum z factor is approximately 0.77 and for temperatures of twice the critical temperature the minimum z factor is 0.937. At high pressures, the z factor increases above 1 where the gas is no longer supercompressible. At these conditions, the specific volume of the gas is becoming so small and the distance between molecules is much smaller, so that the density is affected by the volume occupied by the individual molecules. Hence, the z factor continues to increase above unity as the pressure increases.

The gas deviation factor, z, is commonly determined by measuring the volume of a sample of the natural gas at a specific pressure and temperature, and then measuring the volume of the same quantity of gas at atmospheric pressure and at a temperature sufficiently high so that the hydrocarbon mixture is in the vapor phase. If the gas deviation factor is not measured, it may be estimated from its specific gravity. The method uses a correlation to estimate pseudo-critical temperature and pseudo-critical pressure values from the specific gravity.

Other laws that are pertinent to gases are presented (alphabetically and not by history or preference) in the following paragraphs.

C.2.1 Avogadro's law

Avogadro's law describes the connection between gas volume and number of moles. Thus, under the same conditions of temperature and pressure, equal volumes of different gases contain an equal number of molecules.

If the volume of gas is increased under isothermal and iso-barometric conditions, the number of moles also increases. The ratio between volume and number of moles is therefore a constant. Thus, V (the volume of gas) is proportional to n (the number of moles of the gas) or:

$$V/n = k$$

This law describes how, under the same condition of temperature and pressure, equal volumes of all gases contain the same number of molecules. In order to compare the same substance under two different sets of conditions, the law can be usefully expressed as follows:

$$V_1/n_1 = V^2/n_2$$

The equation shows that, as the number of moles of gas increases, the volume of the gas also increases in proportion. Similarly, if the number of moles of gas is decreased, then the volume also decreases. Thus, the number of molecules or atoms in a specific volume of ideal gas is independent of their size or the molar mass of the gas.

Accordingly, every gas has fixed number of molecules in a particular fixed volume at a given temperature and pressure or under standard conditions and the volume taken by 1 mol of a gas at particular STP condition is the gram molar volume.

C.2.2 Boyle's law

Robert Boyle (1627–1691), during the course of experiments with air, observed the following relation between pressure and volume: if the temperature of a given quantity of gas is held constant, the volume of gas varies inversely with the absolute pressure. This relation, written as an equation, is:

$$p_1/V_1 = p_2/V_2$$

Thus:

$$p_1/p_2 = V_2/V_1$$

In the application of Boyle's law, volume at a second set of pressure conditions is generally desired.

In the current context, the law formulates the relationship between the compression and expansion of a gas at constant temperature which states that the pressure (p) of a given quantity of gas varies inversely with its volume (v) at constant temperature, that is, in equation form the law is expressed as:

$$pv = K_{constant}$$

where k is a constant. Thus, if the temperature is constant, as pressure of a gas increases, the volume decreases:

$$P_1 V_1 = P_2 V_2$$

C.2.3 Charles' law

Approximately 100 years after the discovery of Boyle's law, Jacques A. Charles (1746–1823) and Joseph L. Gay-Lussac (1778–1850) independently discovered the law that is usually called Charles' law. The Charles' law is in two parts:
1. If the pressure on a particular quantity of gas is held constant, then, with any change of state, the volume will vary directly as the absolute temperature. Expressed as an equation,

$$V_1/T_1 = V_2/T_2$$

$$T_1/V_1 = T_2/V_2$$

$$T/V = constant$$

Since the volume at a second set of temperature conditions is desired usually more than any other information:

$$V_2 = V_1 \times T_2/T_1$$

2. If the volume of a particular quantity of gas is held constant, then, with any change of state, the absolute pressure will vary directly as the absolute temperature:

$$P_1/P_2 = T_1/T_2$$

$$T_1/P_1 = T_2/P_2$$

$$T/P = constant$$

Therefore, gases show an inverse relationship between pressure and volume – as the pressure increases, the volume decreases and vice versa. For Boyle's law to be obeyed, the temperature has to remain constant.

The separate relations of Boyle's and Charles' laws may be combined to give:

$$(P_1V_1)/T_1 = (P_2V_2)/T_2$$

This equation is known as Boyle's–Charles' law and as the simple gas law. It is one of the most widely used relations in gas measurement work since it approximately represents the behavior of many gases under conditions close to ordinary atmospheric temperatures and pressures. One can substitute known values in the combined formula and solve for any one unknown value. In cases where one of the parameters, such as temperature, is not to be considered, it may be treated as having the same value on both sides of the formula. However, the equation does not show the relations connecting volumes and masses of gases.

C.2.4 Dalton's law of partial pressure

Most gaseous systems contain a mixture of gases – for example, air is a mixture of nitrogen gas, oxygen gas, xenon gas, carbon dioxide gas, and many others. When working with a mixture of gases, there is often the need to know the total pressure exerted by all of the gases together as well as the portion of the total pressure that is exerted by only one of the gases in the mixture. The portion of the total pressure is that one gas in a mixture of gases contributes to the partial pressure of the gas. Thus, if a container is filled with more than one gas, each gas exerts pressure and the pressure that is exerted by one among the mixture of gases if it occupies the same volume on its own is known the partial pressure.

Thus, the total pressure of a mixture of gases is the sum of all its individual gas pressures (each gas is said to exert a partial pressure). The total pressure will always equal the atmospheric pressure; normally 760 torr at sea level. Each gas in the mixture exerts a partial pressure. Dalton's law of partial pressures states that in a mixture of gases the pressure exerted by each gas is the same pressure it would exert if it alone occupied the container. From which, in a mixture of gases, the total pressure is the sum of the partial pressures of the gases at constant temperature. Thus:

$$P_{total} = P_1 + P_2 + P_3 + P_4 + \cdots + P_n$$

Dalton's law of partial pressures can be combined with the ideal gas equation to solve many kinds of problems involving mixtures of gases. One way of combining them begins by rearranging the ideal gas equation to express the partial pressures of gases in a mixture.

$$PV = nRT \text{ or } P = (nRT)/V$$

C.2.5 Gay-Lussac's law

As the temperature of an enclosed gas increases, the pressure increases at constant volume. Thus:

$$P_1/T_1 = P_2/T_2$$

As a result of these laws, the combined gas law allows you to do calculations for situations in which only the amount of gas is constant. Thus:

$$(P_1V_1)/T_1 = (P_2V_2)/T_2$$

These laws (Boyle's law, Charles' law, Dalton's law, Gay-Lussac's law) can be collectively used as the combined gas law allows an investigator to perform calculations for situations in which only the amount of gas is constant.

C.2.6 Graham's law of diffusion and effusion

Diffusion is the gradual mixing of gases due to the motion of their component particles even in the absence of mechanical agitation such as stirring. The result is a gas mixture with uniform composition. Diffusion is also a property of the particles in liquids and liquid solutions and, to a lesser extent, of solids and solid solutions. The related process, effusion, is the escape of gaseous molecules through a small (usually microscopic) hole, such as a hole in a balloon, into an evacuated space.

Put simply, diffusion is the tendency of molecules to move toward areas of lower concentration until the concentration is uniform throughout.

The phenomenon of effusion has been known for thousands of years, but it was not until the early nineteenth century that quantitative experiments related the rate of effusion to molecular properties. The rate of effusion of a gaseous substance is inversely proportional to the square root of its molar mass. This relationship is referred to as Graham's law, after the Scottish chemist Thomas Graham (1805–1869). The ratio of the effusion rates of two gases is the square root of the inverse ratio of their molar masses:

Put simply, effusion is when a gas escapes through a tiny hole in a container and gases of lower molar mass diffuse and effuse faster than gases of higher molecular weight.

Thus Graham's law can be stated as: the rate of effusion of a gaseous substance (E) is inversely proportional to the square root of the molar mass (M):

$$E_A/E_B = \sqrt{M_B}/\sqrt{M_A}$$

The relationship is based on the postulate that all gases at the same temperature have the same average kinetic energy – the kinetic theory of gases states that temperature, in degrees Kelvin (°K), is directly proportional to the average kinetic energy of the molecules. Thus, the rate at which a molecule, or a mole of molecules, diffuses or effuses is related to the speed at which it moves which is related to the molecular weight of the molecule.

C.2.7 Ideal gas law

In order to behave as an ideal gas, gases could not have any volume and could be attracted to other gas molecules. This is contradictory to the natural order but, under certain conditions, real gases can behave very similarly to an ideal gas.

An ideal gas is an imaginary gas that obeys exactly the following relationship (the ideal gas law):

$$PV = nRT$$

In this equation, P is the absolute pressure of the gas, V is total volume occupied by the gas, n is the number of moles of the gas, R is the ideal gas constants in appropriate units, and T is the absolute temperature of the gas. The ideal gas constant, R, can be used in a variety of forms, thus:

$$= 1.987 \, cal/(gmol)(K)$$
$$= 1.987 \, Btu/(lbmol)(°R)$$
$$= 10.73 \, (psia)(ft^3)/(lbmol)(°R)$$
$$= 8.314 \, (kPa)(m^3)/(kmol)(°K)$$
$$= 8.314 \, J/(gmol)(°K)$$
$$= 82.06 \, (atm)(cm^3)/(gmol)(°K)$$
$$= 0.08206 \, (atm)(L)/(gmol)(°K)$$
$$= 21.9 \, (in.Hg)(ft^3)/(lbmol)(°R)$$
$$= 0.7302 \, (atm)(ft^3)/(lbmol)(°R)$$

The ideal gas law describes the behavior of most gases at pressure and temperature conditions close to atmospheric. Most natural gas engineers and operating personnel, at one time or another, become involved with the erratic behavior of natural gas under pressure. At moderate pressures, the gas tends to compress more than the ideal gas law indicates, particularly for temperatures close to the critical temperature. At high pressures, the gas tends to compress less than the ideal gas law predicts. In most engineering problems, the pressures of interest fall within the moderate range and the real gases are described as super-compressible. To correct for the deviation between the measured or observed volume and that calculated using the ideal gas law, an empirical factor z, called the gas deviation factor or the z-factor, is used. In the literature, this factor is sometimes referred to as the compressibility factor, which can result in confusion with another gas property. In order to avoid ambiguity, this factor will be referred to as the gas deviation factor or z-factor throughout this text.

In a mixture of ideal gases, the partial pressure of a gas component is the pressure that would be exerted by a component if it existed by itself in the same volume as occupied by the mixture and the same temperature of the mixture. Moreover, the sum of the mole fractions of all the components present must equal 1. That is, the partial pressure of any gas in a mixture is the total pressure multiplied by the mole fraction of that gas which is a direct result of the ideal gas law, which assumes that all gas particles behave ideally. Each gas component in a gas mixture exerts an independent pressure (the partial pressure).

The total pressure of a mixture of gases is the sum of the pressures of each individual gas. Thus:

$$P_{total} = P_1 + P_2 + P_3 + \cdots \text{ (etc.)}$$

The partial pressure of an individual gas is equal to the total pressure multiplied by the mole fraction of that gas. Thus, for the gas mixture:

$$P_{total} V_{total} = n_{total} R T_{total}$$

And

$$\frac{P_i V_T}{P_T V_T} = \frac{n_i R T_T}{n_T R T_T}$$

$$P_i = \frac{n_i}{n_T} P_T = y_i P_T$$

C.2.8 Raoult's law

Raoult's law is a law of physical chemistry and thermodynamics which states that the partial pressure of each component of an ideal mixture of liquids is equal to the vapor pressure of the pure component multiplied by the mole fraction of the pure component in the mixture. As a result, the relative lowering of vapor pressure of a dilute solution of nonvolatile solute is equal to the mole fraction of solute in the solution. Mathematically, Raoult's law for a single component in an ideal solution is stated as

$$p = p_i^* x_i$$

In this equation, p_i is the partial pressure of the component i in the gaseous mixture (above the solution), p_i^* is the equilibrium vapor pressure of the pure component, and x_i is the mole fraction of the component I in the mixture (in the solution).

Where two volatile liquids A and B are mixed with each other to form a solution, the vapor phase consists of both components of the solution. Once the components in the solution have reached equilibrium, the total vapor pressure of the solution can be determined by combining Raoult's law with Dalton's law of partial pressures to give:

$$p = p^*_A x_A + p^*_B x_B + p^*_C x_C + \cdots \text{ (etc.)}$$

If a non-volatile solute (zero vapor pressure, does not evaporate) is dissolved into a solvent to form an ideal solution, the vapor pressure of the final solution will be lower than that of the solvent. The decrease in vapor pressure is directly proportional to the mole fraction of solute in an ideal solution:

$$p = p^*_A x_A$$

$$\Delta p = p^*_A - p = p^*_A (1 - x_A) = p^*_A x_B$$

Raoult's law assumes ideal behavior based on the simple microscopic assumption that intermolecular forces between unlike molecules are equal to those between similar molecules: the conditions of an ideal solution. This is analogous to the ideal gas law, which is a limiting law valid when the interactive forces between molecules approach zero, for example as the concentration approaches zero. Raoult's law is valid if the physical properties of the components are identical. The more similar the components are, the more their behavior approaches that described by Raoult's law.

C.3 Bubble point

In the content of natural gas, the bubble point is the temperature (at a given pressure) at which the first bubble of vapor is formed when heating a liquid consisting of two or more components. Insofar as that vapor will probably have a different composition than the liquid, the bubble points (along with the dew point) at different compositions are useful data when designing distillation systems. For a single component, the bubble point and the dew point are the same and are referred to as the boiling point.

In the case of associated gas, the bubble-point pressure is the pressure at which the first bubble of gas comes out of solution. At this point, the crude oil is saturated and cannot hold anymore gas. Above this pressure, the crude oil is undersaturated and acts as a single-phase liquid.

A bubble-point test is a test designed to determine the pressure at which a continuous stream of bubbles is initially seen downstream of a wetted filter under gas pressure. To perform a bubble-point test, gas is applied to one side of a wetted filter, with the tubing downstream of the filter submerged in a bucket of water. The filter must be wetted uniformly such that water fills all the voids within the filter media. When gas pressure is applied to one side of the membrane, the test gas will dissolve into the water, to an extent determined by the solubility of the gas in water. Downstream of the filter, the pressure is lower. Therefore, the gas in the water on the downstream side is driven out of solution. As the applied upstream gas pressure is increased, the diffusive flow downstream increases proportionally. At some point, the pressure becomes great enough to expel the water from one or more passageways establishing a path for the bulk flow of air. As a result, a steady stream of bubbles should be seen exiting the submerged tubing. The pressure at which this steady stream is noticed is referred to as the bubble point.

C.4 Compressibility

Natural gas, like any gas, can be pressurized using a compressor in which the volume of the gas is decreased. Typically, natural gas is compressed using pressure on the order of 2,900–4,300 psi, which gives a 200-fold to 250-fold reduction in the volume of the gas. The compression factor (also known as the compressibility factor or the real gas factor and given the symbol Z) appears in equations governing volumetric metering. Moreover, the conversion of volume at metering conditions to volume at defined reference conditions can properly proceed with an accurate knowledge of Z at both relevant pressure and relevant temperature conditions. When gas is compressed, work is done and thus it gets hotter. It is therefore necessary to cool gas during or after compression. Equally when it is expanded, adiabatically (without heat being added) it gets colder. This latter phenomenon is used to cool gas during treatment to remove liquids.

The compressibility factor (Z), also known as the compression factor, is a useful thermodynamic property for modifying the ideal gas law to account for the real gas behavior. In general, deviation from ideal behavior becomes more significant the closer a gas is to a phase change, the lower the temperature or the larger the pressure. Compressibility factor values are usually obtained by calculation from equations of state (EOS), such as the virial equation which take compound specific empirical constants as input. For a gas that is a mixture of two or more pure gases (e.g., air or natural gas), a gas composition is required before compressibility can be calculated.

The compressibility factor (Z) is represented by the equation:

$$Z = V_m / (V_{m \text{ ideal gas}}) = pV_m / RT$$

Vm is the molar volume, $(V_m)_{\text{ideal gas}} = RT/p$ is the molar volume of the corresponding ideal gas, p is the pressure, T is the temperature, and R is the gas constant.

For engineering applications, the compressibility factor is frequently expressed as:

$$Z = p / (\rho R_{\text{specific}} T)$$

In this equation, ρ is the density of the gas and $R_{specific} = R/M$ which is the specific gas constant with M being the molar mass.

For an ideal gas, the compressibility factor is $Z = 1$ per definition. In many applications, the requirements for accuracy demand that deviations from ideal gas behavior (i.e., real gas behavior) should be taken into account. The value of Z generally increases with pressure and decreases with temperature. At high pressures, molecules are colliding more often which allows repulsive forces between molecules to have a noticeable effect, making the molar volume of the real gas (V_m) greater than the molar volume of the corresponding ideal gas ($(V_m)_{ideal\ gas} = RT/p$), which causes Z to exceed one. When pressures are lower, the molecules are freer to move and, in this case, the attractive forces dominate and $Z < 1$. The closer the gas is to its critical point or its boiling point, the more Z deviates from the ideal case.

The isothermal gas compressibility (cg, also called the bulk modulus of elasticity) of natural gas is useful insofar as it is used extensively in determining the compressible properties of the reservoir. Gas usually is the most compressible medium in the reservoir. However, care should be taken so that it is not confused with the gas deviation factor, Z, which is sometimes called the super compressibility factor:

$$c_g = -\frac{1}{V_g}\left(\frac{\partial V_g}{\partial P}\right)_T$$

where V is the volume, P is the pressure, and T is the absolute temperature. For an ideal gas, the compressibility is defined as:

$$c_g = -\frac{1}{P}$$

For a non-ideal gas, the compressibility is defined by the equation:

$$c_g = \frac{1}{P} - \frac{1}{Z}\left(\frac{\partial Z}{\partial P}\right)_T$$

Gas is difficult to store in the gaseous state outside the reservoir to provide flexibility of supply. Only a small amount of flexibility is provided by the high-pressure gas in pipelines,

C.5 Corrosiveness

Gas streams produced during petroleum recovery, petroleum refining, and natural gas processing, while ostensibly being hydrocarbon in nature, may contain large amounts of corrosive acid gases such as hydrogen sulfide and carbon dioxide resulting in a high potential for corrosion. Although the processing of natural gas is in many respects less complicated than the processing and refining of crude oil, it is equally as necessary to assure that all of the corrosive constituents are removed.

In gas plants, handling product streams from refinery units raises the potential for corrosion from moist hydrogen sulfide and cyanide derivatives. When feedstocks are from the visbreaker, the delayed coker, the fluid coker, or any other thermal cracking unit, corrosion from hydrogen sulfide and deposits of iron sulfide in the high-pressure sections of gas compressors from ammonium compounds is possible. Furthermore, processing opportunity crudes requires refiners to manage greater volumes of corrosive and toxic hydrogen sulfide. In sour-gas streams, the primary corrosion-causing constituents of gas streams are hydrogen sulfide (H_2S) and carbon dioxide (CO_2), with contributions from other corrosive constituents. Streams containing ammonia should be dried before processing. Antifouling additives may be used in absorption oil to protect heat exchangers. Corrosion inhibitors may be used to control corrosion in overhead systems.

As an example, copper corrosion causes difficulties due to the deterioration of copper and copper alloy fittings and connections commonly used in liquefied petroleum gas systems. As little as 1 ppm of hydrogen sulfide can cause a copper strip test failure. Thus, the copper strip corrosion test is an extremely sensitive test that will detect virtually all species of corrosive sulfur, including minute traces of hydrogen sulfide. It is critically important that the product being tested does not contain any additives that may diminish the reaction with the copper strip.

Hydrogen sulfide corrosion results in the formation of black iron sulfide scales and is typified by black water in the separation units. Under-deposit corrosion frequently occurs beneath the scale layer and can result in forming deep, isolated or randomly scattered pits. The prime means of removing or reducing the impact of iron sulfide entering an olamine system are to: (i) prevent the corrosion from occurring initially in the piping by using corrosion inhibitors, (ii) disperse the iron sulfide particles into the water phase so they can be removed by inlet separation equipment, and (iii) remove the iron sulfide from the gas phase upstream of the olamine absorber by use of a suitable filter or by a water wash.

Corrosion by wet carbon dioxide can result in high corrosion rates, but a carbonate film gives some protection and is more protective at higher temperatures. The carbon dioxide content is often not high in refinery streams, except in hydrogen reformer plant systems. In addition, one of the major sources of corrosion on carbon steel vessels in sweetening units is heat-stable materials, which are a product of amine degradation. Oxygen plays a major role in amine degradation – the reaction of oxygen and the amine produces organic acids, such as acetic acid, formic acid and so on.

In addition, the effluent streams from any of the refining processes can contain ammonia (NH_3) and hydrogen sulfide (H_2S) which react to form ammonium bisulfide (NH_4HS), which is highly corrosive to carbon steel and may lead to a catastrophic failure. The severity of ammonium bisulfide-induced corrosion depends upon

(i) the concentration of ammonium bisulfide, (ii) the fluid velocity and turbulence, (iii) wash-water management, as well as (iv) piping configuration and temperature of the system.

C.6 Critical properties

The critical temperature is the temperature above which a substance cannot exist as a liquid, no matter how much pressure is applied. The critical pressure is the pressure required to liquify a vapor at the critical temperature. In the vicinity of the critical point, the physical properties of the liquid and the vapor change dramatically, with both phases becoming ever more similar. At the critical point, only one phase exists and above the critical point there exists a state of matter that is continuously connected with (can be transformed without phase transition into) both the liquid and the gaseous state.

The critical point is the end point of the pressure–temperature curve that designates conditions under which a liquid and the corresponding vapor can coexist. At a higher temperature, the gas cannot be liquefied by pressure alone and at the critical point (defined by the critical temperature and the critical pressure), the phase boundaries vanish. The triple point is the temperature and pressure at which the three phases (gas, liquid, and solid) of a substance coexist in thermodynamic equilibrium.

In the context of natural gas, the gas can be converted to a liquid by compressing the gas at a suitable temperature (Table C.1). However, the gas is more difficult to liquefy as the temperature increases because the kinetic energy of the particles that make up the gas also increases. At the critical temperature, the gas can no longer be liquefied.

C.7 Density and specific gravity

Density is the mass of a substance contained in a unit volume (simply put, density is mass divided by volume). In the SI system of units, the ratio of the density of a substance to the density of water at 15 °C is known as the specific gravity (relative density). Various units of density, such as kg/m^3, lb-mass/ft^3, and g/cm^3, are commonly used. In addition, molar densities or the density divided by the molecular weight is often specified.

The gas density is defined as mass (m) per unit volume (V) and can be calculated from the real gas law:

$$\rho_g = \frac{m}{V} - \frac{pMW_m}{ZRT}$$

Density, a physical property of matter, is a measure of the relative heaviness of hydrocarbons and other chemicals at a constant volume, and each constituents of natural gas has a unique density associated with it. For most chemical compounds (i.e., those that are solid or liquid), the density is measured relative to water (1.00). For gases, the density is more likely to be compared to the density of air (also given the number 1.00 but this is arbitrary and bears no relationship to the density of water). As a comparison, the density of liquefied natural gas (LNG) is approximately from 0.41 to 0.5 kg/L, depending on temperature, pressure and composition; in comparison, the density of water is 1.0 kg/L. In terms of composition, LNG is predominantly methane and it is not surprising that the density of LNG is close to (but not exactly equal to) the density of methane.

The density (or specific gravity) of the various unconventional gases discussed in this chapter may be determined conveniently by any one of several methods and a variety of instruments. Density values (including those of natural gas hydrocarbons) are given at room temperature various in tables unless otherwise indicated by the superscript figure; for example, 2.487^{15} indicates a density of 2.487 g/cm for the substance at 15 °C. A superscript 20 over a subscript 4 indicates a density at 20 °C relative to that of water at 4 °C. For gases, the value of the density is given in grams per liter (g/L).

The specific gravity of a gas is the ratio of the molecular weight of a particular natural gas to that of air. The molecular weight of a gas mixture is the summation of the products of the individual mole fractions and molecular weights of each individual component. Air itself is a mixture of gases such as oxygen (21% v/v), nitrogen (78% v/v), with minor amounts of carbon dioxide, water vapor, and some inactive gases. So the molecular weight of air has been calculated as 28.97. Therefore, the specific gravity of natural gas is

$$sg = M_w / M_{air}$$

In this equation, M_w is the molecular weight of the gas stream and M_{air} is the molecular weight of air.

More specifically, the density of an ideal gas mixture is calculated by simply replacing the molecular weight of the pure component with the average molecular weight of the gas mixture to give

$$\rho_g = \frac{pM_a}{RT}$$

In this equation, ρ_g is the density of the gas mixture, lb/ft^3, Ma is the average molecular weight, p is the absolute pressure, psia, T is the Absolute temperature, and R is the universal gas constant.

The density of low-boiling hydrocarbon derivatives can be determined by several methods including a hydrometer method or by a pressure hydrometer method. The specific gravity (relative density) by itself has little significance compared to its use for higher molecular weight liquid petroleum products and can only give an in-dication of quality characteristics when combined with values for volatility and vapor pressure. It is important for stock quantity calculations and is used in connec-tion with transport and storage.

Another term, specific gravity, is commonly used is relation to the properties of hydrocarbons. The specific gravity of a substance is a comparison of its density to that of water. Both the density of the substance and the density of water should be measured or expressed at the same pressure and temperature. If the behavior of both the gas mixture and the air is described by the ideal gas equation, the specific gravity can then be expressed as in the form:

$$\gamma_g = \frac{\rho_g}{\rho_{air}}$$

Also

$$\gamma_g = \frac{\frac{p_{sc}M_a}{RT_{sc}}}{\frac{p_{sc}M_{air}}{RT_{sc}}}$$

$$\gamma_g = \frac{M_a}{M_{air}} = \frac{M_a}{28.6}$$

γ_g is the is the specific gravity of the gas, ρ_{air} is the density of the air, M_{air} is appar-ent molecular weight of the air which is 28.96, M_a is the apparent molecular weight of the gas, p_{sc} is the standard pressure, psia, and sc is the standard temperature, °R.

The density of any gas compared to the density of air is the vapor density and is an important characteristic of the constituents of natural gas and natural gas con-stituents. Put simply, if the constituents of natural gas are less dense (lighter) than air, they will dissipate into the atmosphere whereas if the constituents of natural gas are denser (heavier) than air, they will sink and be less likely to dissipate into the atmosphere. Of the hydrocarbon constituents of natural gas, methane is the only one that is less dense than air.

The density (or specific gravity) can be calculated, but if it is necessary to mea-sure it several pieces of apparatus are available. For determining the density or spe-cific gravity of liquefied petroleum gas in its liquid state, there are two methods using a metal pressure pycnometer.

The statement is often made that natural gas is lighter than air. This statement often arises because of the continued insistence by engineers and scientists that the properties of a mixture are determined by the mathematical average of the proper-ties of the individual constituents of the mixture. Such mathematical bravado and inconsistency of thought is detrimental to safety and needs to be qualified. In fact, relative to air, methane is less dense than air but the other hydrocarbon constitu-ents of unrefined natural gas (such as ethane, propane, butane, and any higher molecular weight hydrocarbon derivatives) are denser than air. Therefore, should a natural gas leak occur in field operations, especially where the natural gas contains constituents other than methane, only methane dissipates readily into the air whereas the other hydrocarbon constituents that are heavier than air do not readily dissipate into the atmosphere. This poses considerable risk if these constituents of natural gas accumulate or pool at ground level when it has been erroneously assumed that natural gas is lighter than air.

The relative density typically relates to the density of natural as relative to the density of air (Table C.2) although, in some cases, the density of hydrogen may be used for comparison to the density of natural gas. The relative density, as a measure of gas density relative to air at reference conditions this is used for interchangeability specifications to limit the higher hydrocarbon content of the gas. An increased higher hydrocarbon content could lead to combustion problems such as increased carbon monoxide emissions, soot formation, engine knock, or spontaneous ignition on gas turbines even at the same Wobbe index value.

Table C.2: Relative density (specific gravity) of natural gas hydrocarbons relative to air.

Gas	Specific gravity
Air	1.000
Methane (CH_4)	0.5537
Ethane (C_2H_6)	1.0378
Propane (C_3H_8)	1.5219
Butane (C_4H_{10})	2.0061
Pentane (C_5H_{12})	2.487
Hexane (C_6H_{14})	2.973

Caution is advised when using statements like natural gas is lighter than air. This statement often arises because of the continued insistence by engineers and scientists that the properties of a mixture are determined by the mathematical average of the properties of the individual constituents of the mixture. Such mathematical bravado and inconsistency of thought is detrimental to safety and needs to be qualified. The *relative density* (*specific gravity*) is the ratio of the density (mass of a unit volume) of a substance to the density of a given reference material. Specific gravity usually means relative density (of a liquid) with respect to water:

$$\text{Relative density} = [\rho(\text{substance})]/[\rho(\text{reference})]$$

As it pertains to gases, particularly in relation to safety considerations at commercial and industrial facilities in the United States, the relative density of a gas is usually defined with respect to air, in which air is assigned a *vapor density* of one (unity). With this definition, the vapor density indicates whether a gas is denser (greater than one) or less dense (less than one) than air. The vapor density has implications for container storage and personnel safety – if a container can release a dense gas, its vapor could sink and, if flammable, collect until it is at a concentration sufficient for ignition. Even if not flammable, it could collect in the lower floor or level of a confined space and displace air, possibly presenting a smothering hazard to individuals entering the lower part of that space.

Gases can be divided into two groups based upon their vapor density: (i) gases which are heavier than air, and (ii) gases which are as light as air or lighter than air. Gases that have a vapor density greater than one will be found in the bottom of storage containers and will tend to migrate downhill and accumulate in low-lying areas. Gases that have a vapor density which is the same or less than the vapor density of air will disperse readily into the surrounding environment. Additionally, chemicals that have the same vapor density as air (1.0) tend to disperse uniformly into the surrounding air when contained and, when released into the open air, chemicals that are lighter than air will travel up and away from the ground.

Methane is the only hydrocarbon constituent of natural gas that is lighter than air (Table C.2). The higher molecular weight hydrocarbons have a higher vapor density than air and are likely, after a release, to accumulate in low-lying areas and represent a danger to investigator (of the release). However, the other hydrocarbon constituents of unrefined natural gas (such as ethane, propane, butane, and higher molecular weight hydrocarbon derivatives) are denser than air. Therefore, should a natural gas leak occur in field operations, especially where the natural gas contains constituents other than methane, only methane dissipates readily into the air whereas the other hydrocarbon constituents that are heavier than air do not readily dissipate into the atmosphere. This poses considerable risk if these constituents of natural gas accumulate or pool at ground level when it has been erroneously assumed that natural gas is lighter than air.

C.8 Deviation factor

The gas deviation factor is an important gas property and is involved in calculating gas properties such as the formation volume factor, density, compressibility, and viscosity. All these properties are necessary in calculating initial gas-in-place (and, thus, reserves), predicting future gas production, and designing production tubing and pipelines.

A natural gas mixture under reservoir conditions is nonideal and its behavior can be approximated by the real gas law, a general equation of state for gases:

$$pV = ZnRT$$

In this equation, p is pressure in psi, V is the gas volume in ft3, n is the number of moles of the gas, T is absolute temperature in R, and R is the universal gas constant and equals to 10.73 psi ft^3/lb-mol-R. Z is the gas deviation factor or "Z-factor" in some petroleum literature.

Chemical engineers have called it the super-compressibility factor which is defined as the ratio of the real volume (the volume actually occupied by a gas at a given p and T) to the ideal volume (volume it would occupy had it behaved as an ideal gas). It is a measure of how a real gas deviates from ideality.

The gas deviation factor is an important gas property and it is involved in calculating gas properties such as the formation volume factor, density, compressibility, and viscosity. All these properties are necessary in calculating initial gas-in-place (and, thus, reserves), predicting future gas production, and designing production tubing and pipelines (Elsharkawy and Elkamel, 2001).

In thermodynamics, the gas deviation factor is a correction factor which describes the deviation of the behavior of a real gas from the behavior of an ideal gas and is represented by the ratio of the molar volume of a gas to the molar volume of an ideal gas at the same temperature and pressure. It is a useful thermodynamic property for modifying the ideal gas law to account for the real gas behavior. In general, deviation from ideal behavior becomes more significant; the closer a gas is to a phase change, the lower the temperature or the larger the pressure. Also, In the presence of large amounts of nonhydrocarbon gases, the gas deviation factor must be adjusted.

Compressibility factor values are usually obtained by calculation from EOS, such as the virial equation which take compound-specific empirical constants as input. For a gas that is a mixture of two or more pure gases (e.g., air or natural gas), the composition of the gas must be known before compressibility can be calculated. Alternatively, the compressibility factor for specific gases can be read from generalized compressibility charts that plot the compressibility factor (Z) as a function of pressure at constant temperature.

The compressibility factor should not be confused with the compressibility (also known as the coefficient of compressibility or the isothermal compressibility of a material), which is the measure of the relative volume change of a fluid or a solid in response to a pressure change.

C.9 Dew point

The dew point or dew point temperature of a gas is the temperature at which the water vapor or low-boiling hydrocarbon derivatives contained in the gas is transformed into the liquid state. The formed liquid is (condensate) exists as a liquid below the dew point temperature but above the dew point the liquid is a gaseous component of the gas.

The hydrocarbon dew point is often considered to be the most important factor when performing any type of gas sampling. In the simplest terms, hydrocarbon dew point is the point at which the gas components begin to change phase from gaseous to liquid. When phase change occurs, certain components of the gas stream drop out and form liquids thereby making an accurate gas sample impossible to obtain. The hydrocarbon dew point is a function of gas composition and pressure. A hydrocarbon dew point curve is a reference chart that determines the specific pressure and temperature at which condensation occurs. No two hydrocarbon dew point curves are alike due to differing gas compositions. Since dew point can be calculated from composition, direct determination of dew point for a particular liquefied petroleum gas sample is a measure of composition. It is, of course, of more direct practical value and if there are small quantities of higher molecular weight material present, it is preferable to use a direct measurement.

The hydrocarbon dew point is the temperature (at a given pressure) at which the hydrocarbon constituents of any hydrocarbon-rich gas mixture, such as natural gas, will start to condense out of the gaseous phase. The maximum temperature at which such condensation takes place is called the *cricondentherm*. The hydrocarbon dew point is a function of the gas composition as well as the pressure. The hydrocarbon dew point is universally used in the natural gas industry as an important quality parameter, stipulated in contractual specifications and enforced throughout the natural gas supply chain, from producers through gas processing (gas cleaning), transmission, and distribution companies to the consumer.

Thus, if the hydrocarbon dew point is reduced to such a level that retrograde condensation, that is, condensation resulting from pressure drop, cannot occur under the worst conditions likely to be experienced in the gas transmission system. Similarly, the water dew point is reduced to a level sufficient to preclude formation of C_1 to C_4 hydrates in the system. The natural gas after appropriate treatment for acid gas reduction, odorization, and hydrocarbon and moisture dew point adjustment would then be sold within prescribed limits of pressure, calorific value and possibly Wobbe index (cv/sp. gr.).

Typically, in processing, the hydrocarbon dew point is reduced to such a level that retrograde condensation, that is, condensation resulting from pressure drop, cannot occur under the worst conditions likely to be experienced in the gas transmission system. Similarly, the water dew point is reduced to a level sufficient to preclude formation of C_1 to C_4 hydrates in the system. Once the composition of a mixture has been determined, it is possible to calculate various properties such as specific gravity, vapor pressure, calorific value, and dew point. Since dew point can be calculated from composition, direct determination of dew point for a liquefied petroleum gas sample is a measure of composition. If there are small quantities of higher molecular weight material present, it is preferable to use a direct measurement.

The natural gas after appropriate treatment for acid gas reduction, odorization, and hydrocarbon and moisture dew point adjustment, would then be sold within prescribed limits of pressure, calorific value and possibly Wobbe index (cv/sp. gr.).

While the dew point identifies the condition at which vapor first begins to condense to liquid, it provides no information about the quantity of condensation resulting from a small degree of cooling. The condensation rate of liquids in gas transmission lines may vary widely depending on the composition, temperature, and pressure of the system and there needs to be a practical hydrocarbon dew point specification allowing small amounts of liquids that have no significant impact on operations (Bullin et al., 2010).

C.10 Flammability

Flammable chemicals are those chemicals that ignite more easily than other chemicals, whereas those that are harder to ignite or burn less vigorously are combustible. The degree of flammability or combustibility of the constituents of a gas stream in air depends largely upon the chemical composition of the stream which is also related to the volatility of the constituents in the gas stream. Furthermore, volatility is related to the boiling point.

The boiling point (boiling temperature) of a substance is the temperature at which the vapor pressure of the substance is equal to atmospheric pressure. At the boiling point, a substance changes its state from liquid to gas. A stricter definition of boiling point is the temperature at which the liquid and vapor (gas) phases of a substance can exist in equilibrium. When heat is applied to a liquid, the temperature of the liquid rises until the *vapor pressure* of the liquid equals the pressure of the surrounding atmosphere (gases). At this point, there is no further rise in temperature, and the additional heat energy supplied is absorbed as *latent heat* of vaporization to transform the liquid into gas. This transformation occurs not only at the surface of the liquid (as in the case of *evaporation*) but also throughout the volume of the liquid, where bubbles of gas are formed.

The boiling point of a liquid is lowered if the pressure of the surrounding atmosphere (gases) is decreased. On the other hand, if the pressure of the surrounding atmosphere (gases) is increased, the boiling point is raised. For this reason, it is customary when the boiling point of a substance is given to include the pressure at which it is observed, if that pressure is other than standard, that is, 760 mm of mercury or 1 atmosphere (STP).

The boiling points of petroleum fractions are rarely, if ever, distinct temperatures. It is, in fact, more correct to refer to the boiling ranges of the various fractions; the same is true of natural gas. To determine these ranges, the material in question is tested in various methods of distillation, either at atmospheric pressure or at reduced pressure. Thus, the boiling points of the hydrocarbon constituents of natural gas increase with molecular weight and the initial boiling point of natural gas corresponds to the boiling point of the most volatile constituents (i.e., methane).

Purified natural gas is neither corrosive nor toxic, its ignition temperature is high, and it has a narrow flammability range, making it an apparently safe fossil fuel compared to other fuel sources. In addition, purified natural gas (i.e., methane) having a specific gravity (0.60) lower than that of air (1.00) rises if escaping and dissipates from the site of any leak. However, methane is highly flammable, burns easily and almost completely. Therefore, natural gas can also be hazardous to life and property through an explosion. When natural gas is confined, such as within a house or in a coalmine, concentration of the gas can reach explosive mixtures that, if ignited, results in blasts that could destroy buildings.

The flash point of petroleum or a petroleum product, including natural gas, is the temperature to which the product must be heated under specified conditions to give off sufficient vapor to form a mixture with air that can be ignited momentarily by a specified flame. As with other properties, the flash point is dependent on the composition of the gas and the presence of other hydrocarbon constituents. The fire point is the temperature to which the gas must be heated under the prescribed conditions of the method to burn continuously when the mixture of vapor and air is ignited by a specified flame.

From the viewpoint of safety, information about the flash point and fire point is of most significance at or slightly above the maximum temperatures (30–60 °C, 86–140 °F) that may be encountered in storage, transportation, and use of liquid petroleum products, in either closed or open containers. In this temperature range, the relative fire and explosion hazard can be estimated from the flash point. For products with flash point below 40 °C (104 °F), special precautions are necessary for safe handling. Flash points above 60 °C (140 °F) gradually lose their safety significance until they become indirect measures of some other quality. The flash point of a petroleum product is also used to detect contamination. A substantially lower flash point than expected for a product is a reliable indicator that a product has become contaminated with a more volatile product, such as gasoline. The flash point is also an aid in establishing the identity of a petroleum product.

The flammability range is the range of temperature over which natural gas is flammable. The flammable limits (Table C.3) are expressed by the lower explosive limit (LEL) and the upper explosive limit (UEL). The LEL is the concentration of natural gas in the air below which the propagation of a flame will not occur on contact with an ignition source. The lower explosive limit for natural gas is 5% by volume in air and, in most cases, the smell of gas would be detected well before combustion conditions are met. The UEL is the concentration of natural gas in the air above which the propagation of a flame will not occur on contact with an ignition source. The natural gas UEL is 15% by volume in air.

Table C.3: Flammability limits of the constituents of fuel gases.

Gas*	LFL % v/v in air	UFL % v/v in air
n-Butane	1.6	8.4
Butylene (1-butene)	1.7	9.7
Carbon monoxide	12.5	74.2
Carbonyl sulfide	12.5	74
2,2-Dimethyl propane	1.4	7.5
Ethane	3	12.4
Ethylene	2.7	36
n-Heptane	1	7
n-Hexane	1	7.5
Hydrogen	4	75
Hydrogen sulfide	4	46
Isobutane	1.82	9.6
Methane	5	15
Methyl mercaptan	3.9	21.8
n-Octane	1	7
iso-Octane	0.6	6
n-Pentane	1.4	7.8
iso-Pentane	1.3	9.2
Propane	2.1	9.5
Propylene	2	11.1

*Listed alphabetically.

Explosions caused by natural gas leaks occur a few times each year. Individual homes and small businesses are most frequently affected when an internal leak builds up gas inside the structure. Frequently, the blast will be enough to significantly damage a building but leave it standing. Occasionally, the gas can collect in high enough quantities to cause a deadly explosion, disintegrating one or more buildings in the process.

By way of explanation, before a fire or explosion can occur, three conditions must be met simultaneously. A fuel (i.e., combustible gas) and oxygen (air) must exist in certain proportions, along with an ignition source, such as a spark or flame. The ratio of fuel and oxygen that is required varies with each combustible gas or

vapor. The minimum concentration of a particular combustible gas or vapor necessary to support its combustion in air is defined as the LEL for that gas. Below this
level, the mixture is too lean to burn. The maximum concentration of a gas or vapor
that will burn in air is defined as the UEL. Above this level, the mixture is too rich to
burn. The range between the LEL and UEL is known as the flammable range for that
gas or vapor. Typically, the values given for the UEL and the LEL (Table C.4) are
valid only for the conditions under which they were determined (usually room temperature and atmospheric pressure using a 2-inch tube with spark ignition). The
flammability range of most materials expands as temperature, pressure, and container diameter increase.

Table C.4: Lower explosive limits (LEL) and upper
explosive limits (UEL) for various constituents of
gases, gas condensate, and natural gasoline.

Constituent	LEL	UEL
Benzene	1.3	7.9
1,3-Butadiene	2	12
Butane	1.8	8.4
n-Butanol	1.7	12
I-Butene	1.6	10
cis-2-Butene	1.7	9.7
trans-2-Butene	1.7	9.7
Carbon monoxide	12.5	74
Carbonyl sulfide	12	29
Cyclohexane	1.3	7.8
Cyclopropane	2.4	10.4
Diethylbenzene	0.8	
2,2-Dimethylpropane	1.4	7.5
Ethane	3	12.4
Ethyl benzene	1	6.7
Ethylene	2.7	36
Gasoline	1.2	7.1
Heptane	1.1	6.7
Hexane	1.2	7.4

Table C.4 (continued)

Constituent	LEL	UEL
Hydrogen	4	75
Hydrogen sulfide	4	44
Isobutane	1.8	8.4
Isobutylene	1.8	9.6
Methane	5	15
3-Methyl-1-butene	1.5	9.1
Methyl mercaptan	3.9	21.8
Pentane	1.4	7.8
Propane	2.1	9.5
Propylene	2.4	11
Toluene	1.2	7.1
Xylene	1.1	6.6

A further aspect of volatility that receives considerable attention is the vapor pressure of petroleum and its constituent fractions. The *vapor pressure* is the force exerted on the walls of a closed container by the vaporized portion of a liquid. Conversely, it is the force that must be exerted on the liquid to prevent it from vaporizing further. The vapor pressure increases with temperature for any given gasoline, liquefied petroleum gas, or other product. The temperature at which the vapor pressure of a liquid, a pure compound, or a mixture of many compounds equals 1 atm (14.7 psi, absolute) is designated as the boiling point of the liquid.

The vapor density has implications for flammability during storage. If a container can release a dense gas, its vapor could sink and, if flammable, collect until it is at a concentration sufficient for ignition. Even if not flammable, it could collect in the lower floor or level of a confined space and displace air, possibly presenting a smothering hazard to individuals entering the lower part of that space.

Before a fire or explosion can occur, three conditions must be met simultaneously: (i) a fuel gas – a combustible gas – and oxygen (air) must exist in certain proportions, along with an ignition source, such as a spark or flame. The ratio of fuel and oxygen that is required varies with each combustible gas or vapor. The minimum concentration of a combustible gas (or vapor) necessary to support combustion of the gas in air is defined as the *LEL* for that gas. Below this level, the mixture is too lean to burn. The maximum concentration of a gas (or vapor) that will burn in air is defined as the *UEL*. Above this level, the mixture is too rich to burn. The range between the LEL and UEL is known as the flammable range for that gas or vapor.

Thus, the lower and upper limits of flammability indicate the percentage of combustible gas in air below which and above which flame will not propagate. When flame is initiated in mixtures having compositions within these limits, it will propagate and therefore the mixtures are flammable. A knowledge of flammable limits and their use in establishing safe practices in handling gaseous fuels is important, for example, when purging equipment used in gas service, in controlling factory or mine atmospheres, or in handling liquefied gases.

The calculation of flammable limits is accomplished by Le Chatelier's modification of the mixture law, which (expressed in the simplest form) is

$$L = 100/(p_1/N_1 + p_2/N_2 + \cdots + p_n N_n)$$

L is the volume percentage of fuel gas in a limited mixture of air and gas; $p_1, p_2, \ldots, p_n$ are the volume percentages of each combustible gas present in the fuel gas, calculated on an air- and inert-free basis so that $p_1 + p_2 + \ldots + p_n = 100$; and $N_1, N_2, \ldots N_n$ are the volume percentages of each combustible gas in a limit mixture of the individual gas and air. The foregoing relation may be applied to gases with inert content of 10% or less without introducing an absolute error of more than 1% or 2% in the calculated limits.

The rate of flame propagation (also referred to as the burning velocity) in gas–air mixtures is of importance in utilization problems, including those dealing with burner design and rate of energy release. There are several methods that have been used for measuring such burning velocities, in both laminar and turbulent flames. Results by the various methods do not agree, but any one method does give relative values of utility. Maximum burning velocities for turbulent flames are greater than those for laminar flames.

C.11 Formation volume factor

Volumetric factors were introduced in petroleum and natural gas calculations to readily relate the *volume* of reservoir fluids that are obtained at the surface (stock tank fluids) to the volume that the fluid(s) occupied when it was (they were) under reservoir pressure and compressed in the reservoir.

The formation volume factor allows the reservoir engineer to account for all the reservoir fluid as pressure changes, a volume factor (B) is used which is a ratio of the volume of the fluid at reservoir conditions to its volume at atmospheric conditions (usually 60 °F and 14.7 psi). The produced natural gas is measured in standard cubic feet (SCF). The change in volume factor for a measured change in the reservoir pressure allows for simple estimation of the initial gas or oil volume.

Thus, the formation volume factor for gas is the ratio of volume of 1 mol of gas at a given pressure and temperature to the volume of 1 mol of gas at standard conditions (Ps and Ts). The gas formation volume factor is used to relate the volume of gas, as measured at reservoir conditions, to the volume of the gas as measured at standard conditions, that is, 15.5 °C(60 °F) and 14.7 psia. This gas property is then defined as the actual volume occupied by a certain amount of gas at a specified pressure and temperature, divided by the volume occupied by the same amount of gas at standard conditions. In an equation form, the relationship is expressed as

$$B_g = \frac{V_{p,T}}{V_{sc}}$$

Bg is the gas formation volume factor, ft^3/scf., $V_{p,T}$ is the volume of gas at pressure p and temperature, T, in ft^3, V_{sc} is the volume of gas at standard conditions, scf.

The reciprocal of the specific molar volume is the molar density, and thus

$$B_g = \frac{\tilde{v}_g|_{yes}}{\tilde{v}_g|_{sc}} = \frac{\bar{\rho}_g|_{sc}}{\bar{\rho}_g|_{yes}} = \frac{(\rho_g/MW_g)|_{sc}}{(\rho_g/MW_g)|_{yes}}$$

Introducing the definition for densities in terms of compressibility factor,

$$B_g = \frac{\frac{p_{sc}}{RT_{sc}Z_{sc}}}{\frac{p}{RTZ}}$$

Since $Z_{sc} \approx 1$, the relationship is:

$$B_g = \frac{p_{sc}}{T_{sc}}\frac{ZT}{P} = 0.02827\frac{ZT}{P}[RCF/SCF]$$

RCF is the reservoir cubic feet.

Gas formation volume factors can be also expressed in terms of [RB/SCF]. In such a case, 1 RB = 5.615 RCF. Thus:

$$B_g = 0.005035\frac{ZT}{P}[RCF/SCF]$$

The formation volume factor of a liquid **or condensate** (B$_o$) relates the volume of 1 lb-mol of liquid at reservoir conditions to the volume of that liquid once it has gone through the surface separation facility:

$$B_0 = \frac{\text{Volume of 1 lb} - \text{mol of liquid at reservoir conditions, RB}}{\text{Volume of that 1 lb} - \text{mol after going through separation, STB}}$$

The total volume occupied by 1 lb-mol of liquid at reservoir conditions (V$_o$)$_{res}$ can be calculated through the compressibility factor of that liquid, as follows:

$$(V_0)_{yes} = \left(\frac{nZ_0RT}{P}\right)_{yes}$$

where n = 1 lb-mol.

Upon separation, some gas is going to be taken out of the liquid stream feeding the surface facility. If n_{st} is the moles of liquid leaving the stock tank per mole of feed entering the separation facility. The volume that 1 lb-mol of reservoir fluid (including natural gas) is going to occupy after going through the separation facility is given by:

$$(V_0)_{yes} = \left(\frac{n_{st}Z_0RT}{P}\right)_{SC}$$

C.12 Heat of combustion

One of the principal uses of natural gas is as a fuel, and consequently, pipeline gas is normally bought and sold (custody transfer) on the basis of its heating value. Procedures for calculating the heat effect in any chemical reaction are found in standard texts on thermodynamics.

The heat of combustion (energy content) of natural gas is the amount of energy that is obtained from the burning of a volume of natural gas is measured in British thermal units (Btu). The value of natural gas is calculated by its Btu content. One Btu is the quantity of heat required to raise the temperature of one pound of water of 1 °F at atmospheric pressure. A cubic foot of natural gas has an energy content of approximately 1031 Btu, but the range of values is between 500 and 1500 Btu depending upon the composition of the gas. The heat value of gases is generally determined at constant pressure in a flow calorimeter in which the heat released by the combustion of a definite quantity of gas is absorbed by a measured quantity of water or air.

For use as heating agents, the relative merits of gases from different sources and having different compositions can be compared readily on the basis of their heating values. Therefore, the heating value is used as a parameter for determining the price of gas in custody transfer as well as an essential factor in calculating the efficiencies of energy conversion devices such as gas-fired turbines. The heating values of a gas depend not only upon the temperature and pressure, but also upon the degree of saturation with water vapor. However, some calorimetric methods for measuring heating values are based upon the gas being saturated with water at the specified conditions.

Determination of the heating value of a fuel involves two arbitrary but conventional standard states for the water formed in the reaction: (i) all of the water formed is a liquid – gross heating value, frequently called the higher heating value (HHV) and (ii) all of the water formed is a gas (net heating value, frequently called lower heating value (LHV). The gas industry always uses the gross heating value in custody transfer. Obviously, the numerical difference between the two heating values is the heat of condensation of water at the specified conditions. Both states are hypothetical because the heating value is normally calculated at 60 °F and 1 atm (15.6 °C and 14.7 psi), standard conditions for the gas industry, and, thus at equilibrium, the water would be partially liquid and partially vapor. A common practice is also to assume ideal gas behavior, and consequently, the heating values commonly calculated and reported are representative of, but not identical to, the values obtained when the fuel is burned in an industrial or residential furnace.

The heating value of natural gas (or any fuel gas) may be determined experimentally using a calorimeter in which fuel is burned in the presence of air at constant pressure. The products are allowed to cool to the initial temperature and a measurement is made of the energy released during complete combustion. All fuels that contain hydrogen release water vapor as a product of combustion, which is subsequently condensed in the calorimeter. The resulting measurement of the heat released is the HHV, also known as the gross heating value, and includes the heat of vaporization of water. The LHV, also known as the net heating value, is calculated by subtracting the heat of vaporization of water from the measured HHV and assumes that all products of combustion including water remain in the gaseous phase. The US system of measurement uses Btu per pound or Btu per standard cubic foot when expressed on a volume basis. This property is an indicator of the performance and torque potential of the gas for a defined engine configuration.

The principal quality criterion of natural gas is its heating value and the total calorific (heating) value of fuel gas produced or sold in the natural gas range from 900 to 1,200 Btu/standard ft^3. In addition, the gas must be readily transportable through high-pressure pipelines and, therefore, the water content, as defined by the water dew point, must be considered to prevent the formation of ice or hydrates in the pipeline. Likewise, the amounts of the entrained hydrocarbons having higher molecular weight than ethane, as defined by the hydrocarbon dew point, should be considered to prevent accumulation of condensable liquids that may block the pipeline.

Thus, the energy content of natural gas is variable because natural gas has variations in the amount and types of energy gases (methane, ethane, propane, butane) it contains: the more non-combustible gases in the natural gas, the lower the energy (Btu). In addition, the volume mass of energy gases which are present in a natural gas accumulation also influences the Btu value of natural gas. The more carbon atoms in a hydrocarbon gas, the higher its Btu value. It is necessary to conduct the Btu analysis of natural gas are done at each stage of the supply chain. Gas chromatographic process analyzers are used to conduct fractional analysis of the natural gas streams, separating natural gas into identifiable components. The components and their concentrations are converted into a gross heating value in Btu-cubic foot.

In the United States, at retail, natural gas is often sold in units of Therms (th) (1 Therm = 100,000 BTU). Wholesale transactions are generally done in decatherms, thousand decatherms, or million decatherms (MMDth). A MMDth is roughly a billion cubic feet of natural gas.

The gross heats of combustion of crude oil and its products are given with fair accuracy by the equation:

$$Q = 12,400 - 2,100d^2$$

where d is the 60/60 °F specific gravity. Deviation from the formula is generally less than 1%.

Two other terms that need consideration of this point are (i) the gross heating value, and (ii) the net heating value (Table C.5). The *gross heating value* is the total energy transferred as heat in an ideal combustion reaction at a STP in which all water formed appears as liquid. The gross heating is an ideal gas property in a hypothetical state since the water cannot all condense to liquid because some of the water would saturate the carbon dioxide in the products. Thus:

$$Hv^{id} = \sum_i y_i Hv_i^{id}$$

Hv^{id} is the gross heating value per unit volume of ideal gas, MJ/m^3, y_i is the mole fraction in gas phase for component i. Calculation of the ideal energy flow requires multiplication of the gross heating value by the ideal gas volumetric flow rate of gas for the time period.

Table C.5: Gross heating values and net heating values of the constituents of natural gas.

Gas*	Gross heating value		Net heating value	
	(Btu/ft³)**	(Btu/lb)	(Btu/ft³)	(Btu/lb)
Butane	3,225	21,640	2,977	19,976
Ethane	1,783	22,198	1,630	20,295
Hexane	4,667	20,526	4,315	18,976
Hydrogen sulfide	672	7,479		
Methane	1,011	23,811	910	21,433
Natural gas	950	19,500	850	17,500
Pentane	3,981	20,908	3,679	19,322
Propane	2,572	21,564	2,371	19,834

*Listed alphabetically.
**1 Btu/ft³ = 8.9 kcal/m.

To employ a real gas flow to calculate the ideal energy flow requires converting the real gas flow rate to the ideal gas flow rate by dividing by the Z-factor. Thus, Hv^{id}/Z is the ideal gross heating value per unit volume of real gas – the Z-factor must be determined for the natural gas mixture and then divided into the gross heating value of the mixture. Dividing each pure component gross heating value by the pure component Z-factor and then take the molar average leads to incorrect answers.

The net heating value is the total energy transferred as heat in an ideal combustion reaction at a STP in which all water formed appears as vapor. The net heating is an ideal gas property in a hypothetical state (the water cannot all remain vapor because, after the water saturates the CO_2 in the products, the rest would condense). It is a common misconception that the net heating value applies to industrial operations such as fired heaters and boilers. While the flue gases from these operations do not condense, the net heating value does not apply directly because the gases are not at 15 °C (59 °F). When the gases to cool to 15 °C (59 °F), some of the water would condense while the remainder would saturate the gases. It is possible to use either the gross or net heating value in such situations taking care to utilize the hypothetical state properly.

Typically, a gas with a low calorific value is accompanied by high inert gas content and a gas with high calorific value is accompanied by high content of higher molecular weight hydrocarbons ((C_{2+}) content). With respect to the methane number (MN), inert gases are favorable for the MN level, whereas the higher molecular weight hydrocarbons (C_{2+}) reduce the MN.

C.13 Helium content

Helium is an element that occurs in trace amounts in unprocessed natural gas (Rogers, 1921). Most of the helium that occurs in natural gas is proposed to have been formed by the radioactive decay of uranium and thorium in granite-type rocks that are present in the continental crust of the Earth. As a light gas, helium is buoyant and seeks to move upward as soon as it is formed. Very few natural gas fields contain enough to justify a helium recovery process. Generally, a natural gas source must contain at least 0.3% v/v helium to be considered as a potential helium source.

The richest helium accumulations are found where three conditions exist: (i) granitoid basement rocks are rich in uranium and thorium, (ii) the basement rocks are fractured and faulted to provide escape paths for the helium, and (iii) porous sedimentary rocks above the basement faults are capped by an impermeable seal of halite (NaCl) and anhydrite ($CaSO_4$). When all three of these conditions are met, helium might accumulate in the porous sedimentary rock layer.

For large-scale use, helium is extracted from the gas stream by fractional distillation. Since helium has a lower boiling point than any other element, low temperature and high pressure are used to liquefy nearly all the other gases (mostly nitrogen and methane. The resulting crude helium gas is purified by successive exposures to lowering temperatures, in which almost all of the remaining nitrogen and other gases are precipitated out of the gaseous mixture. Activated charcoal is used as a final purification step, usually resulting in 99.995% pure helium. In a final production step, most of the helium that is produced is liquefied by a cryogenic process that allows more facile transportation of the liquid helium.

C.14 Mercury content

Mercury is present as a trace component in many natural gas reservoirs and, as an environmentally hazardous element, mercury has to be removed from the gas stream, preferably on the production site (Mussig and Rothman, 1997). The presence of mercury is not caused by human activities. Mercury in natural gas is not a man-made nuisance; it occurs naturally in certain gas formations and is inevitably produced with the gas. The mercury traces are supposed to originate from volcanic rocks, often underlying the gas reservoirs.

As an environmentally hazardous element, mercury has to be removed from the gas stream, preferably on the production site. After the concentration of the metal has been determined by application of standard methods, the most appropriate removal technique has to be applied and all mercury-contaminated areas such as sludge and soil have to be cleaned and the waste has to be disposed of properly and according to the law.

During production and treatment of the gas, mercury may be released to the environment and also parts of the plant be contaminated. Mercury may be emitted to air such as by way glycol process overheads, leading to soil contamination together with incidental spillage during maintenance around the treatment facilities. Materials, which have been in contact with mercury, such as sludge from dehydration units and activated carbon filters applied in gas treatment units are contaminated. Regardless of its concentration mercury is adsorbed on any metal surfaces, in scales and corrosion products, therefore, during maintenance, revamp and abandonment activities all these parts might be contaminated and have to be cleaned or treated prior scrapping and disposal or they have to be sent for disposal without affecting the environment.

C.15 Methane number

The main parameter for rating the knock resistance of gaseous fuels is theMN, which is analogous to the octane number for gasoline and suitable test methods to determine the MN of a gaseous fuel are being developed (Malenshek and Olsen, 2009).

MN characterizes gaseous fuel tendency to auto-ignition. By convention, this index has the value 100 for methane and 0 for hydrogen. The gaseous fuels are thus compared with a methane–hydrogen binary mixture. Two gases with same value of *MN* have the same resistance against the spontaneous combustion.

Different scales have been used to rate the knock resistance of compressed natural gas (CNG) including the motor octane number (MON) and the MN. The differences in these ratings are the reference fuel blends used for comparison to the natural gas. MN uses a reference fuel blend of methane, with a MN of 100, and hydrogen, with a MN of 0. Correlations have been generated between the reactive hydrogen/carbon ratio (H/C) and the MON and between MON and MN:

$$\text{MON} = -406.14 + 508.04 \times (\text{H}/\text{C}) - 173.55^*(\text{H}/\text{C})2 + 20.17^*(\text{H}/\text{C})3\,\text{MN}$$

$$= 1.624^*\text{MON} - 119.1$$

Thus, if a gas mixture has a MN of 70, its knock resistance is equivalent to that of a gas mixture of 70% methane and 30% hydrogen.

To ensure safe engine operation, the MN must always be at least equal to the methane number requirement (MNR) of the gas engine. The MN required by the engine is affected by design and operating parameters, with the adjustment of the MN requirement being achieved by changing engine operation. Changes in ignition timing, air/fuel ratio, and output are effective measures to reduce the MN requirement.

C.16 Molecular weight

Although, in a strict sense, a gas mixture does not have a unique molecular weight, it behaves as though it does. Thus, the concept of apparent molecular weight (or the average molecular weight) is useful in characterizing a gas mixture. The apparent molecular weight of a gas mixture is a pseudo property of the mixture and is defined as

$$M_a = \Sigma y_i M_i$$

where M_a is the apparent molecular weight of mixture, y_i is the mole fraction of component, and M_i is the molecular weight of component i.

The gas laws can be applied to gas mixtures by simply using apparent molecular weight instead of the single-component molecular weight in the formulas.

C.17 Pseudoreduced properties

For gas mixtures, the gas critical pressure and gas critical temperature are called pseudocritical pressure and pseudocritical temperature, respectively, to be distinguished from those of pure components, and can be calculated as

$$p_{pc} = \sum_{t=1}^{n} y_t p_{ci'}$$

$$T_{pc} = \sum_{t=1}^{n} y_t T_{ci'}$$

where p_{ci} and T_{ci} are critical pressures and temperatures of individual components, respectively. The temperature must be absolute (R or K), which is simply °F + 460 or °C + 273. The pseudoreduced pressure and temperature of the mixture are simply

$$p_{pr} = \frac{p}{p_{pc}},$$

$$T_{pr} = \frac{T}{T_{pc}}.$$

At the standard conditions of $p_{sc} = 14.7$ psi and $T_{sc} = 60\ °F = 520\ R$, the gas deviation factor, Z_{sc}, can be taken as equal to 1.

Pseudocritical properties of gas mixtures can be estimated from the given gas specific gravity if gas composition is not known.

C.18 Residue

The residue in a gas stream is not to be confused with the term residue gas. The residue gas from the NGLs recovery section is the final, purified sales gas (i.e., methane) which is pipelined to the end-user markets.

On the other hand, the standard test definition for residue in a gas stream is the concentration of contaminants boiling above 37.8 °C (100 °F) that may be present in the liquefied petroleum gas. The contaminants are usually compressor oils, lubricants from valves, plasticizers from hoses, corrosion inhibitors, and other crude oil products from pumps, pipelines, and storage vessels that are used in multiple service applications. Contaminants, whatever the source, can be particularly troublesome in liquid withdrawal systems such as those used in internal combustion engine fuel systems where the materials accumulate in the vaporizer and will ultimately plug the fuel system.

C.19 Sulfur content

Sulfur compounds in natural gas are in the form of mercaptans, hydrogen sulfide, and odorants. The first two are naturally occurring at source (gas fields) and have already been reduced by treatment at the gas processing plant.

The manufacturing processes for liquefied petroleum gas are designed so that the majority, if not all, of the sulfur compounds are removed. The total sulfur level is therefore considerably lower than for other crude oil-based fuels and a maximum limit for sulfur content helps to define the product more completely. The sulfur compounds that are mainly responsible for corrosion are hydrogen sulfide, carbonyl sulfide and, sometimes, elemental sulfur. Hydrogen sulfide and mercaptans have distinctive unpleasant odors. A control of the total sulfur content, hydrogen sulfide and mercaptans ensures that the product is not corrosive or nauseating. Stipulating a satisfactory copper strip test further ensures the control of the corrosion.

Corrosive sulfur compounds can be detected by their effect on copper and the form in which the general copper strip corrosion test for petroleum products is applied to liquefied petroleum gas. Hydrogen sulfide can be detected by its action on moist lead acetate paper and a procedure is also used as a measure of sulfur compounds.

C.20 Specifications

Providers of natural gas and their customers (particularly electric utilities) are used to defining quality (property) requirements for natural gas in the form of specification sheets in which the specifications are developed to conform to federal, state, and local regulations governing emissions compliance, operational requirements (such as the type of generating and/or backend clean-up equipment), and/or any other constraints imposed on the generator.

Due to the nature of natural gas and method of transportation, the quality of the delivered gas is often determined by the pipeline company and whether or not the gas is suitable for inclusion in the pipeline transportation system. Utility generators have a need to know the expected range of quality of the fuel being delivered to them and ideally to have some control over the variability of that fuel to assure compliance with regulations, to protect their investment in generating equipment, and to be able to meet the needs of their customers in the most economic manner. Thus, the properties contained in specification sheets (Table C.6) must be in the form of clearly defined listings of gas quality even though theses may vary from pipeline to pipeline.

Table C.6: Examples of the property data (specifications) that may be required by natural gas users.

Property	Symbol	Units	Test method	Minimum value	Maximum value
High heating value	HHV	Btu/scf			
Low heating value	LHV	Btu/scf			
Specific gravity	ρ				
Wobbe index	WI				
Temperature	T	oF			
Water	H_2O	lb/mmscf			
Water dewpoint	D_{H2O}	oF			
Hydrocarbon dewpoint	D_{hc}	oF			
Methane	C_1	%			
Ethane	C_2	%			
Propane	C_3	%			
Iso-butane	$i\text{-}C_4$	%			

Table C.6 (continued)

Property	Symbol	Units	Test method	Minimum value	Maximum value
n-Butane	$n\text{-}C_4$	%			
Iso-pentane	$i\text{-}C_5$	%			
n-Pentane	$n\text{-}C_5$	%			
Hexanes (and heavier)	C_6 (C_6 +)	%			
Heptanes	C_7	%			
Octanes	C_8	%			
Nonanes	C_9	%			
Decanes	C_{10}	%			
Nitrogen	N_2	%			
Carbon dioxide	CO_2	%			
Oxygen	O_2	%			
Hydrogen	H_2	%			
Helium	He	%			
Other gases		%			
Total sulfur	S_{total}	ppm			
Hydrogen sulfide	H_2S	ppm			
Carbonyl sulfide	COS	ppm			
Sulfur in mercaptans	S_{RSH}	ppm			
Iron sulfide	FeS	lbs			
Other sulfur compounds		ppm			
Mercury	Hg	ppb			
Particulate matter		ppm			
Particulate matter		Size: microns			

C.21 Super compressibility factor

The gas super compressibility factor, Z (or the Z-Factor, or the real gas deviation factor), is a function of pressure and temperature that corrects the *ideal gas law* for high-pressure- and high-temperature conditions. In the gas industry, the z-factor correlation for hydrocarbon gases that is defined by the following correlations:

$$P_{pr} = P/P_{pc}$$

$$T_{pr} = T/T_{pc}$$

In these equations, P is the pressure of interest, psi, P_{pc} is the pseudo-critical pressure, psi, T is temperature of interest, degrees Rankine °R, and T_{pc} is the pseudo-critical temperature, degrees Rankine °R.

These properties are pseudo-properties because the pseudo-critical pressure and pseudo-critical temperature are not measured critical properties, but are calculated properties.

C.22 Viscosity

Knowledge of the viscosity of the reservoir fluids is essential for a study of the dynamic or flow behavior of these fluids through pipes, porous media, or, more generally, wherever transport of momentum occurs in fluid motion. The unit of viscosity is g/cm s, or the poise. The kinematic viscosity is the ratio of the absolute viscosity to the density:

$$\frac{\mu}{\rho} = \frac{centipoise}{\frac{g}{cm^3}} = centistokes$$

The viscosity of gases near room temperature is in the centipoise (cP) range, so that is a commonly used unit. Gas viscosity is only marginally dependent on pressure near atmospheric pressure and is primarily a function of temperature and can be modeled in terms of temperature with the input of experimental reference measurements.

The viscosity of natural gas is usually several orders of magnitude lower than oil or water which (fortunately for gas recovery operations) contributes to the higher mobility of the gas the reservoir relative to crude oil or water. The viscosity of gas mixtures at one atmosphere and reservoir temperature can be determined from the gas mixture composition. Thus:

$$\mu_{ga} = \frac{\sum_{i=1}^{N} y_i \mu_i \sqrt{M_{gi}}}{\sum_{i=1}^{N} y_i \sqrt{M_{gi}}}$$

In this equation, μ_{ga} is the viscosity of the gas mixture at the desired temperature and atmospheric pressure, y_i is the mole fraction of the ith component, μ_i is the viscosity of the ith component of the gas mixture at the desired temperature and atmospheric pressure, M_{gi} gas is the molecular weight of the ith component of the gas mixture, and N is the number of components in the gas mixture.

In summary, the viscosity of natural gas is affected by pressure, température, and composition. Contrary to the case for liquids, the viscosity of a gas at low pressures increases as the temperature is raised. At high pressures, gas viscosity decreases as the temperature is raised. At intermediate pressures, gas viscosity may decrease as temperature is raised and then increase with further increase in temperature.

C.23 Volatility and vapor pressure

The vaporization and combustion characteristics of a gas, especially liquefied petroleum gas, are defined for normal applications by volatility, vapor pressure, and, to a lesser extent, specific gravity.

Volatility is expressed in terms of the temperature at which 95% of the sample is evaporated and presents a measure of the least volatile component present. Vapor pressure (also called the saturation pressure – the corresponding temperature is called saturation temperature) is, therefore, a measure of the most extreme low temperature conditions under which initial vaporization can take place. By setting limits to vapor pressure and volatility jointly, the specification serves to ensure essentially single component products for the butane and propane grades. By combining vapor pressure/volatility limits with specific gravity for propane–butane mixtures, essentially two-component systems are ensured. The residue, that is, non-volatile matter, is a measure of the concentration of contaminants boiling above 37.8 °C (100 °F) that may be present in the gas.

In a closed container, the vapor pressure of a pure compound is the force exerted per unit area of walls by the vaporized portion of the liquid. Vapor pressure is also the pressure at which the vapor phase and the liquid phase of a pure chemical are in equilibrium with each other. In an open air under atmospheric pressure, a liquid at any temperature below its boiling point has its own vapor pressure that is less than 1 atm. When the vapor pressure of a compound reaches 1 atm (14.7 psi), the saturation temperature becomes the normal boiling point. Vapor pressure increases with temperature and the highest value of vapor pressure for a substance is its critical pressure in which the corresponding temperature is the critical temperature.

Vapor pressure is an important thermodynamic property of any chemical and it is a measure of the volatility of a fluid (Table C.7). Compounds with a higher tendency to vaporize have higher vapor pressures. More volatile compounds are those that have lower boiling points and are called light compounds. For example, propane (C_3) has boiling point less than that of n-butane (nC_4) and, as a result, it is more volatile. At a fixed temperature, vapor pressure of propane is higher than that of butane. In this case, propane is called the light compound (more volatile) and butane the heavy compound. Generally, more volatile compounds have higher critical pressure and lower critical temperature, and lower density and lower boiling point than those of less volatile (heavier) compounds, although this is not true for the case of some isomeric compounds.

Vapor pressure is a useful parameter in calculations related to hydrocarbon losses and flammability of hydrocarbon vapor in the air. More volatile compounds

Table C.7: Vapor pressure of the hydrocarbon derivatives of various gas streams.

Hydrocarbon	Formula	Molecular weight	Vapor pressure*
Methane	CH_4	16.043	−5,000
Ethane	C_2H_6	30.07	−800
Propane	C_3H_8	44.097	188
n-Butane	C_4H_{10}	58.124	51.54
iso-Butane	C_4H_{10}	58.124	72.39
n-Pentane	C_5H_{12}	72.151	15.575
iso-Pentane	C_5H_{12}	72.151	20.4444
neo-Pentane	C_5H_{12}	72.151	36.66
n-Hexane	C_6H_{14}	86.178	4.96
2-Methylpentane	C_6H_{14}	86.178	6.767
3-Methylpentane	C_6H_{14}	86.178	6.103
neo-Hexane	C_6H_{14}	86.178	9.859
2,3-Dimethylbutane	C_6H_{14}	86.178	7.406
n-Heptane	C_7H_{16}	100.205	1.62
2-Methylhexane	C_7H_{16}	100.205	2.2719
3-Methylhexane	C_7H_{16}	100.205	2.131
3-Ethylpentane	C_7H_{16}	100.205	2.013
2,2-Dimethylpentane	C_7H_{16}	100.205	3.494
2.4-Dimethylpentane	C_7H_{16}	100.205	3.293
3,3-Dimethylpentane	C_7H_{16}	100.205	2.774
n-Octane	C_8H_{18}	114.232	0.537
iso-Octane	C_8H_{18}	114.232	1.709
n-Nonane	C_9H_{20}	128.259	0.1796
n-Decane	$C_{10}H_{22}$	142.286	0.0609
Cyclopentane	C_5H_{10}	70.135	9.914
Methylcyclopentane	C_6H_{12}	84.162	4.503
Cyclohexane	C_6H_{12}	84.162	3.266
Methylcyclohexane	C_7H_{14}	98.189	1.6093
Ethylene	C_2H_4	28.054	

Table C.7 (continued)

Hydrocarbon	Formula	Molecular weight	Vapor pressure*
Propylene	C_3H_6	42.081	227.6
1-Butene	C_4H_8	56.108	62.1
cis-2-Butene	C_4H_8	56.108	45.95
trans-2-Butene	C_4H_8	56.108	49.94
iso-Butene	C_4H_8	56.108	63.64
1-Pentene	C_5H_{10}	70.135	19.117
1,3-Butadiene	C_4H_6	54.092	59.4
Benzene	C_6H_6	78.114	3.225
Toluene	$C_6H_5CH_3$	92.141	1.033
Ethyl benzene	C_8H_{10}	106.168	0.376
o-Xylene	C_8H_{10}	106.168	0.263
m-Xylene	C_8H_{10}	106.168	0.325
p-Xylene	C_8H_{10}	106.168	0.3424

* psia @100 °F.

are more ignitable than heavier compounds. For example, n-butane is added to gasoline to improve its ignition characteristics. Vapor pressure is, therefore, a measure of the most extreme low temperature conditions under which initial vaporization can take place. By setting limits to vapor pressure and volatility jointly, the specification serves to ensure essentially single component products for the butane and propane grades. By combining vapor pressure/volatility limits with specific gravity for propane–butane mixtures, essentially two-component systems are ensured. The residue, that is, non-volatile matter, is a measure of the concentration of contaminants boiling above 37.8 °C (100 °F) that may be present in the gas.

Low-vapor-pressure compounds reduce evaporation losses and chance of vapor lock. Therefore, for a fuel there should be a compromise between low and high vapor pressure. However, one of the major applications of vapor pressure is in the calculation of equilibrium ratios for phase equilibrium calculations. For pure hydrocarbons, values of vapor pressure at the reference temperature of 100 °F (38 °C). For natural gas, the method of Reid is used to measure vapor pressure at 100 °F. *Reid vapor pressure* is approximately equivalent to vapor pressure at 100 °F (38 °C).

Volatility is expressed in terms of the 95% evaporated temperature and is a measure of the amount of least volatile fuel components present in the product. This specification controls the heavy ends in the fuel and is, in effect, a restriction on higher boiling fractions that will not vaporize for use at system temperatures. When coupled with the vapor pressure, single component products for propane and butane and two component products for butane–propane mixtures are assured.

Vapor pressure is an important specification property of commercial propane, special duty propane, propane/butane mixtures, and commercial butane that assures adequate vaporization, safety, and compatibility with commercial appliances. Relative density, while not a typical criterion contained in specifications, is necessary for determination of filling densities and custody transfer. The MON is useful in determining the products' suitability as a fuel for internal combustion engines. Precision and accuracy of compositional data are extremely important when these data are used to calculate various properties of these petroleum products.

Simple evaporation tests in conjunction with vapor pressure measurement give a further guide to composition. In these tests, a liquefied petroleum gas sample is allowed to evaporate naturally from an open graduated vessel. Results are recorded on the basis of volume/temperature changes, such as the temperature recorded when 95% v/v has evaporated, or volume left at a particular temperature. The evaporation characteristic is a measure of the relative purity of the various types of liquefied petroleum gases and is a benefit in the assurance of volatility performance. The test results can be used to indicate the presence of butane and higher molecular weight constituents in propane-type liquefied petroleum gas, and pentane and higher molecular weight constituents in propane/butane-type fuel gases as well as in butane-type fuel gases. The presence of hydrocarbon compounds less volatile than those of which the liquefied petroleum gas is primarily composed is indicated by an increase in the 95% v/v evaporated temperature. When the type and concentration of higher boiling components is required, chromatographic analysis should be used.

C.24 Volume measurement

In the simplest sense, natural gas does not have a definite volume but is affected by (i) the size of the container – the gas expands to completely fill the container in which it is placed, (ii) the pressure, and (iii) the temperature.

More specifically in the current context, natural gas is measured in volume (cubic meters, m^3, or cubic feet, ft^3) at the resource well – $1\ m^3 = 35.31\ ft^3$. One cubic foot of natural gas is the volume of gas contained in 1 cubic foot at STP. Generally, the gas production measured from the reserves ranges in thousands or millions of cubic feet.

As part of the properties of natural gas, the gas processing operators also need to know not only the bulk properties of the material but also the amount of materials held in storage as part of the natural gas inventory. In fact, once the natural gas has been sampled, there is the need to calculate the total volume of the material that remains in storage. Following from this, the amount of natural gas involved in commercial transactions can be calculated with a high degree of accuracy. The

same volume-remaining issue is relevant for several of the unconventional gas resources, such as coal gas, biogas, and landfill gas.

The varying composition of natural gas from different sources requires several
procedures for calculating volume or density of the gas (ASTM D1071: Standard Test
Methods for Volumetric Measurement of Gaseous Fuel Samples, *Annual Book of
Standards*, ASTM International, West Conshohocken, Pennsylvania). In addition,
temperature affects the volume and density of natural gas and measured volumes
must be converted to net quantities at 15.6 °C (60 °F), which is the accepted base
temperature used widely in commercial transactions. Again, obtaining a representative sample of the material that is held in storage or carrier tanks is essential for
converting measured quantities to the standard volume. To assist with volume calculations, petroleum measurement tables that can also be applied to natural gas are
available for use in the calculation of quantities of natural gas as the gas exists
under designated conditions. The tables presented in the test method provide data
that cover typical operating ranges for the reduction of gravity and volume to standard states, for calculation of weight–volume relationships, and interconversion of
a wide variety of units that are of value for commercial sales of the gas or product.

Gas metering and gas meter types

The common method of measuring the gas produced is by metering the volume of
the stream of gas flowing in the pipeline. Since the volume of gas is a function of
pressure and temperature, and then normally measured at international standard
conditions (14.7 psia and 15.7 °C, 14.7 psia and 60.3 °F, 1 atm and 520 °R).

In the Venturi tube meter, the Venturi effect is the reduction in fluid pressure
that results when a fluid flows through a constricted section of a pipe. The elbow
meter is a device used to measure the flow rate of a gas through a pipe. An elbow
meter is a pipe with a right-angle bend, and flow rate is gauged by sensors that measure the difference in pressure between the inner and outer angle of the bend. The
sonic meter (also known as an ultrasonic flow meter) is a type of flow measurement
equipment that measures the velocity of a fluid flowing inside a pipeline by transmitting ultrasonic sound waves across the pipeline.

The rotameter is a device that measures the volumetric flow rate of gas in a
closed tube. It belongs to a class of meters called variable area meters, which measure flow rate by allowing the cross-sectional area the gas travels through to vary,
causing a measurable effect. An orifice meter is basically a type of flow meter used
to measure the rate of flow of gas, using the differential pressure measurement principle. It is the most common technique of gas measurements in pipeline and the gas
flow rate is calculated using the following equation:

$$Q = K(D*P_f)^{05}$$

In this equation, Q is the gas flow rate, K is the orifice constant, D is the differential pressure across the orifice, and P_f is the flowing pressure.

C.25 Water content

The use of natural gas with high water content can result in the formation of liquid water, ice particles or hydrates at low operating temperatures and high pressure which will interfere with consistently smooth flow of fuel into the engine and cause problems such as poor drivability or even engine stoppage.

When natural gas contains water, but the component analysis is on a dry basis, the component analysis must be adjusted to reflect the presence of water. The mole fraction of water in the mixture is estimated from the definition of relative humidity (on a one mole basis):

$$y_w = h^g P_w^\sigma / P = n_w / (1 + n_w)$$

where n is the number of moles, n_w is the number of water moles, y_w is the mole fraction of water, P_w is the partial pressure of water in gas phase, kPa, P is the total pressure, kPa, and h_g is the relative humidity.

Thus, it is a fundamental requirement that liquefied petroleum gas should not contain free water. Dissolved water may give trouble by forming hydrates and giving moisture vapor in the gas phase. Both of these will lead to blockages. Therefore, test methods are available to determine the presence of water using electronic moisture analyzers, dew-point temperature, and length-of-stain detector tubes.

C.26 Wobbe index

The Wobbe index (also called the Wobbe number) gives a measure of the heat input to an appliance through a given aperture at a given gas pressure. Using this as a vertical co-ordinate and the flame speed factor as the horizontal co-ordinate a combustion diagram can be constructed for an appliance, or a whole range of appliances, with the aid of appropriate test gases. The concept behind defining the Wobbe index is to have a measure for the interchangeability of gases, that is, gases with the same Wobbe index (underdefined pressure conditions) generate the same output during combustion. However, a distinction is made between what is referred to as the higher Wobbe index and the lower Wobbe index, depending on whether the superior or the inferior calorific value are used in the formula.

The Wobbe index (*W*) is an important criterion of inter-changeability of gases in the industrial applications (such as engines, boilers, and burners). Gas composition variation does not involve any notable change of air factor and of flame speed when

Wobbe index remains almost constant. The index can be calculated starting from the *HHV* and specific gas density (*d*) by

$$W = HHV/\sqrt{d} \text{ or } W = HHV/d^{1/2}$$

Some typical Wobbe numbers are methane 1,360 Btu/ft^3; ethane 1,740 Btu/ft^3; and propane 2,044 Btu/ft^3.

This parameter is usually used to characterize gas quality. Indeed, two gases with the same Wobbe index deliver the same quantity of heat for the same supply pressure. Thus, for an industrial burner, for example, one maintains heat flow with a constant value by the output control of gas according to the Wobbe index.

The Wobbe index (or Wobbe number = calorific value/(specific gravity)$^{1/2}$ and the flame speed are usually expressed as a factor or an arbitrary scale on which that of hydrogen is 100. This factor can be calculated from the gas analysis. In fact, calorific value and specific gravity can be calculated from compositional analysis).

The Wobbe number gives a measure of the heat input to an appliance through a given aperture at a given gas pressure. Using this as a vertical co-ordinate and the flame speed factor as the horizontal co-ordinate, a combustion diagram can be constructed for an appliance, or a whole range of appliances, with the aid of appropriate test gases. This diagram shows the area within which variations in the Wobbe index of gases may occur for the given range of appliances without resulting in either incomplete combustion, flame lift, or the lighting back of pre-aerated flames. This method of prediction of combustion characteristics is not sufficiently accurate to eliminate entirely the need for the practical testing of new gases. Another important combustion criterion is the gas modulus: $M = P/W$, where P is the gas pressure and W the Wobbe number of the gas. This must remain constant if a given degree of aeration is to be maintained in a pre-aerated burner using air at atmospheric pressure.

Gas turbines can operate with a large range of fuels, but the fuel variation that a particular installation can cope with is limited. The modified Wobbe index (MWI) is used particularly by gas turbine manufacturers because it takes into account the temperature of the fuel. The MWI is the ratio of the LHV to the square root of the product of the specific gravity and the absolute gas temperature:

$$MWI = LHV/\sqrt{SG_{gas} \times T_{gas}}$$

This is equivalent to

$$MWI = LHV/\sqrt{(MW_{gas}/28.96) \times T_{gas}}$$

LHV is the lower heating value of the fuel gas (Btu/scf), SG_{gas} is the specific gravity of the fuel gas relative to air, MW_{gas} is the molecular weight of the fuel gas, T_{gas} is the absolute temperature of the fuel gas (in degrees Rankine), and 28.96 is the molecular weight of dry air.

Any change in the heating value of the gas will require a corresponding change in the fuel's flow rate to the machine, incorporation of temperature effects is important in calculating energy flows in turbines where a large input temperature variation is possible. The allowable MWI range is established to ensure that required fuel nozzle pressure ratios are maintained during all combustion/turbine modes of operation. For older, diffusion-type combustors, the gas turbine control system can typically accommodate variations in the MWI as large as ±15%. But for newer, dry low NOx (DLN) combustors, variations in the MWI of only ±3% could cause problems. Fuel instability can be caused by velocity changes through a precisely sized fuel-nozzle orifice which can cause flame instability, resulting in pressure pulsations and/or combustion dynamics which can, in the worst case, destroy the combustion system.

Also, since the Wobbe index is an indicator of the interchangeability of fuel gases; it (alone or with other analyses) can be used to control blending of fuel gases. Since the Wobbe index and the Btu value of fuel gases make similar curves, either could be used to control blending of fuel gases; thereby, controlling the amount of nitrogen in the blended fuel (Segers et al., 2011). Another important combustion criterion is the gas modulus, M = P/W where P is the gas pressure and W the Wobbe index of the gas. This must remain constant if a given degree of aeration is to be maintained in a pre-aerated burner using air at atmospheric pressure.

Finally, despite the common acceptance of Wobbe index as the main interchangeability parameter, a variety of units and reference temperatures are used across the world.

Appendix D
Gas cleaning processes

Gas processing plants (gas cleaning plants) play a variety of roles in the natural gas industry and the desired quality of the end product dictates the processes required. Natural gas streams contain impurities in varying amounts which can cause serious problems (Table D.1), and, therefore, gas cleaning processes are required to reduce the impurities to minimum (specified) level using a variety of processes (Table D.2).

Table D.1: Types of impurities (alphabetically rather than by occurrence) and examples of the effects of the impurities.

Carbon dioxide (CO_2)
Helium (He)
Hydrogen sulfide (H_2S)
Mercury (Hg)
Nitrogen (N_2)
Oxygen (O_2)
Particulate matter (such as dust)
Sulfur compounds (RSH)
Water (H_2O)

Effect of carbon dioxide
Present in all gas reservoirs
Corrosive acids are formed in the presence of water.
Lowers the heat content of natural gas

Effect of hydrogen sulfide
Present in all gas reservoirs
Will combine with water to produce corrosive acids
Flaring of natural gas containing hydrogen sulfide will result in the formation of toxic sulfur dioxide (SO_2).

Effect of water
Every gas reservoir has certain initial liquid water saturation
Will combine with carbon dioxide and hydrogen sulfide to produce corrosive acids
Will reduce the area available for gas flow
When the temperature of the natural gas falls below a certain limit, it allows the water and gas molecules to form a solid phase known as hydrates.

Hydrate formation conditions
Hydrates are ice-like solids that form when free water and natural gas combine at high pressure and low temperature.
Hydrates consist of several water molecules associated with each hydrocarbon molecules.
Water should be removed from the gas stream to specified minimum limits.

Gas processing (gas cleaning, gas refining) usually involves the use of several integrated unit processes to remove: (i) oil, (ii) water, (iii) elements such as sulfur, helium, and carbon dioxide, and (iv) natural gas liquids (Tables D.3 and D.4). In addition, it is often necessary to install scrubbers and heaters at or near the wellhead that serve

https://doi.org/10.1515/9783110691023-013

Table D.2: General flow chart for gas cleaning and examples of the individual processes*.

Feedstock	Process type	Process name
Wellhead gas	Condensate and water removal	Separator units
Pretreated gas	Acid gas removal	Benfield process Iron oxide process Olamine process Pressure swing adsorption unit Sulfinol process
Water-containing gas	Dehydration	Glycol unit Pressure swing adsorption unit
Nitrogen-containing gas	Nitrogen rejection	Absorption process Adsorption process Cryogenic process
Liquid-containing gas	Liquids recovery	Absorption Demethanizer Turbo-expander
Hydrocarbon recovery	Fractionation	De-ethanizer Depropanizer Debutanizer
Desulfurization	Sweetening units	Merox process Molecular sieves
Tail gas cleaning	Sulfur production	Claus process SCOT process

primarily to remove sand and other large-particle impurities. The heaters ensure that the temperature of the natural gas does not drop too low and form a hydrate with the water vapor content of the gas stream.

Many chemical processes are available for processing or refining natural gas. However, there are many variables in the choice of process of the choice of refining sequence that dictate the choice of process or processes to be employed. In this choice, several factors must be considered: (i) the types and concentrations of contaminants in the gas, (ii) the degree of contaminant removal desired, (iii) the selectivity of acid gas removal required, (iv) the temperature, pressure, volume, and composition of the gas to be processed, (v) the carbon dioxide–hydrogen sulfide ratio in the gas, and (vi) the desirability of sulfur recovery due to process economics or environmental issues.

In addition to hydrogen sulfide and carbon dioxide, gas may contain other contaminants, such as mercaptans (also called thiols, R-SH) and carbonyl sulfide (COS). In fact, the variation in the composition of the gases that require cleaning whether before being assigned to further use requires a variation of process types to ensure that specifications are met and the environment is protected. The presence of these

Table D.3: Common methods for the removal of carbon dioxide and hydrogen sulfide from gas streams.

Carbon dioxide removal	
Water scrubbing	Uses principle of the higher solubility of carbon dioxide in water to separate the carbon dioxide from biogas. The process uses high pressure and removes hydrogen sulfide as well as carbon dioxide. The main disadvantage of this process is that it requires a large volume of water that must be purified and recycled.
Polyethylene glycol scrubbing	This process is more efficient than water scrubbing. It also requires the regeneration of a large volume of polyethylene glycol.
Chemical absorption	Chemical reaction between carbon dioxide and amine-based solvents (olamines).
Carbon molecular sieves	Uses differential adsorption characteristics to separate methane and carbon dioxide, which is carried out at high pressure (pressure swing adsorption). Hydrogen sulfide should be removed before the adsorption process.
Membrane separation	Selectively separates hydrogen sulfide from carbon dioxide from methane. The carbon dioxide and the hydrogen sulfide dissolve while the methane is collected for use.
Cryogenic separation	Cooling until condensation or sublimation of the carbon dioxide.
Hydrogen sulfide removal	
Biological desulfurization	Natural bacteria can convert hydrogen sulfide into elemental sulfur in the presence of oxygen and iron. This can be done by introducing a small amount (2–5%) of air into the headspace of the digester. As a result, deposits of elemental sulfur will be formed in the digester. This process may be optimized by a more sophisticated design where air is bubbled through the digester feed material. It is critical that the introduction of the air be carefully controlled to avoid reducing the amount of biogas that is produced.
Iron/iron oxide reaction	Hydrogen sulfide reacts readily with either iron oxide or iron chloride to form insoluble iron sulfide. The reaction can be exploited by adding the iron chloride to the digester feed material or by passing the biogas through a bed of iron oxide-containing material. The iron oxide media needs to be replaced periodically. The regeneration process is highly exothermic and must be controlled to avoid problems.

Table D.3 (continued)

Carbon dioxide removal	
Activated carbon	Activated carbon impregnated with potassium iodide can catalytically react with oxygen and hydrogen sulfide to form water and sulfur. The activated carbon beds need regeneration or replacement when saturated.
Scrubbing/membrane separation	The carbon dioxide and hydrogen sulfide can be removed by washing with water, glycol solutions, or separated using the membrane technique.

Table D.4: General properties of unrefined refined natural gas.

Property	Unrefined	Refined
Relative molar mass	20	
Carbon content (% w/w)	73	75
Hydrogen content (% w/w)	27	25
Oxygen content (% w/w)	0.4	0
Hydrogen-to-hydrogen atomic ratio	3.5	4.0
Density relative to air @15 °C	1.5	0.6
Boiling temperature (°C/1 atm)	−162	
Autoignition temperature (°C)	540	560
Octane number	120	130
Methane number	69	99
Vapor flammability limits (volume %)	5	15
Flammability limits	0.7	2.1
Lower heating/calorific value (Btu)	900	
Methane concentration (% v/v)	100	80
Ethane concentration (% v/v)	50	0
Nitrogen concentration (% v/v)	15	0
Carbon dioxide concentration (C% v/v)	5	0
Sulfur concentration (ppm, w/w)	5	0

impurities may eliminate some of the sweetening processes since some processes remove large amounts of acid gas but not to a sufficiently low concentration. On the other hand, there are those processes that are not designed to remove (or are incapable of removing) large amounts of acid gases. However, these processes are also capable of removing the acid gas impurities to very low levels when the acid gases are there in low to medium concentrations in the gas.

The focus of this Appendix is the selection of the processes that are an integral part within the concept of production of a pipelineable product (methane) for sale to the consumer.

D.1 Glycol processes

Absorption dehydration involves the use of a liquid desiccant to remove water vapor from the gas. Although many liquids possess the ability to absorb water from gas, the liquid that is most desirable to use for commercial dehydration purposes should possess the following properties: (i) high absorption efficiency, (ii) relatively easy and economic regeneration, (iii) non-corrosive and non-toxic, (iv) no operational problems when used in high concentrations, and (v) no interaction with the hydrocarbon portion of the gas, and no contamination by acid gases.

Glycols are the most widely used absorption liquids as they approximate the properties that meet the commercial application criteria. The glycol derivatives, particularly ethylene glycol, diethylene glycol, triethylene glycol (TEG), and tetraethylene glycol are the most appropriate for satisfying these criteria to varying degrees. Water and the glycols show complete mutual solubility in the liquid phase due to hydrogen–oxygen bonds, and their water vapor pressures are low. A frequently used process is the TEG process. In situations where inhibition is not feasible or practical, dehydration must be used. Both liquid and solid desiccants may be used, but economics frequently favor liquid desiccant dehydration when it meets the required dehydration specification.

In the process, wet natural gas first enters an inlet separator to remove all liquid hydrocarbon derivatives from the gas stream. Then the gas flows to an absorber (contactor) where it is contacted countercurrently and dried by the lean TEG. However, the TEG also absorbs volatile organic compounds (VOCs) that vaporize with the water in the reboiler. Dry natural gas exiting the absorber passes through a gas/glycol heat exchanger, and then into the sales line. The wet (rich) glycol exiting the absorber flows through a coil in the accumulator where it is preheated by hot lean glycol. After the glycol-glycol heat exchanger, the rich glycol enters the stripping column and flows down the packed bed section into the reboiler. Steam generated in the reboiler strips absorbed water and VOCs out of the glycol as it rises up the packed bed. The water vapor and desorbed natural gas are vented from the top of the stripper. The hot regenerated lean glycol flows out of the reboiler into the accumulator (surge tank) where it is cooled via cross exchange with returning rich glycol; it is pumped to a glycol/gas heat exchanger and back to the top of the absorber.

In the TEG process, wet natural gas first enters an inlet separator to remove all liquid hydrocarbon derivatives from the gas stream after which the gas flows to an absorber (contactor) where it is contacted counter-currently and dried by the lean TEG. The TEG also absorbs VOCs that vaporize with the water in the reboiler. The dry natural gas existing the absorber passes through a gas/glycol heat exchanger and then into the sales line.

D.2 Olamine processes

Acid gas constituents (hydrogen sulfide, H_2S, and carbon dioxide, CO_2) present in most natural gas streams are a constant reminder of the need for gas processing, specifically acid gas removal technologies. As currently practiced, acid gas removal processes involve selective absorption of the contaminants into a liquid (such as ethanolamine) that is passed countercurrent to the gas or the chemical reaction of the acid gases with a solid oxide (such as iron oxide). In the case of the use of olamine derivatives (Table D.5), the absorbent is stripped of the gas components (regeneration) and recycled to the absorber. The process design will vary and, in practice, may employ multiple absorption columns and multiple regeneration columns. However, depending upon the application, special solutions such as mixtures of amines; amines with physical solvents such as sulfolane and piperazine; and amines that have been partially neutralized with an acid such as phosphoric acid may also be used.

Different amine derivatives can be selected for use, depending on the composition and operating conditions of the feed gas, different amines can be selected to meet the product gas specification. Amines are categorized as being primary, secondary, and tertiary, depending upon the degree of substitution of the central nitrogen by organic groups. Primary amines react directly with sulfide, carbon dioxide, and COS.

Examples of primary amines include monoethanolamine (MEA) and the proprietary diglycolamine agent (DGA). Secondary amines react directly with hydrogen sulfide and carbon dioxide as well as with COS. The most common secondary amine is diethanolamine (DEA), while diisopropanolamine (DIPA) is another example of a secondary amine that has been used in amine treating systems. Tertiary amines react directly with hydrogen sulfide, react indirectly with carbon dioxide, and react indirectly with COS. The most common examples of tertiary amines are methyldiethanolamine (MDEA) and activated methyldiethanolamine (a-MDEA). Depending upon the application, special solutions such as mixtures of amines; amines with physical solvents such as sulfolane and piperazine; and amines which have been partially neutralized with an acid such as phosphoric acid may also be used.

Table D.5: Olamines used for gas processing.

Olamine	Formula	Derived Name	Molecular weight	Specific gravity	Melting point, °C	Boiling point, °C	Flash point, °C	Relative capacity
Diethanolamine	$(HOC_2H_4)_2NH$	DEA	105.14	1.097	27	217	169	58
Diglycolamine (hydroxyethanolamine)	$H(OC_2H_4)_2NH_2$	DGA	105.14	1.057	−11	223	127	58
Diisopropanolamine	$(HOC_3H_6)_2NH$	DIPA	133.19	0.99	42	248	127	46
Ethanolamine (monoethanolamine)	$HOC_2H_4NH_2$	MEA	61.08	1.01	10	170	85	100
Methyldiethanolamine	$(HOC_2H_4)_2NCH_3$	MDEA	119.17	1.03	−21	247	127	51
Triethanolamine	$(HOC_2H_4)_3NH$	TEA	148.19	1.124	18	335, d	185	41

*Listed alphabetically; d, with decomposition.

In addition, in a refinery other gas streams (process gas) may be added to the natural; gas to be co-processed and these refinery streams will likely contain mercaptan derivatives (RSH), carbon disulfide (CS_2), or COS. Thus, the level of concentration of acid gases in the sour gas is an important consideration for selecting the proper sweetening process. Some processes are applicable for removal of large quantities of acid gas, and other processes have the capacity for removing acid gas constituents to the parts per million (ppm) range. However, whatever the range of non-hydrocarbon constituents in a gas stream, the sweetening process should ensure that the product gas meets pipeline specification or process specifications.

MEA and DEA have found the most general application in the sweetening of natural gas streams. Even though a DEA system may not be as efficient as some of the other chemical solvents are, it may be less expensive to install because standard packaged systems are readily available. In addition, it may be less expensive to operate and maintain.

MEA is a stable compound and in the absence of other chemicals suffers no degradation or decomposition at temperatures up to its normal boiling point and readily reacts with hydrogen sulfide and carbon dioxide, thus:

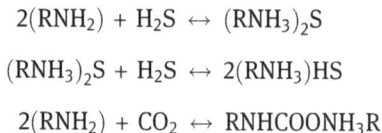

$$2(RNH_2) + H_2S \leftrightarrow (RNH_3)_2S$$

$$(RNH_3)_2S + H_2S \leftrightarrow 2(RNH_3)HS$$

$$2(RNH_2) + CO_2 \leftrightarrow RNHCOONH_3R$$

These reactions are reversible by changing the system temperature. MEA also reacts with COS and carbon disulfide (CS_2) to form heat-stable salts that cannot be regenerated. On the other hand, DEA is a weaker base than MEA and therefore the DEA system does not typically suffer the same corrosion problems. DEA reacts with hydrogen sulfide and carbon dioxide in the following manner:

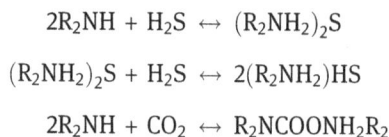

$$2R_2NH + H_2S \leftrightarrow (R_2NH_2)_2S$$

$$(R_2NH_2)_2S + H_2S \leftrightarrow 2(R_2NH_2)HS$$

$$2R_2NH + CO_2 \leftrightarrow R_2NCOONH_2R_2$$

These reactions are reversible. Also, DEA reacts with COS and with carbon disulfide to form compounds that can be regenerated in the stripping column.

D.3 Physical solvent processes

Two of the currently most widely used physical solvent processes for gas cleaning are the Selexol process and the Rectisol process. The process solvent is a mixture of dimethyl ethers of polyethylene glycol [$CH_3(CH_2CH_2O)_nCH_3$] where n is between 3 and 9. The solvent is chemically and thermally stable and has a low vapor pressure that limits its losses to the treated gas. The solvent has a high solubility for carbon dioxide, hydrogen sulfide, and COS. It also has appreciable selectivity for hydrogen sulfide over carbon dioxide.

The principal benefits of physical solvents are: (i) high selectivity for hydrogen sulfide over COS and carbon dioxide, (ii) high loadings at high acid gas partial pressures, (iii) solvent stability, and (iv) low heat requirements because most of the solvent can be regenerated by a simple pressure letdown. The performance of a physical solvent can be easily predicted. The solubility of a compound in the solvent is directly proportional to its partial pressure in the gas phase, hence, the improvement in the performance of physical solvent processes with increasing gas pressure. Physical solvent processes can be configured to take advantage of their high hydrogen sulfide/ carbon dioxide selectivity together with high levels of carbon dioxide recovery.

In the Selexol process, the solvent is composed of a dimethyl ether of polyethylene glycol, which is chemically inert and not subject to degradation. The process also removes COS, mercaptan derivatives, ammonia, hydrogen cyanide, and metal carbonyl derivatives. A variety of flow schemes permit process optimization and energy reduction and the partial pressure of the acid gas is the key driving force. Typical feedstock conditions range between 300 and 2000 psia, with acid composition (carbon dioxide plus hydrogen sulfide) from 5% v/v to more than 60% v/v by volume. The product specifications achievable depend on the application and can range from ppm up to percent levels of acid gas.

The Rectisol process is the most widely used physical solvent gas treating process for acid gas removal using an organic solvent at low temperatures. In general, methanol is used for removal of hydrogen sulfide, COS, and carbon dioxide as well as removal of organic and inorganic impurities. It is possible to produce a clean gas with less than 0.1 ppm sulfur and a carbon dioxide content down to the ppm range. The main advantage over other processes is the use of a cheap, stable and easily available solvent, a very flexible process and low utilities.

The Sulfinol process, developed in the early 1960s, is a combination process that uses a mixture of amines and a physical solvent. The solvent consists of an aqueous amine and sulfolane. The process is used for the removal of hydrogen sulfide, COS, mercaptan derivatives, other organic sulfur compounds, and all or part of the carbon dioxide from natural, synthetic, and refinery gases. The total sulfur compounds in the treated gas can be reduced to ultra-low ppm levels, as required for refinery fuel and pipeline quality gases. An improved application is to selectively remove hydrogen sulfide, COS, mercaptan derivatives and other organic sulfur compounds for pipeline specification, while co-absorbing only part of the carbon dioxide. Deep removal of carbon dioxide removal for liquefied natural gas plants is another application, as well as bulk carbon dioxide removal with flash regeneration of the solvent. The process sequence – Sulfinol/Claus/shell Claus off-gas treating process (SCOT) – can be used advantageously with an integrated Sulfinol system that handles selective hydrogen sulfide removal upstream and the SCOT process that treats the Claus off-gas.

Sulfinol has a good affinity for most of the acid gases and has the ability to release these gases in the regenerator upon pressure reduction and heat application. When operating under suitable conditions it is capable of removing twice as much acid gas as a 20% MEA solution.

The Sulfinol-D process uses DIPA, while Sulfinol-M uses MDEA. The mixed solvents allow for better solvent loadings at high acid gas partial pressures and higher solubility of COS and organic sulfur compounds than straight aqueous amines.

D.4 Metal oxide processes

The use of solids for sweetening gas (typically in batch-type process) is based on adsorption of the acid gases on the surface of the solid sweetening agent, or reaction with some component on that surface. The solids processes are usually best applied to gases containing low-to-medium concentrations of hydrogen sulfide or mercaptan derivatives. The solids processes tend to be highly selective and do not normally remove significant quantities of carbon dioxide. Consequently, the regenerated hydrogen sulfide stream from the process is high purity and, in addition, pressure has relatively little effect on the adsorptive capacity of a sweetening agent.

An example of a hydrogen sulfide scavenger process is the iron sponge process (also called the dry box process) which is the oldest and still the most widely used batch process for sweetening of natural gas and natural gas liquids. Hydrogen sulfide scavengers are appropriate for use at the low concentrations of hydrogen sulfide where conventional chemical absorption and physical solvents are not economical. During recent years, hydrogen sulfide scavenger technology has been expanded with many new materials coming on the market and others being discontinued. Overall, the simplicity of the process, low capital costs, and relatively low chemical (iron oxide) cost continue to make the process an ideal solution for hydrogen sulfide removal. In addition, pressure has relatively little effect on the adsorptive capacity of a sweetening agent. The use of iron sponge process for sweetening sour gas is based on adsorption of the acid gases on the surface of the solid sweetening agent.

The *sponge* consists of wood shavings impregnated with a hydrated form of iron oxide. The wood shavings serve as a carrier for the active iron oxide powder. Hydrogen sulfide is removed by reacting with iron oxide to form ferric sulfide. The process is usually best applied to gases containing low to medium concentrations (300 ppm) of hydrogen sulfide or mercaptans. This process tends to be highly selective and does not normally remove significant quantities of carbon dioxide. As a result, the hydrogen sulfide stream from the process is high purity. The use of iron sponge process for sweetening sour gas is based on adsorption of the acid gases on the surface of the solid sweetening agent followed by chemical reaction of ferric oxide (Fe_2O_3) with hydrogen sulfide:

$$2Fe_2O_3 + 6H_2S \rightarrow 2Fe_2S_3 + 6H_2O$$

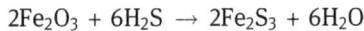

The reaction requires the presence of slightly alkaline water and a temperature below 43 °C (110 °F) and bed alkalinity should be checked regularly, usually on a daily basis. A pH level on the order of 8 to 10 should be maintained through the injection of caustic soda with the water. If the gas does not contain sufficient water vapor, water may need to be injected into the inlet gas stream.

The ferric sulfide produced by the reaction of hydrogen sulfide with ferric oxide can be oxidized with air to produce sulfur and regenerate the ferric oxide:

$$2Fe_2S_3 + 3O_2 \rightarrow 2Fe_2O_3 + 6S$$

$$S + O_2 \rightarrow SO_2$$

The regeneration step, that is, the reaction with oxygen is exothermic and air must be introduced slowly so the heat of reaction can be dissipated. If not and the air is introduced rapidly, there is the potential for the bed to ignite the bed. Some of the elemental sulfur produced in the regeneration step remains in the bed which, after several cycles, the sulfur will form a cake over the ferric oxide thereby decreasing the reactivity of the bed. Typically, after approximately 10 cycles, depending upon the sulfur content of the gas stream, the bed must be removed, and a new bed introduced into the reactor.

The Slurrisweet process uses iron oxide slurry that is similar to those for dry iron oxide processes, except with a higher proportion of the magnetite form of iron oxide (Fe_3O_4). Any foaming and settling problems of the iron oxide particles were solved using a silicon-based defoamer with additives and a dispersant, respectively. Also, corrosion was inhibited by using an epoxy coating on the vessel. Injection of air at 5% mole concentration of hydrogen sulfide extended the batch life and stabilized the spent chemical.

Iron oxide suspensions, like iron oxide slurries, rely upon hydrated ferric oxide as the active regenerable agent. However, iron oxide suspensions react in a basic environment with an alkaline compound, followed by the reaction of the hydrosulfide with iron oxide to form iron sulfide. The iron is then regenerated by aeration. Thus:

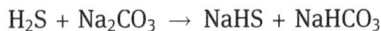

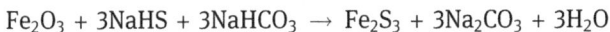

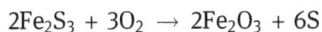

$$H_2S + Na_2CO_3 \rightarrow NaHS + NaHCO_3$$

$$Fe_2O_3 + 3NaHS + 3NaHCO_3 \rightarrow Fe_2S_3 + 3Na_2CO_3 + 3H_2O$$

$$2Fe_2S_3 + 3O_2 \rightarrow 2Fe_2O_3 + 6S$$

Iron oxide suspensions were the precursors to the chelated iron processes.

Adsorption (or solid bed) dehydration is the process where a solid desiccant is used for the removal of water vapor from a gas stream. The solid desiccants commonly used for gas dehydration are those that can be regenerated and, consequently, used over several adsorption–desorption cycles. In fact, there are several solid desiccants which possess the physical characteristic to adsorb water from natural gas but the most popular are (i) alumina, (ii) silica gel, and (iii) silica–alumina gel.

Alumina is a hydrated form of aluminum oxide (Al_2O_3) and is the least expensive adsorbent. It is activated by driving off some of the water associated with it in its hydrated form ($Al_2O_3 \cdot 3H_2O$) by heating. It produces an excellent dew point depression values as low as -100 °F but requires much more heat for regeneration. Also, it is alkaline and cannot be used in the presence of acid gases, or acidic chemicals used for well treating. The tendency to adsorb higher molecular weight hydrocarbon derivatives is high, and it is difficult to remove these during regeneration. It has good resistance to liquids, but little resistance to disintegration due to mechanical agitation by the flowing gas.

Silica gel and silica–alumina gel are granular, amorphous solids manufactured by chemical reaction. Gels manufactured from sulfuric acid and sodium silicate reaction are called silica gels and consist almost solely of silicon dioxide (SiO_2). Alumina gels consist primarily of some hydrated form of aluminum trioxide (Al_2O_3). Silica–alumina gels are a combination of silica and alumina gel. Gels can dehydrate gas to as low as 10 ppm and have the greatest ease of regeneration of all desiccants. They adsorb high molecular weight hydrocarbon derivatives but release them relatively more easily during regeneration. Since these gels are acidic, they can handle sour gases but not alkaline materials such as caustic or ammonia. Although there is no reaction with hydrogen sulfide, sulfur can deposit and block their surface and, therefore, the gels are useful if the content of hydrogen sulfide in the gas stream is less than 5–6% v/v.

The solid desiccants generally are used in dehydration systems consisting of two or more towers and associated regeneration equipment, such as the use of a two-tower system pressure-swing adsorption system. One tower is on-stream adsorbing water from the gas while the other tower is being regenerated and cooled. Hot gas is used to drive off the adsorbed water from the desiccant, after which the tower is cooled with an unheated gas stream. The towers are switched before the on-stream tower becomes water saturated. In this configuration, part of the dried gas is used for regeneration and cooling and is recycled to the inlet separator.

The so-called dry sorption processes are used to scavenge hydrogen sulfide and organic sulfur compounds (mercaptans) from gas streams through reactions with a solid based media. They are typically non-regenerable processes although some are partially regenerable, losing activity upon each regeneration cycle. Most dry sorption processes are governed by the reaction of a metal oxide with hydrogen sulfide to form a metal sulfide compound. For regenerable reactions, the metal sulfide compound can then react with oxygen to produce elemental sulfur and a regenerated metal oxide. The primary metals used for dry sorption processes are iron and zinc.

Zinc oxide (ZnO) has also been used to clean gas tams. For example, at increased temperatures (205–370 °C, 400–700 °F), zinc oxide has a rapid reaction rate, therefore providing a short mass transfer zone, resulting in a short length of unused bed and improved efficiency. An enhanced form of the zinc oxide process (the Puraspec process) can operate more efficiently at reduced temperatures (38–205 °C, 100–400 °F) which is due to an increased porosity and decreased density resulting in higher obtainable sulfur loading per pound of media. In the process, fixed beds of chemical absorbents provide effectively total irreversible selective removal of impurities from wet or dry hydrocarbons without feedstock losses. Radial flow reactor designs are available if low system pressure drop is required. The process displays greater efficiency at the higher operating temperatures (205–370 °C, 400–700 °F).

In a slurry process, the zinc oxide slurry reactor consists of a simple vertical bubble contactor with the gas providing sufficient agitation to keep zinc oxide particles in suspension in addition to a dispersant:

$$ZnO + H_2S \rightarrow ZnS + H_2O$$

$$ZnO + H_2S \rightarrow Zn(OH)(HS)$$

Zinc mercaptide ($Zn(OH)(HS)$) is a minor product and the majority of the hydrogen sulfide is converted to zinc sulfide. Zinc mercaptide will form a sludge in the reactor and can contribute to foaming.

The Sulfa-Check process is used to selectively removes hydrogen sulfide and mercaptan derivatives from natural gas, in the presence of carbon dioxide. The original process is accomplished in a one-step single vessel design using an aqueous solution of sodium nitrite buffered to stabilize the pH above 8. Also, there is enough strong base to raise the pH of the fresh material to 12.5. Removal of hydrogen sulfide is not affected under short contact times since the reaction is almost instantaneous. Sodium hydroxide and sodium nitrite are consumables in the processes and cannot be regenerated. This process is accomplished in a one-step single vessel design using an aqueous solution of sodium nitrite buffered to stabilize the pH above 8. Also, there is enough strong base to raise the pH of the fresh material to 12.5. The reaction with hydrogen sulfide forms elemental sulfur, ammonia, and caustic soda:

$$NaNO_2 + 3H_2S \leftrightarrow NaOH + NH_3 + 3S + H_2O$$

Other reactions forming the oxides of nitrogen do occur (Burnes and Bhatia, 1985) and carbon dioxide in the gas reacts with the sodium hydroxide to form sodium carbonate and sodium bicarbonate. The spent solution is slurry of fine sulfur particles in a solution of sodium and ammonium salts.

The Chemsweet process is a batch process for the removal of hydrogen sulfide from natural gas. The chemicals of choice for the process are a mixture of zinc oxide (ZnO), zinc acetate (($CH3COO)_2Zn$, $ZnAc_2$), water, and a dispersant to keep the zinc oxide particles in suspension. When one part is mixed with five parts of water the acetate dissolves and provides a controlled source of zinc ions that react instantaneously with the bisulfide and sulfide ions that are formed when hydrogen sulfide dissolves in water. In addition to a chemical reaction to remove the hydrogen sulfide, the zinc oxide also replenishes the zinc acetate. Thus:

Sweetening:

$$ZnAc_2 + H_2S \leftrightarrow ZnS + 2HAc$$

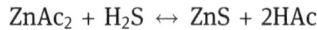

Regeneration:

$$ZnO + 2HAc \leftrightarrow ZnAc_2 + H_2O$$

The overall reaction is

$$ZnO + H_2S \leftrightarrow ZnS + H_2O$$

The presence of carbon dioxide in the natural gas is of little consequence to the process because the pH of the Chemsweet slurry is low enough to prevent significant absorption of the carbon dioxide, even when the ratio of carbon dioxide to hydrogen sulfide is high.

The SulfaTreat process is also a batch-type process for the selective removal of hydrogen sulfide and mercaptans from natural gas. The process is dry, using no free liquids, and can be used for natural gas applications where a batch process is suitable. The SulfaTreat system is a more recent development using iron oxide on a porous solid material. Unlike the iron sponge process, the SulfaTreat material is non-pyrophoric and has a higher capacity than iron sponge on a volumetric or mass basis.

The chemistry of the SulfaTreat I similar to the chemistry of the iron sponge process:

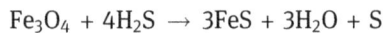

$$Fe_3O_4 + 4H_2S \rightarrow 3FeS + 3H_2O + S$$

$$Fe_3O_4 + 6H_2S \rightarrow 3FeS_2 + 4H_2O + 2H_2$$

$$Fe_2O_3 + 3H_2S \rightarrow Fe_2S_3 + 3H_2O$$

Another fixed-bed dry sorption process makes use of a hydroxide media similar to that of caustic scrubbing processes (Sofnolime process). It is a dry process that claims it has a synergistic mixture of hydroxides in a granular solid. The media is capable of removing hydrogen sulfide, carbon dioxide, COS, sulfur dioxide, and organic sulfur compounds. The process employs the following reactions:

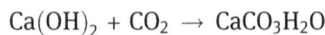

$$2NaOH + H_2S \rightarrow Na_2S + H_2O$$

$$Ca(OH)_2 + CO_2 \rightarrow CaCO_3H_2O$$

The process removes both hydrogen sulfide and carbon dioxide but gas streams with high carbon dioxide content, such as biogas, will exhaust the media rapidly. The packed bed is supported between layers of ceramic balls on the top and bottom, with the flow direction in the upward direction.

While not strictly a metal oxide process, *iron chelate* solutions have been developed with greater acceptability due to the non-hazardous classification of the working solution. Two processes are worthy of note here: (i) the LO-CAT process and (ii) the SulFerrox process. The LO-CAT and SulFerrox process differ mainly on vessel configurations, iron concentrations, proprietary chelates, and additives to optimize the process. The main advantages of chelated iron solutions are the catalytic nature of the reactant, reduced plant footprint, and ability to reclaim elemental sulfur.

The technology involves the use of an iron chelate-type catalyst that converts hydrogen sulfide to elemental sulfur by the reduction of ferric ions to ferrous ions. The ferric ions are regenerated by contact with air. Thus,

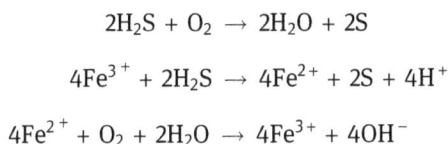

$$2H_2S + O_2 \rightarrow 2H_2O + 2S$$

$$4Fe^{3+} + 2H_2S \rightarrow 4Fe^{2+} + 2S + 4H^+$$

$$4Fe^{2+} + O_2 + 2H_2O \rightarrow 4Fe^{3+} + 4OH^-$$

In addition, the BioDeNOx process is a biological process that removes nitrogen oxides from flue gases. An iron chelate selectively absorbs the nitrogen oxides which are reduced to nitrogen with ethanol in the presence of microorganisms. They use a wet gas scrubber to contact the circulation liquid with flue gas feed and absorb the nitrogen oxides (NO_x). In the sump underneath the scrubber, the absorbed nitrogen oxides are biologically reduced to nitrogen and ethanol is also consumed and, thus, the iron chelate solution is regenerated. The presence of oxygen and acids compounds in the flue gas, such as hydrogen chloride and hydrogen fluoride, oxidizes a part of iron chelate to the ferric (Fe^{3+}) state. Therefore, a purge and makeup of iron chelate is necessary to eliminate this oxidized ferric material. To minimize iron chelate consumption, a nano-filtration can be installed and bleed from the unit is passed through the filter and the chelate is recovered.

Another biological-type process is the THIOPAQ process which involves the biological desulfurization of high pressure natural gas, synthesis gas, fuel gas streams, acid gas from amine regeneration and treatment of spent caustic. The process is the oxidation of hydrogen sulfide is oxidized to elemental sulfur using sulfur bacteria (Thiobacilli) which are naturally occurring bacteria are not genetically modified. In the process, the gas stream gas is sent to a caustic scrubber in which the hydrogen sulfide reacts to produce sodium sulfide which is converted to elemental sulfur and caustic by the bacteria when air is supplied in the bioreactor. The sulfur particles are covered with a (bio) macropolymer layer, which maintains the sulfur in a mil-like suspension that does not cause fouling or plugging. The sulfur slurry is produced can be concentrated to a cake containing 60% w/w dry matter. This cake can be used directly for agricultural purposes, or as feedstock for sulfuric acid manufacturing or, alternately, the biological sulfur slurry can be purified further by melting to high-quality sulfur to meet Claus sulfur specifications.

D.5 Methanol-based processes

Methanol is probably one of the most versatile solvents in the natural gas processing industry. Historically, methanol was the first commercial organic physical solvent and has been used for hydrate inhibition, dehydration, gas sweetening and
liquids recovery. Most of these applications involve low temperature where methanol's physical properties are advantageous compared with other solvents that exhibit
high viscosity problems or even solids formation. Operation at low temperatures tends
to suppress the most significant disadvantage of methanol, high solvent loss. Furthermore, methanol is relatively inexpensive and easy to produce making the solvent an
attractive alternate for gas processing applications.

The use of methanol has been further exploited in the development of the Rectisol process either alone or as toluene–methanol mixtures are used to remove hydrogen sulfide and slip carbon dioxide to the overhead product more selectively. Toluene
has an additional advantage insofar as COS is more soluble in toluene than in methanol. The Rectisol process was primarily developed to remove both carbon dioxide and
hydrogen sulfide (along with other sulfur-containing species) from gas streams resulting from the partial oxidation of coal, oil, and petroleum residua. The ability of methanol to absorb these unwanted components made it the natural solvent of choice.
Unfortunately, at cold temperatures, methanol also has a high affinity for hydrocarbon
constituents of the gas streams. For example, propane is more soluble in methanol
than is carbon dioxide. There are two versions of the Rectisol process – the two stage
and the once through. The first step of the two-stage process is desulfurization before
shift conversion; the concentrations of hydrogen sulfide and carbon dioxide are about 1
and 5% by volume, respectively. Regeneration of the methanol following the desulfurization of the feed gas produces high sulfur feed for sulfur recovery. The once-through
process is only applicable for high-pressure partial-oxidation products. The once-
through process is also applicable when the hydrogen sulfide to carbon dioxide
content is unfavorable, in the neighborhood of 1:50.

Recently, a process using methanol has been developed in which the simultaneous capability to dehydrate, to remove acid gas, and to control hydrocarbon dew
point. The IFPEXOL-1 process is used for water removal and hydrocarbon dew point
control; the IFPEXOL-2 process is used for acid gas removal. The novel concept behind the IFPEXOL-1 process is to use a portion of the water-saturated inlet feed to
recover the methanol from the aqueous portion of the low temperature separator.
That approach has solved a major problem with methanol injection in large facilities, the methanol recovery via distillation. Modifications to the process include
water washing the hydrocarbon liquid from the low temperature separator to enhance the methanol recovery. The IFPEXOL-2 process for acid gas removal is similar
to an amine-type process except for the operating temperatures. The absorber operates below −20 °F to minimize methanol losses, and the regenerator operates at about
90 psi. Cooling is required on the regenerator condenser to recover the methanol.

This process usually follows the IFPEXOL-1 process so excessive hydrocarbon absorption is not as great a problem.

D.6 Alkali washing processes

Alkali washing processes typically fall under the general banner of scrubber-based processes (such as chemical scrubbers and gas scrubbers) which are a diverse group of gas cleaning processes that can be used to remove some particulate matter and/or gases from industrial exhaust streams. Traditionally, the term scrubber is used to refer to devices that use a liquid to wash remove contaminants from gas streams. More recently, the term has also been used to describe systems that inject a dry reagent or a slurry into a contaminated gas stream to wash out acid gases.

Wet scrubbers can also be used for heat recovery from hot gases by flue-gas condensation. In this option, water from the scrubber drain is circulated through a cooler to the nozzles at the top of the scrubber. The hot gas enters the scrubber at the bottom and, if the gas temperature is above the dew point of the water, it is initially cooled by evaporation of the water drops. Further cooling causes water vapor to condense thereby adding to the amount of circulating water.

Caustic scrubbing for hydrogen sulfide removal with caustic scrubbing is only economical when small amounts if hydrogen sulfide are present and suitable means of disposing the spent solution are available. The chemistry is simple and to some extent, depends on the concentration of hydrogen sulfide in the gas stream efficient. Thus,

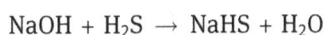

$$NaOH + H_2S \rightarrow NaHS + H_2O$$

$$2NaOH + H_2S \rightarrow Na_2S + H_2O$$

$$2NaOH + CO_2 \rightarrow Na_2CO_3 + H_2O$$

Carbonate washing is a mild alkali process for emission control by the removal of acid gases (such as carbon dioxide and hydrogen sulfide) from gas streams and uses the principle that the rate of absorption of carbon dioxide by potassium carbonate increases with temperature. It has been demonstrated that the process works best near the temperature of reversibility of the reactions:

$$K_2CO_3 + CO_2 + H_2O \rightarrow 2KHCO_3$$

$$K_2CO_3 + H_2S \rightarrow KHS + KHCO_3$$

In the Benfield process, acid gases are scrubbed from the feed in an absorber column using potassium carbonate solution with Benfield additives to improve performance and avoid corrosion.

Water washing, in terms of the outcome, is analogous to washing with potassium carbonate, and it is also possible to carry out the desorption step by pressure reduction. The absorption is purely physical and there is also a relatively high absorption of hydrocarbon derivatives, which are liberated at the same time as the acid gases.

The hot potassium carbonate process has been utilized successfully for bulk removal of carbon dioxide from a number of gas mixtures. It has been used for sweetening natural gases containing both carbon dioxide and hydrogen sulfide. The process is not suitable for sweetening gas mixtures containing little or no carbon dioxide, as the potassium bisulfide should be difficult to regenerate if carbon dioxide is not present.

The process using potassium phosphate is known as phosphate desulfurization, and it is used in the same way as the Girbotol process to remove acid gases from liquid hydrocarbon derivatives as well as from gas streams. The treatment solution is a water solution of tripotassium phosphate (K_3PO_4), which is circulated through an absorber tower and a reactivator tower in much the same way as the ethanolamine is circulated in the Girbotol process; the solution is regenerated thermally.

Other processes include the Alkazid process, which removes hydrogen sulfide and carbon dioxide using concentrated aqueous solutions of amino acids. The hot potassium carbonate process decreases the acid content of natural and refinery gas from as much as 50% to as low as 0.5% and operates in a unit similar to that used for amine treating.

The Giammarco–Vetrocoke process is used for hydrogen sulfide and/or carbon dioxide removal. In the hydrogen sulfide removal section, the reagent consists of sodium or potassium carbonates containing a mixture of arsenite derivatives and arsenate derivatives; the carbon dioxide removal section utilizes hot aqueous alkali carbonate solution activated by arsenic trioxide or selenous acid or tellurous acid.

The Catacarb process employs a modified potassium salt solution containing a stable and nontoxic catalyst and corrosion inhibitor. In the process, a catalyst is used to activate a carbonate solution in the absorption and desorption of carbon dioxide, thus overcoming the above disadvantage of carbonate scrubbing. Several other catalysts and inhibitors are also used in this process – the choice depends on the composition of the gas stream to be treated. This process is also capable of removing trace amounts of other acid gases such as COS, carbon disulfide, and mercaptan derivatives.

The Merox process is used to treat end product streams by rendering any mercaptan sulfur compounds inactive. This process can be used for treating liquefied petroleum gas, natural gasoline, and higher molecular weight fractions. The method of treatment is the extraction reaction of the sour feedstock containing mercaptans (RSH) with caustic soda (NaOH) in a single, multi-stage extraction column using high efficiency trays. The extraction reaction is shown by the following equation:

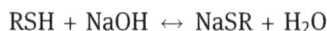

$$RSH + NaOH \leftrightarrow NaSR + H_2O$$

After extraction, the extracted mercaptans in the form of sodium mercaptide derivatives (NaSR) are catalytically oxidized to water insoluble disulfide derivatives (RSSR):

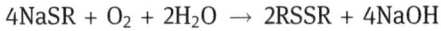

$$4NaSR + O_2 + 2H_2O \rightarrow 2RSSR + 4NaOH$$

The disulfide oil is decanted and sent to fuel or to further processing in a hydrotreater. The regenerated caustic is then recirculated to the extraction column.

The Sulfa-Check process uses sodium nitrite ($NaNO_2$) as the basis for the caustic process. Gas streams with elevated oxygen levels with the Sulfa-Check process will produce some NOx in the gas stream.

D.7 Membrane processes

The use of membranes has several advantages compared to the other technologies such as absorption and adsorption processes. For instance, the possibility of having a large membrane area in a small module volume due to high packing density allows lower space and weight requirements for membrane units for the same production, and consequently reduced investment. Moreover, the process is more environmentally friendly compared to the absorption process due to the non-use of chemicals. In addition to the previously mentioned advantages, membranes also have the advantage of reduced operating costs, no moving parts and being suitable for remote locations. Its tolerance of motion makes membrane technology promising for offshore and subsea applications.

Several carbon dioxide selective membrane systems for carbon dioxide removal from natural gas have been installed onshore or topside on platforms. For small fields with low carbon dioxide content in the gas stream, membrane systems work quite effectively, and the carbon dioxide content in the product can be quite close to the pipeline specification (<2%). However, for bigger fields and high carbon dioxide content in the feed, the membrane treated streams still have a high carbon dioxide content (e.g., >6%) and need further treatment. For small fields membrane technologies are economically competitive compared to absorption, while for large fields absorption is still more favorable. However, due to limitations in weight and size on platforms, membrane processes can be more advantageous for offshore or subsea operations.

In another process, a membrane-based process for upgrading natural gas that contains C_{3+} hydrocarbon derivatives and/or acid gas is described. The conditioned natural gas can be used as fuel for gas-powered equipment, including compressors, in the gas field or the processing plant. Optionally, the process can be used to produce natural gas liquids.

D.8 Molecular sieve processes

Molecular sieves can be used for removal of sulfur compounds from gas streams. Hydrogen sulfide can be selectively removed to meet 4 ppm v/v specification. The sieve bed can be designed to dehydrate and sweeten simultaneously. In addition, molecular sieve processes can be used for removal of carbon dioxide from gas streams.

Molecular sieves are highly selective for the removal of hydrogen sulfide (as well as other sulfur compounds) from gas streams and over continuously high absorption efficiency. They are also an effective means of water removal and thus offer a process for the simultaneous dehydration and desulfurization of gas. However, gas that has excessively high water content may require upstream dehydration. The molecular sieve process is similar to the iron oxide process. Regeneration of the bed is achieved by passing heated clean gas over the bed.

As the temperature of the bed increases, it releases the adsorbed hydrogen sulfide into the regeneration gas stream. The sour effluent regeneration gas is sent to a flare stack, and up to 2% of the gas seated can be lost in the regeneration process. A portion of the natural gas may also be lost by the adsorption of hydrocarbon components by the sieve. In this process, unsaturated hydrocarbon components, such as olefins and aromatics, tend to be strongly adsorbed by the molecular sieves. The molecular sieves are susceptible to poisoning by such chemicals as glycols and require thorough gas cleaning methods before the adsorption step. Alternatively, the sieve can be offered some degree of protection by the use of *guard beds* in which a less expensive catalyst is placed in the gas stream before contact of the gas with the sieve, thereby protecting the catalyst from poisoning.

D.9 Sulfur recovery processes

Sulfur is present in natural gas principally as hydrogen sulfide and, in crude oil, as sulfur-containing compounds which are converted to hydrogen sulfide during processing. The hydrogen sulfide, together with some or all of any carbon dioxide present, is removed from the natural gas or refinery gas by means of one gas treating processes. The side stream from acid gas treating units consists mainly of hydrogen sulfide/or carbon dioxide. Carbon dioxide is usually vented to the atmosphere but sometimes is recovered for carbon dioxide floods. Hydrogen sulfide could be routed to an incinerator or flare, which would convert the hydrogen sulfide to sulfur dioxide. The release of hydrogen sulfide to the atmosphere may be limited by environmental regulations. There are many specific restrictions on these limits, and the allowable limits are revised periodically. In any case, environmental regulations severely restrict the amount of hydrogen sulfide that can be vented or flared in the regeneration cycle.

Sulfur recovery from sulfur-containing gas streams typically involves application of the Claus process using the reaction between hydrogen sulfide and sulfur dioxide (produced in the Claus process furnace from the combustion of hydrogen sulfide with air and/or oxygen) yielding elemental sulfur and water vapor:

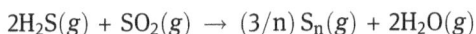

$$2H_2S(g) + SO_2(g) \rightarrow (3/n)\,S_n(g) + 2H_2O(g)$$

Therefore, higher conversions for this exothermic, equilibrium-limited reaction call for low temperatures which lead to low reaction rates and require the use of a catalyst. The catalytic conversion is usually carried out in a multi-stage fixed-bed adsorptive reactors process, to counteract the severe equilibrium limitations at high conversions.

The chemistry of the Claus process involves partial oxidation of hydrogen sulfide to sulfur dioxide and the catalytically promoted reaction of hydrogen sulfide and sulfur dioxide to produce elemental sulfur. The reactions are staged and are as follows:

Thermal stage:

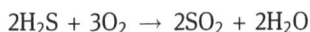

$$2H_2S + 3O_2 \rightarrow 2SO_2 + 2H_2O$$

Thermal and catalytic stage:

$$SO_2 + 2H_2S \rightarrow 3S + 2H_2O$$

The off-gas leaving a Claus plant is referred to as tail gas and, in the past was burned to convert the unreacted hydrogen sulfide to sulfur dioxide, before discharge to the atmosphere, which has a much higher toxic limit. However, the increasing standards of efficiency required by the pressure from environmental protection has led to the development of a large number of Claus tail gas clean-up units, based on different concepts, in order to remove the last remaining sulfur species.

The oxygen-blown Claus process was originally developed to increase capacity at existing conventional Claus plants and to increase flame temperatures of gases having low hydrogen sulfide content. The process has also been used to provide the capacity and operating flexibility for sulfur plants where the feed gas is variable in flow and composition such as often found in refineries.

In the Selectox process, the first stage thermal reactor (the furnace) is replaced the Claus with a catalytic oxidation step for the conversion of hydrogen sulfide in dilute acid–gas streams to liquid sulfur. The catalytic oxidation reactor can operate at much lower temperatures than the furnace and maintain a more stable flame temperature with gas streams having low hydrogen sulfide content. The Recycle Selectox process is an all-catalytic process in which there are no flames at any point in the process. A special catalyst bed replaces the acid gas burner in a conventional Claus plant and the catalyst occupies the top few inches of the first bed, where it promotes the selective oxidation of hydrogen sulfide to sulfur dioxide. The remainder of the bed is filled with Claus catalyst where the Claus reaction occurs to approximately

80 percent completion. The highly exothermic nature of these reactions requires that the feed gas be monitored for the concentration of hydrogen sulfide to avoid overheating.

Finally, liquid sulfur flowing from the Claus plant to the sulfur pit contains typically 250 to 350 ppm v/v of hydrogen sulfide and the sulfur is degassed using an active gas liquid contacting system to release dissolved gas. In this operations, sulfur from the pit is pumped into the degassing tower where it is contacted countercurrently with hot compressed air over a fixed catalyst bed. The degassed sulfur is returned to the product section of the sulfur pit.

The SuperClaus process consists of a thermal stage followed by three or four catalytic reaction stages with sulfur removed between stages by condensers. The first two or three reactors are filled with standard Claus catalyst while the last reactor is filled with the selective oxidation catalyst. In the thermal stage, the acid gas is burned with a less-than-stoichiometric amount of controlled combustion air so that the tail gas leaving the last Claus reactor typically contains 0.5–0.9 vol.% of hydrogen sulfide. There are two main principles are applied in operating the SuperClaus process: (i) operating the Claus plant with excess hydrogen sulfide to suppress the sulfur dioxide content in the Claus tail gas, and (ii) selective oxidation of the remaining hydrogen sulfide by the process catalyst selectively converts the hydrogen sulfide in the presence of water vapor and excess oxygen to elemental sulfur.

The DynaWave wet gas scrubber, which is a reverse jet scrubber that performs desulfurization in a wet gas environment, can be used as a follow-on to the SuperClaus process and the and the incinerator. In the process, the scrubbing liquid is injected, through a non-restrictive jet nozzle, counter current to the inlet incinerator flue gas. Liquid, containing caustic reagent, collides with the down-coming gas to create the region of extreme turbulence (the froth zone) with a high rate of mass transfer. Quench, sulfur dioxide removal, and removal of particulate matter occur in this zone. The clean, saturated gas and charged liquid continue through a separation vessel. The saturated gas continues through the vessel to mist removal devices. The liquid descends into the vessel sump for recycle back to the reverse jet nozzle. In the vessel sump, oxidation air is used to convert sodium sulfite (Na_2SO_3) to sodium sulfate (Na_2SO_4).

In the Clauspol process, the Claus tail gas is contacted counter-currently with an organic solvent in a low pressure drop packed column. Hydrogen sulfide and sulfur dioxide are absorbed in the solvent and react to form liquid elemental sulfur according to the Claus reaction, which is promoted by an inexpensive dissolved catalyst. The solvent is pumped around the contactor, and the heat of reaction is removed through a heat exchanger to maintain a constant temperature above the sulfur melting point. Due to the limited solubility of sulfur in the solvent, pure liquid sulfur separates from the solvent and is recovered from settling section at the bottom of the contactor. This process allows sulfur recovery up to 99.8% and the recovery level can be customized by adapting the size of the contactor.

The Selectox process is based on replacing the Claus first-stage thermal reactor (the furnace) with a catalytic oxidation step. A catalytic oxidation reactor can operate at much lower temperatures than the furnace and maintain a more stable flame temperature with lean hydrogen sulfide feeds.

Two versions of the process are offered. The first option is a once through option, treating acid gases with up to 5% v/v hydrogen sulfide and in which sulfur recovery is on the order of 84–94% w/w for hydrogen sulfide in the feed gas of 2–5% v/v, respectively. In the second option, which is a recycle version that handles hydrogen sulfide concentrations on the order of 5–100% v/v, the gas is recycled from the Selectox reactor condenser to cool the reactor outlet temperature not to exceed 370 °C (700 °F). The temperature limit is set so that carbon steel can be used for the reactor vessel.

In the Beavon Stretford reactor (BSR)/Selectox tail gas process, the gases are first hydrogenated to hydrogen sulfide in the BSR, then they proceed to another Selectox reactor stage. Sulfur recoveries up to 99.3% w/w have been reported for BSR/Selectox.

The wet oxidation processes are based on reduction–oxidation (redox) chemistry to oxidize the hydrogen sulfide to elemental sulfur in an alkaline solution containing an oxygen carrier. Vanadium and iron are the two oxygen carriers that are used. The best example of a process using the vanadium carrier is the Stretford process. The most prominent examples of the processes using iron as a carrier are the LO-CAT process and the SulFerox process.

The Stretford process using vanadium finds little use now because of the toxic nature of the vanadium solution and iron-based processes are more common.

Both the LO-CAT process and the SulFerox process are essentially the same in principle. The SulFerox process differs from the LO-CAT in that the oxidation and the regeneration steps are carried out in separate vessels and sulfur is recovered from the filters, melted, and sent to sulfur storage. Also, the SulFerox process uses a higher concentration of iron chelates (about 2–4% by weight vs 0.025–0.3% by weight for the LO-CAT process). Both processes are capable of up to 99+% sulfur recovery. However, using the processes for Claus tail gas treating requires hydrolysis of all the sulfur dioxide in the tail gas to hydrogen sulfide because the sulfur dioxide will react with the buffering base potassium hydroxide (KOH) and form potassium sulfate (K_2SO_4) which will consume the buffering solution and quickly saturate it.

Tail gas treating involves the removal of the remaining sulfur compounds from gases remaining after sulfur recovery. Tail gas from a typical Claus process, whether a conventional Claus or one of the extended versions of the process, usually contains small but varying quantities of COS, carbon disulfide, hydrogen sulfide, and sulfur dioxide as well as sulfur vapor. In addition, there may be hydrogen, carbon monoxide, and carbon dioxide in the tail gas.

In order to remove the rest of the sulfur compounds from the tail gas, all of the sulfur bearing species must first be converted into hydrogen sulfide which is then absorbed into a solvent and the clean gas vented or recycled for further processing.

The reduction of COS, carbon disulfide, sulfur dioxide, and sulfur vapor in Claus tail gas to hydrogen sulfide is necessary when sulfur recovery of 99.9+% is required. Usually the sulfur recovery level is set by the allowable emissions of sulfur from the tail gas incinerator. In addition, the reduction of COS is done on raw synthesis gas when the downstream acid gas removal process is unable to remove COS to a sufficient extent to meet sulfur emissions regulations from combustion of the cleaned fuel gas. These sulfur compounds are reduced to hydrogen sulfide by hydrogenation or by hydrolysis, at a raised temperature, over a catalytic bed.

In these processes, elemental sulfur and sulfur dioxide are reduced mainly via hydrogenation, while COS and carbon disulfide are mainly hydrolyzed to hydrogen sulfide. Sulfur and sulfured dioxide are virtually completely converted to hydrogen sulfide when an excess of hydrogen is present.

The SCOT process was developed in the early 1970s and consists of a combination of a catalytic hydrogenation/hydrolysis step and an amine scrubbing unit. In the process, tail gas from the Claus sulfur recovery unit is heated in an in-line burner before entering the hydrogenation reactor, where all sulfur species are converted to hydrogen sulfide (H_2S). Hydrogenation reactor effluent is then cooled by generating low-pressure steam, followed by additional cooling by cooling water exchange. Residual hydrogen sulfide in the cooled tail gas is removed with amine in a counter-current packed absorber. The treated tail gas from the absorber top is incinerated before being vented to the atmosphere. The rich solvent from the amine absorber is pumped to the regenerator after heat exchange against the hot lean solvent from the regenerator. Acid gases are stripped from the solvent in the trayed regenerator via a steam reboiler. The hot lean solvent from the regenerator bottom is pumped back to the absorber after being heat exchanged with rich solvent and cooling water to lower its temperature. Acid gas from the amine regenerator overhead is recycled back to the Claus plant for sulfur recovery. The reactions are

$$SO_2 + 3H_2 \rightarrow 2H_2O + H_2S$$

$$S_8 + 8H_2 \rightarrow 8H_2S$$

$$COS + H_2O \rightarrow CO_2 + H_2S$$

$$CS_2 + 2H_2O \rightarrow CO_2 + H_2S$$

When carbon monoxide is also present in the tail gas then the following reactions could also take place:

$$SO_2 + 3CO \rightarrow COS + 2CO_2$$

$$S_8 + 8CO \rightarrow 8COS$$

$$H_2S + CO \rightarrow COS + H_2$$

$$H_2O + CO \rightarrow CO_2 + H_2$$

The last reaction (shift reaction) is very rapid, and the presence of CO does not seem to favor the first three reactions.

D.10 Process selection

The selection of an optimum process will depend on conditions and composition of the inlet gas, cost of fuel and energy, product specifications, and relative product values. For example, some processes remove both hydrogen sulfide and carbon dioxide; other processes are designed to remove hydrogen sulfide only. It is important to consider the process selectivity for, say, hydrogen sulfide removal compared to carbon dioxide removal that ensures minimal concentrations of these components in the product, thus the need for consideration of the carbon dioxide to hydrogen sulfide in the gas stream.

Decisions in selecting a gas treating process can many times be simplified by gas composition and operating conditions. High partial pressures (50 psi) of acid gases enhance the probability of using a physical solvent. The presence of significant quantities of higher molecular weight hydrocarbon derivatives in the feed discourages using physical solvents. Low partial pressures of acid gases and low outlet specifications generally require the use of amines for adequate treating.

In general, the batch and amine processes are used for over 90% of all onshore wellhead applications. Amine processes are preferred when the lower operating cost, the chemical cost for this process is prohibitive, and justifies the higher equipment cost. The key is the sulfur content of the feed gas where below 20-pound sulfur per day, batch processes are more economical, and over 100-pound sulfur per day amine solutions are preferred.

Appendix E
Examples of ASTM test methods
for natural gas and gas condensate

D56   Standard Test Method for Flash Point by Tag Closed Cup Tester
D92   Standard Test Method for Flash and Fire Points by Cleveland Open Cup Tester
D93   Standard Test Methods for Flash Point by Pensky-Martens Closed Cup Tester
D129   Standard Test Method for Sulfur in Petroleum Products (General High Pressure
        Decomposition Device Method)
D240   Standard Test Method for Heat of Combustion of Liquid Hydrocarbon Fuels by Bomb
        Calorimeter
D287   Standard Test Method for API Gravity of Crude Petroleum and Petroleum Products
        (Hydrometer Method)
D323   Standard Test Method for Vapor Pressure of Petroleum Products (Reid Method)
D1018   Standard Test Method for Hydrogen in Petroleum Fractions
D1070   Standard Test Methods for Relative Density of Gaseous Fuels
D1071   Standard Test Methods for Volumetric Measurement of Gaseous Fuel Samples
D1072   Standard Test Method for Total Sulfur in Fuel Gases by Combustion and Barium Chloride
        Titration
D1142   Standard Test Method for Water Vapor Content of Gaseous Fuels by Measurement of
        Dew-Point Temperature
D1217   Standard Test Method for Density and Relative Density (Specific Gravity) of Liquids by
        Bingham Pycnometer
D1265   Standard Practice for Sampling Liquefied Petroleum (LP) Gases, Manual Method
D1266   Standard Test Method for Sulfur in Petroleum Products (Lamp Method)
D1267   Standard Test Method for Gage Vapor Pressure of Liquefied Petroleum (LP) Gases
        (LP-Gas Method)
D1550   Standard ASTM Butadiene Measurement Tables
D1657   Standard Test Method for Density or Relative Density of Light Hydrocarbons by Pressure
        Hydrometer
D1826   Standard Test Method for Calorific (Heating) Value of Gases in Natural Gas Range by
        Continuous Recording Calorimeter
D1835   Standard Specification for Liquefied Petroleum (LP) Gases
D1837   Standard Test Method for Volatility of Liquefied Petroleum (LP) Gases
D1838   Standard Test Method for Copper Strip Corrosion by Liquefied Petroleum (LP) Gases
D1945   Standard Test Method for Analysis of Natural Gas by Gas Chromatography
D1946   Standard Practice for Analysis of Reformed Gas by Gas Chromatography
D1988   Standard Test Method for Mercaptans in Natural Gas Using Length-of-Stain Detector
        Tubes
D2156   Standard Test Method for Smoke Density in Flue Gases from Burning Distillate Fuels
D2157   Standard Test Method for Effect of Air Supply on Smoke Density in Flue Gases from
        Burning Distillate Fuels
D2158   Standard Test Method for Residues in Liquefied Petroleum (LP) Gases
D2163   Standard Test Method for Determination of Hydrocarbons in Liquefied Petroleum (LP)
        Gases and Propane/Propene Mixtures by Gas Chromatography
D2384   Standard Test Methods for Traces of Volatile Chlorides in Butane-Butene Mixtures

https://doi.org/10.1515/9783110691023-014

D2420	Standard Test Method for Hydrogen Sulfide in Liquefied Petroleum (LP) Gases (Lead Acetate Method)
D2421	Standard Practice for Interconversion of Analysis of C5 and Lighter Hydrocarbons to Gas-Volume, Liquid-Volume, or Mass Basis
D2427	Standard Test Method for Determination of C2 through C5 Hydrocarbons in Gasolines by Gas Chromatography
D2504	Standard Test Method for Noncondensable Gases in C2 and Lighter Hydrocarbon Products by Gas Chromatography
D2505	Standard Test Method for Ethylene, Other Hydrocarbons, and Carbon Dioxide in High-Purity Ethylene by Gas Chromatography
D2593	Standard Test Method for Butadiene Purity and Hydrocarbon Impurities by Gas Chromatography
D2598	Standard Practice for Calculation of Certain Physical Properties of Liquefied Petroleum (LP) Gases from Compositional Analysis
D2650	Standard Test Method for Chemical Composition of Gases by Mass Spectrometry
D2713	Standard Test Method for Dryness of Propane (Valve Freeze Method)
D2779	Standard Test Method for Estimation of Solubility of Gases in Petroleum Liquids
D2789	Standard Test Method for Hydrocarbon Types in Low Olefinic Gasoline by Mass Spectrometry
D3120	Standard Test Method for Trace Quantities of Sulfur in Light Liquid Petroleum Hydrocarbons by Oxidative Microcoulometry
D3246	Standard Test Method for Sulfur in Petroleum Gas by Oxidative Microcoulometry
D3429	Standard Test Method for Solubility of Fixed Gases in Low-Boiling Liquids
D3588	Standard Practice for Calculating Heat Value, Compressibility Factor, and Relative Density of Gaseous Fuels
D3827	Standard Test Method for Estimation of Solubility of Gases in Petroleum and Other Organic Liquids
D3828	Standard Test Methods for Flash Point by Small Scale Closed Cup Tester
D3956	Standard Specification for Methane Thermophysical Property Tables
D3984	Standard Specification for Ethane Thermophysical Property Tables
D4051	Standard Practice for Preparation of Low-Pressure Gas Blends
D4084	Standard Test Method for Analysis of Hydrogen Sulfide in Gaseous Fuels (Lead Acetate Reaction Rate Method)
D4150	Standard Terminology Relating to Gaseous Fuels
D4362	Standard Specification for Propane Thermophysical Property Tables
D4423	Standard Test Method for Determination of Carbonyls in C4 Hydrocarbons
D4424	Standard Test Method for Butylene Analysis by Gas Chromatography
D4468	Standard Test Method for Total Sulfur in Gaseous Fuels by Hydrogenolysis and Rateometric Colorimetry
D4486	Standard Test Method for Kinematic Viscosity of Volatile and Reactive Liquids
D4650	Standard Specification for Normal Butane Thermophysical Property Tables
D4651	Standard Specification for Isobutane Thermophysical Property Tables
D4784	Standard Specification for LNG Density Calculation Models
D4810	Standard Test Method for Hydrogen Sulfide in Natural Gas Using Length-of-Stain Detector Tubes
D4888	Standard Test Method for Water Vapor in Natural Gas Using Length-of-Stain Detector Tubes
D4891	Standard Test Method for Heating Value of Gases in Natural Gas and Flare Gases Range by Stoichiometric Combustion

D4984	Standard Test Method for Carbon Dioxide in Natural Gas Using Length-of-Stain Detector Tubes
D5134	Standard Test Method for Detailed Analysis of Petroleum Naphtha through n-Nonane by Capillary Gas Chromatography
D5234	Standard Guide for Analysis of Ethylene Product
D5273	Standard Guide for Analysis of Propylene Concentrates
D5274	Standard Guide for Analysis of 1,3-Butadiene Product
D5287	Standard Practice for Automatic Sampling of Gaseous Fuels
D5303	Standard Test Method for Trace Carbonyl Sulfide in Propylene by Gas Chromatography
D5305	Standard Test Method for Determination of Ethyl Mercaptan in LP-Gas Vapor
D5454	Standard Test Method for Water Vapor Content of Gaseous Fuels Using Electronic Moisture Analyzers
D5504	Standard Test Method for Determination of Sulfur Compounds in Natural Gas and Gaseous Fuels by Gas Chromatography and Chemiluminescence
D5623	Standard Test Method for Sulfur Compounds in Light Petroleum Liquids by Gas Chromatography and Sulfur Selective Detection
D5842	Standard Practice for Sampling and Handling of Fuels for Volatility Measurement
D5954	Standard Test Method for Mercury Sampling and Measurement in Natural Gas by Atomic Absorption Spectroscopy
D6159	Standard Test Method for Determination of Hydrocarbon Impurities in Ethylene by Gas Chromatography
D6228	Standard Test Method for Determination of Sulfur Compounds in Natural Gas and Gaseous Fuels by Gas Chromatography and Flame Photometric Detection
D6273	Standard Test Methods for Natural Gas Odor Intensity
D6350	Standard Test Method for Mercury Sampling and Analysis in Natural Gas by Atomic Fluorescence Spectroscopy
D6450	Standard Test Method for Flash Point by Continuously Closed Cup (CCCFP) Tester
D6667	Standard Test Method for Determination of Total Volatile Sulfur in Gaseous Hydrocarbons and Liquefied Petroleum Gases by Ultraviolet Fluorescence
D6849	Standard Practice for Storage and Use of Liquefied Petroleum Gases (LPG) in Sample Cylinders for LPG Test Methods
D6897	Standard Test Method for Vapor Pressure of Liquefied Petroleum Gases (LPG) (Expansion Method)
D6968	Standard Test Method for Simultaneous Measurement of Sulfur Compounds and Minor Hydrocarbons in Natural Gas and Gaseous Fuels by Gas Chromatography and Atomic Emission Detection
D7314	Standard Practice for Determination of the Heating Value of Gaseous Fuels using Calorimetry and On-line/At-line Sampling
D7551	Standard Test Method for Determination of Total Volatile Sulfur in Gaseous Hydrocarbons and Liquefied Petroleum Gases and Natural Gas by Ultraviolet Fluorescence
D7607	Standard Test Method for Analysis of Oxygen in Gaseous Fuels (Electrochemical Sensor Method)
D7756	Standard Test Method for Residues in Liquefied Petroleum (LP) Gases by Gas Chromatography with Liquid, On-Column Injection
D7833	Standard Test Method for Determination of Hydrocarbons and Non-Hydrocarbon Gases in Gaseous Mixtures by Gas Chromatography

Appendix F
Examples of the gas processors association test methods for natural gas

GPA 1167   GPA Glossary – Definition of Words and Terms Used in the Gas Processing Industry
GPA 2100   Qualitative Determination of Carbonyl Sulfide in Propane Using Length of Stain Tubes
GPA 2103   Tentative Method for Analysis of Natural Gas Condensate Mixtures Containing Nitrogen and Carbon Dioxide by Gas Chromatography
GPA 2108   GPA Fractionation Grade Product Specifications
GPA 2140   Liquefied Petroleum Gas Specifications and Test Methods
GPA 2145   Table of Physical Constants of Paraffin Hydrocarbons and Other Components of Natural Gas Liquids Industries
GPA 2166   Obtaining Natural Gas Samples for Analysis by Gas Chromatography
GPA 2172   Calculation of Gross Heating Value, Relative Density and Compressibility Factor for Natural Gas Mixtures from Compositional Analysis
GPA 2174   Obtaining Liquid Hydrocarbons Samples for Analysis by Gas Chromatography
GPA 2177   Analysis of Demethanized Hydrocarbon Liquid Mixtures Containing Nitrogen and Carbon Dioxide by Gas Chromatography
GPA 2186   Method for the Extended Analysis of Hydrocarbon Liquid Mixtures Containing Nitrogen and Carbon Dioxide by Temperature Programmed Gas Chromatography
GPA 2188   Method for Determination of Ethyl Mercaptan in LP-Gas Using Length of Stain Tubes
GPA 2194   Low-Pressure Field Method for Determining Ethyl Mercaptan Odorant in LP-Gas Using Length of Stain Tubes
GPA 2198   Selection, Preparation, Validation, Care and Storage of Natural Gas and Natural Gas Liquids Reference Standard Blends
GPA 2199   Determination of Specific Sulfur Compounds by Capillary Gas Chromatography and Sulfur Chemiluminescence Detection
GPA 2261   Method of Analysis for Natural Gas and Similar Gaseous Mixtures by Gas Chromatography
GPA 2286   Tentative Method for the Extended Analysis for Natural Gas and Similar Gaseous Mixtures by Temperature Programmed Gas Chromatography
GPA 2377   Test for Hydrogen Sulfide and Carbon Dioxide in Natural Gas Using Length of Stain Tubes
GPA 8117   A Simplified Vapor Pressure Correlation for Commercial NGLs
GPA 8173   Standard for Converting Mass of Natural Gas Liquids or Vapors to Equivalent Liquid Volumes
GPA 8182   Standard for Mass Measurement of Natural Gas Liquids
GPA 8186   Measurement of Liquid Hydrocarbons by Truck Scales
GPA 8195   Tentative Standard for Converting Net Vapor Space Volumes to Equivalent Liquid Volumes.
GPA 8217   Temperature Correction for the Volume of NGL.

https://doi.org/10.1515/9783110691023-015

About the author

Dr. James G. Speight has doctorate degrees in chemistry, geological sciences, and petroleum engineering. He is the author of more than 90 books in petroleum science, fossil fuel science, petroleum engineering, environmental sciences, and ethics.

He has more than 50 years of experience in areas associated with (i) the properties, recovery, and refining of reservoir fluids, conventional petroleum, heavy oil, extra heavy oil, tar sand bitumen, and oil shale; (ii) the properties and refining of natural gas, gaseous fuels; (iii) the production and properties of chemicals from crude oil, coal, and other sources; (iv) the properties and refining of biomass, biofuels, biogas, and the generation of bioenergy; and (v) the environmental and toxicological effects of energy production and fuels use. His work has also focused on safety issues, environmental effects, environmental remediation, and safety issues as well as reactors associated with the production and use of fuels and biofuels.

Although he has always worked in private industry with emphasis on contract-based work leading to commercialization of concepts, Dr. Speight was (among other appointments) a visiting professor in the College of Science, University of Mosul (Iraq), and has also been a visiting professor in chemical engineering at the Technical University of Denmark (Lyngby, Denmark) and the University of Trinidad and Tobago (Point Lisas, Trinidad).

In 1996, Dr. Speight was elected to the Russian Academy of Sciences and awarded the Academy's Gold Medal of Honor that same year for outstanding contributions to the field of petroleum sciences. In 2001, he received the Scientists without Borders Medal of Honor of the Russian Academy of Sciences and was also awarded the Einstein Medal for outstanding contributions and service in the field of geological sciences. In 2005, the Academy awarded Dr. Speight the Gold Medal – Scientists without Frontiers, Russian Academy of Sciences, in recognition of his continuous encouragement of scientists to work together across international borders. In 2007, Dr. Speight received the Methanex Distinguished Professor award at the University of Trinidad and Tobago in recognition of excellence in research. In 2018, he received the American Excellence Award for Excellence in Client Solutions from the United States Institute of Trade and Commerce, Washington, DC.

https://doi.org/10.1515/9783110691023-016

Index

https://doi.org/10.1515/9783110691023-017

www.ingramcontent.com/pod-product-compliance
Lightning Source LLC
Chambersburg PA
CBHW080125220326
41598CB00032B/4960